AF183013

CELL SIGNALLING IN PROKARYOTES AND LOWER METAZOA

CELL SIGNALLING IN PROKARYOTES AND LOWER METAZOA

Edited by

Ian Fairweather

School of Biology and Biochemistry
The Queen's University of Belfast
Belfast, Northern Ireland

KLUWER ACADEMIC PUBLISHERS
DORDRECHT / BOSTON / LONDON

Library of Congress Cataloging-in-Publication Data is available

ISBN 1-4020-1739-1

Published by Kluwer Academic Publishers,
PO Box 17, 3300 AA Dordrecht, The Netherlands.

Sold and distributed in North, Central and South America
by Kluwer Academic Publishers,
101 Philip Drive, Norwell, MA 02018, USA

In all other countries, sold and distributed
by Kluwer Academic Publishers,
PO Box 322, 3300 AH Dordrecht, The Netherlands

Printed on acid-free paper

All rights reserved
© 2004 Kluwer Academic Publishers
No part of the material protected by this copyright may be reproduced or utilized in any form or by any means, electronic, mechanical, including photocopying, recording or by any information storage and retrieval system, without written permission from the copyright owners.

Printed in the Netherlands

CONTENTS

Preface vii

Chapter 1 1
G proteins and MAP kinase cascades in the pheromone response of fungi
Ann Kays, and Katherine A. Borkovich

Chapter 2 27
Prokaryotic intercellular signalling – Mechanistic diversity and unified themes
Clay Fuqua and David White

Chapter 3 73
Signal transduction mechanisms in protozoa
Fernando L. Renaud, José de Ondarza, Pierangelo Luporini, Michael J. Marino, and Judy van Houten

Chapter 4 91
Signalling systems in cnidaria
Werner Müller

Chapter 5 115
Neuropeptides in cnidarians
Cornelis J.P. Grimmelikhuijzen, Michael Williamson, and Georg N. Hansen

Chapter 6 141
Signalling mechanisms in platyhelminths
Ian Fairweather

Chapter 7 195
Control of *Caenorhabditis elegans* behaviour and development by G proteins big and small
Carol A. Bastiani, Melvin I. Simon, and Paul W. Sternberg

Chapter 8 243
Electrophysiological and pharmacological studies on excitable tissues in nematides
Robert J. Walker, Candida M. Rogers, Christopher J. Franks, and Lindy Holden-Dye

Chapter 9 303
Evidence for an annelid neuroendocrine system
Michel Salzet, Didier Vieau, and Christophe Breton

Chapter 10 325
Ion channels of microbes
Christopher P. Palmer, Ann Batiza, Xin-Liang Zhou, Stephen H. Loukin, Yoshiro Saimi, and Ching Kung

Chapter 11 347
Bacterial signal transduction: Two-component signal transduction as a model for therapeutic intervention
Lenore A. Pelosi, Kwasi A. Ohemeng, and John F. Barrett

Index 403

PREFACE

Cell signalling lies at the heart of many biological processes and currently is the focus of intense research interest. In multicellular organisms, it is central to how different types of cell communicate with each other and how they detect and respond to extracellular changes. Intercellular communication is vital to single-celled organisms as well, allowing them to respond to environmental cues and signals. As a parasitologist, I became interested in cell signalling through studies on the nervous system of flatworms and nematodes. The idea for the book, however, stemmed from a course in Endocrinology I taught at Queen's. It struck me that there is a perception in the academic community that, in terms of cell signalling, invertebrates barely exist below the level of molluscs and arthropods and that much of the understanding of signalling mechanisms and integration of pathways has come from research on specific cell types, such as lymphocytes and cardiomyocytes, in mammals. More simple organisms are largely ignored. Yet there is much of interest to be learnt from examining the ways in which the basic principles of signalling mechanisms were laid down in unicellular organisms, then developed in different directions as body patterns became more complex and different types of cells and tissues needed to find new ways to communicate with each other. Separate nervous and endocrine systems evolved, as did vascular systems, and signalling molecules were appropriated for new uses. The book deliberately focuses on the early stages in the evolution of communication systems, drawing information together in a way that has not been attempted before.

The book begins – as signalling did – with intercellular communication in fungi, bacteria and protozoans. These organisms needed to be able to respond to environmental signals and to be able to secrete extracellular signals (pheromones) in order to co-ordinate their activity. The book then follows the development of the nervous system in the cnidarians, platyhelminths and nematodes through to the annelids, where a separate endocrine system is present. This opened up the possibility of signalling molecules being used for intercellular communication within organisms, as neurotransmitters, neuromodulators, neuromuscular transmitters, neurohormones and morphogens, or "hormones". Many of the molecules are similar to those in more advanced organisms and the evolutionary conservation of signalling molecules and pathways will be highlighted. Nor are ion channels the sole preserve of higher organisms with nervous systems, either, as the chapter on ion channels in microbes will demonstrate. Simple organisms have considerable value as models for studying developmental processes and unravelling the underlying signalling pathways and this will be emphasised. Many of the groups covered contain important pathogens or parasites and the potential for manipulating signalling pathways for therapeutic intervention will be highlighted. Adaptation to a parasitic way of life has opened up the possibility of even more complex signalling processes, with bidirectional communication between the parasite and its host. Consequently, while the groups covered in the book may be considered to be relatively "simple" with regard to their morphology, they are certainly far from simple in terms of their complement of signalling molecules and processes. Many of the elements typical of more advanced and more widely studied organisms are already present in lower groups, and so it is to them that one must look for the initial, exciting steps in how cells and tissues learnt to talk to each other.

This book should be of value to students and researchers in a wide variety of disciplines: endocrinology, neurobiology, cell signalling, microbiology, parasitology, veterinary science and clinical science. It will serve to put cell signalling in a wider, evolutionary context. I hope that it will attract interest in some groups which have been neglected in the past.

I have been particularly lucky with the authors who have contributed to the book, not only because of their expertise, but because of the way in which their Chapters have fitted in with the overall themes of the book. I must thank them for their patience in what has been a long gestation period. Thanks, too, to other academics not in the book who made many useful suggestions about potential authors in areas outside my own. To those potential authors who agreed to contribute to the book yet in the end failed to do so, which led to delays in publication, I extend my disappointment. The book would have been more balanced and the better with their Chapters. I would like to extend my sincere gratitude to the staff at Kluwer who have kept faith with the project. Particular appreciation goes to Esther Verdries who has done a superb job in putting the book together. Finally to my son, Simon, whose computer knowledge and advice has been invaluable to me as I moulded the Chapters into a common format.

Ian Fairweather

CHAPTER 1

G PROTEINS AND MAP KINASE CASCADES IN THE PHEROMONE RESPONSE OF FUNGI

ANN KAYS and KATHERINE A. BORKOVICH

Department of Microbiology and Molecular Genetics, University of Texas-Houston Medical School, Houston, Texas 77030, USA

Summary

Development in fungal systems occurs in response to environmental cues and external stimuli. Heterotrimeric G protein coupled receptors (GPCRs) provide these systems with the ability to receive and transmit external signals into the cell. Once the signal has been internalized, mitogen activated protein kinase (MAPK) and/or cAMP-dependent protein kinase (PKA) cascades amplify and integrate various stimuli. Studies of heterotrimeric G proteins in yeast and filamentous fungi reveal remarkable evolutionary conservation in the signal transduction pathways of lower eukaryotic and mammalian cells. Because of the ease of genetic and biochemical manipulation, fungi have proven to be an invaluable system for dissecting the complex regulatory networks involved in higher eukaryotic signalling. In this chapter, we will examine cell-cell communication in fungi by addressing the pheromone response signal transduction pathway. The similarities and differences observed between the *Saccharomyces cerevisiae* and *Schizosaccharomyces pombe* yeast pheromone response pathways will be used as a paradigm for discussing sexual development in the filamentous species *Cryptococcus neoformans*, *Magnaporthe grisea*, *Neurospora crassa*, and *Ustilago maydis*.

1. Introduction

1.1. HETEROTRIMERIC G PROTEINS

Heterotrimeric G proteins are composed of a GTPase α subunit and a βγ dimer (Dessauer, Posner and Gilman, 1996). In the basal state, the βγ moiety is associated with the GDP-bound Gα subunit at the cytoplasmic domains of seven transmembrane α helical receptors, referred to as G protein coupled receptors (GPCRs). A variety of external stimuli, such as pheromones, hormones, and odourants are bound by GPCRs, resulting in the exchange of GDP for GTP on the Gα subunit and dissociation of the Gα-GTP and βγ moiety. Gα-GTP and the free βγ complex are then able to activate downstream effectors, such as enzymes and ion channels. The initiating stimulus is terminated by hydrolysis of GTP to GDP by the Gα subunit. The otherwise slow GTPase activity of Gα is accelerated by 'Regulators of G-protein Signalling' (RGS) proteins (Berman and Gilman, 1998; Neer, 1997). The GDP-Gα reassociates with βγ to await the next cycle of activation.

Gα proteins range in size from 39–52 kDa and belong to a family of GTP

I. Fairweather (ed.), Cell Signalling in Prokaryotes and Lower Metazoa, 1–26.
© 2004 *Kluwer Academic Publishers. Printed in Great Britain.*

binding proteins that includes Ras (Dessauer, Posner and Gilman, 1996). The Gα subunit is composed of two domains, a GTPase and an α-helical domain, that are connected by two linkers. Most Gα subunits undergo post-translational lipid modifications that tether the protein to the membrane (Wedegaertner, Wilson and Bourne, 1995). The Gβ subunit is approximately 36 kDa and interacts with the Gα subunit (Dessauer, Posner and Gilman, 1996; Gautam *et al.*, 1998). The NH_2-terminus of Gβ is an amphipathic α-helix followed by seven repeating units of $\sim$43 residues called a WD motif. Crystal structure analysis indicates that WD motifs form seven blades that contribute to an overall organization resembling a propeller, referred to as a β superbarrel. The Gγ subunit is much smaller (6–9 kDa) compared to the Gα and Gβ proteins and is known to undergo post-translational lipid modifications as well (Downes and Gautam, 1999; Gautam *et al.*, 1998). Lipid modification of Gγ is not required for interacting with Gβ, but functions in membrane targeting and, therefore, facilitates complex formation with Gα. The Gγ protein contacts loops in the fifth and sixth blades of the β propeller and forms a coiled coil with the Gβ α-helical domain.

G proteins have been shown to regulate adenylyl cyclase activity in many systems, including fungi (Dessauer, Posner and Gilman, 1996; Ivey, Yang and Borkovich, 1999). Adenylyl cyclase catalyzes the production of the second messenger cAMP from ATP. cAMP-dependent protein kinase (PKA) is an internal cAMP receptor and is composed of two catalytic and two regulatory subunits (Francis and Corbin, 1999). Binding of cAMP by the regulatory subunits leads to their dissociation from the catalytic subunits. The freed catalytic subunits phosphorylate downstream protein targets, which in many cases include transcription factors. PKA activity is negatively regulated by hydrolysis of cAMP to AMP by cAMP phosphodiesterase (Beavo, 1995).

1.2. MITOGEN ACTIVATED PROTEIN KINASES

Mitogen activated protein kinase (MAPK) cascades transmit signals from extracellular stimuli to the nucleus to alter gene expression. Frequently, the activating signal is delivered to the cascade by a Gα subunit or the Gβγ dimer (Crespo *et al.*, 1995). The structure of MAPK cascades is conserved among eukaryotic cells and is composed of three sequentially phosphorylated proteins: MAPK kinase kinase (MAPKKK), MAPK kinase (MAPKK), and MAPK. Binding of a scaffolding protein to all three components of the MAPK cascade sequesters or allows proteins to interact, maintains specificity of the cascade, and/or monitors cellular localization (Garrington and Johnson, 1999; Widmann *et al.*, 1999).

The cascade is initiated by activation of MAPKKK through interacting with heterotrimeric and/or small GTP-binding proteins or through MAPKKK kinase phosphorylation (Crespo *et al.*, 1994; Siow *et al.*, 1997). Dimerization and subcellular localization are also implicated in regulation of MAPKKK.

MAPKKK is a serine/threonine kinase composed of an N-terminal autoinhibitor and a C-terminal catalytic domain. The amino acid sequence of MAPKKK indicates regulatory motifs not present in MAPKK and MAPK, such as Pleckstrin homology domains, leucine-zipper dimerization motifs, and tyrosine and serine/threonine phosphorylation sites. The activated MAPKKK phosphorylates serine and/or threonine residues of the next module in the cascade, the MAPKK (Widmann *et al.*, 1999).

Activated MAPKK recognizes specific MAPK tertiary structures and phosphorylates threonine and tyrosine residues at a T-X-Y motif (where X is P, G, or E) in the MAPK activation loop (Gartner, Namyth and Ammerer, 1992). MAPKK is called a dual specificity kinase because it phosphorylates threonine and tyrosine, and both must be phosphorylated to achieve maximal MAPK activation. As the last kinase in the cascade, the phosphorylated MAPK frequently acts on transcription factors in the nucleus. MAPK has also been shown to phosphorylate other substrates on serine and threonine residues in close proximity to proline residues (Widmann *et al.*, 1999).

2. *Saccharomyces cerevisiae*

Pheromone response pathways facilitate communication between haploid fungal cells of opposite mating type to mediate mating (Banuett, 1998; see Fig. 1). Peptide pheromones secreted by one cell bind to receptors on a cell of opposite mating type to activate downstream pathways. The pheromone response frequently results in the transcription of specific mating genes, cell

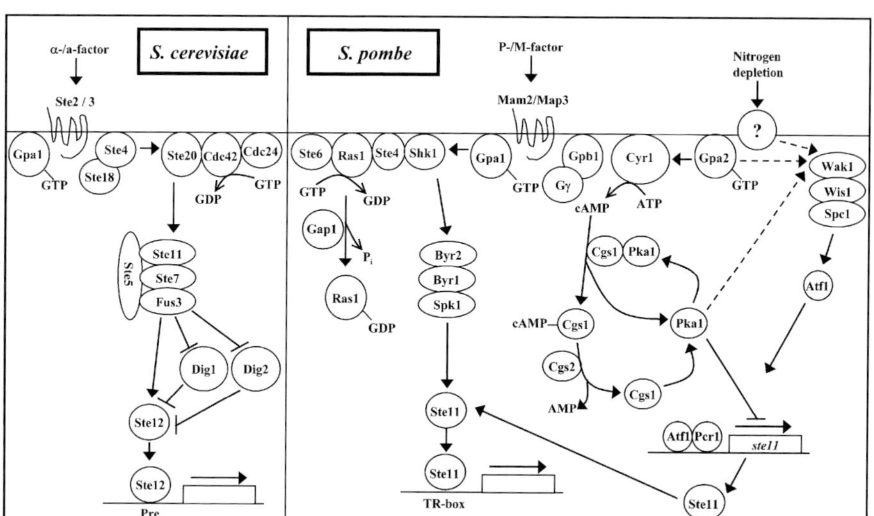

Figure 1. Proteins involved in the pheromone response pathways of *S. cerevisiae* and *S. pombe*. Participation of both heterotrimeric G proteins and MAP kinases is indicated. See text for details.

cycle arrest, and changes in cellular morphology. The *S. cerevisiae* pheromone response pathway is the most well understood and studies on this species underlie the current understanding of G proteins and MAPK cascades in all systems. Mating in *S. cerevisiae* occurs between two haploid cells of **a** and **α** mating type, determined by the *MAT* genes that are expressed during vegetative growth (Banuett, 1998). The *Mα1* and *Mα2* genes encode the peptide pheromone a-factor that is secreted by **α** cells. The a-factor peptide pheromone is encoded by *MFA1* and *MFA2*. Following translation, a-factor is carboxymethylated and farnesylated at the COOH-terminus prior to being secreted by **a** cells (Kurjan and Herskowitz, 1982; Michaelis and Herskowitz, 1988; Singh *et al.*, 1983).

The α-factor GPCR Ste2p is expressed by **a** cells and binds α-factor, while a-factor is recognized by the a-factor GPCR Ste3p, produced by **α** cells. Consistent with early hypotheses, Ste2p exists as an oligomer (Blumer, Reneke and Thorner, 1988; Overton and Blumer, 2000). Furthermore, oligomerization of Ste2p is necessary for successful signal transduction to occur (Blumer, Reneke and Thorner, 1988; Overton and Blumer, 2000). Complexes of Ste2p may activate multiple G proteins and thus amplify the pheromone signal.

The Ste2p and Ste3p GPCRs activate the same heterotrimeric G proteins in both mating types: Gpa1p (Gα), Ste4p (Gβ) and Ste18p (Gγ). Evidence suggests that Gpa1p-GDP is complexed with the Ste4p-Ste18p dimer and docked at Ste2p or Ste3p in the resting state (Coria, Savinon-Tejeda and Birnbaumer, 1995). Study of Ste4p and Ste18p has demonstrated that the Gβγ dimer is essential for receptor coupling (Blumer and Thorner, 1990). Pheromone binding results in a conformational change, allowing Ste4p-Ste18p to activate the downstream pheromone response MAPK cascade (Klein, Reuveni and Levitzki, 2000). Deletion of *GPA1* causes constitutive activation of the pheromone pathway and subsequent cell death due to cell cycle arrest (Dietzel and Kurjan, 1987; Miyajima *et al.*, 1987; Kurjan and Dietzel, 1988). Loss of *STE4* or *STE18* is not lethal, but results in sterility.

Gpa1p is post-translationally myristoylated and palmitoylated (Whiteway and Thomas, 1994; Hirschman and Jenness, 1999; Manahan *et al.*, 2000). Loss of palmitoylation in Gpa1p produces a cell that can partially activate the pheromone response pathway independently of pheromone (Song and Dohlman, 1996; Manahan *et al.*, 2000). This defective Gpa1p may be deficient in membrane anchoring and, thus, not complexed with Gβγ, leading to partial activation of the MAPK cascade. The Ste18p subunit is post-translationally modified with farnesyl and palmitoyl lipid groups. Lipid modification of Ste18p is necessary for maintaining membrane association following dissociation from Gpa1p, but is not necessary for association with Gβ (Hirschman and Jenness, 1999; Manahan *et al.*, 2000). Mutation of the Ste18p lipid modification residues causes the release of the Gβγ moiety from the plasma membrane and disruption of the pheromone response pathway (Whiteway and Thomas, 1994; Pryciak and Huntress, 1998; Hirschman and Jenness, 1999; Mahanty *et al.*, 1999).

The Ste4p-Ste18p dimer transmits the pheromone signal to the three sequential kinases Ste11p (MAPKKK), Ste7p (MAPKK) and Fus3p (MAPK) by first interacting with the Ste5p scaffolding protein. During vegetative growth, Ste5p is located in the cytoplasm and a pool of Ste5p shuttles through the nucleus (Mahanty *et al.*, 1999). Exposure of cells to pheromone increases migration of Ste5p from the nucleus to the plasma membrane, where it can interact with the Gβγ dimer. This migration is dependent on the importin β-like protein Ste21p (Gorlich *et al.*, 1997). The area of Ste5p localization at the membrane is the future site of mating projection (shmoo) formation (Mahanty *et al.*, 1999). *ste5* mutants that are unable to shuttle through the nucleus do not interact with Ste4p, suggesting that the scaffolding protein is modified in the nucleus to facilitate membrane targeting (Mahanty *et al.*, 1999). Deletion of Gβ disrupts Ste5p translocation to the membrane and pathway activation, but Ste5p recruitment is not affected by deletion of MAPK components (Pryciak and Huntress, 1998). Taken together, this suggests that Gβ is only responsible for triggering migration of Ste5p to the membrane and is not responsible for directly activating the MAPK cascade in *S. cerevisiae*. Although extensive evidence supports Ste5p translocation to the membrane following pheromone exposure, it is not yet known whether the MAPK cascade components are also transported to the membrane.

Recruitment of Ste5p to the membrane places the scaffolding protein in closer proximity to the membrane localized proteins Ste20p, Cdc24p and Cdc42p. These three proteins have been shown to function upstream of the MAPKKK Ste11p, but downstream of Ste5p (Pryciak and Huntress, 1998). Recruitment of Ste20p to the membrane occurs prior to pheromone exposure (Peter *et al.*, 1996; Pryciak and Huntress, 1998). The kinase activity of Ste20p is independent of Ste4p, an observation that further supports the hypothesis that Gβ is not a direct activator of Ste20p. Ste20p is a serine/threonine protein kinase that belongs to the p21 Activated Kinase (PAK) family of Cdc42/Rac-dependent kinases, and is a MAPKKK kinase to activate the downstream cascade (Leberer *et al.*, 1997). The Ste20p kinase domain is located at the C-terminus, while the N-terminal region contains the Cdc42/Rac Interactive Binding site (CRIB domain) that is bound by Cdc42p-GTP (Burbelo, Dreschesel and Hall, 1995). Cdc42p is a small GTP binding protein from the Rho family that exchanges GDP for GTP to become activated, and later hydrolyzes GTP to return to the basal state. The pheromone pathway leads to activation of the cell polarity pathway to facilitate polarized growth of the cell (shmoo formation), a morphogenetic pathway that is regulated by Cdc42p (Cabib, Drgonova and Drgon, 1998; Madden and Snyder, 1998). The putative Guanine nucleotide Exchange Factor (GEF) Cdc24p is also required for polarized growth and is thought to mediate GTP cycling on Cdc42p (Burbelo, Dreschel and Hall, 1995; Johnson, 1999).

The specific role of these membrane-localized proteins in the pheromone response pathway is currently under debate. Recent reports have analyzed the importance of the *STE20* CRIB domain in mating (Peter *et al.*, 1996; Leberer *et al.*, 1997; Pryciak and Huntress, 1998). Cdc42 has been shown to bind specifically to the CRIB domain in a GTP-dependent manner (Peter *et al.*, 1996). Initial reports indicated that the deletion of the Ste20p Cdc42 binding domain does not significantly affect mating (Peter *et al.*, 1996; Leberer *et al.*, 1997). However, a second study with the same *ste20* mutant demonstrates that this strain is more defective in the pheromone response pathway than previously reported (Pryciak and Huntress, 1998). Loss of *CDC42* produces a cell that is reduced in mating and, conversely, an activating *cdc42* mutation increases mating efficiency (Akada *et al.*, 1996). Results from such studies with *CDC42* and the contradictory reports regarding the functional significance of the Ste20p CRIB domain in the pheromone response pathway suggest a subtle role for Cdc42p. For example, Cdc42p, Ste4p-Ste18p, Ste20p and Cdc24p may interact to form a stable scaffolding complex to recruit Ste5p to the membrane. Such a scenario would allow efficient activation of both the pheromone MAPK cascade and the Cdc42p-mediated cell polarity pathway. Another possibility is that Cdc24p and Cdc42p may allow Gβγ and Ste20p to participate in an unidentified cellular event, without having a direct role in cascade activation (Johnson, 1999).

The Ste5p scaffolding protein binds Ste11p, Ste7p, and Fus3p using different domains (Cairns, Ramer and Kornberg, 1992; Marcus *et al.*, 1994; Whiteway *et al.*, 1995). The Ste11p MAPKKK is thought to be phosphorylated by Ste20p, although dependence of Ste20p kinase activity on pheromone activation has not been demonstrated. Activation of Ste11p causes the sequential phosphorylation and activation of Ste7p and Fus3p (Gartner, Namyth and Ammerer, 1992). Ste11p and Ste7p are components of both the pheromone and filamentous growth MAPK cascades; therefore, sequestration of the pheromone response MAPK cascade proteins by Ste5p facilitates pathway specificity (Gustin *et al.*, 1998).

Specificity of the pheromone MAPK cascade is also determined by the last component of the cascade, the MAPK. Deletion of the MAPKs *FUS3* and *KSS1* results in sterility; therefore, these two MAPKs were originally thought to have redundant functions (Elion, Brill and Fink, 1991). During the pheromone response in *fus3*Δ mutants, Kss1p activates both the filamentation and pheromone response pathways. Kss1p is unable to substitute for Fus3p when a catalytically inactive Fus3p is expressed (Madhani, Styles and Fink, 1997). These findings support the hypothesis that Fus3p preferentially participates in the pheromone response pathway. Differentiation of the two pathways was further complicated because the transcription factor Ste12p was thought to be downstream of both Fus3p and Kss1p (Madhani, Styles and Fink, 1997). Activation of the filamentation pathway results in Ste12p associating with Tec1p to bind sites in the promoter (filamentation response elements, or FREs) of genes regulated by this pathway. Inactive (dephosphorylated) Kss1p inhibits the filamentation pathway by binding Ste12p-Tec1p to physically interfere with promoter binding. Fus3p is thought to phosphorylate Ste12p, which can then bind pheromone response elements (PREs) in the promoter of target genes to induce transcription. The Dig1p and Dig2p proteins are negative regulators of Ste12p and activation of Fus3p results in inhibition of Dig1p and Dig2p (Olson *et al.*, 2000).

Phosphorylated Fus3p is also hypothesized to activate the bifunctional protein Far1p. The function of Far1p is dictated by cellular location: cell cycle arrest in the nucleus and cell polarity in the cytoplasm (Blondel *et al.*, 1999). The rate of export from the nucleus determines Far1p functional specificity and export is triggered by pheromone exposure (Butty *et al.*, 1998; Blondel *et al.*, 1999). Far1p was isolated as a multi-copy suppressor of *fus3*Δ, and Fus3p was shown to phosphorylate and interact with Far1p (Elion, Satterberg and Kranz, 1993; Peter *et al.*, 1993). Cell cycle arrest occurs as a result of Far1p-mediated inhibition of cyclin-dependent kinases (Peter and Herskowitz, 1994; Gartner *et al.*, 1998). Interestingly, Cdc24p and Far1p have been shown to interact independently of pheromone. Cytoplasmic Far1p participates in the cell polarity pathway by forming a complex with Cdc24p and the Ste4p-Ste18p dimer (Butty *et al.*, 1998; Nern and Arkowitz, 1999). Since Cdc24p localizes at the site of polarized growth, it is assumed that the entire Far1p protein complex is also found at this site. Localization of this protein complex is thought to facilitate cell fusion by signalling the site of shmoo formation to the actin cytoskeleton (Blondel *et al.*, 1999).

3. *Schizosaccharomyces pombe*

In the fission yeast, *Schizosaccharomyces pombe* mating occurs between two haploid cells of opposite mating type, h^- or M, and h^+ or P (Banuett, 1998; Davey, 1998; see Fig. 1). Nitrogen limitation causes reduction in intracellular cAMP levels, leading to expression of the pheromone response pathway components. Depletion of both nitrogen and glucose is communicated intracellularly through the action of heterotrimeric G proteins. Internalization of the decreased nutrients signal causes activation of the Gα subunit, Gpa2 (Isshiki *et al.*, 1992). Activated Gpa2 is responsible for positively stimulating adenylyl cyclase (Cyr1), which is also dependent on the cyclase-associated protein, Cap1 (Kawamukai *et al.*, 1991; Kawamukai *et al.*, 1992). Deletion of *gpa2*, *cyr1* or *cap1* results in inappropriate activation of the pheromone response pathway in nitrogen-rich medium (Maeda, Mochizuki and Yamamoto, 1990; Isshiki *et al.*, 1992). No cAMP is detected in *cyr1*Δ cells and *gpa2*Δ strains have decreased intracellular cAMP levels (Maeda, Mochizuki and Yamamoto, 1990; Isshiki *et al.*, 1992). Furthermore, mating is reduced by overexpression of *cyr1* or by mutation of *gpa2* to yield a constitutively-activated protein that is unable to hydrolyze GTP (Kawamukai *et al.*, 1991; Isshiki *et al.*, 1992). Although Gpa2 has been demonstrated to influence adenylyl cyclase activity, it is not known if the effect is direct or indirect. Intracellular cAMP levels are maintained by coordinating the activities of adenylyl cyclase and the cAMP-phosphodiesterase, Cgs2

(Maeda, Mochizuki and Yamamoto, 1990; Devoti *et al.*, 1991). Null mutations of *cgs2* result in increased cellular cAMP levels and inhibition of mating.

By analogy to other systems, the PKA regulatory subunit, Cgs1, is thought to be bound to the PKA catalytic subunit, Pka1. cAMP binding by Cgs1 would result in dissociation of the complex, freeing Pka1 to phosphorylate downstream targets (Devoti *et al.*, 1991; Maeda *et al.*, 1994). In rich medium, intracellular cAMP levels and Pka1 activity are high and the pheromone response pathway is inhibited; therefore, Pka1 has an inhibitory effect on sexual development. Increasing PKA catalytic activity by deletion of *cgs1* or overexpression of *pka1* inhibits mating, while deletion of *pka1* to decrease PKA activity causes inappropriate mating (Devoti *et al.*, 1991). Overexpression of a second cAMP-dependent protein kinase catalytic subunit, *sck1*, suppresses the defects of *pka1*Δ cells (Jin *et al.*, 1995). Deletion of *sck1* does not have a significant effect on mating, again suggesting that Pka1 negatively regulates sexual development.

Pka1 activates two different downstream pathways in response to distinct nutritional cues. A decrease in glucose results in a sudden significant decrease in intracellular cAMP which activates the gluconeogenesis pathway through induction of fructose-1,6-bisphosphatase, encoded by *fbp1* (Hoffman and Winston, 1991; Nocero *et al.*, 1994). Nitrogen starvation causes a slow reduction in cAMP levels and PKA activity to allow transcription of the *ste11* transcription factor gene and activation of the pheromone response pathway.

The transcription factor Ste11 is a HMG (high mobility group) DNA binding protein and is essential to mating (Sugimoto *et al.*, 1991). Ste11 regulates expression of components of the pheromone response pathway and, therefore, null mutations in *ste11* block mating at a point upstream of the pheromone response pathway. (Sugimoto *et al.*, 1991). Expression of *ste11*-dependent genes is achieved by binding of Ste11 to TR-box motifs (TTTCTTTGTT), so named due to the high number of thymines. Transcription of *ste11* is positively regulated by a MAPK cascade that is activated in response to osmotic, oxidative, and nutrient stresses. This stress-mediated cascade consists of Wak1 (MAPKKK), Wis1 (MAPKK), and Spc1 (also called Sty1; MAPK) (Stettler *et al.*, 1996; Aiba *et al.*, 1998; Nguyen *et al.*, 2000). At this time, little is known about how the cell detects stress or how the signal is transmitted to the MAPK cascade. Deletion of *spc1* results in reduced mating efficiency due to defective induction of *ste11*. Activated Spc1 phosphorylates an Activating Transcription Factor, Atf1, and *atf1*Δ strains are sterile due to an inability to respond to nitrogen starvation or induce *ste11* expression (Takeda *et al.*, 1995; Shiozaki and Russel, 1996). Phosphorylated Atf1 acts as a dimer with Pcr1, the *S. pombe* cAMP Response element binding protein, to induce *ste11* transcription (Watanabe and Yamamoto, 1996). The 5′ promoter region of *ste11* has the consensus site for Atf1-Pcr1 binding. The pathways mediated by Atf1-Pcr1 and cAMP are the major positive and negative regulators of *ste11* transcription, respectively.

Mating type in *S. pombe* is determined by two different groups of genes present at the *mat* locus. In M cells, *mat-Mm* and *mat-Mc* genes are encoded at the *mat* locus, while in P cells the *mat-Pm* and *mat-Pc* genes are present (Dooijes *et al.*, 1993). The *mat* genes are not expressed in vegetative cells and Ste11-mediated induction of these genes is necessary for pheromone and pheromone receptor transcription. Under nitrogen limiting conditions, *mat-Mm* and *mat-Pm* are expressed at a low level and pheromone exposure induces full expression of these genes. The *mat-Mm* and *mat-Pm* genes are both necessary for meiosis to occur following mating (Aono *et al.*, 1994). In M cells, the *mat-Mc* gene product binds ACAAT motifs, called M-boxes, to facilitate Ste11 binding at weak TR-boxes and subsequent transcription of M cell-specific genes (Kjaerulff *et al.*, 1997). Mat-Pc is thought to activate expression of the P cell pheromone receptor, Map3, by interacting with the transcription factor Map1 (Nielsen, Friis and Kjaerulff, 1996).

The pheromone response pathway is initiated by binding of a diffusible peptide hormone to a specific membrane receptor. P-factor is encoded by *map2* and secreted by P cells (Imai and Yamamoto, 1994). Mature P-factor is composed of 23 residues and does not undergo post-translational modifications, in contrast to *S. cerevisiae* α-factor. M cells produce M-factor, which is similar to the *S. cerevisiae* a-factor, and is encoded by the *mfm1*, *mfm2*, and *mfm3* genes (Davey, 1992; Kjaerulff, Davey and Nielsen, 1994). Biologically active M-factor is post-translationally carboxymethylated and farnesylated at the C-terminus (Davey, 1992). P-factor is bound by the P-factor receptor, Mam2, a GPCR similar to *S. cerevisiae* Ste2p that is expressed by M cells (Kitamura and Shimoda, 1991). Map3, a GPCR located in the plasma membrane of P cells, is similar to *S. cerevisiae* Ste3p and binds secreted M-factor (Tanaka *et al.*, 1993).

Disruption of the Gβ subunit gene *gpb1* results in accelerated mating and sporulation and, therefore, is thought to act as a negative regulator of the pheromone response pathway (Kim *et al.*, 1996). At this time, direct interaction between Gpb1 and Gpa1 has not been demonstrated and no Gγ subunit has been identified. Gpa1-GTP has been implicated in activation of the downstream pheromone response MAPK cascade, because deletion of *gpa1* results in a loss of response to pheromone (Obara *et al.*, 1991). However, it is not known how Gpa1 activates the pheromone response MAPKKK Byr2.

Activation of the pheromone MAPK cascade also requires Ras1, Ste4, Ste6 and Shk1. Ras1 binds and hydrolyzes GTP to GDP to cycle between activated and inactivated forms, respectively (Marshall, 1996). GTP cycling on Ras1 is facilitated by Ste6, a putative GEF, and Gap1, a GTPase Activating Protein, which accelerates the rate of GTP hydrolysis by Ras1. Ras1 localizes at, and recruits cytosolic Byr2 to, the plasma membrane (Bauman, Cheng and Albright, 1998). Ste4, a protein containing a leucine zipper, is only expressed during nitrogen starvation (Barr *et al.*, 1996). The leucine zipper of Ste4 interacts with the sterile alpha-motif (SAM) in the Byr2 regulatory region (Tu *et al.*, 1997; Bauman and Albright, 1998). The regulatory and catalytic domains of Byr2 display autoinhibition and binding of Ste4 and Ras1 to the regulatory domain may be required to interfere in this interaction. The autoinhibitory interaction of the Byr2 regulatory and catalytic domains is also thought to be antagonized by the PAK-like kinase Shk1. A specific interaction between Byr2 and Shk1 has not been defined, but Shk1 is thought to act as a MAPKKK kinase to phosphorylate Byr2 (Tu *et al.*, 1997). Phosphorylation of Byr2 is presumed to result in the sequential phosphorylation of Byr1 (MAPKK) and Spk1 (MAPK). Phosphorylated Spk1 is expected to activate mating-specific transcription factors and to have more than one downstream phosphorylation target. Ste11 is considered the most likely target of Spk1 because of the central role of Ste11 during sexual development and the observation that genes induced by pheromone exposure contain the TR-box for Ste11 binding.

4. *Cryptococcus neoformans*

C. neoformans is a dimorphic basidiomycete that alternates between budding yeast cells and filamentous growth (Alspaugh, Perfect and Heitman, 1998; Wang and Heitman, 1999). Nitrogen starvation causes the switch to filamentous growth and induces sexual development. Mating is dependent on the exchange of peptide pheromones between opposite mating partners, **α** and **a**, and results in the formation of conjugation tubes and cell fusion. Following cell fusion, the dikaryotic cell produces a mycelium composed of filaments that are linked by fused clamp connections. Differentiation occurs at the hyphal tips to produce rounded specialized structures called basidia. Inside basidia, nuclear fusion, meiosis, and mitosis occur to produce four haploid progeny spores. The progeny spores (basidiospores) form on the basidial surface by repeated budding and mitosis (Kwon-Chung, 1976). Basidiospores subsequently germinate and begin growth as a budding yeast cell. Sterility is defined as little or no filaments or spores produced following cell fusion.

Asexual sporulation (haploid fruiting) occurs independently of a mating partner in *MAT***α** cells during severe nitrogen and water depletion or in the presence of pheromone secreted by *MAT***a** cells (Wickes *et al.*, 1996; Wang, Perfect and Heitman, 2000). In haploid fruiting, cells undergo filamentation and formation of basidia and basidiospores that germinate and continue growth as budding yeast cells. One distinction between filamentation in haploid and dikaryotic cells is that clamp connections are not present in filaments of haploid fruiting cells (Sia, Lengeler and Heitman, 2000). Due to the mating type specificity involved, haploid fruiting is thought to function in fertilization by dispersing increased numbers of spores (Yue *et al.*, 1999).

A Gα, *GPA1*, and a Gβ, *GPB1* have been identified in *C. neoformans* (Tolkacheva *et al.*, 1994; Wang, Perfect and Heitman, 2000). The *gpa1*Δ displays little filamentous growth but, given enough time, mating occurs with production of a reduced number of spores (Alspaugh, Perfect and Heitman, 1997; Wang, Perfect and Heitman, 2000). Exogenous cAMP (2 mM) suppresses the mating defect of *gpa1*Δ and the zygote cell filaments and sporulates. The addition of cAMP does not affect wild-type mating and does not bypass the nitrogen starvation requirement (Alspaugh, Perfect and Heitman, 1997). Interestingly, the *gpa1*Δ mutant is still able to produce conjugation tubes in the presence of pheromone secreted from a wild-type cell of opposite mating type (Wang, Perfect and Heitman, 2000), suggesting the defect is downstream of pheromone sensing. The *gpb1*Δ mutant is sterile and does not produce filaments, basidia, or spores, even after prolonged incubation, and does not undergo haploid fruiting (Wang, Perfect and Heitman, 2000). In the presence of pheromone, *gpb1*Δ cells do not form conjugation tubes and 10 mM exogenous cAMP has no effect on the mating defect, suggesting that GPB1 participates in pheromone response independently of cAMP. Yeast two hybrid analysis also indicates that GPA1 and GPB1 do not physically interact (Wang, Perfect and Heitman, 2000). Taken together, these results suggest that GPA1 participates in sexual development by monitoring the cellular nutrient status through a cAMP-dependent pathway and that GPB1 is responsible for downstream activation in the pheromone response pathway. A MAPK homologue, CPK1, has recently been identified in *C. neoformans*. Expression of *CPK1* suppresses the *gpb1*Δ mating and haploid fruiting defects (Wang, Perfect and Heitman, 2000). Expression of *CPK1* in the *gpa1*Δ strain background has no effect on the sexual development defect. Genes encoding a protein similar to *S. cerevisiae* Ste20p have been identified, with different alleles in the two mating types (Sia, Lengeler and Heitman, 2000). Due to its similarity to *S. cerevisiae* Ste20p, *C. neoformans* STE20 may also function in activation of the pheromone response MAPK cascade.

A protein similar to the *S. cerevisiae* Ste12p transcription factor has been identified in *C. neoformans* and is specific to *MAT*α cells (Wickes, Edman and Edman, 1997). Overexpression of *STE12* results in haploid fruiting and the *ste12*Δ strain has slightly reduced frequency of mating and basidia formation early in mating (Wickes, Edman and Edman, 1997; Yue *et al.*, 1999; Chang *et al.*, 2000). However, further incubation results in *ste12*Δ mated cultures appearing like wild type (Yue *et al.*, 1999; Chang *et al.*, 2000). The *gpb1*Δ haploid fruiting defect is suppressed by overexpression of *STE12*, but there is no effect on mating (Yue *et al.*, 1999). As a result, STE12 is thought to participate in pathogenic development and not in mating (Chang *et al.*, 2000).

A Ras gene, *RAS1*, has been isolated from *C. neoformans* and is post-translationally modified with a lipid group, suggesting it is membrane bound (Tanaka *et al.*, 1999). After coculturing with the opposite mating type, no mating filaments are observed and no spores containing the mutation are recovered from the *ras1*Δ mutant (Alspaugh *et al.*, 2000). The *ras1*Δ mutant is unable to undergo cell fusion or establish a stable dikaryotic cell and, therefore, the mating defect occurs early (Alspaugh *et al.*, 2000). Overexpression of *GPB1* or the addition of 50 mM cAMP suppresses the mating defect (Alspaugh *et al.*, 2000). Expression of a *RAS1* gene with an activating mutation has no effect on the mating defects of *gpa1*Δ and *gpb1*Δ strains. These results suggest that RAS1 functions independently of GPA1 upstream of GPB1 in the pheromone response pathway or may participate in nutrient sensing in an upstream or parallel pathway.

5. *Neurospora crassa*

Neurospora crassa is a saprophytic filamentous ascomycete. Nitrogen starvation causes sexual development in *N. crassa* by inducing formation of sexual structures (protoperithecia) containing female gametes (Raju and Perkins, 1994). Male gametes can be any vegetative cell, usually conidia. For mating to occur, a male and female gamete of opposite mating type, **A** and **a**, must come into contact. Contact is achieved through growth of a specialized hyphal tube (trichogyne) from the protoperithecium towards the male conidium (Bistis, 1981, 1983). There is evidence for pheromone production by both males and females. The substance produced by males directs the females to grow towards them, while the female pheromone causes an apparent growth arrest in the male conidium, inhibiting germination. After fusion between the trichogyne and conidium, the conidial nucleus moves up the trichogyne tube by an unknown mechanism. Fertilization and subsequent meiotic and mitotic divisions take place once the conidial nucleus reaches the female gametes at the base of the protoperithecium. The whole structure then matures into a perithecium, containing asci each with eight ordered ascospores.

The genes that specify mating type map to a single locus whose sequence varies between the two mating types (Nelson, 1996). The region surrounding the mating type locus contains genes that regulate sexual development and heterokaryon incompatibility (Glass *et al.*, 1988). The **A** mating type region contains three predicted proteins: MT A-1, MT A-2 and MT A-3 (Ferreira, Saupe and Glass, 1996). MT A-1 is similar to *S. cerevisiae* MATα1 and is required for mating. MT A-2, an amphipathic α-helical protein, and MT A-3, a HMG-box protein, are essential for postfertilization functions. The **a** mating type region encodes a single protein, MT a-1, which is required for mating and heterokaryon incompatibility (Staben and Yanofsky, 1990). The N-terminus of MT a-1 contains a HMG domain which is similar to that of *S. pombe* Mat-Mc. Interestingly, the MT a-1 protein binds DNA sequences with a conserved CAAAT core, similar to that recognized by *S. pombe* Ste11 (Philley and Staben, 1994).

All three of the Gα subunits cloned from *N. crassa* – *gna-1*, *gna-2* and *gna-3* – have been shown to regulate sexual development. Deletion of *gna-1* results in formation of abnormal protoperithecia and perithecia and female sterility, traits which are exacerbated by the loss of *gna-2* (Ivey *et al.*, 1996; Baasiri *et al.*, 1997). Sexual development in Δ*gna-3* mutants is impaired during homozygous crosses, resulting in the production of inviable ascospores (Kays *et al.*, 2000). Exogenous cAMP has no effect on the mating defects of Δ*gna-1* and Δ*gna-3* mutants, although intracellular cAMP levels are lower in Δ*gna-3* strains during sexual development. GNA-1/GNA-2 and GNA-3 are hypothesized to mediate the pheromone response pathway and ascospore maturation, respectively; therefore, Gα proteins in *N. crassa* differentially regulate various aspects of sexual development. A Gβ (GNB-1) and Gγ (GNG-1) subunit have also been identified in *N. crassa*; preliminary results suggest that GNB-1 plays an indirect role in mating (Yang, Poole, and Borkovich, unpublished observations; Krystofova and Borkovich, unpublished observations).

Components of the *N. crassa* pheromone MAPK cascade have not been identified at this time, but ample evidence supports the hypothesis that this developmental stage is largely cAMP-independent. The adenylyl cyclase mutant, *cr-1*, has been shown to contain no intracellular cAMP or adenylyl cyclase activity and to display serious growth defects, resulting in a high occurrence of spontaneous suppressor mutations (Terenzi, Flawia and Torres, 1974; Terenzi *et al.*, 1976). However, *cr-1* is able to progress through sexual development with only a spore maturation delay, indicating that sexual development can occur in the absence of cAMP in *N. crassa.*

6. *Magnaporthe grisea*

The filamentous ascomycete *Magnaporthe grisea* can be induced to undergo sexual development by transferring cultures to a complex minimal medium containing plant material (oatmeal agar in most studies) (Kronstad *et al.*, 1998). Fertilized structures (perithecia) have been observed in *M. grisea*, but specialized unfertilized female structures have not been described (Shen, Bobrowicz and Ebbole, 1999). Within the perithecium, cell fusion, meiosis and two rounds of mitosis occur to yield eight haploid progeny spores encased in an ascus. The perithecium contains many asci and once the ascospores are mature they are discharged from the neck of the perithecium (Shen, Bobrowicz and Ebbole, 1999).

Mating in *M. grisea* is predicted to occur between two haploid cells of opposite mating type, *MAT1-1* and *MAT1-2* (Kang, Chumley, and Valent, 1994). Genes encoding pheromones have recently been identified in *M. grisea* (Shen, 1999). The sequence of the *MF1-1* putative pheromone gene, identified in the *MAT1-1* mating type, predicts a 26-residue peptide containing consensus sites for post-translational lipid modification. The *MF2-1* gene is expressed in MAT1-2 cells and encodes a protein predicted to be cleaved into four individual MF2-1 factors. Analysis of *MF1-1* and *MF2-1* indicates that expression is cell-type specific (Shen, Bobrowicz and Ebbole, 1999).

Three Gα subunits, *magA*, *magB* and *magC*, have been identified and characterized in *M. grisea* (Liu and Dean, 1997). Deletion of *magA* and *magC* results in defective ascospores that do not germinate, while the *magB* mutant is completely unable to form perithecia (Liu and Dean, 1997). These phenotypes suggest that MAGB functions early in mating and MAGA and MAGC participate in ascospore formation downstream of MAGB. Thus, *MAGB* appears to possess functions similar to *N. crassa gna-1* and *gna-2*, while *MAGA* and *MAGC* are more similar to *N. crassa gna-3*.

Mutation of the adenylyl cyclase gene, *mac1*, impairs *M. grisea* mating, resulting in the loss of perithecia formation (Choi and Dean, 1997). The similar phenotypes of the *mac1* and *magB* strains suggest that they participate in the same or parallel pathways. MAGB may sense nutrient availability by modulating adenylyl cyclase activity upstream of the pheromone response pathway, as observed in *S. pombe*. In a different scenario, MAGB could be responding to pheromone by activating adenylyl cyclase and, subsequently, PKA activates the downstream pheromone response MAPK cascade. A gene encoding a PKA catalytic subunit, *CPKA*, has been identified in *M. grisea* (Mitchell and Dean, 1995). Measurement of cAMP dependent kinase activity in *cpka* strains indicates that no remaining activity is detected, supporting the hypothesis that CPKA is the major PKA (Adachi and Hamer, 1998). Surprisingly, deletion of *CPKA* has no effect on vegetative and sexual development (Mitchell and Dean, 1995). This last observation can be explained by two hypotheses. Either a second unidentified catalytic subunit compensates for the deletion of *CPKA*, as observed in *S. pombe*, or the PKA pathway could be one of two or more pathways mediating development.

The two MAPKs identified in *M. grisea* are most similar to MAPKs from *S. cerevisiae* (Xu and Hamer, 1996; Xu, Staiger and Hamer, 1998). The *PMK1* (Pathogenicity MAP Kinase 1) protein sequence is most identical to Fus3p from *S. cerevisiae* (Xu and Hamer, 1996) and *PMK1* suppresses the mating defect of *S. cerevisiae fus3Δ kss1Δ* mutants. Deletion of *pmk1* has no effect on sexual development in *M. grisea*, suggesting that PMK1 does not function in the pheromone response pathway. Expression of the *M. grisea MPK1* MAPK in *S. cerevisiae slt2Δ* mutants suppresses the autolysis defect of the cell integrity pathway mutants. Furthermore, deletion of *MPK1* in *M. grisea* has a pleiotropic effect on vegetative and sexual development. The *mpk1Δ* strains undergoes progressive autolysis beginning in the older portion of a culture, displays reduced aerial hyphae and conidia formation and are female sterile, producing no ascospores (Xu, Staiger and Hamer, 1998).

7. *Ustilago maydis*

Pathogenesis in the dimorphic hemibasidiomycete, *Ustilago maydis* is developmentally regulated, occurring only after mating. During asexual development, *U. maydis* grows as saprophytic elongated unicellular sporidia that divide by budding (Bölker, 1998; Kronstad *et al.*, 1998). Sexual development is regulated by two loci, the biallelic ***a*** and multiallelic ***b***, that regulate expression of pheromones and pheromone receptors and formation of dikaryotic hyphae, respectively (Kronstad and Staben, 1997). Mating occurs between two haploid sporidia of opposite mating type, which is determined by the ***a*** locus (Banuett and Herskowitz, 1989; Bölker, Urban and Kahman, 1992). Products of the

a locus facilitate recognition of a mating partner and formation of conjugation tubes that fuse at the tips through the exchange of pheromones. Pheromone binding results in the formation of the infectious dikaryotic hyphae, only if the mated cells contain different alleles of the ***b*** locus (Banuett and Herskowitz, 1989; Spellig *et al.*, 1994). The infectious hyphae invade maize and initiate hyphal proliferation within the plant to form tumours. The dikaryotic filaments differentiate unicellular diploid teliospores that are subsequently released in distinctive clouds of black spores when the tumor bursts, explaining why the name "corn smut disease" is commonly used to describe *U. maydis* infections of maize. The released teliospores undergo meiosis to yield four haploid cells that form the promycelium and bud sporidia.

The ***a*** mating type locus has been shown to have two alleles, *a1* and *a2* (Bölker, Urban and Kahmann, 1992). In contrast to other fungi, these mating type alleles encode a lipopeptide pheromone and a pheromone receptor. The *a2* allele also contains two genes, *lga2* and *rga2*, of undetermined function and a region that appears to have encoded a second pheromone gene at one time (Urban, Kahmann and Bölker, 1996b). The multiallelic b locus has been demonstrated to have four alleles, *b1*, *b2*, *b3* and *b4* (Kronstad and Leong, 1990; Schulz *et al.*, 1990). The four alleles have an allele-specific region in the N-terminus. The remaining region is highly conserved and has been shown to encode two homeodomain proteins, bE and bW. These two homeodomain proteins dimerize to regulate development of dikaryotic filaments (Schulz *et al.*, 1990). Expression of the genes encoded at both the ***a*** and ***b*** locus is induced by pheromone exposure (Urban, Kahmann and Bölker, 1996a).

A pheromone response factor, *prf1*, has been identified and deletion of *prf1* results in sterility and avirulence (Hartmann, Kahmann and Bölker, 1996). The sterility of *prf1*Δ mutants has been attributed to its impaired expression of basal and induced levels of both pheromone and pheromone receptor. Furthermore, loss of a putative MAPK binding site in *prf1* causes reduced dikaryotic hyphae formation when the mutant is crossed to wild type. Although no filamentation is observed during a homokaryotic cross, this mutation does not result in complete sterility, as compared to *prf1*Δ strains (Müller *et al.*, 1999). These observations suggests that a MAPK activates Prf1, which then binds pheromone response elements to mediate infectious filamentous growth by regulating expression of genes encoded at the b locus.

As in *N. crassa* and *M. grisea*, more than two Gα subunits have also been identified in *U. maydis*: *gpa1*, *gpa2*, *gpa3*, and *gpa4* (Regenfelder *et al.*, 1997). Mutants deleted for *gpa1*, *gpa2* and *gpa4* are able to mate and form dikaryotic hyphae that display wild-type virulence; therefore, these Gα proteins are not essential for pathogenic and sexual development. In contrast, deletion of *gpa3* affects both asexual and sexual development. *gpa3* mutants produce less dikaryotic hyphae. This defect results from an inability to respond to pheromone, as deduced by the observation that *gpa3*Δ cells do not express pheromone-inducible genes (Regenfelder *et al.*, 1997). Homokaryotic crosses are unable to infect and form tumours; however, crosses of *gpa3*Δ to wild type are able to produce a few tumours on maize. Exogenous cAMP suppresses the pathogenic and mating defects of *gpa3*Δ and the addition of cAMP to wild-type cells bypasses the need for pheromone exposure to stimulate pheromone gene expression (Krüger *et al.*, 1998). Mutational activation of *gpa3* results in pheromone-independent mating (Regenfelder *et al.*, 1997). A mutant defective in the gene encoding adenylyl cyclase, *uac1*, grows constitutively as hyphae, forms filaments, and aggregates during growth in liquid medium (Barrett, Gold and Kronstad, 1993). Despite similar appearance of growth in *uac1* mutants and wild type, the *uac1* strain is nonpathogenic (Barrett, Gold and Kronstad, 1993). High levels of cAMP (25 mM) are required to observe budding growth in *uac1*Δ mutants; this same concentration causes multiple budding in wild type (Gold *et al.*, 1994). Deletion of *uac1* in a strain carrying an activating mutation in *gpa3* leads to constitutive filamentation, similar to that observed for the Δ*uac1* mutant, suggesting that Gpa3 functions upstream of Uac1 (Krüger *et al.*, 1998). Therefore, Gpa3 is thought to regulate the pheromone response and maize infection through a cAMP-dependent pathway.

A suppressor of Δ*uac1*, *Ustilago* bypass of cyclase (*ubc1*), has been shown to encode the PKA regulatory subunit (Gold *et al.*, 1994). Both the *uac1 ubc1* double mutant and the *ubc1* strain display multiple budding due to frequent misplaced bud site selection; therefore, bud site selection and budding is believed to be cAMP-dependent. Homokaryotic crosses between the double mutant or Δ*ubc1* strains result in reduced dikaryotic mycelial formation, but both strains are able to form normal mycelia during heterokaryotic crosses with wild type. Mutation of *ubc1* causes attenuated virulence and indicating that Ubc1 is required for maize infection. Two PKA catalytic subunits, *uka1* and *adr1*, have been identified, with Adr1 contributing the majority of catalytic activity (Dürrenberger, Wong and Kronstad, 1998). Deletion of *uka1* has no effect on mating, morphogenesis and pathogenesis; however, loss of *adr1* has a pleiotropic effect in *U. maydis*. *adr1*Δ mutants are similar to *uac1*Δ strains, in that they are covered in aerial hyphae and aggregate in liquid media. Furthermore, coinjection of wild type and *adr1*Δ into maize results in significantly reduced virulence and no infection symptoms are visible after coinjection of compatible *adr1*Δ strains. The ability of *adr1*Δ to mate cannot be evaluated on plates due to the constitutive filamentous growth phenotype of the mutant. Taken together, these studies suggest that the cAMP-dependent pathway regulating mating and virulence occurs through Gpa3-Uac1-Adr1, although activation of Uac1 by Gpa3 has not yet been been demonstrated.

One MAPKK, Fuz7, and two MAPKs, Kpp2 and Ubc3, have been identified in *U. maydis* (Banuett and Herskowitz, 1994; Mayorga and Gold, 1999, Müller *et al.*, 1999). Mutational analysis has shown that Fuz7 regulates mating by mediating formation of conjugation tubes and filamentation (Banuett and Herskowitz, 1994). Fuz7 is required for pathogenesis and has a role in tumour formation and teliospore germination. Epistatic studies indicate that *fuz7* and *gpa3* do not participate in the same pathway (Regenfelder *et al.*, 1997). The predicted amino acid sequence of Kpp2 indicates that it is most similar to

M. grisea Pmk1 and *C. albicans* CEK1 (Müller *et al.*, 1999). During confrontation assays, the *kpp2Δ* strain is unresponsive and exhibits reduced pheromone and *prf1* expression. Kpp2 also functions in formation of dikaryotic hyphae following cell fusion and filamentation. Exogenous cAMP has no effect on the *kpp2Δ* phenotypes and epistasis studies indicate that *kpp2* functions independently of the cAMP. Furthermore, over expression of *prf1* has no effect on *kpp2Δ*, but a *kpp2Δ* strain with a constitutively-expressed b allele displays improved formation of tumours. The *ubc3* gene was identified as a suppressor of the *uac1Δ* mutation and the encoded protein is similar to the *S. cerevisiae* pheromone response MAPK Fus3p (Mayorga and Gold, 1999). Ubc3 is essential for filamentation and formation of dikaryotic hyphae; therefore, Δ*ubc3* cells are defective in pheromone response and are significantly reduced in virulence. During mating, *ubc3Δ* mutants are defective in forming dikaryotic hyphae due to an inability to respond to pheromone. Mating is possible between two *ubc3Δ* cells of opposite mating type only when they come in contact, eliminating the need for a conjugation tube. *ubc3Δ* strains are capable of infecting maize, but have reduced virulence. However, the Δ*uac1* Δ*ubc3* mutant is significantly reduced in virulence, indicating that deletion of the two genes has an additive effect on pathogenesis (Mayorga and Gold, 1999). The *uac1Δ ubc3Δ* and *uac1Δ fuz7Δ* strains display similar phenotypes, suggesting that the Fuz7-Ubc3 cascade is downstream of PKA (Mayorga and Gold, 1999).

8. Conclusions

Analysis of the pheromone response pathways of *S. cerevisiae* and *S. pombe* has provided an expanded knowledge of the regulation and transmission of G protein ssignalling pathways in eukaryotes. In both systems, diffusible peptide pheromones are secreted and bound to GPCRs expressed by cells of opposite mating type to initiate the pheromone response pathway. The signal is internalized by dissociation of heterotrimeric G proteins, which in turn activate the pheromone pathway-specific MAPK cascade. Completion of the pheromone response pathway results in cell cycle arrest at the G1 phase, although the arrested cells continue to grow. The cell produces a shmoo projection that searches out its mating partner and undergoes cell fusion to form a zygote cell. How the signal is transmitted from the G protein to the cascade, as well as the role of membrane-localized proteins (such as Ras, Rho, and Pak), remain active areas of debate in both yeasts; however, in both systems the MAPKKK must be recruited to the membrane during the response. Evidence of membrane localization and complex formation by these proteins suggests that their purpose is to form a scaffolding structure to facilitate maintenance of the MAPK cascade. This mechanism is similar to that suggested for the role of Ras, Rho and Pak proteins during activation of MAPK cascades in mammalian systems (Lopez-Ilasaca, 1998).

Although the overall mode of response is similar, differences do exist between

S. cerevisiae and *S. pombe*. The *S. cerevisiae* MAPK cascade is activated by the Gβγ dimer, while the *S. pombe* MAPK cascade is regulated by the GTP-bound Gα subunit. Different G proteins are responsible for activating the cascades, but the opposing free G protein subunit(s) (Gα in *S. cerevisiae* and Gβγ in *S. pombe*) appears to serve an inhibitory role in each pathway. In contrast to *S. pombe*, the *S. cerevisiae* pheromone MAPK components are bound by the Ste5p scaffolding protein, maintaining the three modules in close physical association. This scaffolding system allows efficient sequential phosphorylation and prevents crosstalk between the pheromone and filamentous MAPK pathways. No such scaffolding protein has been discovered for either of the MAPK cascades in *S. pombe*; however, the possibility that one of the cascade components acts as a scaffolding module has not been explored.

The most notable difference between the two systems is the requirement for nitrogen starvation conditions to initiate sexual development in *S. pombe*. *S. cerevisiae* only requires nitrogen starvation (and a non-fermentable carbon source) to undergo meiosis, once a diploid cell has already been formed by mating. The requirement for nitrogen depletion significantly complicates mating in *S. pombe*. A stress-sensing MAPK cascade and a cAMP-dependent pathway operate upstream of the pheromone pathway to monitor nutrient availability during sexual development. Therefore, the apparent absence of scaffolding proteins in the MAPK cascades of *S. pombe* may allow more fluidity in signalling. A lack of physical association between the modules may facilitate easy integration of signals to interrupt the sequential phosphorylation cascade as needed. This system of co-dependent signalling networks would provide *S. pombe* with the mechanism to continually monitor nutrient availability and relay any changes to the pheromone response pathway.

Many components in the pheromone response pathway in filamentous fungi have been identified by their similarity to *S. cerevisiae* proteins. However, sexual development in filamentous fungi appears more similar to *S. pombe* than to *S. cerevisiae*. Like *S. pombe*, all filamentous fungal species require defined nutrient conditions to initiate mating and nutrient availability is monitored through a cAMP-dependent pathway and MAPK cascade. Deletion of the G proteins required for nutrient monitoring or the pheromone response results in impaired mating. Mating is also impaired if intracellular cAMP levels are reduced, often through mutational inactivation of the adenylyl cyclase gene. Interestingly, the *M. grisea* homolog of *S. cerevisiae* Fus3p (PMK1) is able to suppress the mating defects of a *S. cerevisiae fus3Δ kss1Δ* mutant, but does not appear to participate in *M. grisea* mating. The cell wall integrity MAPK homologue, MPS1, does appear to positively regulate perithecial formation in *M. grisea*, supporting the prediction that a MAPK cascade mediates the pheromone response pathway in *M. grisea*. The saprophytic fungus, *N. crassa* displays the *S. pombe* nutrient requirement for induction of sexual development; however, *N. crassa* apparently does not monitor the cellular nutrient status through a cAMP-dependent pathway. Thus, filamentous fungal systems all apparently possess a conserved MAPK cascade in the pheromone response pathway. The advent of genome sequencing projects promises that all MAPK components will soon be defined in these species. Future investigations will reveal the circuitry that integrates nutritional and pheromone cues.

The corn smut pathogen, *U. maydis* is unique among the organisms discussed here, in that it is the only species that has coupled mating and virulence. The pheromone response pathway does not appear to be regulated by a G protein; instead, the Gpa3 Gα somehow activates adenylyl cyclase and PKA, which then signals to the downstream Fuz7-Ubc3 pheromone response MAPK cascade. Gβ and Gγ subunits have not been identified in this system; therefore, the possibility that Gβγ functions in the pheromone response pathway cannot be ruled out.

Homologues of the different signalling components regulating mating also participate in virulence of fungal pathogens, such as *C. neoformans* and *M. grisea.* Therefore, it appears that the pathways regulating mating have been conserved between filamentous fungi, with as yet unknown accessory proteins determining whether a species is a pathogen or saprophyte. Future studies that include comparison of the signal transduction systems in the organisms presented here will undoubtedly elucidate fundamental paradigms for environmental sensing in all fungal species.

References

Adachi, K., and Hamer, J.E. (1998) Divergent cAMP signaling pathways regulate growth and pathogenesis in the rice blast fungus *Magnaporthe grisea. Plant Cell* **10**, 1361–1373.

Aiba, H., Kawaura R., Yamamoto E., Yamada, H., Takegawa, K., and Mizuno, T. (1998) Isolation and characterization of high-osmolarity-sensitive mutants of fission yeast. *Journal of Bacteriology* **180**, 5038–5043.

Akada, R., Kallal, L., Johnson, D.I., and Kurjan, J. (1996) Genetic relationships between the G protein βγ complex, Ste5p, Ste20p, and Cdc42p. Investigation of effector roles in the yeast pheromone response pathway. *Genetics* **143**, 103–117.

Alspaugh, J.A., Perfect, J.R., and Heitman, J. (1997) *Cryptococcus neoformans* mating and virulence are regulated by the G-protein α subunit GPA1 and cAMP. *Genes and Development* **11**, 3206–3217.

Alspaugh, J.A., Perfect, J.R., and Heitman, J. (1998) Signal transduction pathways regulating differentiation and pathogenicity of *Cryptococcus neoformans. Fungal Genetics and Biology* **25**, 1–14.

Alspaugh, J.A., Cavallo, L.M., Perfect, J.R., and Heitman, J. (2000) *RAS1* regulates filamentation, mating and growth at high temperature of *Cryptococcus neoformans. Molecular Microbiology* **36**, 352–365.

Aono, T., Yanai, H., Miki, F., Davey, J., and Shimoda, C. (1994) Mating pheromone-induced expression of the *mat1-Pm* gene of *Schizosaccharomyces pombe*: identification of signalling components and characterization of upstream controlling elements. *Yeast* **10**, 757–770.

Baasiri, R.A., Lu, X., Rowley, P.S., Turner, G.E., and Borkovich, K.A. (1997) Overlapping functions for two G protein α subunits in *Neurospora crassa. Genetics* **174**, 137–145.

Banuett, F. (1998) Signalling in the yeasts: an informational cascade with links to the filamentous fungi. *Microbiology and Molecular Biology Reviews* **62**, 249–274.

Banuett, F., and Herskowitz, I. (1989) Different *a* alleles of *Ustilago maydis* are necessary for maintenance of filamentous growth but not for meiosis. *Proceedings of the National Academy of Sciences, USA* **86**, 5878–5882.

Banuett, F., and Herskowitz, I. (1994) Identification of *fuz7*, a *Ustilago maydis* MEK/MAPKK homolog required for **a**-locus-dependent and -independent steps in the fungal life cycle. *Genes and Development* **8**, 1367–1378.

Barr, M.M., Tu, H., Van Aelst, L., and Wigler, M. (1996) Identification of Ste4 as a potential regulator of Byr2 in the sexual response pathway of *Sichizosaccharomyces pombe. Molecular and Cellular Biology* **16**, 5597–5603.

Barrett, K.J., Gold, S.E., and Kronstad, J.W. (1993) Identification and complementation of a mutation to constitutive filamentous growth in *Ustilago maydis. Molecular Plant – Microbe Interactions* **6**, 274–283.

Bauman, P., and Albright, C.F. (1998) Functional analysis of domains in the Byr2 kinase. *Biochimie* **80**, 621–625.

Bauman, P., Cheng, Q.C., and Albright, C.F. (1998) The Byr2 kinase translocates to the plasma membrane in a Ras1-dependent manner. *Biochemical and Biophysical Research Communications* **244**, 468–474.

Beavo, J.A. (1995) Cyclic nucleotide phosphodiesterases: functional implications of multiple isoforms. *Physiological Reviews* **75**, 725–748.

Berman, D.M., and Gilman, A.G. (1998) Mammalian RGS proteins: barbarians at the gate. *Journal of Biological Chemistry* **273**, 1269–1272.

Bistis, G.N. (1981) Chemotropic interactions between trichogynes and conidia of opposite mating-type in *Neurospora crassa. Mycologia* **73**, 959–975.

Bistis, G.N. (1983) Evidence for diffusible, mating-type-specific trichogyne attractants in *Neuropora crassa. Experimental Mycology* **7**, 292–295.

Blondel, M., Alepuz, P.M., Huang, L.S., Shahan, S., Ammerer, G., and Peter, M. (1999) Nuclear export of Far1p in response to pheromones requires the export receptor Msn5p/Ste21p. *Genes and Development* **13**, 2284–2300.

Blumer, K.J., and Thorner, J. (1990) β and γ subunits of a yeast guanine nucleotide-binding protein are not essential for membrane association of the α subunit but are required for receptor coupling. *Proceedings of the National Academy of Sciences, USA* **87**, 4363–4367.

Blumer, K.J., Reneke, J.E., and Thorner, J. (1988) The *STE2* gene product is the ligand-binding componet of the α-factor receptor of *Saccharomyces cerevisiae. Journal of Biological Chemistry* **263**, 10836–10842.

Bölker, M. (1998) Sex and Crime: heterotrimeric G proteins in fungal mating and pathogenesis. *Fungal Genetics and Biology* **25**, 143–156.

Bölker, M., Urban, M., and Kahmann, R. (1992) The *a* mating type locus of *U. maydis* specifies cell signaling components. *Cell* **68**, 441–450.

Burbelo, P.D., Drechesel, D., and Hall, A. (1995) A conserved binding motif defines numerous candidate target proteins for both Cdc42 and Rac GTPases. *Journal of Biological Chemistry* **270**, 29071–29074.

Butty, A.C., Pryciak, P.M., Huang, L.S., Herskowitz, I., and Peter, M. (1998) The role of Far1p in linking the heterotrimeric G protein to polarity establishment proteins during yeast mating. *Science* **282**, 1511–1516.

Cabib, E., Drgonova, J., and Drgon, T. (1998) Role of small G proteins in yeast cell polarization and wall biosynthesis. *Annual Review of Biochemistry* **67**, 307–333.

Cairns, B.R., Ramer, S.W., and Kornberg, R.D. (1992) Order of action of components in the yeast pheromone response pathway revealed with a dominant allele of the STE11 kinase and the multiple phosphorylation of the STE7 kinase. *Genes and Development* **6**, 1305–1318.

Chang, Y.C., Wickes, B.L., Miller, G.F., Penoyer, L.A., and Kwon-Chung, K.J. (2000) *Cryptococcus neoformans STE12α* regulates virulence but is not essential for mating. *Journal of Experimental Medicine* **191**, 871–881.

Choi, W., and Dean, R.A. (1997) The adenylate cyclase gene *MAC1* of *Magnaporthe grisea* controls appressorium formation and other aspects of growth and development. *Plant Cell* **9**, 1973–1983.

Coria, R., Savinon-Tejeda, A.L., and Birnbaumer, L. (1995) *STE2/SCG1*-dependent inhibition of *STE4*-induced growth arrest by mutant *STE4*Δ C6 in the yeast pheromone response pathway. *FEBS Letters* **367**, 122–126.

Crespo, P., Xu, N., Simonds, W.F., and Gutkind, J.S. (1994) Ras-dependent activation of MAP kinase pathway mediated by G-protein βγ subunits. *Nature* **369**, 418–420.

Crespo, P., Cachero, T.G., Xu, N., and Gutkind, J.S. (1995) Dual effect of β-adrenergic receptors on mitogen-activated protein kinase. *Journal of Biological Chemistry* **270**, 25259–25265.

Davey, J. (1992) Mating pheromones of the fission yeast *Schizosaccharomyces pombe*: purification and structural characterization of M-factor and isolation and analysis of two genes encoding the pheromone. *EMBO Journal* **11**, 951–960.

Davey, J. (1998) Fusion of a fission yeast. *Yeast* **14**, 1529–1566.

Dessauer, C.W., Posner, B.A., and Gilman, A.G. (1996) Visualizing signal transduction: receptors, G proteins, and adenylate cyclases. *Clinical Sciences* **91**, 527–537.

Devoti, J., Seydoux, G., Beach, D., and McLeod, M. (1991) Interaction between *ran1*$^+$ protein kinase and cAMP-dependent protein kinase as negative regulators of fission yeast meiosis. *EMBO Journal* **10**, 3759–3768.

Dietzel, C., and Kurjan, J. (1987) Pheromonal regulation and sequence of the *Saccharomyces cerevisiae* SST2 gene: a model for desensitization to pheromone. *Molecular and Cellular Biology* **7**, 4169–4177.

Dooijes, D., van de Wetering, M., Knippels, L., and Clevers, H. (1993) The *Schizosaccharomyces pombe* mating-type gene *mat-Mc* encodes a sequence-specific DNA-binding high mobility group box protein. *Journal of Biological Chemistry* **268**, 24813–24817.

Downes, G.B., and Gautam, N. (1999) The G protein subunit gene families. *Genomics* **62**, 544–552.

Dürrenberger, F., Wong, K., and Kronstad, J.W. (1998) Identification of a cAMP-dependent protein kinase catalytic subunit required for virulence and morhogenesis in *Ustilago maydis*. *Proceedings of the National Academy of Sciences, USA* **95**, 5684–5689.

Elion, E.A., Brill, J.A., and Fink, G.R. (1991) FUS3 represses CLN1 and CLN2 and in concert with KSS1 promotes signal transduction. *Proceedings of the National Academy of Sciences, USA* **88**, 9392–9396.

Elion, E.A., Satterberg, B., and Kranz, J.E. (1993) FUS3 phosphorylates multiple components of the mating signal transduction cascade: evidence for STE12 and FAR1. *Molecular Biology of the Cell* **4**, 495–510.

Ferreira, A.V., Saupe, S., and Glass, N.L. (1996) Transcriptional analysis of the *mtA* idiomorph of *Neurospora crassa* identifies two genes in addition to *mtA-1*. *Molecular and General Genetics* **250**, 767–774.

Francis, S.H., and Corbin, J.D. (1999) Cyclic nucleotide-dependent protein kinases: intracellular receptors for cAMP and cGMP action. *Critical Reviews in Clinical Laboratory Sciences* **36**, 275–328.

Garrington, T.P., and Johnson, G.L. (1999) Organization and regulation of mitogen-activated protein kinase signaling pathways. *Current Opinion in Cell Biology* **11**, 211–218.

Gartner, A., Namyth, K., and Ammerer, G. (1992) Signal transduction in *Saccharomyces cerevisiae* requires tyrosine and threonine phosphorylation of FUS3 and KSS1. *Genes and Development* **6**, 1280–1292.

Gartner, A., Jovanovic, A., Jeoung, D.I., Bourlat, S., Cross, F.R., and Ammerer, G. (1998) Pheromone-dependent G1 cell cycle arrest requires Far1 phosphorylation, but may not involve inhibition of Cdc28-Cln2 kinase, *in vivo*. *Molecular and Cellular Biology* **18**, 3681–3691.

Gautam, N., Downes, G.B., Yan, K., and Kisselev, O. (1998) The G-protein βγ complex. *Cellular Signalling* **10**, 447–455.

Glass, N.L., Vollmer, S.J., Staben, C., Grotelueschen, J., Metzenberg, R.L., and Yanofsky, G. (1988) DNAs of the two mating-type alleles of *Neuropora crassa* are highly dissimilar. *Science* **241**, 570–573.

Gold, S., Duncan, G., Barrett, K., and Kronstad, J. (1994) cAMP regulates morphogenesis in the fungal pathogen *Ustilago maydis*. *Genes and Development* **8**, 2805–2816.

Gorlich, D., Dabrowski, M., Bischoff, F.R., Kutay, U., Bork, P., Hartmann, E., Prehm, S., and Izaurralde, E. (1997) A novel class of RanGTP binding proteins. *Journal of Cell Biology* **138**, 65–80.

Gustin, M.C., Albertyn, J., Alexander, M., and Davenport, K. (1998) MAP kinase pathways in the yeast *Saccharomyces cerevisiae. Microbiology and Molecular Biology Reviews* **2**, 1264–1300.

Hartmann, H.A., Kahmann, R., and Bölker, M. (1996) The pheromone response factor coordinates filamentous growth and pathogenicity in *Ustiliago maydis. EMBO Journal* **15**, 1632–1641.

Hirschman, J.E., and Jenness, D.D. (1999) Dual lipid modification of the yeast Gγ subunit Ste18p determines membrane localization of Gβγ. *Molecular and Cellular Biology* **19**, 7705–7711.

Hoffman, C.S., and Winston, F. (1991) Glucose repression of transcription of the *Schizosaccharomyces pombe flb1* gene occurs by a cAMP signaling pathway. *Genes and Development* **5**, 561–571.

Imai, Y., and Yamamoto, M. (1994) The fission yeast mating pheromone P-factor: its molecular structure, gene structure, and ability to induce gene expression and G1 arrest in the mating partner. *Genes and Development* **8**, 328–338.

Isshiki, T., Mochizuki, N., Maeda, T., and Yamamoto, M. (1992) Characterization of a fission yeast gene, *gpa2*, that encodes a Gα subunit involved in the monitoring of nutrition. *Genes and Development* **6**, 2455–2462.

Ivey, F.D., Yang, Q., and Borkovich, K.A. (1999) Positive regulation of adenylyl cyclase activity by a Gαi homolog in *Neurospora crassa. Fungal Genetics and Biology* **26**, 48–61.

Ivey, F.D., Hodge, P.N., Turner, G.E., and Borkovich, K.A. (1996) The Gαi homologue *gna-1* controls multiple differentiation pathways in *Neurospora crassa. Molecular Biology of the Cell* **7**, 1283–1297.

Jin, M., Fujita, M., Culley, B.M., Apoliario, E., Yamamoto, M., Maundrell, K., and Hoffman, C.S. (1995) *sck1*, a high copy number suppressor of defects in the cAMP-dependent protein kinase pathway in fission yeast, encodes a protein homologous to the *Saccharomyces cerevisiae* SCH9 kinase. *Genetics* **140**, 457–467.

Johnson, D.I. (1999) Cdc42: An essential Rho-type GTPase controlling eukaryotic cell polarity. *Microbiology and Molecular Biology Reviews* **63**, 54–105.

Kang, S., Chumley, F.G., and Valent, B. (1994) Isolation of the mating-type genes of the phytopathogenic fungus *Magnaporthe grisea* using genomic subtraction. *Genetics* **138**, 289–296.

Kawamukai, M., Ferguson, K., Wigler, M., and Young, D. (1991) Genetic biochemical analysis of the adenylyl cyclase of *Schizosaccharomyces pombe. Cell Regulation* **2**, 155–164.

Kawamukai, M., Gerst, J., Field, J., Riggs, M., Rodgers, L., Wigler, M., and Young, D. (1992) Genetic and biochemical analysis of the adenylyl cyclase-associated protein, cap, in *Schizosaccharomyces pombe. Molecular Biology of the Cell* **3**, 167–180.

Kays, A.M., Rowley, P.S., Baasiri, R.A., and Borkovich, K. (2000) Regulation of conidiation and adenylyl cyclase levels by the Gα protein GNA-3 in *Neurospora crassa. Molecular and Cellular Biology* **20**, 7693–7705.

Kim, D.U., Park, S.K., Chung, K.S., Choi, M.U. and Yoo, H.S. (1996) The G protein β subunit Gpb1 of *Schizosaccharomyces pombe* is a negative regulator of sexual development. *Molecular and General Genetics* **252**, 20–32.

Kitamura, K., and Shimoda, C. (1991) The *Schizosaccharomyces pombe mam2* gene encodes a putative pheromone receptor which has a significant homology with the *Saccharomyces cerevisiae* Ste2 protein. *EMBO Journal* **10**, 3743–3751.

Kjaerulff, S., Davey, J., and Nielsen, O. (1994) Analysis of the structural genes encoding M-factor in the fission yeast *Schizosaccharomyces pombe*: identification of a third gene, *mfm3*. *Molecular and Cellular Biology* **4**, 3895–3905.

Kjaerulff, S., Dooijes, D., Clevers, H., and Nielsen, O. (1997) Cell differentiation by interaction of two HMG-box proteins: Mat1-Mc activates M cell-specific genes in *S. pombe* by recruiting the ubiquitous transcription factor Ste11 to weak binding sites. *EMBO Journal* **16**, 4021–4033.

Klein, S., Reuveni, H., and Levitzki, A. (2000) Signal transduction by a nondissociable heterotrimeric yeast G protein. *Proceedings of the National Academy of Sciences, USA* **97**, 3219–3223.

Kronstad, J.W., and Leong, S.A. (1990) The *b* mating-type locus of *Ustilago maydis* contains variable and constant regions. *Genes and Development* **4**, 1384–1395.

Kronstad, J.W., and Staben, C. (1997) Mating type in filamentous fungi. *Annual Review of Genetics* **31**, 245–276.

Kronstad, J., De Maria, A.D., Funnell, D., Laidlaw, R.D., Lee, N., and Ramesh, M.M.D.M. (1998) Signaling via cAMP in fungi: interconnections with mitogen-activated protein kinase pathways. *Archives of Microbiology* **170**, 395–404.

Krüger, J., Loubradou, G., Regenfelder, E., Hartmann, A., and Kahmann, R. (1998) Crosstalk between cAMP and pheromone signalling pathways in *Ustilago maydis. Molecular and General Genetics* **260**, 193–198.

Kurjan, J., and Herskowitz, I. (1982) Structure of a yeast pheromone gene (MFα): a putative α-factor precursor contains four tandem copies of mature α-factor. *Cell* **30**, 933–943.

Kurjan, J., and Dietzel, C. (1988) Analysis of the role of SCG1, a Gα homolog, and SST2 in pheromone response and desensitization in yeast. *Cold Spring Harbor Symposia on Quantitative Biology* **53**, 577–584.

Kwon-Chung, K.J. (1976) Morphogenesis of *Filobibasidiella neoformans*, the sexual state of *Cryptococcus neoformans. Mycologia* **68**, 943–946.

Leberer, E., Wu, C., Leeuw, T., Fourest Lieuvin, A., Segall, J.E., and Thomas, D.Y. (1997) Functional characterization of the Cdc42p binding domain of yeast Ste20p protein kinase. *EMBO Journal* **16**, 83–97.

Liu, S., and Dean, R.A. (1997) G protein α subunit genes control growth, development, and pathogenicity of *Magnaporthe grisea. Molecular Plant – Microbe Interactions* **10**, 1075–1086.

Lopez-Ilasaca, M. (1998) Signaling from G-protein-coupled receptors to mitogen-activated protein (MAP)-kinase cascade. *Biochemical Pharmacology* **56**, 269–277.

Madden, K., and Snyder, M. (1998) Cell polarity and morphogenesis in budding yeast. *Annual Review of Microbiology* **52**, 687–744.

Madhani, H.D., Styles, C.A., and Fink, G.R. (1997) MAP kinases with distinct inhibitory functions impart signaling specificity during yeast differentiation. *Cell* **91**, 673–684.

Maeda, T., Mochizuki, N., and Yamamoto, M. (1990) Adenylyl cyclase is dispensable for vegetative cell growth in the fission yeast *Schizosaccharomyces pombe. Proceedings of the National Academy of Sciences, USA* **87**, 7814–7818.

Maeda, T., Watanabe, Y., Kunitomo, H., and Yamamoto, M. (1994) Cloning of the *pka1* gene encoding the catalytic subunit of the cAMP-dependent protein kinase in *Schizosaccharomyces pombe. Journal of Biological Chemistry* **269**, 9632–9637.

Mahanty, S.K., Wang, Y., Farley, F.W., and Elion, E.A. (1999) Nuclear shuttling of yeast scaffold Ste5 is required for its recruitment to the plasma membrane and activation of the mating MAPK cascade. *Cell* **98**, 501–512.

Manahan, C.L., Patnana, M., Blumer, K.J., and Linder, M.E. (2000) Dual lipid modification motifs in Gα and Gγ subunits are required for full activity of the pheromone response pathway in *Saccharomyces cerevisiae. Molecular Biology of the Cell* **11**, 957–968.

Marcus, S., Polverino, A., Barr, M., and Wigler, M. (1994) Complexes between STE5 and components of the pheromone-responsive mitogen-activated protein kinase module. *Proceedings of the National Academy of Sciences, USA* **91**, 7762–7766.

Marshall, C.J. (1996) Ras effectors. *Current Opinion in Cell Biology* **8**, 197–204.

Mayorga, M.E., and Gold, S.E. (1999) A MAP kinase encoded by the *ubc3* gene of *Ustilago maydis* is required for filamentous growth and full virulence. *Molecular Microbiology* **34**, 485–497.

Michaelis, S., and Herskowitz, I. (1988) The a-factor pheromone of *Saccharomyces cerevisiae* is essential for mating. *Molecular and Cellular Biology* **8**, 1309–1318.

Mitchell, T.K., and Dean, R.A. (1995) The cAMP-dependent protein kinase catalytic subunit is required for appressorium formation and pathogenesis by the rice blast pathogen *Magnaporthe grisea. Plant Cell* **7**, 1869–1878.

Miyajima, I., Nakafuku, M., Nakayama, N., Brenner, C., Miyajima, A., Kaibuchi, K., Arai, K., Kaziro, Y., and Matsumoto, K. (1987) GPA1, a haploid-specific essential gene, encodes a yeast homolog of mammalian G protein which may be involved in mating factor signal transduction. *Cell* **50**, 1011–1019.

Müller, P., Aichinger, C., Feldbrügge, M., and Kahmann, R. (1999) The MAP kinase Kpp2 regulates mating and pathogenic development in *Ustilago maydis. Molecular Microbiology* **34**, 1007–1017.

Nelson, M.A. (1996) Mating systems in ascomycetes: a romp in the sac. *Trends in Genetics* **12**, 69–74.

Neer, E.J. (1997) Intracellular signalling: turning down G-protein signals. *Current Biology* **7**, R31–R33.
Nern, A., and Arkowitz, R.A. (1999) A Cdc24p-Far1p-Gβγ protein complex required for yeast orientation during mating. *Journal of Cell Biology* **138**, 65–80.
Nguyen, A.N., Lee, A., Place, W., and Shiozaki, K. (2000) Multistep phosphorelay proteins transmit oxidative stress signals to the fission yeast stress-activated protein kinase. *Molecular Biology of the Cell* **11**, 1169–1181.
Nielsen, O., Friis, T., and Kjaerulff, S. (1996) The *Schizosaccharomyces pombe map1* gene encodes an SRF/MCM1-related protein required for P-cell specific gene expression. *Molecular and General Genetics* **253**, 387–392.
Nocero, M., Isshiki, T., Yamamoto, M., and Hoffman, C.S. (1994) Glucose repression of *fpb1* transcription in *Schizosaccharomyces pombe* is partially regulated by adenylate cyclase activation by a G protein α subunit encoded by *gpa2* (*git8*). *Genetics* **138**, 39–45.
Obara, T., Nakafuku, M., Yamamoto, M., and Kaziro, Y. (1991) Isolation and characterization of a gene encoding a G-protein α subunit from *Schizosaccharomyces pombe*: involvement in mating and sporulation pathways. *Proceedings of the National Academy of Sciences, USA* **88**, 5877–5881.
Olson, K.A., Nelson, C., Tai, G., Hung, W., Yong, C., Astell, C., and Sadowski, I. (2000) Two regulators of Ste12p inhibit pheromone-responsive transcription by separate mechanisms. *Molecular and Cellular Biology* **20**, 4199–4209.
Overton, M.C., and Blumer, K.J. (2000) G-protein-coupled receptors function as oligomers *in vivo*. *Current Biology* **10**, 341–344.
Peter, M., and Herskowitz, I. (1994) Direct inhibition of the yeast cyclin-dependent kinase Cdc28-Cln by Far1. *Science* **265**, 1228–1231.
Peter, M., Gartner, A., Horecka, J., Ammerer, G., and Herskowitz, I. (1993) FAR1 links the signal transduction pathway to the cell cycle machinery in yeast. *Cell* **73**, 747–760.
Peter, M., Neiman, A.M., Park, H.O., van Lohnizen, M., and Herskowitz, I. (1996) Functional analysis of the interaction between the small GTP binding protein Cdc42 and the Ste20 protein kinase in yeast. *EMBO Journal* **15**, 7046–7059.
Philley, M.L., and Staben, C. (1994) Functional analyses of the *Neurospora crassa* MT *a-1* mating type polypeptide. *Genetics* **137**, 715–722.
Pryciak, P.M., and Huntress, F.A. (1998) Membrane recruitment of the kinase cascade scaffold protein Ste5 by the Gβγ complex underlies activation of the yeast pheromone response pathway. *Genes and Development* **12**, 2684–2697.
Raju, N.B., and Perkins, D.D. (1994) Diverse programs of ascus development in pseudohomothallic species of *Neurospora, Gelasinospora*, and *Podospora*. *Developmental Genetics* **15**, 104–118.
Regenfelder, E., Spellig, T., Hartmann, A., Lauenstein, S., Bolker, M., and Kahmann, R. (1997) G proteins in *Ustilago maydis*: transmission of multiple signals? *EMBO Journal* **16**, 1934–1942.
Schulz, B., Banuett, F., Dahl, M., Schlesinger, R., Schafer, W., Martin, T., Herskowitz, I., and Kahmann, R. (1990) The *b* alleles of *U. maydis*, whose combinations program pathogenic development, code for polypeptides containing a homeodomain-related motif. *Cell* **60**, 295–306.
Shen, W.C., Bobrowicz, P., and Ebbole, D.J. (1999) Isolation of pheromone precursor genes of *Magnaporthe grisea*. *Fungal Genetics and Biology* **27**, 253–263.
Shiozaki, K., and Russel, P. (1996) Conjugation, meiosis, and the osmotic stress response are regulated by Spc1 kinase through Atf1 transcription factor in fission yeast. *Genes and Development* **10**, 2276–2288.
Sia, R.A., Lengeler, K.B., and Heitman, J. (2000) Diploid strains of the pathogenic basidiomycete *Cryptococcus neoformans* are thermally dimorphic. *Fungal Genetics and Biology* **29**, 153–163.
Singh, A., Chen, E.Y., Lugovoy, J.M., Chang, C.N., Hitzeman, R.A., and Seeburg, P.H. (1983) *Saccharomyces cerevisiae* contains two discrete genes coding for the α-factor pheromone. *Nucleic Acids Research* **11**, 4049–4063.
Siow, Y.L., Kalmar, G.B., Sanghera, J.S., Tai, G., Oh, S.S., and Pelech, S.L. (1997) Identification of two essential phosphorylated threonine residues in the catalytic domain of Mekk1. *Journal of Biological Chemistry* **272**, 7586–7594.

Song, J., and Dohlman, H.G. (1996) Partial constitutive activation of pheromone responses by a palmitoylation-site mutant of a G protein α subunit in yeast. *Biochemistry* **35**, 14806–14817.

Spellig, T., Bölker, M., Lottspeich, F., Frank, R.W., and Kahmann, R. (1994) Pheromones trigger filamentous growth in *Ustilago maydis*. *EMBO Journal* **13**, 1620–1627.

Staben, C., and Yanofsky, C. (1990) *Neurospora crassa* **a** mating-type region. *Proceedings of the National Academy of Sciences, USA* **87**, 4917–4921.

Stettler, S., Warbrick, E., Prochnik, S., Mackie, S., and Fantes, P. (1996) The wis1 signal transduction pathway is required for expression of cAMP-repressed genes in fission yeast. *Journal of Cell Science* **109**, 1927–1935.

Sugimoto, A., Iino, Y., Maeda, T., Watanabe, Y., and Yamamoto, M. (1991) *Schizosaccharomyces pombe ste11⁺* encodes a transcription factor with an HMG motif that is a critical regulator of sexual development. *Genes and Development* **5**, 1990–1999.

Takeda, T., Toda, T., Kominami, K., Kohnosu, A., Yanagida, M., and Jones, N. (1995) *Schizosaccharomyces pombe atf1⁺* encodes a transcription factor required for sexual development and entry into stationary phase. *EMBO Journal* **14**, 6193–6208.

Tanaka, K., Davey, J., Imai, Y., and Yamamoto, M. (1993) *Schizosaccharomyces pombe map3⁺* encodes the putative M-factor receptor. *Molecular and Cellular Biology* **13**, 80–88.

Tanaka, K., Nambu, H., Katoh, Y., Kai, M., and Hidaka, Y. (1999) Molecular cloning of homologs of *RAS1* and *RHO1* genes from *Cryptococcus neoformans*. *Yeast* **15**, 1133–1139.

Terenzi, H.F., Flawia, M.M., and Torres, H.N. (1974) A *Neurospora crassa* morphological mutant showing reduced adenylate cyclase activity. *Biochemical and Biophysical Research Communications* **58**, 990–996.

Terenzi, H.F., Flawia, M.M., Tellez-Inon, M.T., and Torres, H.N. (1976) Control of *Neurospora crassa* morphology by cyclic adenosine 3′,5′-monophosphate and dibutyryl cyclic adenosine 3′,5′-monophosphate. *Journal of Bacteriology* **126**, 91–99.

Tolkacheva, T., McNamara, P., Piekarz, E., and Courchesne, W. (1994) Cloning of a *Cryptococcus neoformans* gene, *GPA1*, encoding a G-protein α-subunit homolog. *Infection and Immunity* **62**, 2849–2856.

Tu, H., Barr, M., Dong, D.L., and Wigler, M. (1997) Multiple regulatory domains on the Byr2 protein kinase. *Molecular and Cellular Biology* **17**, 5876–5887.

Urban, M., Kahmann, R., and Bölker, M. (1996a) The biallelic *a* mating type locus of *Ustilago maydis*: remnants of an additional pheromone gene indicate evolution from a multiallelic ancestor. *Molecular and General Genetics* **250**, 414–420.

Urban, M., Kahmann, R., and Bölker, M. (1996b) Identification of the pheromone response element in *Ustilago maydis*. *Molecular and General Genetics* **251**, 31–37.

Wang, P., and Heitman, J. (1999) Signal transduction cascades regulating mating, filamentation, and virulence in *Cryptococcus neoformans*. *Current Opinion in Micobiology* **2**, 358–362.

Wang, P., Perfect, J.R., and Heitman, J. (2000) The G-protein β subunit GPB1 is required for mating and haploid fruiting in *Cryptococcus neoformans*. *Molecular and Cellular Biology* **20**, 352–362.

Watanabe, Y. and Yamamoto, M. (1996) *Schizosaccharomyces pombe pcr1⁺* encodes a CREB/ATF protein involved in regulation of gene expression for sexual development. *Molecular and Cellular Biology* **16**, 704–711.

Wedegaertner, P.B., Wilson, P.T., and Bourne, H.R. (1995) Lipid modifications of trimeric G proteins. *Journal of Biological Chemistry* **270**, 503–506.

Whiteway, M.S., and Thomas, D.Y. (1994) Site-directed mutations altering the CAAX box of Ste18, the yeast pheromone-response pathway Gγ subunit. *Genetics* **137**, 967–976.

Whiteway, M.S., Wu, C., Leeuw, T., Clark, K., Fourestlieuvin, A., Thomas, D.Y., and Leberer, E. (1995) Association of the yeast pheromone response G protein βγ subunits with the MAP kinase scaffold Ste5p. *Science* **269**, 1572–1575.

Wickes, B.L., Edman, U., and Edman, J.C. (1997) The *Cryptococcus neoformans STE12α* gene: a putative *Saccharomyces cerevisiae STE12* homologue that is mating type specific. *Molecular Microbiology* **26**, 951–960.

Wickes, B.L., Mayorga, M.E., Edman, U., and Edman, J.C. (1996) Dimorphism and haploid fruiting in *Cryptococcus neoformans*: Association with the a-mating type. *Proceedings of the National Academy of Sciences, USA* **93**, 7327–7331.

Widmann, C., Gibson, S., Jarpe, M.B., and Johnson, G.L. (1999) Mitogen-activated protein kinase: conservation of a three-kinase module from yeast to human. *Physiological Reviews* **79**, 143–180.

Xu, J.R., and Hamer, J.E. (1996) MAP kinase and cAMP signaling regulate infection structure formation and pathogenic growth in the rice blast fungus *Magnaporthe grisea. Plant Cell* **10**, 2696–2706.

Xu, J.R., Staiger, C.J., and Hamer, J.E. (1998) Inactivation of the mitogen-activated protein kinase Mps1 from the rice blast fungus prevents penetration of host cells but allows activation of plant defense responses. *Proceedings of the National Academy of Sciencs, USA* **95**, 12713–12718.

Yue, C., Cavallo, L.M., Alspaugh, J.A., *et al.* (1999) The STE12α homolog is required for haploid filamentation but largely dispensable for mating and virulence in *Cryptococcus neoformans. Genetics* **153**, 1601–1615.

CHAPTER 2

PROKARYOTIC INTERCELLULAR SIGNALLING
Mechanistic diversity and unified themes

CLAY FUQUA and DAVID WHITE

Department of Biology, Indiana University, Bloomington, Indiana 47405, USA

Summary
Although most bacteria are single-cell creatures, they are far from solitary, independent organisms. Indeed, many bacteria in the natural environment are in close physical proximity in dense populations where they are involved in a desperate competition for food and space. In these environments, intercellular signalling is common and perhaps unavoidable. Although we have only begun to understand the range of interactions affected by signalling, the diversity of those few systems currently being explored is striking. Cell signalling can involve soluble signal molecules such as pheromones which diffuse between cells, cell-associated signals which require intimate contact, or both. Cell-to-cell communication via diffusible signals often takes place at high cell densities, as part of an adaptive response to impending nutrient depletion and crowding. Thus far, the majority of cell-associated signals have been identified in microbes that exhibit complex structural development patterns. However, as new approaches to identify cell bound signals are developed, the list of signalling mechanisms that require cell-to-cell contact is likely to increase. For both diffusible and cell-contact-dependent signalling, high population densities are often required. The various niches where one finds the highest concentrations of bacteria are the surfaces of tissues of fungi, plants and animals, as well as the surfaces of soil particles and other inanimate objects where nutrients may be concentrated. In this Chapter we will attempt to encapsulate the current understanding of prokaryotic intercellular communication, focussing first on the better understood diffusible signalling pathways, and following with a discussion of cell-associated mechanisms.

1. Introduction and general themes

The importance of intercellular signalling in metazoan organisms is relatively easy to appreciate (see other Chapters in this text). The structural and developmental complexity inherent to multicellular life requires that different cell types be capable of communication with each other. Even relatively simple eukaryotic organisms such as the cellular slime molds and budding yeasts possess mechanisms by which cells in a population communicate and coordinate their actions. The requirement for intercellular signalling is less obvious for the prokaryotic organisms, such as the bacteria and the archeae. In fact, prokaryotic microbes have long been considered the paradigm of unicellularity. The metabolic plasticity of prokaryotes in response to changes in environmental conditions is well documented and equally well accepted. However, beyond the normal competition for resources inherent in any given environment, prokaryotic cells were

I. Fairweather (ed.), Cell Signalling in Prokaryotes and Lower Metazoa, 27–71.
© 2004 *Kluwer Academic Publishers. Printed in Great Britain.*

generally not considered to utilize any deliberate means of communication between cells.

These longstanding views have been shattered by the recent evidence for a variety of intercellular signalling systems in prokaryotes. In fact, the idea of intercellular signalling among prokaryotes is not new. As early as 1965, studies on the development of genetic competence in *Streptococcus pneumoniae* revealed the activity of a microbial hormone (now known to be an oligopeptide) important for the timing of competence (Tomasz, 1965). Likewise, in 1967, a soluble signalling molecule called A-factor was identified as essential for the production of the antibiotic streptomycin by the soil microbe *Streptomyces griseus* (Kholkov *et al.*, 1967). What is new, however, is recognition of the prevalence of intercellular signalling in prokaryotes (Table 1). Cell-to-cell communication has gone from a few isolated examples in unusual microbes, to observations that many, if not all, bacteria exhibit some form of intercellular signalling. This signalling usually regulates significant features of microbial physiology and allows coordination of bacterial behaviour in ways that challenge the prokaryotic unicellular paradigm (for visual examples of signalling microbes see Fig. 1). While there have been no reports of intercellular signalling in the archeae, prokaryotes phylogenetically distinct from the bacteria, it is almost certain that it exists, and simply awaits discovery. This Chapter will review the mechanisms of bacterial cell-cell signalling that are currently recognized, with emphasis on the molecular mechanisms of signalling (for more comprehensive coverage of bacterial intercellular signalling see Dunny and Winans, 1999).

1.1. DIFFUSIBLE AND CELL-ASSOCIATED SIGNALS

For the purposes of discussing the diversity of prokaryotic signalling, we will draw a distinction between those signalling mechanisms that function through the release of soluble chemical cues that we shall refer to as pheromones (see below), and those that rely on cell-to-cell contact. In metazoan cell biology there are many examples of each type of signalling and several that blur the margin between them. For prokaryotes, however, the examples of diffusible signals currently far outnumber the cell-associated signals. This disproportion may reflect the historical bias of microbiologists towards studies on pure, dispersed culture, but more likely is an accurate representation of prokaryotic signalling. Cell-associated signals in metazoan organisms are most often involved in instances where cell position or ordered stratification of cell types (*eg.* tissue or organ development) is required. While there are microbial examples where positional information is required, they are certainly far less common than in metazoan organisms. This is not meant to imply that the diffusible signals are in any way more important than those that remain associated with the cell, but rather that they may be more numerous.

Pheromones are any chemical substances secreted into the environment that elicit a physiological, behavioural, or developmental response from organisms of the same species. They have been described for insects, fungi and protozoans

Table 1. Examples of intercellular signalling by bacteria

Bacteria	**Pheromone**	**Process Regulated**
Gram-positive		
Bacillus subtilis	Oligopeptides	DNA uptake (competence), sporulation
Enterococcus faecalis	Oligopeptides	DNA exchange (conjugation)
Lactic acid bacteria	Oligopeptides	Bacteriocin production
Staphylococcus aureus	Oligopeptides	Animal pathogenesis
Streptococcus pneumoniae	Oligopeptides	DNA uptake, pathogenesis
Streptomyces griseus	γ-butyrolactones	Antibiotic production, secondary metabolism
Gram-negative		
Vibrio fischeri	acyl HSLs	Bioluminescence
Vibrio harveyi	acyl HSL, AI-2 type signal	Bioluminescence
Agrobacterium tumefaciens	acyl HSL	DNA exchange (conjugation)
Pseudomonas aeruginosa	acyl HSLs, quinolones	Animal pathogenesis
Myxococcus xanthus	amino acids	Fruiting body development
Ralstonia solanacearum	acyl HSLs, fatty acid methyl esters (FAME)	Plant pathogenesis
Xanthomonas campestris pv. *campestris*	diffusible signal factor (DSF), diffusible factor (DF)	Plant pathogenesis
Salmonella typhimurium	AI-2 like factor	Animal pathogenesis
Proteus mirabilis	glutamine	Swarming motility
Anabaena spp. PCC7120	Oligopeptide	Cell differentiation

Adapted from Dunny and Winans (1999); see text for references.

(Kubota *et al.*, 1973; Miyake and Beyer, 1973; Gooday, 1974; Shorey and McKelvey, 1977; Newell, 1978). The intercellular diffusible signals that bacteria use have also been referred to as pheromones. In some ways bacterial pheromones are analogous to hormones, which are intercellular signals secreted by cells in a multicellular animal, except that the interacting bacterial cells are not part of a single organism, but rather part of a population of organisms whose individual survival depends upon cell-to-cell signalling, as this Chapter will make clear.

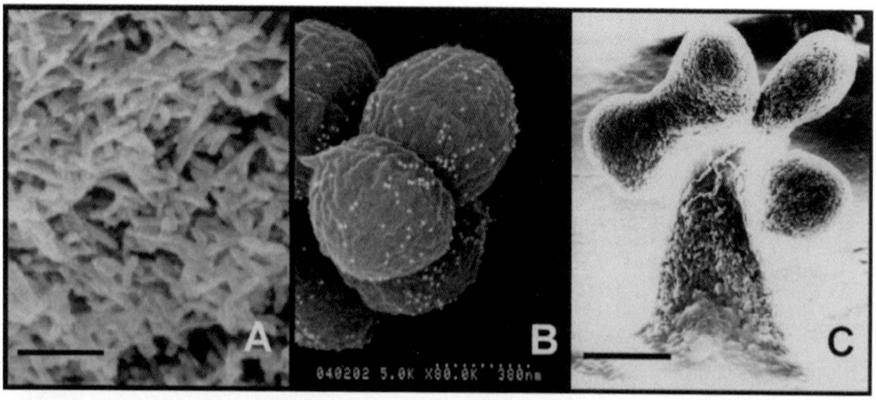

Figure 1. Micrographic examples of bacterial signalling. (A) Biofilm of *Pseudomonas aeruginosa* (produces the acyl HSLs 3-oxo-C_{12}-HSL and C_4-HSL; scale bar 5 μm; (Stickler *et al.*, 1998), (B) Cluster of *Enterococcus faecalis* cells induced with the peptide pheromone cCF10 stained with anti-AS (aggregation substance, bright flecks on cells) antibody (courtesy of Helmut Hirt and Stanley Erlandsen, University of Minnesota), (C) Fruiting body of *Stigmatella aurantica*; scale bar 20 μm.

1.2. QUORUM SENSING AND AUTOINDUCTION

A recurrent theme in prokaryotic intercellular signalling is the utilization of signal molecules to gauge and respond to population density. In fact, there are several evolutionarily distinct systems which function in this capacity, regulating cellular physiology and development with respect to the number and density of bacteria in a given environment. A general term used to describe this microbial census-taking process is "quorum sensing", in which the bacteria wait to attain a certain population density, a bacterial quorum, prior to expressing some aspect of their behaviour (Fuqua, Winans and Greenberg, 1994, 1996). The generic quorum sensing mechanism involves release of a diffusible cue into the environment that, at low population density, is rapidly diluted (Fig. 2). Increasing population density, the inevitable outcome of successive rounds of binary fission by the bacteria, or the result of new cells entering into the population, results in a relative elevation of the signal level. Eventually, signal accumulation is perceived via receptor proteins of the responding bacteria (Fig. 2). Often, these are the very bacteria that produced the signal (*ie.* intraspecies signalling), although there are now examples of interspecies signalling (Bassler, Greenberg and Stevens, 1997). The term autoinduction has historically been used to describe perception of a self-produced signal (Nealson, 1977). Although intraspecies quorum sensing requires an autoinduction circuit, autoinduction does not necessarily imply quorum-sensing. There are instances in which the self-produced autoinducer may serve a function distinct from quorum sensing, such as amplification of a weak upstream signal. Likewise, interspecies signalling between different bacteria, sometimes called alloinduction, does not always reflect quorum sensing, and in fact may not require

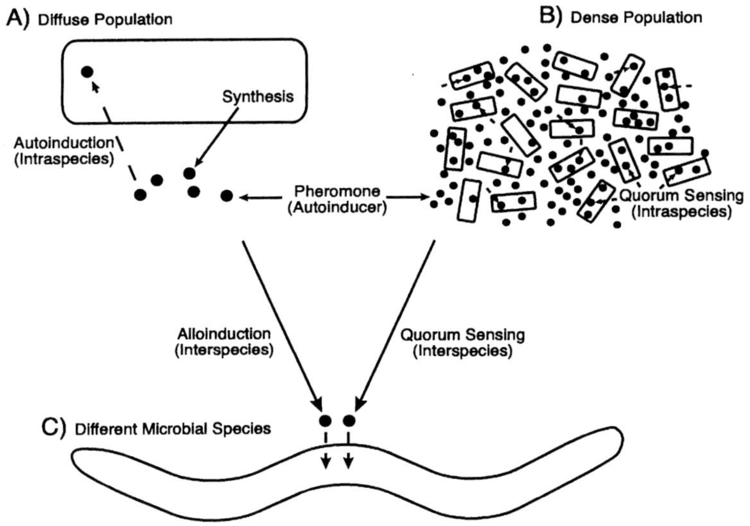

Figure 2. Autoinduction, alloinduction and quorum sensing. Rectangles represent a signal-producing bacterium and the curved shape represents a different bacterium responding to, but not producing the signal. The signal(s) is shown as black dots. Solid arrows indicate signal release while dashed arrows indicate perception of the signal (signals are depicted entering the cell, although many signal molecules are perceived externally by transmembrane sensors, see text). The three panels depict (A) autoinduction of a single cell, irrespective of population density, (B) intraspecies quorum-sensing at high population density, and (C) perception of the signal by a different species as simple alloinduction and interspecies quorum-sensing.

autoinduction. An example would be a quorum-sensing bacterium that monitors the population density of competitor bacteria by sensing the accumulation of a metabolic waste product excreted into the environment (Fig. 2).

1.3. DIVERSITY IN PROKARYOTIC DIFFUSIBLE SIGNAL MOLECULES

Bacteria produce a range of signal molecules which regulate a diversity of functions (Fig. 3 and Table 1). The bacteria are often divided into two distinct classes: Gram-negative and Gram-positive. Gram-negative bacteria have a cell wall composed of a phospholipid bilayer membrane, inside of a thin layer of rigid peptidoglycan, and an outer membrane composed of phospholipid and lipopolysaccharide. In contrast, Gram-positive cell walls have the internal bilayer and a thick external layer of peptidoglycan, as well as negatively charged techoic acids.

Broadly considered, it is found that gram-positive microorganisms have evolved to utilize peptide-based signals, while the more common signal molecules in gram-negative bacteria are lipids or lipid derivatives. Two large classes of signal molecules that reflect this trend are the oligopeptides, produced by a range of gram-positive bacteria, and the lipoidal acylated homoserine lactones

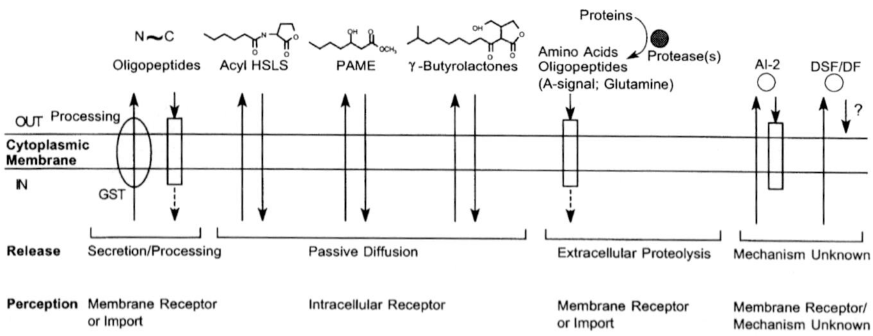

Figure 3. Summary of pheromone-based signalling. The basic aspects of each signalling system are presented. Arrows extending away from the cell indicate release of the signal; arrows towards or into cells indicate perception. Dashed arrows indicate possible transport into cells. Some of the oligopeptide pheromones are perceived externally while others are imported and perceived internally. Additionally, a number of these oligopeptides are modified or further processed (see text and Table 2). Question marks indicate ambiguity in the mechanism of signal release and/or perception. GSP – general secretion pathway.

(acyl HSLs), now identified in many gram-negative organisms (Fig. 3, Table 1 and Table 3). There are, however, numerous exceptions to this overall trend. For example, the filamentous cyanobacterium, *Anabaena* (a so-called gram-variable microbe) utilizes oligopeptide signal molecules, and the gram-positive filamentous *Streptomyces* spp. are known to utilize butyrolactone signals, structurally similar to the acyl HSLs.

Molecular signals also identified in diverse gram-negative bacteria and thought to play roles in cell-to-cell communication are a subset of amino acids in *Myxococcus xanthus*, aliphatic hydroxy ketones in *Stigmatella aurantica*, fatty acid methyl esters in *Ralstonia solanacearum*, the amino acid glutamine in *Proteus mirabilis*, the diffusible factors DSF and DF (structures undefined) of *Xanthomonas campestris*, and the so-called "autoinducer two" (AI-2, structure undefined) that controls luminescence in *Vibrio harveyi* (Fig. 3 and Table 1).

Each of the examples cited above, as well as others, utilizes distinct signalling mechanisms, many of which we will detail later in this Chapter. However, the unifying theme is that all of these bacteria employ signal molecules that are in some manner released, or produced in the external environment, and subsequently perceived by cells of the same and/or different species.

Perception of the signal is usually correlated with increased expression of specific groups of genes. Most often, the signals are used to determine cell density, and in some way adapt to the elevated population. Considering the diversity of signalling mechanisms we shall discuss below, there is clearly a strong selective advantage for bacteria to be able to regulate their individual physiology, as well as their community development, via shared information pathways.

2. Specific Signalling Systems

2.1. PEPTIDE PHEROMONES

Oligopeptide pheromones, commonly found in the Gram-positive bacteria, are small peptides ranging in size from 5 to 40 amino acids. They are processed from larger, ribosomally synthesized peptides prior to or during export (Table 2).

2.1.1. *Location of Peptide Pheromone Receptors*

In order to have an effect, the peptide pheromone must bind to a receptor. In some cases the receptor is on the cell surface, and the pheromone does not enter the cell. Binding of the pheromone to the receptor generates a transmembrane signal that initiates a signal cascade within the cell, often involving a phosphorelay signalling system described later (Fig. 4). Other peptide pheromones must enter the cell in order to have an effect. The pheromones that enter the cells do so via membrane-associated transporters, usually oligopeptide permeases, and subsequently bind to a cytoplasmic receptor, thereby initiating the signal transduction pathway. The cytoplasmic receptor might be an enzyme, or a transcription factor that regulates the transcription of specific genes, and may be activated or inactivated by pheromone binding.

Table 2. Examples of oligopeptide pheromones

Bacteria	**Peptide Name**	**Amino acids Mature (Precursor)**	**Modification**
Lactococcus lactis	Nisin A	34 (57)	lanthionine and methyl-lanthionine rings, dihydroalanine and dihydrobutyrate residues
Streptococcus pneumoniae	CSP (ComC)	17 (41)	Unmodified
Bacillus subtilis	ComX	10 (55)	Lipid – likely farnesylation of Trp residue
	CSF	5 (40)	Unmodified
Staphylococcus aureus	AgrD (AIP)	8 (45)	Cyclization – internal thioester (Cys4 - Met8)
Enterococcus faecalis	CF10	7 (23)	Unmodified
	iCF10	7 (21)	Unmodified
Anabaena sp. PCC 7120	PatS	>5 (17)	Unmodified

Adapted from Dunny and Leonard (1997); see text for references.

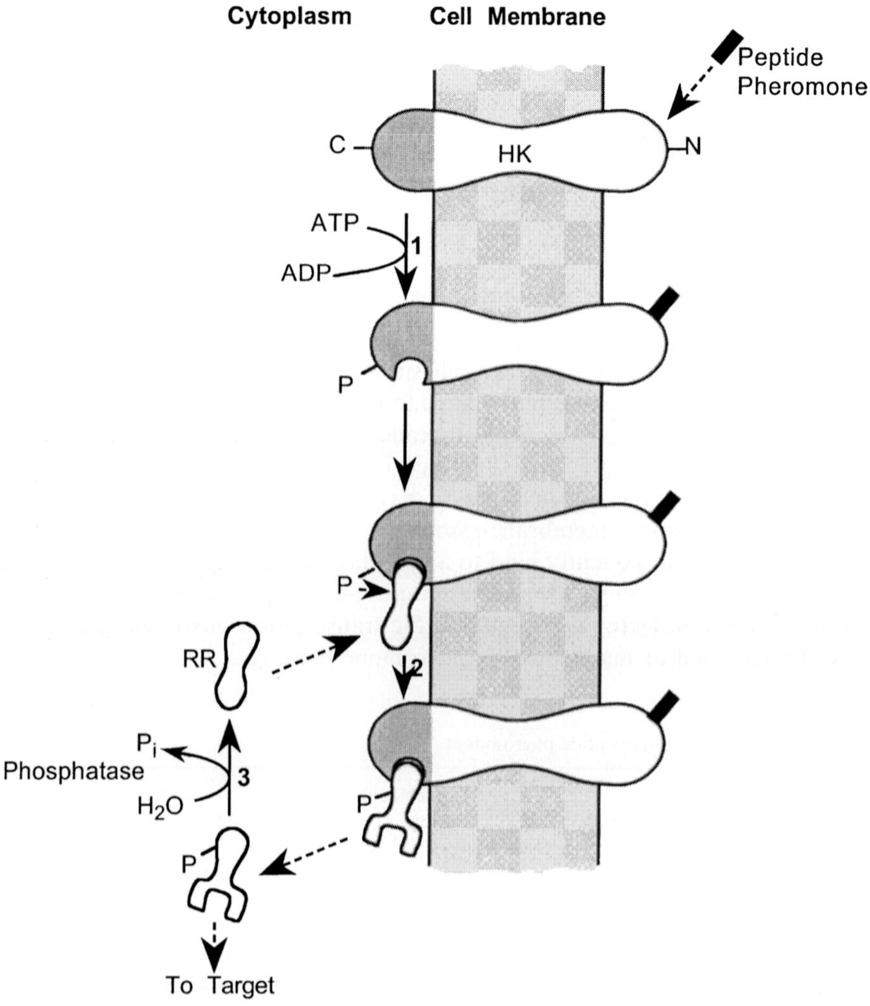

Figure 4. Model of prokaryotic phosphorelay systems. A transmembrane sensor/kinase (HK) has an N-terminal portion on the outside surface, and a C-terminal portion on the cytoplasmic side. The peptide pheromone binds to the HK and induces a conformational change that induces autophosphorylation. The phosphoryl group is transferred to a cytoplasmic response regulator protein (RR). Autophosporylation and subsequent phosphotransfer is depicted as ATP ⟶ ADP in Rxn 1. RR~P commonly binds DNA and stimulates transcription of specific genes. One or more phosphatases dephosphorylate RR~P, controlling its activity. Adapted from White (2000).

2.1.2. *Intracellular Targets of Signalling*

In the systems described in this Chapter, the target of most pheromone signalling is the control of gene expression, and ultimately microbial behaviour. The end result of the signal cascade initiated by most of the well-studied peptide pheromones is the enhancement of transcription of specific genes, and in some cases the stimulation of translation of specific mRNAs. Pheromones alter gene activity by affecting the activity of transcription factors that regulate the expression of specific genes (see Dunny and Leonard, 1997, for a review). The effect of peptide pheromones on the transcription factors is rarely direct, and is usually the outcome of a signalling pathway begun by the pheromone. However, as will be described later, some peptide pheromones that enter the cell do interact directly with transcription factors (Clewell, 1999).

2.1.3. *A Description of Two-Component Phosphorelay Systems*

Pheromones that bind to surface receptors generally use a two-component phosphorelay system to affect gene expression. Two-component phosphorelay systems involved in signal transduction are widespread among bacteria (Fig. 4 and Hock and Silhavy, 1995). Briefly, these systems comprise an enzyme called a histidine kinase (HK), which autophosphorylates upon receiving a signal (Rxn 1). This phosphoryl group is then transferred to a protein called a response regulator (RR) (Rxn 2). In most cases the phosphorylated response regulator (RR$\sim$P) stimulates or inhibits the transcription of specific genes. There are two ways in which the levels of RR-P can be increased by a signal: stimulation of the HK, and inhibition of a phosphatase that removes the phosphate from RR-P (Rxn 3). The common feature in all "two component" systems is a phosphorelay (in fact, often more than two components), during which the phosphoryl group is sequentially transferred from one protein to another until a response regulator protein is phosphorylated. Multiple proteins along the phosphorelay are generally thought to provide regulatory input points for integration of additional environmental signals. Once phosphorylated, the response regulator protein affects a target, usually modulating gene expression. A single bacterium has many different phosphorelays with exquisite specificity for their signals and targets, and which are used to respond to many different environmental conditions. The environmental signals include the levels of specific carbon sources, inorganic nitrogen supply, inorganic phosphate supply, potassium supply, osmolarity of the extracellular media, oxygen levels, and cell-density cues.

(Rxn 1) Histidine kinase (HK)	signal + HK + ATP $\longrightarrow$ HK-P + ADP
(Rxn 2) Response regulator (RR)	RR + HK-P $\longrightarrow$ RR-P (active) + HK
(Rxn 3) Phosphatase	RR-P + H_2O $\longrightarrow$ RR + Pi

2.1.4. *Pheromones that Initiate Two-Component Signalling Cascades*

Examples of two-component systems that respond to peptide pheromones include sporulation in *Bacillus subtilis*, the development of competence in *B. subtilis* and *Streptococcus pneumoniae*, and the regulation of pathogenicity by *Staphylococcus aureus*. *Myxococcus xanthus* uses amino acids as pheromones to stimulate multicellular development, and this too appears to involve the mediation of two-component systems. These systems will be described later. A general model for how an extracellular peptide pheromone might use a two-component system to stimulate the transcription of specific genes is diagrammed in Fig. 4. In this case it is the peptide that interacts with the histidine kinase (HK), either directly or indirectly. Alternatively, pheromones that enter the cell may inhibit the activity of phosphatases, and in this way also contribute to increased transcription of specific genes. All of the models involving two-component systems described later are based upon the generic scheme outlined in Fig. 4, and therefore we will return to this Figure.

Competence and Endosporulation in B. subtilis. *Bacillus subtilis* is a gram-positive bacterium commonly found in soil. It is a well-studied developmental model system for competence and sporulation. Competence is defined as the ability to bind and transport exogenous DNA into the cell, and to recombine the DNA into the genome of the recipient cell. It is one of the ways that genes are transferred between closely related bacteria in the natural environment. Endospore formation is a process by which a cell divides asymmetrically, giving rise to two compartments, one of which ultimately becomes a spore, whereas the other one lyses. Competence and endospore formation can be viewed as alternative developmental fates. They both tend to occur at the end of exponential growth when cell densities are high. Whether one or the other occurs is regulated in part by peptide pheromones secreted by the cells into the extracellular medium.

Competence. There are two competence-inducing peptide pheromones that accumulate in the extracellular media at high cell densities (Lazazzera and Grossman, 1997). These pheromones are ComX and CSF (competence and sporulation factor). ComX is a 10 amino acid peptide processed from the C-terminal end of a 55 amino acid precursor peptide (Table 2). Between 5 and 10nM of extracellular ComX induces a 50% maximal response. The mechanism of processing and export are not known. The precursor protein does not have a leader sequence required for export through the General Secretory Pathway (GSP), nor does it possess the double glycine motif required for processing and export in the ABC transporter pathway. The processing and export of ComX are presumed to require specialized proteins. Interestingly, one of the amino acids in ComX, a tryptophan residue, is modified by lipid attachment likely to be a C15 polyisoprenoid moiety (farnesylation) (Magnuson, Solomon and Grossman, 1994). Similar modification of fungal pheromones is well-documented, except that it is usually a cysteine residue that is alkylated (Caldwell, Naider and Becker, 1995).

The response to ComX requires a two-component system consisting of a histidine kinase called ComP with six transmembrane segments and a cytoplasmic response regulator called ComA (Fig. 5). According to this model, ComX binds to ComP on the outer membrane surface and thereby initiates a phosphorelay whereby ComP autophosphorylates and this phosphoryl group it transferred to ComA. ComA-P is a transcriptional activator that stimulates transcription of genes in the competence pathway.

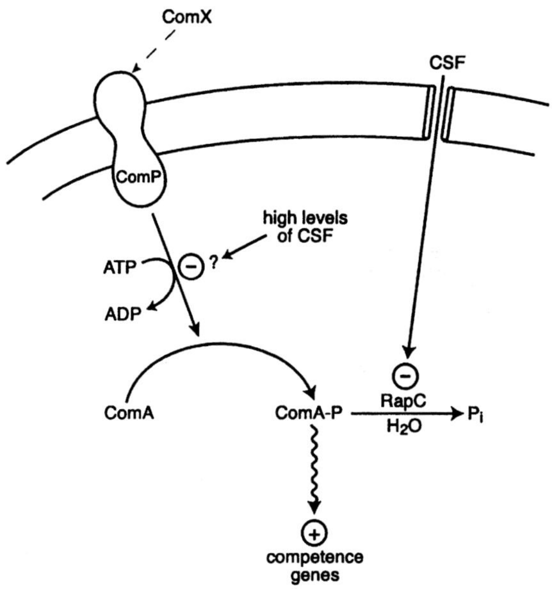

Figure 5. Model for regulation of competence by oligopeptide pheromones in *B. subtilis*. Stimulation of competence by ComX and CSF. ComX binds to the sensor/kinase ComP and stimulates it to autophosphorylate; as a result the phosphoryl group is transferred to the response regulator ComA. Autophosporylation and phosphotransfer is depicted as ATP ⟶ ADP. ComA-P activates the transcription of genes in the competence pathway. CSF enters through an oligopeptide permease and inactivates the RapC phosphatase which dephosphorylates ComA-P.

CSF is a 5 amino acid peptide that is processed from the C-terminal end of a secreted 40 amino acid peptide and accumulates in the medium as the cells grow to a high cell density (Table 2 and Solomon, Lazazzera and Grossman, 1996). It has been suggested that a peptide of between 11 and 25 amino acids is processed from the 40 amino acid precursor and is exported using the GSP pathway. Processing of the CSF precursor to the 5 amino acid peptide is not understood, but an extracellular peptidase, perhaps bound to the outer cell surface, may be involved.

Low concentrations (1–5nM) of CSF stimulate the competence pathway. CSF enters the cell via an oligopeptide permease and inhibits a phosphatase (Rap C phosphatase) that dephosphorylates ComA-P (Solomon, Lazazzera and Grossman, 1996). Thus, both ComX and CSF increase the amounts of ComA-P. However, as explained next, as the extracellular concentrations of CSF surpass a minimum threshold (>20nM), then competence is inhibited and sporulation is stimulated.

Sporulation. The extracellular peptide pheromone that stimulates the sporulation pathway is CSF, which as previously described also stimulates the competence pathway (Fig. 6). In addition, there are as yet undefined sporulation factors that are also integrated with the effect of CSF. Whether CSF stimulates

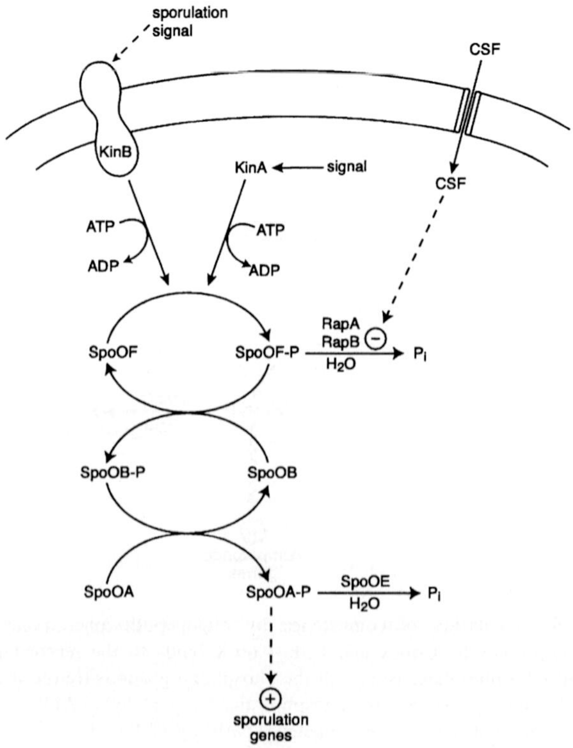

Figure 6. Stimulation of *Bacillus subtilis* sporulation by CSF. There are two histidine kinases: cytoplasmic KinA and membrane-associated KinB. In response to a sporulation signal, KinB autophosphorylates and the phosphoryl group is transferred to a response regnlator protein SpoOF. Autophosphorylation and phosphotransfer is depicted as ATP ⟶ ADP. The phosphoryl group is passed from SpoOF to SpoOB, and in turn is transferred SpoOA. SpoOA-P is a transcriptional activator for sporulation genes. In addition to the histidine kinases, three phosphatases regulate the levels of SpoOA-P. RapA and RapB dephosphorylate SpoOF-P. and SpoOE dephosphorylates SpoOA-P. The peptide pheromone CSF enters the cell through the oligopeptide permease and inactivates RapB. RapA is regulated in an independent manner. Adapted from White (2000).

competence or sporulation depends upon its extracellular concentration. If the extracellular concentration of CSF is sufficiently high (>20nM), then sporulation, rather than competence, is stimulated (Lazazzera, Solomon and Grossman, 1997). It has been suggested that high intracellular levels of CSF inhibit the histidine kinase ComP, resulting in lower levels of the phosphorylated ComA-P (Fig. 5). The result would be decreased expression of ComA-P dependent genes in the competence pathway.

If this occurs and conditions are appropriate for sporulation, then the cells will sporulate rather than become competent. CSF also stimulates sporulation by enhancing phosphotransfer (Fig. 6). Evidence suggests that CSF inhibits a phosphatase (RapB phosphatase) that dephosphorylates SpoOF-P, which is part of a phosphorelay cascade that phosphorylates SpoOA, the primary positive transcriptional regulator for *spo* genes (Fig. 6) (Perego, 1997). The undefined sporulation signals apparently also stimulate phosphorylation of SpoOA through activation of the HKs that initiate the phosphorelay.

Competence in Streptococcus pneumoniae. *S. pneumoniae* is a gram-positive bacterium that is found in the upper respiratory tract of many healthy people and is a common cause of middle ear infections, bacterial meningitis in children and pneumonia. As with *B. subtilis*, competence in *S. pneumoniae* is stimulated by a peptide pheromone that activates a two-component regulatory system. The signal transduction pathway is similar to the generalized pattern drawn in Fig. 4. The peptide pheromone that stimulates the competence pathway is a competence inducing factor called CSP, a 17 amino acid peptide made from a larger precursor protein and likely processed during export via an ABC-type permease (ComAB) (Håvarstein, Coomaraswamy and Morrison, 1995; Håvarstein, Diep and Nes, 1995). CSP accumulates in the extracellular media in proportion to cell density and binds to a receptor on the surface of target cells. The receptor is called ComD, and it is the histidine kinase portion of a two-component system (Håvarstein, Hakenbeck and Morrison, 1996). The binding of CSP to ComD stimulates autophosphorylation of ComD and the subsequent phosphorylation of the response regulator protein called ComE (Ween, Gaustad and Håvarstein, 1999). The phosphorylated form of ComE stimulates the transcription of genes in the competence pathway (for a model see Håvarstein and Morrison, 1999).

Virulence Genes in Staphylococcus aureus. *S. aureus* is a gram-positive pathogenic bacterium that is responsible for a variety of diseases ranging from skin diseases such as impetigo and infection of hair follicles (boils) to more serious infections of the lungs, heart, and blood. The bacteria produce a variety of virulence factors that contribute to pathogenesis. The expression of many of the virulence genes is dependent upon regulatory genes called the *agr* system (accessory gene regulation). Two of the gene products of the *agr* system, AgrC and AgrA, comprise a two-component regulatory system, AgrC being a membrane-bound histidine kinase and AgrA a cytoplasmic response regulator (see Fig. 4 for a generic scheme). AgrC is activated by an extracellular peptide pheromone of 8 amino acids called AIP for *agr* autoinducing peptide (Table 3 and Ji, Beavis and Novick, 1995). AIP is processed from the larger AgrD gene product and is internally cyclized, containing a thioester bond between the cysteine residue at position 4 and the carboxy terminus (Ji, Beavis and Novick, 1997). The maturation of AIP requires the AgrB gene product. In the current model, external AIP binds to AgrC on the cell surface and stimulates autophosphorylation. The phosphoryl group of the phosphorylated AgrC is passed to the cytoplasmic RR AgrA (see Novick, 1999, for a review). AgrA-P does not appear to bind to DNA, but may stimulate expression by interacting with a DNA-binding transcription factor.

Table 3. Examples of acyl HSL quorum sensors

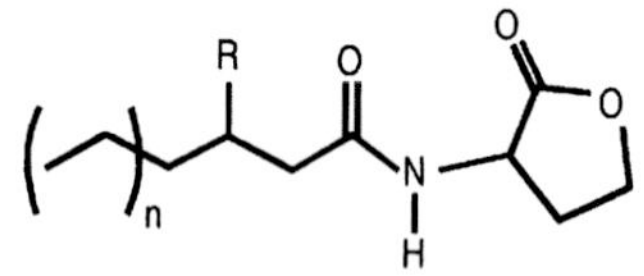

Bacteria	**Regulators**[1]	Chain length (n)[2]	β R-group[3]	Abbreviation
Vibrio fischeri	LuxR-LuxI	6 (1)	=O	3-oxo-C_6-HSL
	AinR-AinS[4]	8 (2)	−H	C_8-HSL
Vibrio harveyi	LuxN-LuxLM[4]	4 (0)	−OH	3-OH-C_4-HSL
Pseudomonas aeruginosa	LasR-LasI	12 (8)	=O	3-oxo-C_{12}-HSL
	RhlR-RhlI	4 (0)	−H	C_4-HSL
Agrobacterium tumefaciens	TraR-TraI	8 (4)	=O	3-oxo-C_8-HSL
Rhodobacter sphaeroides	CerR-CerI	14 (10)	−H	Δ7-C_{14}-HSL[5]
Vibrio anguillarum	VanR-VanI	10 (6)	=O	3-oxo-C_{10}-HSL
Yersenia enterocolitica	YenR-YenI	6 (2)	−H, =O	C_6-HSL 3-oxo-C_6-HSL

For details and references see Fuqua, Winans and Greenberg (1996); Fuqua and Eberhard (1999).
[1] Quorum-sensing regulatory proteins, except where indicated proteins are members of the LuxR-LuxI family, R – receptor/regulator, I/S/LM – acyl HSL synthases.
[2] Corresponds to n in diagram.
[3] Chemical moiety at β position, R in diagram.
[4] AinR and LuxN are homologous to two component sensor kinases, AinS and LuxLM are acyl HSL synthases, but not homologous to LuxI.
[5] An unsaturated bond exists between pos. 7 and 8 on the acyl chain.

Bacteriocin Production. Many bacteria that live in animals secrete bacteriocins, which are antibiotic proteins or peptides that inhibit or kill other bacteria that may be competitors for food or space. For example, *Lactococcus lactis* secretes a 34 amino acid bacteriocin called nisin that inhibits the growth of many different gram-positive bacteria (Table 2 and Mattick and Hirsch, 1944). Nisin is considered a class I bacteriocin, synthesized as a larger gene product that is processed and post-translationally modified with internal lanthionine bonds (intramolecular thioether bridges) and dehydroalanine, as well as dehydrobutyrine residues (Kleerebezem, de Vos and Kuipers, 1999). The production of nisin by *L. lactis* is stimulated by extracellular nisin, which in this context acts like a peptide pheromone (Kuipers *et al.*, 1995). When nisin accumulates in the extracellular media, it binds to a membrane-bound histidine kinase (NisK) and activates it (see Fig. 4 for general scheme). The activated NisK autophosphorylates and then this phosphoryl group is transferred to the response regulator called NisR. Phosphorylated NisR is suggested to activate the transcription of genes that result in the biosynthesis of nisin (see Kleerebezem, de Vos and Kuipers, 1999, for a review). Thus, in this case the bacteria use the bacteriocin to signal the rest of the population to make more nisin. There are, however, variations on this theme. Lactic acid bacteria that synthesize class II bacteriocins (unmodified peptide antibiotics) secrete peptide pheromones that activate bacteriocin production but are not themselves bacteriocins (Nes and Eijsink, 1999).

Heterocyst Formation in a Cyanobacterium. Species of the genus *Anabaena* are filamentous cyanobacteria (blue-green alga) that form nitrogen-fixing cells called heterocysts. The heterocysts, which form in wild-type organisms only when the medium is devoid of fixed nitrogen, are non-dividing cells located at approximately 10-cell intervals in the filament. They supply fixed nitrogen to the rest of the cells (called vegetative cells) in the filament, which are responsible for the growth and the production of new filaments. In *Anabaena* sp. PCC 7120, conditions of nitrogen deprivation cause proheterocyst cells to synthesize a 17 amino acid peptide called PatS which is trimmed, retaining the C-terminal end (at least the terminal 5 amino acid residues) that comprises the oligopeptide signalling molecule called PatS (Table 2 and Yoon and Golden, 1998). PatS diffuses into adjacent cells and prevents them from developing into heterocysts. Mutant strains unable to produce PatS make numerous heterocysts unevenly spaced even when grown in the presence of nitrogen. In order to account for the fact that there is a distinct pattern (one in every 10 cells) of heterocysts in wild type filaments, a model has been proposed postulating that PatS diffuses through the filament between the plasma membrane and the outer portion of the cell wall (the periplasmic space). This creates a gradient originating at the proheterocyst and dissipating laterally along the filament. This type of signalling differs from those discussed previously in that the signal acts specifically on cells within the same filament, and not on cells outside the filament. In this respect, it is more like a hormone produced by a multicellular animal. The signal transduction system is not known, but a phosphorelay has been suggested based upon other known peptide signalling systems.

Enterococcus Signalling. Not all peptide pheromones bind to histidine kinase/receptors on the cell surface and stimulate two-component signalling systems, or enter the cell like CSF to prevent the dephosphorylation of a response regulator. One peptide signalling system that does not appear to involve a two-component pathway is responsible for mating and plasmid transfer in *Enterococcus* (Dunny and Leonard, 1997).

E. faecalis is a gram-positive bacterium that is part of the normal flora in the human intestine. It can infect the urinary tract and other parts of the body, and is a frequent cause of hospital-acquired infections (nosocomial infections). A strain of *E. faecalis* harbouring a particular plasmid can transfer a copy of the plasmid to a strain which lacks the plasmid via a process called conjugation or mating (Dunny, Brown and Clewell, 1978). Such plasmid transfer is stimulated by peptide pheromones 7-8 amino acids long produced by the recipient strain (Table 2 and Clewell, 1999). In contrast to the pheromone systems previously described, stimulation by the enterococcal pheromones does not appear to involve two-component systems. The enterococci can carry several plasmids belonging to different families, the transfer of each one being responsive to a cognate peptide pheromone (Dunny *et al.*, 1979). As a consequence of a cell reacting to a peptide pheromone, several genes required for mating and the transfer of the cognate plasmid to a recipient cell are transcribed. Once a cell receives a plasmid, the production of pheromone for that particular plasmid is reduced, although it can continue to secrete pheromones for other families of plasmids. In addition, the conjugal plasmids encode production of inhibitor pheromones. The inhibitors are also oligopeptides, released in excess to the endogenously generated inducing pheromone and thereby preventing self-activation (Ike *et al.*, 1983). Mutant donor cells that cannot synthesize the inhibitor peptides are activated by their own pheromone signal and are not responsive to recipients.

A model has been proposed for the function of the peptide pheromone cAD1 that stimulates the transfer of plasmid pAD1 (Clewell, 1999). The pheromone is transported into the cell by a membrane-associated permease. Once inside the cell, the peptide binds to a negative transcription regulator called TraA. This inactivates the negative transcription regulator and thus enhances the transcription of genes that promote mating and plasmid transfer. One of the genes whose transcription is enhanced encodes TraE1, a positive regulator of the transcription of genes required for mating. Thus, the pheromone controls conjugation by stimulating the synthesis of a positive transcription factor for the expression of genes required for mating and plasmid transfer. It must be stated, however, that the molecular control mechanisms are not the same for all of the pheromones and their cognate plasmids. For example, in the case of the plasmid pCF10 and its cognate pheromone, regulation includes control at the translational level (Bensing, Manias and Dunny, 1997). It is not clear why *E. faecalis* has evolved such different mechanisms to transduce oligopeptide signals, rather than utilize a two-component phosphorelay as in other Gram-positive bacteria. One unique requirement in the *E. faecalis* system is the necessity to inhibit pheromone release, and hence self-activation in donors as well as recipients that receive the plasmid. If one considers the donors and recipients as discrete populations, mating pheromone signalling is in fact an excellent example of alloinduction (see section 1.2. and Fig. 2). A complete discussion of pheromones in enterococci can be found in Clewell (1999).

2.2. INTERCELLULAR SIGNALLING IN MYXOBACTERIAL DEVELOPMENT

Myxobacteria are unique among the bacteria in that they form multicellular fruiting bodies (reviews of the biology of myxobacteria can be found in Dworkin and Kaiser, 1993; Brun and Shimkets, 2000). Myxobacteria are usually found growing on decaying vegetation on forest floors, or on the surface of pellets of dung excreted by rabbits and other herbivores. The life cycle of a well-studied myxobacterium, *Myxococcus xanthus*, is depicted in Fig. 7. The cells live on solid surfaces as dense populations called swarms wherein the individual cells move by gliding motility.

2.2.1. *Developmental Biology of the Myxobacteria*

During vegetative growth and cell division the population of cells form a cohesive swarm. Under conditions of nutrient depletion at high density on a solid surface, the individual cells glide into numerous aggregation centres and construct multicellular fruiting bodies that are visible with the naked eye. Within the fruiting bodies the cells differentiate into resting cells called myxospores. *M. xanthus* fruiting bodies are mounds of myxospores. Fruiting bodies

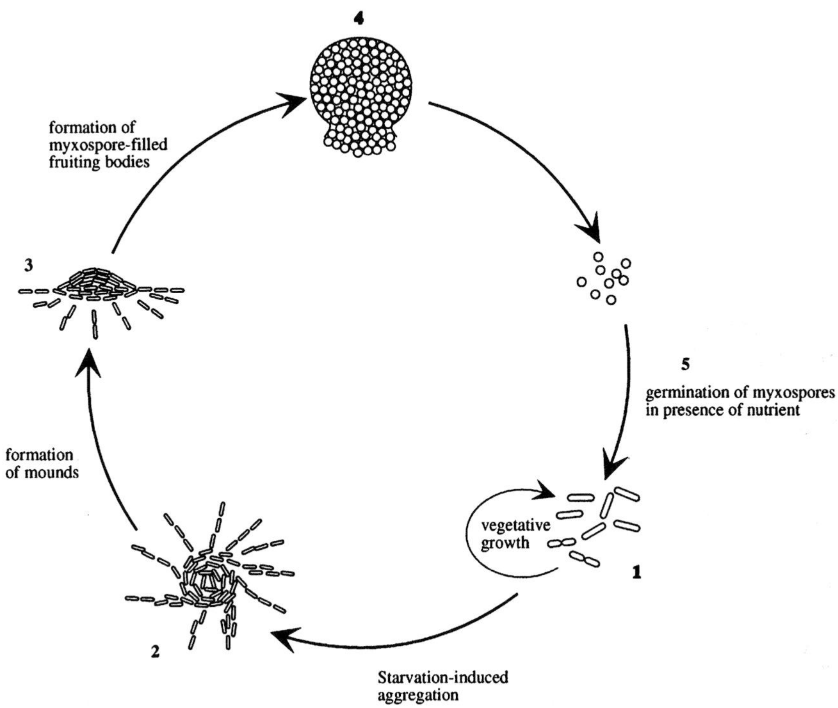

Figure 7. Life cycle of *Myxococcus xanthus*. 1. Vegetative growth. Cells grow in dense swarms on solid surfaces. 2. Aggregation. When nutrients are depleted, cells glide into aggregation centres, each one consisting of many thousands of cells. 3. Mound formation. Each aggregation centre becomes a mound of cells as bacteria continue to accumulate. 4. The mound develops into a fruiting body when the cells differentiate into resting cells called myxospores surrounded by a coat (capsule). 5. Nutrient availability causes the myxospores to germinate into vegetative cells, returning to a new growth phase. Depletion of nutrients again results in new fruiting body formation within 24 hrs. Adapted from White (2000).

are dispersed to new locations, probably by sticking to the bristles of wandering insects. When fresh nutrients are available, the myxospores germinate and a new vegetative swarm emerges from the fruiting body. Thus, at no time does an individual cell glide away from the swarm and begin a new colony. These bacteria are truly interactive, and are often referred to as "social bacteria", having properties of both unicellular and multicellular organisms.

2.2.2. *A-Signalling in the Aggregation Phase of Fruiting Body Formation*

In order for the myxobacteria to initiate fruiting body formation, a diffusible intercellular signal is produced within the first hour after starvation (signalling based upon cell-cell contact is a separate event that occurs later in fruiting body development: see section 2.9.1.). The pheromone produced by *M. xanthus* is composed of a subset of amino acids that must accumulate above a threshold concentration of 10 μM (Kuspa, Plamann and Kaiser, 1992). The amino acids that are used as signals by *M. xanthus* are collectively called the A-signal. *M. xanthus* does not secrete the amino acids as such, but rather secretes proteases that degrade extracellular proteins to the amino acids used as the A-signal (Fig. 8 and Plamann, Kuspa and Kaiser, 1992). The A-signal is produced in proportion to the cell density as a response to starvation by the extracellular proteases, which degrade surface proteins. When myxobacteria are induced to form fruiting bodies in the laboratory in media devoid of protein, the substrate for the proteases is not known, but presumably is derived from the myxobacteria cells themselves (Plamann and Kaplan, 1999). Thus, two conditions must be met in order for *M. xanthus* to aggregate and form fruiting bodies – nutrient depletion and high cell density.

Mutants blocked in A-signal generation are arrested within the first two hours of development and form loose aggregates that do not mature into fruiting bodies and form few myxospores (La Rossa *et al.*, 1983). It is known that the addition of A-signal stimulates the transcription of genes that are active primarily during development. A current model is that A-signal either directly or indirectly stimulates a membrane-bound histidine kinase called SasS (Fig. 8 and Yang and Kaplan, 1997). As a result, SasS is autophosphorylated and the phosphoryl group is transferred to an RR called SasR. SasR-P is an activator for a group of developmental genes. A negative regulator called SasN is required for normal A-signal signalling and appears to block the phospotransfer during vegetative growth through interaction with SasS (Fig. 8 and Xu, Yang and Kaplan, 1998). This multicomponent pathway is likely to allow the integration of additional environmental conditions with the response to A-signal.

Another myxobacterium, *Stigmatella aurantiaca* secretes a lipoidal pheromone called stigmalone (Stephens, Hegeman and White, 1982; Hull, Berkessel and Plaga, 1998; Plaga, Stamm and Schairer, 1998). When the levels of extracellular stigmalone are sufficiently high, the cells form aggregates that mature into fruiting bodies. Whereas *Myxococcus* fruiting bodies are simply mounds of

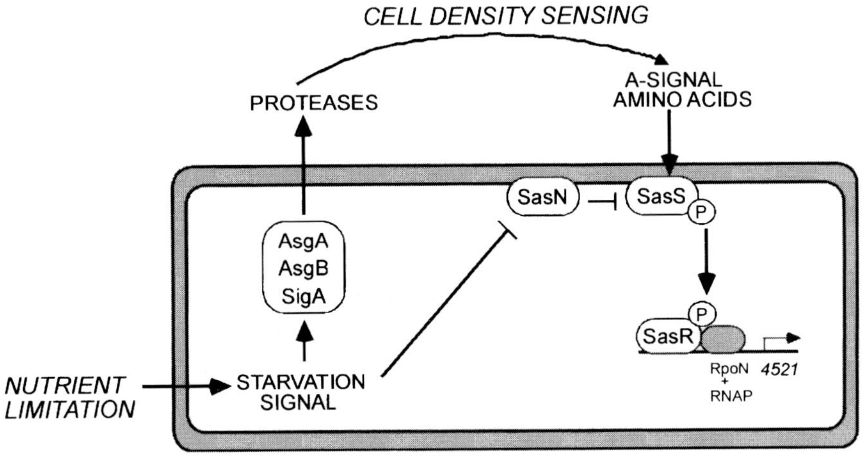

Figure 8. The A-signal response pathway in *Myxococcus xanthus*. An individual *M. xanthus* cell is represented during early development to illustrate the relationship between the regulatory elements and signals that control expression of developmental genes such as *4521*. Nutrient limitation initiates a signalling cascade that results in the activation of a histidine kinase (AsgA) and a transcriptional regulator (AsgB). SigA is the major σ factor required for transcription. As a consequence of this signal generating pathway, specific genes are transcribed that result in the secretion of proteases which are responsible for producing the A-signal (a subset of amino acids). The A-signal response pathway includes SasS, a membrane-bound histidine kinase. SasS~P is suggested to serve as a source of phosphate for SasR, a response regulator. Phospho-SasR activates the transcription of developmental genes, such as *4521*. RpoN and RNAP are the σ^{54} RNA polymerase holoenzyme. SasN prevents the expression of developmental genes during growth, but is inactivated by the starvation signal. (Figure kindly supplied by H. Kaplan.)

myxospores, *Stigmatella* fruiting bodies have a much more complex shape (see Fig. 1C). They resemble certain fungi, in consisting of a stalk supporting several sporangia that house the myxospores. The signalling pathway for the response to stigmalone is not known. *Myxococcus* does not respond to stigmalone, and the use of different pheromones in these closely related myxobacteria might be one factor in ensuring that mixed aggregates consisting of both types of myxobacteria do not form.

2.3. ACYLATED HOMOSERINE LACTONES

2.3.1. *Luminescence Regulation in Marine Vibrios*

Acyl HSLs were first identified as signalling molecules in several bioluminescent marine *Vibrio* species: *V. harveyi* colonizes the intestines of certain fishes and is prevalent in sea water, and *V. fischeri* is a symbiont that inhabits the "light organs" of specific fishes and squids and produces light for its animal host (Nealson, Platt and Hastings, 1970; Nealson, 1977). These *Vibrio* species produce light in standard laboratory culture, but luminescence is significantly delayed until the late stages of culture growth. Light production can be aberrantly induced in fresh cultures by the addition of conditioned medium from dense, bioluminescent cultures, implicating the presence of a bacterially-derived factor(s) that signals the younger cells to begin producing light. A great deal of work over many years has led to a reasonable understanding of the bioluminescence control in these microbes (see Fuqua and Eberhard, 1999). *V. fischeri* synthesizes 3-oxo-hexanoyl-homoserine lactone (3-oxo-C_6-HSL) via an enzyme called LuxI (Eberhard *et al.*, 1981; Schaefer *et al.*, 1996). This signal molecule is freely diffusible across the bacterial envelope. At low population densities, such as when *V. fischeri* is free in seawater, the acyl HSL effectively exits the cell down a steep concentration gradient, resulting in a low intracellular concentration which is insufficient to induce luminescence. Colonization of the animal host light organ by *V. fischeri* results in clonal growth of the population and eventually a high cell density that is paralleled by a concominant increase in the 3-oxo-C_6-HSL concentration. As the diffusion gradient across the cellular envelope is reduced, less acyl HSL is released from the cell and the intracellular concentration of pheromone increases, fostering interaction with a cytoplasmic acyl HSL receptor called LuxR. LuxR binds to 3-oxo-C_6-HSL, in turn associates with a DNA regulatory element upstream of the bioluminescence (*lux*) genes, activates expression of the downstream operon, and turns on the lights. For this microbe, the high population density provided by the light organ environment leads to accumulation of the pheromone, and subsequent activation of *lux* genes. The dramatic difference in the *V. fischeri* populations in the light organ compared to those cells in the free-living state results in a symbiosis-specific regulation of light production. However, artificially high population densities, such as those provided in the laboratory, also effectively activate the acyl HSL regulatory system. Surprisingly, light regulation in the closely-related *V. harveyi*, while sharing a population density sensing component and utilizing an acyl HSL, functions via an entirely distinct mechanism (see section 2.3.7. below).

2.3.2. *Acyl HSLs are a Common Intercellular Signal in Gram-Negative Bacteria*

Although the original observations of luminescence regulation in the marine vibrios were made in the early 1970s, it was not until 1992 that other bacteria were found to also produce acyl HSLs (Bainton *et al.*, 1992). Since that time there has been a tremendous explosion of information on acyl HSLs from over 20 different genera of gram-negative bacteria (Fuqua, Winans and Greenberg, 1996; Greenberg, 1997). The acyl HSLs synthesized by these bacteria share the same general structure as the *V. fischeri* pheromone, but may vary in length (with an even number of alkyl groups ranging from 4 to 14), modification at the 3rd (β) position (fully reduced, hydroxylated, or carrying a carbonyl group) and the presence or absence of unsaturated bonds in the acyl chain (see Table 3 and Fuqua and Eberhard, 1999). In almost all cases, the acyl HSL is synthesized by an enzyme that shares sequence homology with LuxI (Fig. 9). Perception of the acyl HSL and subsequent transcriptional regulation of target genes is

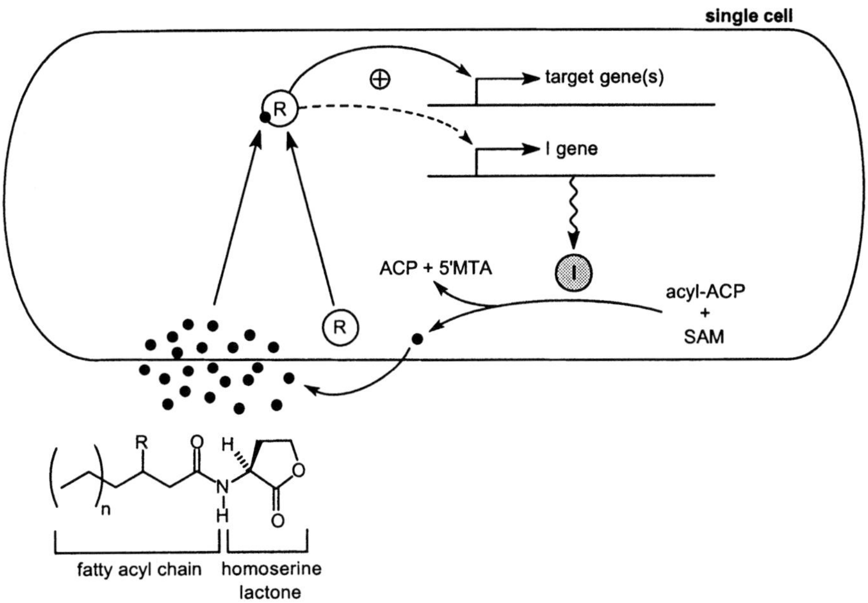

Figure 9. General model for acyl HSL signalling. LuxR type protein and LuxI type protein are represented as spheres labelled with R and I, respectively; acyl HSL is a filled circle. Filled arrows indicate noncovalent association and catalysis; dashed arrow implies that activation of I gene expression is variable between different bacteria. Squiggles indicate transcription and translation of the I gene. On the generalized acyl HSL structure, R can be H, OH, or O and n equals the number of additional two carbon groups beyond the minimal 4 carbon length depicted for the acyl chain; 0–5.

almost always mediated by a LuxR homologue. The target functions controlled by acyl HSLs and LuxI-LuxR regulatory circuits are strikingly diverse (for examples see Table 1 and Fuqua, Winans and Greenberg, 1996). More often than not, the genes regulated by the acyl HSL are either directly or indirectly involved in interactions between the microbe and host organisms.

There are several examples in which the acyl HSL signals play a global regulatory role, directly or indirectly controlling the expression of many different genes (Whiteley, Lee and Greenberg, 1999). It is generally thought that acyl HSLs provide information on the intraspecies population dynamics in a given environment. There are also reports of interspecies communication, or crosstalk, via acyl HSLs from several different bacteria (Allison *et al.*, 1993; Wood *et al.*, 1997).

2.3.3. *Synthesis and Release of Acyl HSLs*

Several LuxI-type proteins have been shown to utilize the ubiquitous methyl group donor S-adenosylmethionine (SAM) as a substrate for the homoserine portion of the acyl HSL (see Fig. 9 and Hanzelka and Greenberg, 1996; Moré *et al.*, 1996; Schaeffer *et al.*, 1996; Parsek *et al.*, 1999). The methionine moiety in SAM is ligated to the appropriate acyl chain donated by an acyl-acyl carrier protein (ACP) fatty acid biosynthetic intermediate. For the RhlI acyl HSL synthase of *Pseudomonas aeruginosa*, the reaction occurs in a sequential manner with the binding of SAM followed by ligation with the acyl chain (in this case C_4-ACP). There are exactly ten amino acid residues that are absolutely conserved among the LuxI family of proteins and there is evidence that of these, all the charged amino acid side chains are necessary for enzymatic activity (Hanzelka *et al.*, 1997; Parsek, Schaeffer and Greenberg, 1997). However, beyond these analyses, there is very little understanding of how different LuxI-type proteins recognize different acyl-ACP conjugates to synthesize the appropriate acyl HSL (for a detailed review see Fuqua and Eberhard, 1999).

The 3-oxo-C_6-HSL that regulates *lux* genes in *V. fischeri* is freely diffusible across the bacterial envelope, and therefore passive diffusion is considered the primary mode by which the signal is released into the environment (Kaplan and Greenberg, 1985). By analogy to the Lux system, and because exogenous addition of synthetic acyl HSLs provides rapid and dose-responsive activation of quorum-sensing systems, it is generally assumed that most acyl HSLs are released from, and perhaps re-enter, cells by passive diffusion (Fig. 9). However, acyl HSLs with significantly longer acyl chains, and thus greater hydrophobicity, might partition to a greater extent into the membrane bilayers. In fact, several studies have shown that the release of 3-oxo-C_{12}-HSL in *P. aeruginosa* is enhanced by the activity of a membrane efflux system thought to be involved in the scavenging of hydrophobic molecules from the lipid bilayer (Evans *et al.*, 1998; Pearson, Van Delden and Iglewski, 1999).

2.3.4. *Mechanisms of Acyl HSL-Dependent Gene Regulation*

Members of the LuxR family of proteins are usually transcriptional activators, although there is evidence that the LuxR-homologues from several plant pathogenic species of *Erwinia* are repressors (Beck von Bodman, Majerczak and Coplin, 1998; Nasser *et al.*, 1998). All LuxR proteins can be subdivided into two major functional regions, based on sequence similarities and biochemical analysis of several representative members of the family. The N-terminal half of the protein contains a conserved region of approximately 40 amino acid residues where mutations abolish the binding of 3-oxo-C_6-HSL for LuxR itself (Hanzelka and Greenberg, 1995; Stevens and Greenberg, 1999). Acyl HSLs function at concentrations ranging from 1×10^{-6}–1×10^{-9}M, although this varies among different bacteria. Presumably, the threshold concentration of acyl HSL is set by the binding affinity of each acyl HSL for its cognate receptor.

All LuxR homologues contain a helix-turn-helix (HTH) motif, that is required for DNA binding. The HTH and flanking sequences are particularly conserved and place the LuxR-type regulators into the larger FixJ-NarL family of prokaryotic transcription factors that all share similar DNA binding motifs (Da Re *et al.*, 1994). In fact, a LuxR protein, missing its amino terminal 162 residues and containing only the C-terminal DNA binding motif, is sufficient for binding to *lux* gene promoter elements (Choi and Greenberg, 1992a; Stevens, Dolan and Greenberg, 1994). This amino-terminal truncation LuxR protein binds in a constitutive manner, similar to several other members of the larger FixJ-NarL family, suggesting that the amino terminal region functions to inhibit DNA binding activity. This has led to a model for LuxR, and by extension other LuxR-type activators, where acyl HSL binding releases inhibition of the DNA binding region by the amino terminal region, which in turn facilitates promoter binding and subsequent transcriptional activation. However, there is also evidence that multimerization (probably dimerization) is stimulated by acyl HSL association for several LuxR-type activators (Choi and Greenberg, 1992b; Zhu and Winans, 1998; Luo and Farrand, 1999). It remains unclear how multimerization integrates with the apparent inhibitory effect of the amino terminal region on the DNA binding activity. Even less well understood is the mechanism by which the LuxR repressors, such as EsaR from *Erwinia stewartii*, function (Beck von Bodman, Majerczak and Coplin, 1998). Acyl HSL binding in these cases should result in release of the protein from its DNA binding sites and induction of target gene expression. Clearly, the conformational alterations that occur upon acyl HSL binding to the repressors must be dramatically different from those that occur in the activators. However, there are no obvious, consistent differences between the amino acid residues conserved among the repressors (Stevens and Greenberg, 1999).

LuxR-type proteins bind to specific DNA sequences upstream or within the promoters of the genes they regulate. Often, short inverted repeat sequences of 18–20 bp are associated with these regulated promoters. These sequences share significant primary sequence similarity with the sequence bound by LuxR upstream of the *V. fischeri lux* genes and are therefore collectively called *lux*-type boxes (Fuqua and Winans, 1994; Fuqua, Winans and Greenberg, 1996). Disruption of the sequences usually abolishes target gene regulation (Devine, Shadel and Baldwin, 1989; Fuqua and Winans, 1996; Rust, Pesci and Iglewski, 1996; Luo and Farrand, 1999). LuxR itself, TraR from *A. tumefaciens* and LasR from *P. aeruginosa* have been demonstrated to bind to *lux*-type boxes *in vitro* (Stevens, Dolan and Greenberg, 1994; You *et al.*, 1996; Zhu and Winans, 1999). There are also multiple examples of genes that are directly regulated by LuxR-type proteins that possess no obvious or highly degenerate lux-type boxes. Therefore, *lux*-type DNA sequence elements are common, but not absolutely required for all acyl HSL regulated genes.

2.3.5. *Acyl HSLs in Microbial Bioflims*

In the *V. fischeri* Lux regulatory system, 3-oxo-C_6-HSL accumulates dramatically when the bacteria colonize the tissue of the light organ (Boettcher and Ruby, 1995). However, for most other quorum-sensing bacteria, no such specialized inducing environment exists. The high population densities which lead to acyl HSL accumulation in other microbes are almost certainly those environments associated with surfaces, both inanimate or biological. Bacteria grow in surprisingly structured surface aggregates called biofilms on many, if not all, non-sterile surfaces (see Fig. 1 and Costerton *et al.*, 1995; Palmer and White, 1997). Biofilms can cover large surface areas or be concentrated in smaller aggregates commonly called microcolonies. Individual cells in biofilms are commonly embedded in exopolymeric matrix material and concentrated in large numbers. It is arguable that biofilms are the structures for which the acyl HSL quorum sensors, as well as other quorum-sensing systems, evolved to function. Indeed, acyl HSLs have been detected in natural biofilms from several sources (McLean *et al.*, 1997; Stickler *et al.*, 1998). Moreover, for the well-studied biofilm former, *P. aeruginosa* mutants that cannot synthesize 3-oxo-C_{12}-HSL, one of two major acyl HSLs this microbe produces, form biofilms that lack any of the complex structures of wild-type biofilms, and are easily removed from surfaces (Davies *et al.*, 1998). These observations suggest that not only do acyl HSLs function within biofilms, but also that they are involved in biofilm formation. The ubiquity of acyl HSLs as biofilm signal molecules is currently an area of active study, but it is unlikely that acyl HSLs play this role in all biofilms (*eg.* those formed by predominantly gram-positive microbes). Other intercellular signalling systems probably perform this function along with, or in place of, acyl HSLs in biofilms of different bacteria. Understanding the role of acyl HSLs and other signalling molecules in these complex microbial communities is an exciting area of future work.

2.3.6. *Modulation of Acyl HSL Activity by Additional Signal Molecules*

Bacterially Produced Signal Molecules. Several recent studies have identified additional signal molecules that are recognized by LuxR-type receptors. The simplest example of this is found in *V. fischeri*, in which an alternative acyl HSL synthase called AinS is responsible for synthesizing C_8-HSL (Kuo, Blough and Dunlap, 1994; Hanzelka *et al.*, 1999). Mutants that cannot synthesize C_8-HSL due to a lesion in *ainS* activate expression of the *lux* genes at lower population density than the wild-type strain, suggesting that the C_8-HSL may competitively inhibit the LuxR protein from interacting with its cognate acyl HSL 3-oxo-C_6-HSL (Kuo, Callaghan and Dunlap, 1996).

More surprising are the reports of non-acyl HSL molecules that affect the activity of Lux-type systems. Diketopiperazines, or cyclic dipeptides, are compounds produced by a number of different bacteria that, at high concentrations, can effectively supplant acyl HSL activity for several LuxR-type proteins (Holden *et al.*, 1999). In combination with acyl HSLs, the diketopiperazines act as competitive inhibitors of the acyl HSL response. The physiological relevance of this observation is not yet clear, but it does have implications for the biochemical specificity of LuxR-type proteins for acyl HSLs. Another possible example exists in *P. aeruginosa*, where it has also been shown that a diffusible molecule, 2-heptyl-3-hydroxy-4-quinolone (called the *P. aeruginosa* quinolone signal), related to the family of quinolone antibiotics (exhibiting no antibiotic activity of its own), can induce the activity of the LasR protein in the absence of its cognate acyl HSL, 3-oxo-C_{12}-HSL (Pesci *et al.*, 1999). The integration of the quinolone signal into the already complex regulatory circuitry of *P. aeruginosa* quorum sensing is currently under investigation.

Signal Interference and Interception by Eukaryotes. There is at least one convincing example of eukaryotic interference with acyl HSL signalling. The red alga, *Delisea pulchra* synthesizes large amounts of compounds generally called furanones, which are released from secretory organs in the fronds of the alga (Givskov *et al.*, 1996). Several furanones have been shown to potently inhibit acyl HSL-based signalling pathways in *V. fischeri* and *S. liquefaciens* by competing with acyl HSLs for the binding site on the LuxR-type protein (Manefield *et al.*, 1999). It remains unclear whether this inhibition of acyl HSL signalling has evolved specifically to interfere with LuxR-type signalling. The furanones have broad spectrum activity on metazoan marine surface colonizers, as well as microorganisms (de Nys *et al.*, 1995). The general question whether it is common for eukaryotic hosts to evolve systems by which to block the communication lines of their microbial colonizers is currently an active area of research.

Acyl HSLs as Immunoactive Compounds. There is also evidence for direct interaction of acyl HSL signals and the mammalian immune system. The *P. aeruginosa* acyl HSL 3-oxo-C_{12}-HSL has immunomodulatory effects in animal infection models (Dimango *et al.*, 1995; Telford *et al.*, 1998). There is as yet conflicting information on the response of the animal immune system to acyl HSL, and it is by no means clear whether the immune recognition of acyl HSLs benefits the host or the infecting microbe. There are numerous examples of microbial toxins (*ie.* superantigens) that function by hyperinducing the mammalian inflammatory response, and it is possible that acyl HSLs can directly damage infected tissues in a similar manner. It is also possible, perhaps even more likely, that the mammalian immune system has developed the ability to eavesdrop on a common signalling pathway of its microbial assailants and thereby more effectively eradicate them.

2.3.7. *Acyl HSL Signalling without LuxR and LuxI-Type Proteins*

Bioluminescence in *V. harveyi* is regulated by population density via two diffusible signals, one of which is 3-OH-C_4-HSL (Cao and Meighen, 1989). The LuxL and LuxM proteins are required for synthesis of the acyl HSL signal (Bassler *et al.*, 1993), but share no homology with LuxI-type proteins. Futhermore, perception of 3-OH-C_4-HSL involves the activity of a two-component-type sensor kinase called LuxN, that engages a phosphorelay system (Freeman and Bassler, 1999; Freeman, Lilley and Bassler, 2000; and see section 2.1.4.). Remarkably, *V. fischeri* also possesses a homologue of LuxM called AinS, which catalyzes synthesis of C_5-HSL (Hanzelka *et al.*, 1999). Catalysis occurs by the same basic mechanism as for LuxI homologues and utilizes similar precursors. There is a gene encoding a two-component sensor kinase, tentatively designated AinR, linked to the *ainS* gene, that seems a likely candidate to regulate some other set of target functions in *V. fischeri*, although these have yet to be identified (Gilson, Kuo and Dunlap, 1995). The observations that acyl HSLs can be synthesized and perceived by evolutionarily different proteins is a striking indication of the utility of these molecules as intercellular signals.

2.4. SIGNALLING VIA AI-2-TYPE FACTORS

2.4.1. *A Second Quorum-Sensing Signal in Vibrio harveyi*

V. harveyi lux genes are regulated by a dual-channel, diffusible signalling system (for review see Bassler and Silverman, 1995). As mentioned above, one of these signals is 3-OH-C_4-HSL (Cao and Meighen, 1989). This signal is thought to act in gauging intraspecies population density. Surprisingly, however, *V. harveyi* mutants that cannot synthesize the acyl HSL still exhibit cell density-dependent regulation of bioluminescence. This is due to a second quorum-sensing system that generates an as yet undefined, diffusible signal molecule(s) called autoinducer 2 (AI-2) (Bassler, Wright and Silverman, 1994). AI-2 is perceived via the LuxQ two-component sensor kinase, and engages the same phosphorelay system that 3-OH-C_4-HSL activates (Bassler, Wright and Silverman, 1994). Therefore, in mutants that cannot synthesize the acyl HSL, AI-2 provides the signal. Likewise, strains that carry mutations in LuxQ, and hence cannot perceive AI-2, maintain their population density responsiveness via the acyl HSL. Double mutants defective for synthesis of both signals cannot activate *lux* genes. Using a mutant *V. harveyi* strain that responds only to AI-2, Bassler *et al.* found that a range of different bacteria produce factors capable of activating the AI-2-dependent system suggesting that this signal molecule may provide interspecies information to *V. harveyi* and other bacteria (Bassler, Greenberg and Stevens, 1997).

2.4.2. *AI-2-Type Signals Are Produced by Diverse Bacteria*

Recent work suggests that AI-2-like factors are synthesized by a broad range of bacteria. AI-2 is a small molecule(s), but does not appear to be an acyl HSL, a γ-butyrolactone (see section 2.5.), or an oligopeptide (Surette and Bassler, 1998). Additionally, in contrast to these other signalling systems, AI-2 and related signals are actively degraded in late-phase cultures and are most active towards the end of logarithmic growth (Surette and Bassler, 1999). Synthesis of AI-2 in *V. harveyi* requires the *luxS* gene. Although homologues of *luxS* have been identified in the genome sequences of over 18 different bacteria, the biochemical function of the *luxS* gene product remains unclear (Surette, Miller and Bassler, 1999). *V. harveyi* has evolved the capacity to detect the AI-2 signal(s) as a means of regulating bioluminescence, most likely using this to gauge its own population density, as well as that of other bacteria in its environment. However, the question still remains whether the AI-2 signal(s) produced by *luxS* homologues in other bacteria also serves a signalling function.

There is evidence that AI-2-type signalling may play a role in regulating the transcription of genes encoding a Type III secretion system (the *sec/esc* genes of the *LEE* pathogenicity island) in enteropathogenic and enterohemorrhagic *E. coli* strains (Sperandio *et al.*, 1999). The Type III secretion system is involved in attachment of these pathogenic *E. coli* strains to intestinal cells during infection, linking AI-2 signalling to host interactions. No additional target genes have been reported, nor have the presumptive regulatory proteins, analogous to the *V. harveyi* LuxQ sensor kinase, been isolated. Genome sequences of several of the *luxS*-harbouring strains do not reveal any particularly strong homologues to LuxQ. However, many two-component-type sensor kinases with unknown function have been identified and these, or other regulatory proteins, may yet prove to be the AI-2 response elements.

2.5. γ-BUTYROLACTONES: REGULATORS OF ANTIBIOTIC SYNTHESIS AND SECONDARY METABOLISM IN *STREPTOMYCES*

One of the first hints at intercellular signalling in prokaryotes came from the pioneering work of Kholkov on the filamentous, fungus-like microbes of the genus *Streptomyces* (Kholkov *et al.*, 1967). The streptomycetes are gram-positive soil bacteria that grow as a branching, filamentous, nonseptate mycelium, superficially resembling fungi. Older parts of the colony send up aerial filaments called hyphae or sporophores. Septa form at regular intervals in the aerial hyphae, and each septated compartment develops into a spore called a conidium (pl. conidia) (Fig. 10). Members of the genus *Streptomyces* are ubiquitous soil microbes and are well recognized as the primary source of pharmaceutically important antibiotics. It is easy to imagine that *Streptomyces* has evolved antibiotic production as a mechanism to compete with other soil microbes for limiting nutrients. In laboratory culture, antibiotics are typically synthesized in the late stages of growth and represent a product of so-called secondary metabolism (metabolism occurring after the primary stage of growth in laboratory culture which results in the secretion of products not required for growth).

2.5.1. *Mechanism of A-factor Signalling*

Sporulation and synthesis of the aminoglycoside streptomycin by *Streptomyces griseus* is delayed until late stages of culture growth and its synthesis depends on a diffusible signal molecule called A-factor (for a detailed review see Horinouchi, 1999). A-factor is a γ-butyrolactone, specifically 2-isocaprolyl-3*R*-hydroxymethyl-γ-butyrolactone, which is effective at a concentration of 1nM (Kholkov *et al.*, 1967; Hara and Beppu, 1982). The AfsA protein is a probable A-factor synthase, utilizing fatty acid biosynthetic precursors and glycerol derivatives as substrates (Ando *et al.*, 1997). A-factor shares structural similarity with the acyl HSLs and, as such, is also freely diffusible across the bacterial envelope. A-factor is thought to coordinate antibiotic synthesis in adjacent

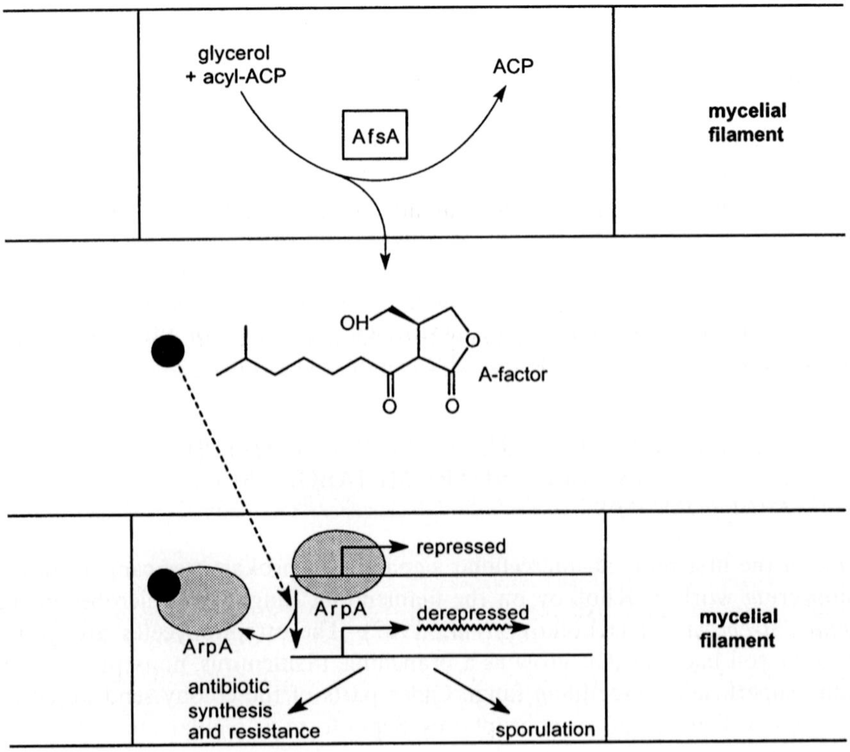

Figure 10. *Streptomyces griseus* signalling between mycelia. Exchange of A-factor between two adjacent *S. griseus* filaments. Filled circle indicates A-factor. Intrafilament A-factor signalling is not depicted but is thought to occur in the same fashion. See text for details.

cells along a filament, as well as between different filaments in the same population (Fig. 10). Resistance genes are also under the control of A-factor, and it seems reasonable that coordination of antibiotic synthesis and resistance gene expression might ensure that members of a *S. griseus* population will not suffer growth inhibition due to bursts of antibiotic production in adjacent parts of the population.

A-factor binds to the cytoplasmic ArpA protein, a repressor that inhibits several genes in the absence of the signal molecule (Fig. 10 and Miyake *et al.*, 1989; 1990). The ArpA amino acid sequence is not similar to LuxR-type or LuxN-type proteins, despite the similarity in signal molecule structure as well as the common events that lead to regulated gene expression. ArpA possesses an HTH motif in its N-terminal half. A-factor binding, presumably to a site in the carboxy terminal portion of ArpA, causes derepression in which the protein releases its DNA binding site resulting in target gene induction (including streptomycin biosynthetic and resistance genes, and other genes involved in microbial development (Horinouchi, 1999).

2.5.2. *γ-Butyrolactones in Other Streptomyces spp.*

Although A-factor is the best characterized system, similar γ-butyrolactone molecules have been identified in several other species of *Streptomyces* and shown to regulate antibiotic synthesis and cell differentiation (Horinouchi, 1999). Virginiamycin synthesis in *S. virginiae* is regulated by a group of several γ-butyrolactones called the virginiae butanolides (VBs), in much the same way as A-factor functions in *S. griseus* (Yamada *et al.*, 1987). Despite a VB receptor protein that shares homology with ArpA, there is apparently no cross-recognition between *Streptomyces* γ-butyrolactones from different species, due to a high degree of specificity (Miyake *et al.*, 1989; Okamoto, Nakamura and Yamada, 1995). Therefore, it is likely that interspecies quorum-sensing via the butyrolactones rarely, if ever, occurs between streptomycetes.

2.6. UNUSUAL SIGNAL MOLECULES IN PLANT PATHOGENS

2.6.1. *Regulation of Exopolysaccharide Synthesis in the Plant Pathogen Ralstonia solanacearum*

The primary virulence factor for *R. solanacearum* on host plants is an acidic, heteropolymeric exopolysaccharide (EPS I). Expression of the genes directing EPS I production (*eps*) is controlled by a complex regulatory network and a series of environmental signals. Included among these is a diffusible signal molecule produced by the bacteria themselves, that has been identified as 3-hydroxypalmitic acid methyl ester (3-OH-PAME, Flavier, Schell and Denny, 1997). 3-OH-PAME is effective at less than 5nM concentration to activate expression of *eps* genes. The signal molecule accumulates during culture growth and acts to integrate population density into *eps* gene regulation via intraspecies quorum sensing. Remarkably, this signal can be transmitted through the vapour phase between two cultures that are separated from each other by an air interface (Clough, Schell and Denny, 1994). It is unclear whether this vapour phase signalling is relevant *in situ*. Synthesis of 3-OH-PAME requires the activity of the PhcB protein, the sequence of which contains a SAM binding motif, suggesting that biosynthesis may utilize fatty acid precursors and SAM in a manner analogous to acyl HSL biosynthesis.

Response to 3-OH-PAME requires the regulatory proteins PhcS, PhcR and PhcA. PhcS is a two-component-type HK that is likely to bind the 3-OH-PAME signal (see Fig. 4 for a general scheme). PhcR is a response regulator protein that lacks the DNA binding function and probably inhibits the activity of the transcriptional activator PhcA (Clough *et al.*, 1997). Perception of 3-OH-PAME by PhcS is thought to relieve the inhibition of PhcA by PhcR, and elevate expression of *eps* genes. In addition to the *eps* genes, PhcA also regulates expression of the *solR* gene encoding a LuxR-type acyl HSL receptor (Flavier *et al.*, 1997). SolR requires the presence of acyl HSL signals (likely to be C_8-HSL and C_6-HSL) produced by the SolI protein, a *R. solanacearum* LuxI homologue (Flavier *et al.*, 1997). SolR and SolI have no impact on *eps* gene regulation and their role in plant pathogenesis is unclear. A gene of unknown function, *aidA*, genetically linked to the *solR solI* gene cluster, is regulated by SolR and should provide an avenue by which the target functions can be identified.

2.6.2. *Diffusible Signal Molecules in the Plant Pathogen Xanthomonas campestris*

X. campestris pv. campestris causes black rot disease on cruciferous plants. As with *R. solanacearum*, production of an EPS, in this case called xanthan gum, is the primary cause of disease symptoms. Production of extracellular enzymes, such as endoglucanases and polygalacturonate lyases, probably adds to the damage of host plant tissue.

DSF Regulates EPS Production and Extracellular Enzyme Synthesis. Xanthan gum synthesis and extracellular enzyme activity are regulated in a complex manner that includes the effect of a self-produced diffusible extracellular factor called DSF (Barber *et al.*, 1997). DSF is a small molecule of unknown identity that, like the acyl HSLs, can be extracted into the solvent ethyl acetate. DSF synthesis is dependent on the *rpfB* and *rpfF* genes, the products of which share homology with proteins involved in fatty acid metabolism. Several lines of evidence led Barber *et al.* to speculate that DSF is synthesized by ligation of acyl-CoA derivatives (intermediates of degradative fatty acid metabolism in bacteria) with another unidentified molecule (Barber *et al.*, 1997). DSF activity is degraded in late-stage cultures, analogously to AI-2 type signalling factors. Xanthan gum and extracellular enzyme production initiates in late-stage cultures, parallel with accumulation of DSF. However, addition of solvent-extracted DSF to early-phase cultures of *X. campestris* pv. campestris does not alter the expression pattern of these virulence factors. This suggests that there are other environmental parameters that influence regulation of virulence factors in *X. campestris* pv. campestris and are integrated with increasing levels of DSF.

DF Regulates Pigment Production and Xanthan Gum Synthesis. A second signal called diffusible factor (DF) also influences xanthan gum synthesis. DF was discovered during analysis of a mutant *X. campestris* pv. campestris strain that synthesizes significantly less of the pigment xanamonadin than wild-type strains, due to a lesion in the *pigB* gene (Poplawsky and Chun, 1997). Cultivation of the *pigB* mutant adjacent to a wild-type strain rescued this phenotype, suggesting cross-feeding of a diffusible factor eventually characterized as DF. The *pigB* mutants are also significantly reduced for xanthan gum synthesis. DF is ethyl acetate-extractable, low molecular weight (<500 Dal) and heat stable. Chemical analysis suggests that the compound may be a γ-butyrolactone, although this remains unproven. Despite some common features, DSF and DF are not identical. Biosynthetic mutants of each have different, at times often opposite, phenotypes. Moreover, DF and DSF have different migration rates by thin layer chromatography (Poplawsky and Chun, 1998). Lastly, the *rpfB* and *rpfF* gene products share no similarity with that of *pigB*. Therefore, it appears that *X. campestris* utilizes at least two distinct diffusible signals to differentially regulate aspects of its interaction with plants. For both the DSF and the DF signals, the pathways that lead to response are not well understood and are subjects of current investigation.

2.7. GLUTAMINE SIGNALLING IN *PROTEUS*

Members of the genus *Proteus* are capable of a type of surface motility called swarming, by which the entire colony spreads via movement of cells across a surface. Swarming motility has been most extensively studied in *P. mirabilis*, and entails differentiation from a standard, rod shaped cell (called a swimmer cell, approx. 2 μm) with several thin flagella, into an elongated (up to 80 μm in length), multinucleoid form called a swarmer cell that is covered with thousands of flagella (see Belas, 1997). In the laboratory, swarmer cells form when *Proteus* is cultivated on agar surfaces of the appropriate growth medium. Swarming growth moves from the point of inoculum outwards in concentric waves that reflect the differentiation of *P. mirabilis* swimmer cells into swarmers, and subsequent concerted movement away from the point of inoculation. After several hours of colony expansion, the swarmers stop moving and differentiate back into swimmer cells, and the growth ring consolidates by vegetative multiplication. At a specific time following consolidation, the swimmers begin differentiating into swarmer cells and a new wave of swarming occurs. This cycle repeats itself until the entire plate is covered with growth. Swarming motility in the host animal probably allows the bacteria to persist in the high flow rate environment of the urinary tract.

Although there are many factors that influence swarming motility, intercellular signalling plays an important role. In *P. mirabilis*, the amino acid glutamine has been identified as a potent and specific signal for swarming motility (Allison *et al*., 1993). Although it has yet to be unequivocally shown, glutamine signalling may act as an intraspecies quorum-sensing mechanism in much the same way as A-signalling via amino acids in *M. xanthus*. *P. mirabilis* produces several extracellular proteases that may degrade proteins (microbially-derived or found in the environment) into free amino acids, and glutamine released during this process may act as a population density cue (Zhao *et al*., 1999). The specific signalling pathways through which the glutamine-response is transduced are not yet defined.

Although in *P. mirabilis* there is no evidence for acyl HSL-dependent control of swarming, in *Serratia liquefaciens*, another swarming bacterium, a LuxI-LuxR-type regulatory system is involved in regulating this process (Eberl *et al*., 1996). Clearly, intercellular signalling plays an important role in the coordinated behaviour of swarming bacteria and several distinct mechanisms of cell-to-cell communication have evolved.

2.8. EMERGING DIFFUSIBLE SIGNALLING SYSTEMS

The examples described in the preceding pages represent a diversity of signalling mechanisms utilized by bacteria. However, these are only those systems about which a substantial amount of mechanistic information is available. It is certain that we have simply "scratched the surface" of signalling in the prokaryotes, and that there are multiple other systems at work that await discovery.

2.8.1. *Novel Factors Produced in Late Stationary Phase Cultures*

There are several other systems in which it is clear that intercellular signalling is occurring, but the signals and most regulatory components remain undefined. Baca-DeLancey *et al.* utilized a pool of random, promoter-probe (carrying a promoterless *lacZ* reporter gene) transposon mutants to identify *E. coli* genes induced by late-stage culture fluids. The genes that were identified were designated *cma*, for conditioned-media-activated, and sequence analysis suggests they encode components of amino acid metabolism (Baca-Delancey *et al.*, 1999). The extracellular factor(s) responsible for induction is distinct from the AI-2-like signal(s) identified by Surette and Bassler, is stable to alkalinization and heat treatment, and is present in stationary phase cultures (Surette and Bassler, 1998). The chemical structure of the *cma*-inducing factor remains to be elucidated.

A similar approach with the urinary tract pathogen, *Providencia stuartii* identified *cma* genes, that are also presumptively involved in amino acid turnover in the cell (Rather *et al.*, 1999). The inducing factor(s) identified is only present in cultures grown in rich medium. A *P. stuartii* gene designated *aarA* is required for production of the *cma* inducing factor in this microbe. The AarA gene product was previously identified as necessary for activation of a peptidoglycan/aminoglycoside 2′-*N*-acetyltransferase gene, possibly affecting bacterial cell wall composition. AarA is predicted to be a membrane protein and may be involved in trafficking of the signal molecule(s). In both of these studies, the signal molecules are produced in late-stage cultures, suggestive of a quorum-sensing mechanism, although other late-stage growth regulation is also possible.

Stationary phase supernatant extracts prepared from cultures of the marine bacterium, *Vibrio* sp. S14 and added to early exponential phase cultures of the same microbe alter the expression pattern of over 100 different gene products, as assessed by two-dimensional protein electrophoresis (Srinivasan *et al.*, 1998). This response overlaps with that of carbon starvation and therefore shares some influence over the response to nutrient limitation. The signal molecule(s) present in these extracts does not appear to be an acyl HSL, although the extracts had an inducing effect on a *V. harveyi* bioluminescent reporter system, and may be analogous to the AI-2 signal (see section 3.4.2. above).

One important question that emerges from these and similar studies concerns how a signal molecule is defined. It is possible that some or all of the signalling activities detected in late-stage culture supernates represent metabolic waste products. Can these be considered as true intercellular signals? As the downstream regulatory components are elucidated for response to the extracellular signals, the answer to this question may become clear. In general, for those systems in which intercellular signalling has been well-studied, there are dedicated response pathways in the cells, including specific receptors that are completely or partially involved in transducing the signal into alterations in physiology. The downstream response elements in the cases such as the late-stage culture signals of *E. coli*, *P. stuartii* and *Vibrio* sp. S14 strain are being actively investigated.

2.9. CELL-CELL SIGNALLING THAT REQUIRES CONTACT

One might imagine that since bacteria of the same species are frequently in contact with each other within populations growing on solid surfaces, and are also in contact with tissue surfaces on host animals, that signalling requiring cell-cell contact probably takes place. Examples of contact-dependent communication between bacteria are not as prevalent as pheromone-based systems. A component of this apparent inequality is semantic. For instance, the process of bacterial conjugation has been the subject of study for many years and might be considered a form of contact-dependent signalling. In most forms of conjugation the donor and recipient cell first fuse membranes to form a mating pair. The donor then transfers genetic information in the form of plasmid or chromosomal DNA to the recipient. It is clear that recipients are not passive in this process and therefore conjugation represents a stabilized, physical interaction between bacteria. Other examples also blur the boundary between information exchange and other processes. There are now numerous examples of contact-dependent signal exchange between bacteria and eukaryotic cells. One important example is that of the Type III secretion systems of pathogenic bacteria. These secretion systems have evolved to act as a contact-dependent syringe for injection of cell-damaging toxins and enzymes from the bacteria directly into the cytoplasm of eukaryotic cells, radically altering their regulatory balance (Galan and Collmer, 1999).

For the purposes of our discussion, however, we are restricting our focus to those signalling pathways between bacteria where there is a discrete transduction pathway that enables a specific response to the signal. We acknowledge that this distinction is not always clear and certainly inadequate to describe the diversity of microbial interactions.

2.9.1. *Fruiting Body Formation and Gliding in Myxococcus xanthus*

The examples of contact-dependent signalling we will cover are all from the social myxobacteria, where fruiting body formation requires coordination of cell alignment and location. Signalling between cells that touch each other during fruiting body formation is essential for this process. As described earlier, myxobacteria must signal each other in order to aggregate, form fruiting bodies and sporulate normally. Signalling via cell contact is as important as pheromone signalling for development in the myxobacteria. *M. xanthus* must also engage in cell contact-dependent signalling in order to exhibit a flagellar-independent surface locomotion called gliding motility (Shimkets and Kaiser, 1999).

C-Signalling and Development. One signalling pathway requiring cell-to-cell contact relies on the so-called C-signal. Mutants defective in C-signalling have altered motility patterns, delayed development and construct abnormal aggregates instead of fruiting bodies. Furthermore, they sporulate at a reduced efficiency and reduce or abolish expression of genes that are expressed late during wild-type development (Kroos and Kaiser, 1987). *M. xanthus* cells are rod-shaped, and cells that are aligned end-to-end appear to stimulate each other via the C-signal to complete the developmental program. A model for C-signalling is shown in Fig. 11. It has been suggested that the C-signal is a protein on the surface of the ends of the cells that binds to a receptor at the end of the second cell, resulting in activation of C-signal-dependent pathways. Cells unable to signal other cells have a mutation in a gene called *csgA* (Shimkets, Gill and Kaiser, 1983). Based upon amino acid sequence and other information, it has been suggested that CsgA is a short-chain alcohol dehydrogenase (Baker, 1994; Lee *et al.*, 1995). CsgA is not strongly expressed during vegetative growth and increases markedly during development. C-signalling results in the activation of the transcription factor FruA. The FruA protein either directly or indirectly activates transcription of the *dev* operon, which contains genes required for sporulation and aggregation. Examination of the gliding motility of single *csgA* mutant cells to which C-signal has been added, has revealed that the C-signal stimulates cells to glide faster, increases the duration of the gliding interval, and decreases the number of stops/min (Jelsbak

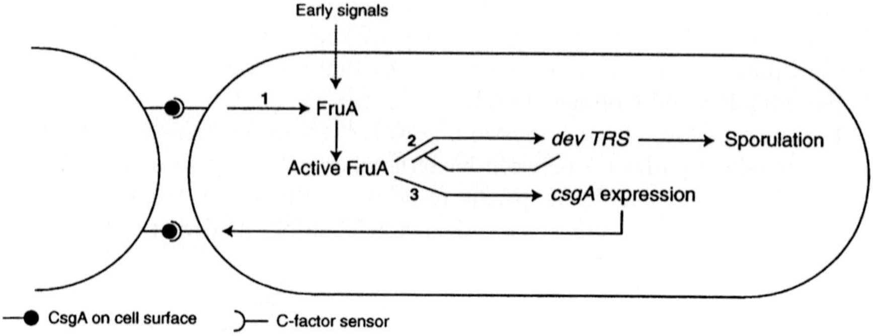

Figure 11. C-signalling in *M. xanthus*. Positive control is indicated by arrows, negative control by lines ending in a bar. Upon receiving the C-signal, a cell activates a transcription factor called FruA. FruA activates the transcription of the operon *devTRS* whose products are required for sporulation. FruA also stimulates the transcription of *csgA* which encodes the C-factor, providing positive feedback. There is also negative feedback on *devTRS* expression. Adapted from Kaiser (1999).

and Søgaard-Andersen, 1999). As a consequence, the C-signal caused the cells to glide further per unit time. It was suggested that this encourages the cells to move into aggregation centres which is an initial step in fruiting body formation. None of these effects were seen in cells that had mutations in the Frz motility system which controls the frequency of reversals in the direction of gliding (see below), suggesting that the Frz proteins are required in order to respond to the C-signal. The addition of C-signal causes the methylation of one of the Frz proteins (FrzCD), and the signalling pathway resulting in methylation requires FruA (see Fig. 11 and Ward and Zusman, 2000, for a review).

Signalling Required for Gliding Motility. Individual cells in a myxobacterial population must coordinate their gliding relative to one another. The coordination is required to keep the swarm intact as a communally feeding entity, as well as to provide the necessary contact-mediated signals for aggregation and fruiting body development. Gliding motility in myxobacteria is coordinated by several cell contact-mediated signalling systems.

Gliding motility in *M. xanthus* is controlled by two sets of genes: adventurous (A) and social (S) (Shimkets, 1999). Mutants in system A genes (A^-S^+) can glide only as groups, indicating that system A genes are required for single cell motility. Mutants in system S genes (A^+S^-) show a motility pattern in which they glide as single cells or small clusters, suggesting that system S genes are required for group motility.

Both single cell and group motility are significantly stimulated when the cell density is increased, reflecting the fact that cell-to-cell communication stimulates gliding motility in both systems. The signalling systems that stimulate single cell and group motility are not well understood, but appear to involve cell-cell contact in both the A and S motility systems (Hodgkin and Kaiser, 1977).

The signalling molecules have not been identified, but presumably they are bound to the cell surface or to fibrils or pili that extend from the cell surface. Genetic analysis and extracellular complementation studies of mutants have suggested that there are five signalling molecules for the A-motility system. Very little is known about the transduction pathways for the A-motility signals.

Gliding using the S motility system requires pili that are present at the cell poles. Pili are short protein fibrils that protrude from the surfaces of most gram-negative bacteria. In general, they mediate attachment to other bacteria, animal, plant or fungal cells, and are very important for colonization of surfaces. The *M. xanthus* polar pili that are required for S motility belong to the class called type IV pili (Wu and Kaiser, 1995). It is believed that a signal required for S motility is transmitted from cell to cell via the polar pili. Mutants with lesions in the gene *tglA*, which encodes a protein homologous to eukaryotic targeting proteins, are defective in the assembly of pilin (the major protein component of pili) into pili and consequently cannot glide via the S motility system. However, these mutants can be induced to transiently produce pili and glide if they make end-to-end contact with wild-type cells. It has been suggested that the TglA protein is localized at the poles of the cells where the type IV pili are located, and that the protein stimulates pilus assembly at the poles of recipient cells that are properly aligned with the poles of the stimulatory cells (Wall and Kaiser, 1998). It has also been suggested that when properly aligned with respect to one another, the wild-type cells can transfer TglA to the mutant cells (Shimkets, 1999). It is of interest that type IV pili have been shown to be involved in several different processes in a wide variety of other gram-negative bacteria, including the adherence of pathogenic bacteria to host cells and twitching motility (McNamara and Donnenberg, 2000).

Regulation of Motility Patterns. The signalling pathways used by myxobacteria in regulating the direction of motility to construct aggregates and fruiting bodies is under intense investigation. Two intercellular signalling pathways, the Frz (frizzy) and Dif (defective in fruiting) pathways, have been discovered. Both of these have protein components that are homologous to the chemotaxis genes (*che* genes) in *E. coli* and appear to be involved in sensing intercellular signals (however, the Dif and Frz systems are genetically distinct from one another). Mutations in the Frz system result in several different phenotypes. The mutant cells form abnormal aggregates that are characterized by "tangled, swirling patterns" as opposed to the mounds of cells normally formed, and they form abnormal fruiting bodies (Ward and Zusman, 2000). Probably related to this phenotype is the fact that most *frz* mutants reverse their direction of motility much less frequently than wild-type (some *frz* mutants show hyper-reversability). The Frz system is also required for adaptation to a lipid that can serve as a chemoaltractant, phosphatidylethanolamine (PE), and is stimulated by C-factor (Søgaard-Andersen and Kaiser, 1996; Kearns and Shimkets, 1998). Mutants in the Dif system are defective in S motility. They form small mounds of cells that do not mature into fruiting bodies (Yang *et al.*, 1998).

The Role of Fibrils in Signalling Via Cell Contact. Protruding from the cell surface of *M. xanthus* are thin appendages called fibrils (distinct from polar pili) that are composed of protein attached to a polysaccharide backbone. The absence of fibrils results in loss of cohesion, a defect in S motility, and inability to form fruiting bodies. The addition of isolated fibrils to fribrilless mutants restores the wild-type phenotype (Chang and Dworkin, 1994). The fibrils are formed under conditions of high-cell density, perhaps reflecting the fact that their production is stimulated by cell-cell contact. The cells use the fibrils to attach to one another, and it has been suggested that they are involved in contact-mediated cell-cell signalling (Chang and Dworkin, 1996; Smith and Dworkin, 1997; Dworkin, 1999).

3. Conclusions

In this Chapter we have introduced the diversity of intercellular signalling mechanisms employed by bacteria. There are many different signal molecules that play important roles in microbial behaviour. The primary considerations for diffusible signals are how the signal is released from the source cell(s), what dictates accumulation of the signal, and perception of the signal by the responding cells. The lipoidal pheromones often can cross the cellular envelope by passive diffusion, as they exit and enter cells. The oligopeptide pheromones must be actively secreted from cells or generated from external substrates. Perception of the oligopeptide depends on transmembrane sensors or transport of the oligopeptide into the cytoplasm of responding cells. Cell-associated signalling is less well understood, but clearly important aspects of signal exchange are the anchoring and presentation of the signal by the donating cell, interactions between cell surfaces of interacting cells, including specific receptors, and the relative alignment between the signal-donating partner and the responding cell.

A major goal of the many laboratories that study bacterial intercellular signalling is to gain a sufficient understanding of the underlying mechanisms in order that we may begin to combat infectious disease in plants and animals by subversion or corruption of signalling molecules. Likewise, enhancement of beneficial bacterial interactions could be stimulated by deliberate application of bacterial signalling molecules or mimetics. The general approach of modifying microbial activity via misinformation has great promise. In contrast to the harsh selection imposed by antimicrobial agents such as antibiotics, where resistant cells are given the ultimate competitive advantage over sensitive cells, it seems reasonable to hope that manipulation of bacterial signalling will provide control over microbial behaviour with less stringent selective pressure. In fact, there are several recent reports of the manipulation of host-associated microbial populations by altering the normal intercellular signalling pathways, suggesting that this area has significant potential (Ji, Beavis and Novick, 1997; Fray *et al.*, 1999).

Acknowledgements

Eric White prepared the drawings in this Chapter (unless otherwise indicated) for which we are extremely grateful. We also thank Heidi Kaplan, Dave Stickler, Helmut Hirt, Stanley Erlandsen, and Gary Dunny for providing Figures, as well as Heidi Kaplan and Bronwyn Ramey for critical reading of the manuscript. Research in the author's laboratories is supported by the National Science Foundation Grant No. MCB-9974863 (C.F.) and the Indiana University Department of Biology (D.W. and C.F.).

References

Allison, C., Lai, H.-C., Gygi, D., and Hughes, C. (1993) Cell differentiation of *Proteus mirabilis* is initiated by glutamine, a specific chemoattractant for swarming cells. *Molecular Microbiology* **8**, 53–60.

Ando, N., Malsumori, N., Sakuda, S., Beppu, T., and Horinouchi, S. (1997) Involvement of AfsA in A-factor biosynthesis as a key enzyme. *Journal of Antibiotics* **50**, 847–852.

Baca-Delancey, R., South, M.M.T., Ding, X., and Rather, P.N. (1999) *Escherichia coli* genes regulated by cell-to-cell signalling. *Proceedings of the National Academy of Sciences, USA* **96**, 4610–4614.

Bainton, N.J., Stead, P., Chhabra, S.R., Bycroft, B.W., Salmond, G.P.C., Stewart, G.S.A.B., and Williams, P. (1992) *N*-(3-oxohexanoyl)-L-homoserine lactone regulates carbapenem antibiotic production in *Erwinia carotovora. Biochemical Journal* **288**, 997–1004.

Baker, M. (1994) *Myxococcus xanthus* C factor, a morphogenic paracrine signal, is similar to *Escherichia coli* 3-oxoacyl-[acyl-carrier-protein] reductase and human 17β-hydroxysteroid dehydrogenase. *Biochemical Journal* **301**, 311–312.

Barber, C.E., Tang, J.L. Feng, J.X., Pan, M.Q., Wilson, T.J.G., Slater, H., Dow, J.M., Williams, P. and Daniels, M.J. (1997) A novel regulatory system required for pathogenicity of *Xanthomonas campestris* is mediated by a small diffusible signal molecule. *Molecular Microbiology* **24**, 555–566.

Bassler, B.L., and Silverman, M.R. (1995) Intercellular communication in marine *Vibrio* species: density-dependent regulation of tbe expression of bioluminescence. In: J.A. Hoch and T.J. Silhavy (eds.) *Two-Component Signal Transduction.* ASM Press, Washington D.C., pp. 431–445.

Bassler, B.L., Wright, M. and Silverman, M.R. (1994) Multiple signalling systems controlling expression of luminescence in *Vibrio harveyi*: sequence and function of genes encoding a second sensory pathway. *Molecular Microbiology* **13**, 273–286.

Bassler, B.L., Greenberg, E.P., and Stevens, A.M. (1997) Cross-species induction of luminescence in the quorum-sensing bacterium *Vibrio harveyi. Journal of Bacteriology* **179**, 4043–4045.

Bassler, B.L., Wright, M., Showalter, R.E., and Silverman, M.R. (1993). Intercellular signalling in *Vibrio harveyi*: sequence and function of genes regulating expression of luminescence. *Molecular Microbiology* **9**, 773–786.

Beck von Bodman, S., Majerczak, D.R., and Coplin, D.L. (1998) A negative regulator mediates quorum-sensing control of expolysaccharide production in *Pantoea stewartii* subsp. *stewartii. Proceedings of the National Academy of Sciences, USA* **95**, 7687–7692.

Belas, R. (1997) *Proteus mirabilis* and other swarming bacteria. In: J.A. Shapiro and M. Dworkin (eds.) *Bacteria as Multicellular Organisms.* Oxford Press, New York, pp. 183–219.

Bensing, B.A., Manias, D.A., and Dunny, G.M. (1997) Pheromone cCF10 and plasmid pCF10-encoded regulatory molecules act post-transcriptionally to activate expression of downstream conjugation functions. *Molecular Microbiology* **24**, 285–294.

Boettcher, K.J., and Ruby, E.G. (1995) Detection and quantification of *Vibrio fischeri* autoinducer from the symbiotic squid light organs. *Journal of Bacteriology* **177**, 1053–1058.

Brun, Y.V., and Shimkets, L.J. (2000). *Prokaryotic Development.* ASM Press, Washington D.C.

Caldwell, G.A., Naider, F., and Becker, J.M. (1995) Fungal lipopeptide mating pheromones: A model system for the study of prenylation. *Microbiological Reviews* **59**, 406–422.

Cao, J.-G., and Meighen, E.A. (1989) Purification and structural identification of an autoinducer for the luminescence system of *V. harveyi. Journal of Biological Chemistry* **264**, 21670–21676.

Chang, B., and Dworkin, M. (1994) Isolated fibrils rescue cohesion and development in the Dsp mutant of *Myxococcus xanthus. Journal of Bacteriology* **176**, 7190–7196.

Chang, B.-Y., and Dworkin, M. (1996) Mutants of *Myxococcus xanthus dsp*, defective in fibril binding. *Journal of Bacteriology* **178**, 697–700.

Choi, S.H., and Greenberg, E.P. (1992a) Genetic dissection of DNA binding and luminescence gene activation by the *Vibrio fischeri* LuxR protein. *Journal of Bacteriology* **174**, 4064–4069.

Choi, S.H., and Greenberg, E.P. (1992b) Genetic evidence for multimerization of LuxR, the transcriptional activator of *Vibrio fischeri* luminescence. *Molecular Marine Biology and Biotechnology* **1**, 408–413.

Clewell, D. (1999) Sex pheromone systems in enterococci. In: G.M. Dunny and S.C. Winans (eds.) *Cell-Cell Signalling in Bacteria.* ASM Press, Washington D.C., pp. 47–65.

Clough, S.J., Schell, M.A., and Denny, T.P. (1994) Evidence for involvement of a volatile extracellular factor in *Pseudomonas solanacearum* virulence gene expression. *Molecular Plant-Microbe Interactions* **3**, 286–292.

Clough, S.J., Lee, K.-E., Schell, M.A., and Denny, T.P. (1997) A two-component system in *Ralstonia (Pseudomonas) solanacearum* modulates production of PhcA-regulated virulence factors in response to 3-hydroxypalmitic acid methyl ester. *Journal of Bacteriology* **179**, 3639–3648.

Costerton, J.W., Lewandowski, Z., Caldwell, D.E., Korber, D.R., and Lappin-Scott, H.M. (1995) Microbial biofilms. *Annual Review of Microbiology* **49**, 711–745.

Da Re, S., Bertagnoli, S., Fourment, J., Reyrat, J.-M., and Kahn, D. (1994) Intramolecular signal transduction within the FixJ transcriptional activator: in vitro evidence for the inhibitory effect of the phosphorylatable regulatory domain. *Nucleic Acids Research* **9**, 1555–1561.

Davies, D.G., Parsek, M.R., Pearson, J.P., Iglewski, B.H., Costerton, J.W., and Greenberg, E.P. (1998) The involvement of cell-to-cell signals in the development of a bacterial biofilm. *Science* **280**, 295–298.

de Nys, R., Steinberg, P.D., Willemsen, P., Dworjanyn, S.A., Gabelish, C.L., and King, R.J. (1995) Broad spectrum effects of secondary metabolites from the red alga *Delisea pulchra* in antifouling assays. *Biofouling* **8**, 259–271.

Devine, J.H., Shadel, G.S., and Baldwin, T.O. (1989) Identification of the operator of the *lux* regulon from the *Vibrio fischeri* strain ATCC7744. *Proceedings of the National Academy of Sciences, USA* **86**, 5688–5692.

Dimango, E., Zar, H.J., Bryan, R., and Prince, A. (1995) Diverse *Pseudomonas aeruginosa* gene products stimulate respiratory epithelial cells to produce interleukin-8. *Journal of Clinical Investigation* **5**, 2204–2210.

Dunny, G.M., and Leonard, B.A.B. (1997) Cell-cell communication in gram-positive bacteria. *Annual Review of Microbiology* **51**, 527–564.

Dunny, G.M., and Winans, S.C. (1999) *Cell-Cell Signalling in Bacteria.* ASM Press, Washington, D.C.

Dunny, G.M., Brown, B.L., and Clewell, D.B. (1978) Induced cell aggregation and mating in *Streptococcus faecalis*: evidence for a bacterial sex pheromone. *Proceedings of the National Academy of Sciences, USA* **75**, 3479–3483.

Dunny, G.M., Graig, R.A., Carron, R.L., and Clewell, D.B. (1979) Plasmid transfer in *Streptococcus faecalis*: production of multiple sex pheromones by recipients. *Plasmid* **2**, 454–465.

Dworkin, M. (1999) Fibrils as extracellular appendages of bacteria: their role in contact-mediated cell-cell interactions in *Myxococcus xanthus. BioEssays* **21**, 590–595.

Dworkin, M., and Kaiser, D. (1993) *Myxobacteria II*. ASM Press, Washington, D.C.

Eberhard, A., Burlingame, A.L., Eberhard, C., Kenyon, G.L., Nealson, K.H., and Oppenheimer, N.J. (1981) Structural identification of autoinducer of *Photobacterium fischeri. Biochemistry* **20**, 2444–2449.

Eberl, L., Winson, M.K., Steinberg, C., Stewart, G.S.A.B., Christiansen, G., Chhabra, S.R., Bycroft, B.W., Williams, P., Molin, S., and Givskov, M. (1996) Involvement of *N*-acyl-L-homoserine lactone autoinducers in control of multicellular behavior of *Serratia liquefaciens. Molecular Microbiology* **20**, 127–136.

Evans, K., Passador, L., Srikumar, R., Tsang, E., Nezezon, J., and Poole, K. (1998) Influence of the MexAB-OprM multidrug efflux system on quorum sensing in *Pseudomonas aeruginosa. Journal of Bacteriology* **180**, 5443–5447.

Flavier, A.B., Schell, M.A., and Denny, T.P. (1997) Identification of 3-hydroxypalmitic acid methyl ester as a novel autoregulator controlling virulence in *Ralstonia solanacearum. Molecular Microbiology* **26**, 251–259.

Flavier, A.B., Ganova-Raeva, L.M., Schell, M.A., and Denny, T.P. (1997) Hierarchical autoinduction in *Ralstonia solanacearum*: control of acyl-homoserine lactone production by a novel autoregulatory system responsive to 3-hydroxypalmitic acid methyl ester. *Journal of Bacteriology* **179**, 7089–7097.

Fray, R.P., Throup, J.P., Daykin, M., Wallace, A., Williams, P., Stewart, G.S.A.B., and Grierson, D. (1999) Plants genetically modified to produce *N*-acylhomoserine lactones communicate with bacteria. *Nature Biotechnology* **17**, 1017–1020.

Freeman, J.A., and Bassler, B.L. (1999) Sequence and function of LuxU: a two-component phosphorelay protein that regulates quorum sensing in *Vibrio harveyi. Journal of Bacteriology* **181**, 899–906.

Freeman, J.A., Lilley, B.N., and Bassler, B.L. (2000) A genetic analysis of the functions of LuxN: a two-component hybrid sensor kinase that regulates quorum sensing in *Vibrio harveyi*. *Molecular Microbiology* **35**, 139–149.

Fuqua, W.C., and Winans, S.C. (1994) A LuxR-LuxI type regulatory system activates *Agrobacterium* Ti plasmid conjugal transfer in the presence of a plant tumor metabolite. *Journal of Bacteriology* **176**, 2796–2806.

Fuqua, C., and Winans, S.C. (1996) Conserved *cis*-acting promoter elements are required for density-dependent transcription of *Agrobacterium tumefaciens* conjugal transfer genes. *Journal of Bacteriology* **178**, 435–440.

Fuqua, C., and Eberhard, A. (1999) Signal generation in autoinduction systems: synthesis of acylated homoserine lactones by LuxI-type proteins. In: S.C. Winans and G.M. Dunny (eds.) *Cell-Cell Signalling in Bacteria*. ASM Press, Washington, D.C., pp. 211–230.

Fuqua, C., Winans, S.C., and Greenberg, E.P. (1996) Census and consensus in bacterial ecosystems: The LuxR-LuxI family of quorum-sensing transcriptional regulators. *Annual Review of Microbiology* **50**, 591–624.

Fuqua, W.C., Winans, S.C., and Greenberg, E.P. (1994) Quorum sensing in bacteria: The LuxR/LuxI family of cell density-responsive transcriptional regulators. *Journal of Bacteriology* **176**, 269–275.

Galan, J.E., and Collmer, A. (1999) Type III secretion machines: bacterial devices for protein delivery into host cells. *Science* **284**, 1322–1328.

Gilson, L., Kuo, A., and Dunlap, P.V. (1995) AinS and a new family of autoinducer synthesis proteins. *Journal of Bacteriology* **177**, 6946–6951.

Givskov, M., de Nys, R., Manefield, M., Gram, L., Maximilien, R., Eberl, L., Molin, S., Steinberg, P.D., and Kjelleberg, S. (1996) Eukaryotic interference with homoserine lactone-mediated prokaryotic signalling. *Journal of Bacteriology* **178**, 6618–6622.

Gooday, G.W. (1974) Fungal sex pheromones. *Annual Review of Biochemistry* **43**, 35–39.

Greenberg, E.P. (1997) Quorum sensing in gram-negative bacteria. *ASM News* **63**, 371–377.

Hanzelka, B.L., and Greenberg, E.P. (1995) Evidence that the N-terminal region of the *Vibrio fischeri* LuxR protein constitutes an autoinducer-binding domain. *Journal of Bacteriology* **177**, 815–817.

Hanzelka, B.L., and Greenberg, E.P. (1996) Quorum sensing in *Vibrio fischeri*: evidence that *S*-adenosylmethionine is the amino acid substrate for autoinducer synthesis. *Journal of Bacteriology* **178**, 5291–5294.

Hanzelka, B.L., Stevens, A.M., Parsek, M.R., Crone, T.J., and Greenberg, E.P. (1997) Mutational analysis of the *Vibrio fischeri* LuxI polypeptide: critical regions of an autoinducer synthase. *Journal of Bacteriology* **179**, 4882–4887.

Hanzelka, B.L., Parsek, M.R., Val, D.L., Dunlap, P.V., Cronan, J.E.J., and Greenberg, E.P. (1999) Acylhomoserine lactone synthase activity of the *Vibrio fischeri* AinS protein. *Journal of Bacteriology* **181**, 5766–5770.

Hara, O., and Beppu, T. (1982) Induction of streptomycin inactivating enzyme by A-factor in *Streptomyces griseus*. *Journal of Antibiotics* **35**, 1208–1215.

Håvarstein, L.S., and Morrison, D.A. (1999) Quorum sensing and peptide pheromones in streptococcal competence for genetic transformation. In: G.M. Dunny and S.C. Winans (eds.) *Cell-Cell Signalling in Bacteria*. ASM Press, Washington D.C., pp. 9–36.

Håvarstein, L.S., Coomaraswamy, G., and Morrison, D.A. (1995) An unmodified heptadecapeptide pheromone induces competence for genetic transformation in *Streptococcus pneumoniae*. *Proceedings of the National Academy of Sciences, USA* **92**, 11140–11144.

Håvarstein, L.S., Diep, D.B., and Nes, I.F. (1995) A family of bacteriocin ABC transporters carry out proteolytic processing of their substrates concomitant with export. *Molecular Microbiology* **16**, 229–240.

Håvarstein, L.S., Hakenbeck, R., Nes, I.F., and Morrison, D.A. (1996) Identification of the streptococcal competence pheromone receptor. *Molecular Microbiology* **21**, 863–869.

Hoch, J.A., and Silhavy, T.J. (1995) *Two-Component Signal Transduction*. ASM Press, Washington, D.C.

Hodgkin, J., and Kaiser, D. (1977) Cell-to-cell stimulation of movement in nonmotile mutants of *Myxococcus*. *Proceedings of the National Academy of Sciences, USA* **74**, 2938–2942.

Holden, M.T.G., Chhabra, S.R., de Nys, R., Stead, P., Bainton, N.J., Hill, P.J., Manfield, M., Kumar, N., Labatte, M., England, D., Rice, S., Givskov, M., Salmond, G.P.C., Stewart, G.S.A.B., Bycroft, B.W., Kjelleberg, S., and Williams, P. (1999) Quorum-sensing crosstalk: isolation and chemical characterization of cyclic dipeptides from *Pseudomonas aeruginosa* and other Gram-negative bacteria. *Molecular Microbiology* **33**, 1254–1266.

Horinouchi, S. (1999) Gamma-butyrolactones that control secondary metabolism and cell differentiation in *Streptomyces*. In: S.C. Winans and G.M. Dunny (eds.) *Cell-Cell Signalling in Bacteria*. ASM Press, Washington, D.C., pp. 193–207.

Hull, W.E., Berkessel, A., and Plaga, W. (1998) Structure elucidation and chemical synthesis of stigmalone, a novel type of prokaryotic pheromone. *Proceedings of the National Academy of Sciences, USA* **95**, 11268–11273.

Ike, Y., Craig, R.C., White, B.A., Yagi, Y., and Clewell, D.B. (1983) Modification of *Streptococcus faecalis* sex pheromones after acquisition of plasmid DNA. *Proceedings of the National Academy of Sciences, USA* **80**, 5369–5373.

Jelsbak, L., and Søgaard-Andersen, L. (1999) The cell surface-associated intercellular C-signal induces behavioral changes in individual *Myxococcus xanthus* cells during fruiting body morphogenesis. *Proceedings of the National Academy of Sciences, USA* **96**, 5031–5036.

Ji, G., Beavis, R.C., and Novick, R.P. (1995) Cell density control of staphylococcal virulence mediated by an octapeptide pheromone. *Proceedings of the National Academy of Sciences, USA* **92**, 12055–12059.

Ji, G., Beavis, R., and Novick, R.P. (1997) Bacterial interference caused by autoinducing peptide variants. *Science* **276**, 2027–2030.

Kaiser, D. (1999) Cell fate and organogenesis in bacteria. *Trends in Genetics* **15**, 273–277.

Kaplan, H.B., and Greenberg, E.P. (1985) Diffusion of autoinducer is involved in regulation of the *Vibrio fischeri* luminescence system. *Journal of Bacteriology* **163**, 1210–1214.

Kearns, D.B., and Shimkets, L.J. (1998) Chemotaxis in a gliding bacterium. *Proceedings of the National Academy of Sciences, USA* **95**, 11957–11962.

Kholkov, A.S., Tovarova, I.I., Borisova, L.N., Pliner, S.A., Schevchenko, L.A., Kornitskaya, E.Y., Ivkina, N.S., and Rapoport, I.A. (1967) A-factor responsible for the biosynthesis of streptomycin by a mutant strain of *Actinomyces streptomycini*. *Doklady Akademii Nauk SSSR* **1771**, 232–235.

Kleerebezem, M., de Vos, W.M., and Kuipers, O.P. (1999) The lantibiotics nisin and subtilin act as extracellular regulators of their own biosynthesis. In: G.M. Dunny and S.C. Winans (eds.) *Cell-Cell Signalling in Bacteria*. ASM Press, Washington D.C., pp. 159–174.

Kroos, L., and Kaiser, D. (1987) Expression of many developmentally regulated genes depends on a sequence of cell interactions. *Genes and Development* **1**, 840–854.

Kubota, T., Tokonoyama, T., Tsukuda, Y., Koyama, H., and Miyake, A. (1973) Isolation and structure determination of blepharismin, a conjugation initiation gamone in the ciliate *Blepharisma*. *Science* **170**, 400–402.

Kuipers, O.P., Beerthuyzen, M.M., de Ruyter, P.G., Luesink, E.J., and de Vos, W.M. (1995) Autoregulation of nisin biosynthesis in *Lactococcus lactis* by signal transduction. *Journal of Biological Chemistry* **270**, 27299–27304.

Kuo, A., Blough, N.V., and Dunlap, P.V. (1994) Multiple *N*-acyl-L-homoserine lactone autoinducers of luminescence genes in the marine symbiotic bacterium *Vibrio fischeri*. *Journal of Bacteriology* **176**, 7558–7565.

Kuo, A., Callaghan, S.M., and Dunlap, P.V. (1996) Modulation of luminescence operon expression by *N*-octanoyl-L-homoserine lactone in *ainS* mutants of *Vibrio fischeri*. *Journal of Bacteriology* **178**, 971–976.

Kuspa, A., Plamann, L., and Kaiser, D. (1992) Identification of heat-stable A-factor from *Myxococcus xanthus*. *Journal of Bacteriology* **174**, 3319–3326.

LaRossa, R., Kuner, J., Hagen, D., Manoil, C., and Kaiser, D. (1983) Developmental cell interactions of *Myxococcus xanthus*: analysis of mutants. *Journal of Bacteriology* **153**, 1394–1404.

Lazazzera, B.A., and Grossman, A.D. (1997) A regulatory switch involving a Clp ATPase. *BioEssays* **19**, 455–458.

Lazazzera, B.A., Solomon, J.M., and Grossman, A.D. (1997) An exported peptide functions intracellularly to contribute to cell density signalling in *Bacillus subtilis*. *Cell* **89**, 917–925.

Lee, B.-U., Lee, K., Mendez, J., and Shimkets, L.J. (1995) A tactile sensory system of *Myxococcus xanthus* involves and extracellular NAD(P)+-containing protein. *Genes and Development* **9**, 2964–2973.

Luo, Z.-Q., and Farrand, S.K. (1999) Signal-dependent DNA binding and functional domains of the quorum-sensing activator TraR as identified by repressor activity. *Proceedings of the National Academy of Sciences,* **96**, 9009–9014.

Magnuson, R., Solomon, J., and Grossman, A.D. (1994) Biochemical and genetic characterization of a competence pheremone from *B. subtilis*. *Cell* **77**, 207–216.

Manefield, M., de Nys, R.,. Kumar, N., Read, R., Givskov, M., Steinberg, P., and Kjelleberg, S. (1999) Evidence that halogenated furanones from *Delisea pulchra* inhibit acylated homoserine lactone (AHL)-mediated gene expression by displacing the AHL signal from its receptor protein. *Microbiology* **145**, 283–291.

Mattick, A.T.R., and Hirsch, A. (1944) A powerful inhibitory substance produced by group N streptococci. *Nature* **154**, 551.

McLean, R.J.C., Whiteley, M., Stickler, D.J., and Fuqua, W.C. (1997) Evidence of autoinducer activity in naturally occurring biofilms. FEMS Microbiology Letters **154**, 259–263.

McNamara, B.P., and Donnenberg, M.S. (2000) Evidence for specificity in type IV pilus biogenesis by enteropathogenic *Escherichia coil*. *Microbiology* **146**, 719–729.

Miyake, A., and Beyer, J. (1973) Cell interaction by means of soluble factors (gamones) in conjugation of *Blepharisma intermedium*. *Experimental Cell Research* **76**, 15–24.

Miyake, K., Horinouchi, S., Yoshida, M., Chiba, N., Mori, K., Nogawa, N., Morikawa, N., and Beppu, T. (1989) Detection and properties of A-factor binding protein from *Streptomyces griseus*. *Journal of Bacteriology* **171**, 4298–4302.

Miyake, K., Kuzuyama, T., Horinouchi, S., and Beppu, T. (1990) The A-factor-binding protein of *Streptomyces griseus* negatively controls streptomycin production and sporulation. *Journal of Bacteriology* **172**, 3003–3008.

Moré, M.I., Finger, L.D., Stryker, J.L., Fuqua, C., Eberhard, A., and Winans, S.C. (1996) Enzymatic synthesis of a quorum-sensing autoinducer through use of defined substrates. *Science* **272**, 1655–1658.

Nasser, W., Bouillant, M.L., Salmond, G., and Reverchon, S. (1998) Characterization of the *Erwinia chrysanthemi expI-expR* locus directing the synthesis of two *N*-acyl-homoserine lactone signal molecules. *Molecular Microbiology* **29**, 1391–1405.

Nealson, K.H. (1977) Autoinduction of bacterial luciferase: occurrence, mechanism, and significance. *Archives of Microbiology* **112**, 73–79.

Nealson, K.H., Platt, T., and Hastings, J.W. (1970) Cellular control of the synthesis and activity of the bacterial luminescent system. *Journal of Bacteriology* **104**, 313–322.

Nes, I.F., and Eijsink, V.G.H. (1999) Regulation of group II peptide bacteriocin synthesis by quorum-sensing mechanisms. In: G.M. Dunny and S.C. Winans (eds.) *Cell-Cell Signalling in Bacteria*. ASM Press, Washington D.C., pp. 175–192.

Newell, P.C. (1978) Progress on slime mold morphology. *Nature* **271**, 302–303.

Novick, R.P. (1999) Regulation of pathogenicity in *Staphylococcus aureus* by a peptide-based density-sensing system. In: G.M. Dunny and S.C. Winans (eds.) *Cell-Cell Signalling in Bacteria*. ASM Press, Washington D.C., pp. 129–146.

Okamoto, S., Nakamura, K., and Yamada, Y. (1995) Virginiae butanolide binding protein for *Streptomyces virginiae*: evidence that Vbr is not the virginiae butanolide binding protein and reidentification of the true binding protein. *Journal of Biological Chemistry* **270**, 12319–12326.

Palmer, R.J., and White, D.C. (1997) Developmental biology of biofilms: implications for treatment and control. *Trends in Microbiology* **5**, 435–440.

Parsek, M.R., Schaefer, A.L., and Greenberg, E.P. (1997) Analysis of random and site-directed mutations in *rhlI*, a *Pseudomonas aeruginosa* gene encoding an acylhomoserine lactone synthase. *Molecular Microbiology* **26**, 301–310.

Parsek, M.R., Val, D.E., Hanzelka, B.L., Cronan, J.E.J., and Greenberg, E.P. (1999) Acyl homoserine lactone quorum-sensing signal generation. *Proceedings of the National Academy of Sciences, USA* **96**, 4360–4365.

Pearson, J.P., Van Delden, C., and Iglewski, B.H. (1999) Active efflux and diffusion are involved in transport of *Pseudomonas aeruginosa* cell-to-cell signals. *Journal of Bacteriology* **181**, 1203–1210.

Perego, M. (1997) A peptide export-import control circuit modulating bacterial development regulates protein phosphatases of the phosphorelay. *Proceedings of the National Academy of Sciences, USA* **94**, 8612–8617.

Pesci, E.C., Milbank, J.B.J., Pearson, J.P., McKnight, S., Kende, A.S., Greenberg, E.P., and Iglewski, B.H. (1999) Quinolone signalling in the cell-to-cell communication system of *Pseudomonas aeruginosa. Proceedings of the National Academy of Sciences, USA* **96**, 11229–11234.

Plaga, W., Stamm, I., and Schairer, H.U. (1998) Intercellular signalling in *Stigmatella aurantica*: purification and characterization of stigmalone, a myxobacterial pheromone. *Proceedings of the National Academy of Sciences, USA* **95**, 11263–11267.

Plamann, L., and Kaplan, H.B. (1999) Cell density sensing during early development in *Myxococcus xanthus.* In: G.M. Dunny and S.C. Winans (eds.) *Cell-Cell Signalling in Bacteria.* ASM Press, Washington D.C., pp. 67–82.

Plamann, L., Kuspa, A., and Kaiser, D. (1992) Proteins that rescue A-signal-defective mutants of *Myxococcus xanthus. Journal of Bacteriology* **174**, 3311–3318.

Poplawsky, A.R., and Chun, W. (1997) *pigB* determines a diffusible factor needed for extracellular polysaccharide slime and xanthomonadin production in *Xanthomonas campestris* pv. campestris. *Journal of Bacteriology* **179**, 439–444.

Poplawsky, A.R., and Chun, W. (1998) Regulation of epiphytic fitness factor production in *Xanthomonas campestris* pv. campestris. *Molecular Plant-Microbe Interactions* **11**, 466–475.

Rather, P.N., Ding, X., Baca-DeLancey, R., and Siddiqui, S. (1999) *Providencia stuartii* genes activated by cell-to-cell signalling and identification of a gene required for production or activity of an extracellular factor. *Journal of Bacteriology* **181**, 7185–7191.

Rust, L., Pesci, E.C., and Iglewski, B.H. (1996) Analysis of the *Pseudomonas aeruginosa* elastase (*lasB*) regulatory region. *Journal of Bacteriology* **178**, 1134–1140.

Schaefer, A.L., Val, D.L., Hanzelka, B.L., Cronan, J.E., and Greenberg, E.P. (1996) Generation of cell-to-cell signals in quorum sensing: acyl homoserine activity of purified *Vibrio fischeri* LuxI protein. *Proceedings of the National Academy of Sciences, USA* **93**, 9505–9509.

Shimkets, L. (1999) Intercellular signalling during fruiting-body development of *Myxococcus xanthus. Annual Review of Microbiology* **53**, 525–549.

Shimkets, L.J., and Kaiser, D. (1999) Cell contact-dependent C signalling in *Myxococcus xanthus.* In G.M. Dunny and S.C. Winans (eds.) *Cell-Cell Signalling in Bacteria.* ASM Press, Washington D.C., pp. 83–97.

Shimkets, L.J., Gill, R.E., and Kaiser, D. (1983) Developmental cell interactions in *Myxococcus xanthus* and the *spoC* locus. *Proceedings of the National Academy of Sciences, USA* **80**, 1406–1410.

Shorey, H.H., and McKelvey, J.J.J. (1977) *Chemical Control of Insect Behavior.* John Wiley and Sons, New York.

Smith, D.R., and Dworkin, M. (1997) A mutation that affects fibril protein, development, cohesion and gene expression in *Myxococcus xanthus. Microbiology* **143**, 3683–3692.

Søgaard-Andersen, L., and Kaiser, D. (1996) C-factor, a cell-surface-associated intercellular signalling protein, stimulates the cytoplasmic Frz signal transduction system in *Myxococcus xanthus. Proceedings of the National Academy of Sciences, USA* **93**, 2625–2679.

Solomon, J.M., Lazazzera, B.A., and Grossman, A.D. (1996) Purification and characterization of an extracellular peptide factor that affects two different developmental pathways in *B. subtilis. Genes and Development* **10**, 2014–2024.

Sperandio, V., Mellies, J.L., Nguyen, W., Shin, S., and Kaper, J.B. (1999) Quorum sensing controls expression of the type III secretion gene transcription and protein secretion in enterohemorrhagic and enteropathogenic *Escherichia coli. Proceedings of the National Academy of Sciences, USA* **96**, 15196–15201.

Srinivasan, S., Ostling, J., Charlton, T., de Nys, R., Takayama, K., and Kjelleberg, S. (1998) Extracellular signal molecule(s) involved in the carbon starvation response of marine *Vibrio* sp. strain S14. *Journal of Bacteriology* **180**, 210–219.

Stephens, K., Hegeman, G.D., and White, D. (1982) Pheromone produced by the myxobacterium *Stigmatella aurantiaca. Journal of Bacteriology* **149**, 739–747.

Stevens, A.M., and Greenberg, E.P. (1999) Transcriptional activation by LuxR, In S.C. Winans and G.M. Dunny (eds.) *Cell-Cell Signalling in Bacteria.* ASM Press, Washington, D.C., pp. 231–242.

Stevens, A.M., Dolan, K.M., and Greenberg, E.P. (1994) Synergistic binding of the *Vibrio fischeri* LuxR transcriptional activator domain and RNA polymerase to the *lux* promoter region. *Proceedings of the National Academy of Sciences, USA* **91**, 12619–12623.

Stickler, D.J., Morris, N.S., McLean, R.J.C., and Fuqua, C. (1998) Biofilms on indwelling urethral catheters produce quorum-sensing molecules *in situ* and *in vitro. Applied and Environmental Microbiology* **64**, 3486–3490.

Surette, M.G., and Bassler, B.L. (1998) Quorum sensing in *Escherichia coli* and *Salmonella typhimurium. Proceedings of the National Academy of Sciences, USA* **95**, 7046–7050.

Surette, M.G., and Bassler, B.L. (1999) Regulation of autoinducer production by *Salmonella typhimurium. Molecular Microbiology* **31**, 585–595.

Surette, M.G., Miller, M.B., and Bassler, B.L. (1999) Quorum sensing in *Escherichia coli, Salmonella typhimurium,* and *Vibrio harveyi*: a new family of genes responsible for autoinducer production. *Proceedings of the National Academy of Sciences, USA* **96**, 1639–1644.

Telford, G., Wheeler, D., Williams, P., Tomkins, P.T., Appleby, P., Sewell, H., Stewart, G.S.A.B., Bycroft, B.W., and Pritchard, D.I. (1998) The *Pseudomonas aeruginosa* quorum-sensing signal molecule *N*-(3-oxododecanoyl)-L-homoserine lactone has immunomodulatory activity. *Infection and Immunity* **66**, 36–42.

Tomasz, A. (1965) Control of the competent state in *Pneumococcus* by a hormone-like cell product: an example of a new type of regulatory mechanism in bacteria. *Nature* **208**, 155–159.

Wall, D., and Kaiser, D. (1998) Alignment enhances the cell-to-cell transfer of pilus phenotype. *Proceedings of the National Academy of Sciences, USA* **95**, 3054–3058.

Ward, M.J., and Zusman, D. (2000) Developmental aggregation and fruiting body formation in the gliding bacterium *Myxococcus xanthus.* In: Y.V. Brun and L.J. Shimkets (eds.) *Prokaryotic Development.* ASM Press, Washington D.C., pp. 243–262.

Ween, O., Gaustad, P., and Håvarstein, L.S. (1999) Identification of DNA binding sites for ComE, a key regulator of natural competence in *Streptococcus pneumoniae. Molecular Microbiology* **33**, 817–827.

White, D. (2000) *The Physiology and Biochemistry of Prokaryotes*, 2nd ed. Oxford Press, New York.

Whiteley, M., Lee, K.M. and Greenberg, E.P. (1999) Identification of genes controlled by quorum sensing in *Pseudomonas aeruginosa. Proceedings of the National Academy of Sciences, USA* **96**, 13904–13909.

Wood, D.W., Gong, F., Daykin, M.M., Williams, P., and Pierson III, L.S. (1997) *N*-acyl-homoserine lactone-mediated regulation of phenazine gene expression by *Pseudomonas aureofaciens* 30-84 in the wheat rhizosphere. *Journal of Bacteriology* **179**, 7663–7670.

Wu, S.S., and Kaiser, D. (1995) Genetic and functional evidence that type IV pili are required for social gliding motility in *Myxococcus xanthus. Molecular Microbiology* **18**, 547–558.

Xu, D., Yang, C., and Kaplan, H.B. (1998) *Myxococcus xanthus sasN* encodes a regulator that prevents developmental gene expression during growth. *Journal of Bacteriology* **180**, 6215–6223.

Yamada, Y., Sugamura, K., Kondo, K., Yanagimoto, M., and Okada, H. (1987) The structure of inducing factors for virginianycin production in *Streptomyces virginiae. Journal of Antibiotics* **40**, 496–504.

Yang, C., and Kaplan, H. (1997) *Myxococcus xanthus sasS* encodes a regulator that prevents developmental gene expression during growth. *Journal of Bacteriology* **179**, 7759–7767.

Yang, Z.Y., Geng, Y., Xu, D., Kaplan, H.B., and Shi, W. (1998) A new set of chemotaxis homologues is essential for *Myxococcus xanthus* social motility. *Molecular Microbiology* **30**, 1123–1130.

Yoon, H.-S., and Golden, J.W. (1998) Heterocyst pattern formation controlled by a diffusible peptide. *Science* **282**, 935–938.

You, Z., Fukushima, J., Ishiwata, T., Chang, B., Kurata, M., Kawamoto, S., Williams, P., and Okuda, K. (1996) Purification and characterization of LasR as a DNA-binding protein. *FEMS Microbiology Letters* **142**, 301–307.

Zhao, H., Li, X., Johnson, D.E., and Mobley, H.L.T. (1999) Identification of protease and *rpoN*-associated genes of uropathogenic *Proteus mirabilis* by negative selection in a mouse model of ascending urinary tract infection. *Microbiology* **145**, 185–195.

Zhu, J., and Winans, S.C. (1998) Activity of the quorum-sensing regulator TraR of *Agrobacterium tumefaciens* is inhibited by a truncated, dominant defective TraR-like protein. *Molecular Microbiology* **27**, 289–297.

Zhu, J., and Winans, S.C. (1999) Autoinducer binding by the quorum-sensing regulator TraR increases affinity for target promoters *in vitro* and decreases TraR turnover rates in whole cells. *Proceedings of the National Academy of Sciences, USA* **96**, 4832–4837.

CHAPTER 3

SIGNAL TRANSDUCTION MECHANISMS IN PROTOZOA

FERNANDO L. RENAUD[1], JOSÉ DE ONDARZA[2], PIERANGELO LUPORINI[3], MICHAEL J. MARINO[4], and JUDY VAN HOUTEN[5]

[1]*University of Puerto Rico, Rio Piedras Campus, San Juan, PR;* [2]*Plattsburgh State University of New York, Plattsburgh, NY;* [3]*University of Camerino, Camerino, Italy;* [4]*Emory University School of Medicine, Atlanta, GA;* [5]*University of Vermont, Burlington, VT*

Summary

Although protozoa are unicellular eukaryotes, they have developed signalling mechanisms to modulate fundamental activities such as acquisition of nutrients and reproduction. However, our knowledge of signal transduction events in protozoa has lagged behind advances made in this field in more complex organisms. Nevertheless, evidence discussed in this review suggests that signal transduction events take place in protozoa by mechanisms that are similar to those present in higher eukaryotes, namely, binding of signalling molecules by specific receptors. This binding then results in the activation of specific signalling pathways. The latter have been studied in more detail in ciliates, in particular, in the opioid mechanism in *Tetrahymena* and *Stentor*, chemosensing in *Paramecium* and pheromone secretion in *Euplotes*. These data suggest that the complex signalling pathways present in metazoa originated in evolution from unicellular eukaryotes.

1. Introduction

Protozoan ciliates are amazing unicellular eukaryotes that have been able to adapt to every aquatic environment, from hot springs to the freezing waters of polar regions (Fenchel, 1980). Some of them have also become sessile (peritrichs and suctorians), or chosen a symbiotic lifestyle with the colonization of different niches in animal hosts, such as the intestine of herbivorous mammals (entodiniomorphs). Ciliates have solved the basic problems of life by mechanisms that appear akin to mechanisms present in more complex, advanced organisms. They secrete and can respond to a wide variety of signals in their environment that are crucial in either being attracted to a source of food, or a mate, or in being repelled by potentially noxious substances. However, the study of signal transduction in ciliates and other protozoa has lagged behind compared to that in mammalian organisms, in spite of the enticing evolutionary questions posed by studies of protozoa. Responses to signals by higher eukaryotes are mediated by signal transduction cascades that have been extensively studied. Most of these pathways involve a receptor that binds the signalling molecule specifically and with high affinity; the receptor-ligand complex then interacts with either a G protein (Kaziro *et al.*, 1991; Sprang, 1997), or activates a

I. Fairweather (ed.), Cell Signalling in Prokaryotes and Lower Metazoa, 73–90.
© 2004 *Kluwer Academic Publishers. Printed in Great Britain.*

cytoplasmic tyrosine kinase (Van der Geer, Hunter and Linberg, 1994). Activated G proteins then activate effectors such as adenylate cyclase or a membrane channel (De Vivo and Iyengar, 1994), whereas activated tyrosine kinases catalyze activation of substrate proteins that may also have kinase activity (van der Geer, Hunter and Lindberg, 1994). The result in any case is amplification of the signal by a cascade-like mechanism. In protozoa some progress has been made recently in characterizing signal transduction mechanisms, as will be seen below. However, in spite of similarities, some important differences are likely to be found when contrasting these mechanisms with those present in metazoa. In the latter, chemical messengers or hormones are carried from one cell to another by means of bodily fluids, such as the blood, and these extracellular environments are kept mostly constant by homeostatic mechanisms. However, protozoa live in environments that are amenable to significant changes in temperature, pH and chemical composition. Therefore, signal transduction mechanisms in these cells should reflect an adaptation to this environmental peculiarity.

2. Chemical Signals and their Receptors

A wide variety of signalling molecules have been reported in ciliates, and these may be classified in three main categories: hormones and neurotransmitters first reported in vertebrates; chemotactic agents such as some proteins and small molecules; and chemical messengers that are unique to ciliates. Among the vertebrate signalling molecules, biogenic amines such as serotonin and catecholamines (LeRoith and Roth, 1984), and peptide hormones such as insulin and β-endorphin (LeRoith *et al.*, 1980, 1982) have been reported to function in such a manner. However, for the most part, these reports have not been followed by thorough studies to confirm their identity, and to determine their biological relevance (see Renaud *et al.*, 1991, for a review). Chemotactic agents have received greater attention, and a number of small biological molecules, whose presence in the environment could be indicative of the presence of bacteria or other sources of food, have been studied. They include compounds such as glutamate, biotin and acetic acid (Van Houten, 1998). Finally, another category of chemical messengers appears to be unique to ciliates and is represented by protein pheromones that have been discovered in *Blepharisma* and *Euplotes* for their effects as cell mating-inducers (see Luporini *et al.*, 1995, for a review). A brief discussion of the mechanism of action of some of these signalling compounds and their receptors is found below.

2.1. EFFECTS OF OPIOIDS ON *TETRAHYMENA* AND *STENTOR*

Opioids were first reported to have specific effects in *Amoeba*, where both morphine and opioid peptides were reported to inhibit pinocytosis by a mechanism that appears to be receptor-mediated, since it is naloxone-reversible

(Josefsson and Johansson, 1979). Similar effects have been reported on phagocytosis in *Tetrahymena* (De Jesús and Renaud, 1989) and mechanoreception in *Stentor* (Marino and Wood, 1993), where some of the components of this opioid mechanism have started to be elucidated. The initial report of these effects was intriguing, since, as opposed to chemotactic stimuli, opioids are not likely to be encountered in the protozoan natural environment. Therefore, the signalling molecule may be surmised to be an endogenous opioid. A β-endorphin-like protein was reported earlier in *Tetrahymena* by means of radioimmunoassays and binding displacement experiments (LeRoith *et al.*, 1982), but this report was not pursued or confirmed by further characterization. Recently, proteins with β-endorphin-like immunoreactivity have been detected by means of Western blots in *Tetrahymena* extracts, using an anti-β-endorphin antibody: bands of 12 and 9 kDa were detected in intracellular extracts, whereas a band of 7 kDa was detected in the extracellular medium, suggesting perhaps a precursor-product relationship (Fig. 1). However, the molecular mass of these proteins does not resemble that of any of the opioid peptides reported so far, or of their precursors; for example, mammalian β-endorphin has a molecular mass of 3.5 kDa in contrast to 7 kDa for the putative secreted opioid detected in *Tetrahymena.* Nevertheless, these extracts are functional in several opioid mechanisms: in *Tetrahymena* both cellular and extracellular extracts inhibit phagocytosis by a naloxone-reversible mechanism (Renaud *et al.*, 1995), thus

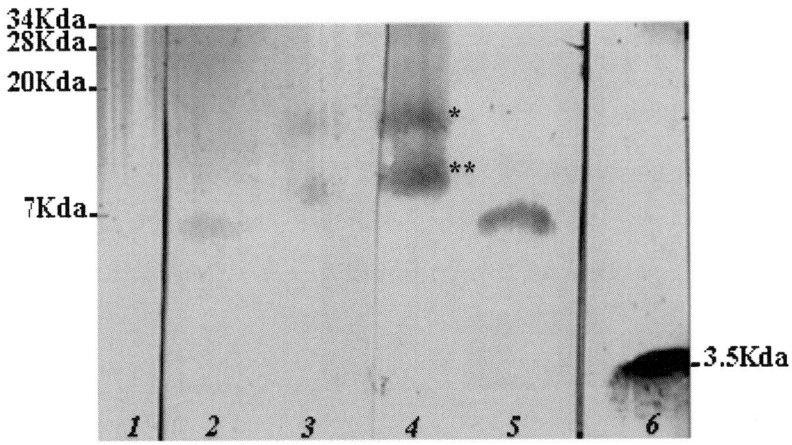

Figure 1. Detection of β-endorphin-like immunoreactivity in *Tetrahymena.* Cell extracts and aliquots from the culture medium were processed as described in Renaud *et al.* (1995). Proteins were separated using SDS-urea gels (Kyte and Rodriguez, 1983), followed by immunoblotting using a monoclonal anti-β-endorphin antibody (Boehringer Mannheim). Under these conditions, bands of 12 kDa (single asterisk) and 9 kDa (double asterisk) are detected in intracellular extracts (lane 4); whereas a band of 7 kDa is detected in the culture medium (lane 5). Control human β-endorphin resulted in a band of the expected mass, 3.5 kDa (lane 6). If the monoclonal antibody was pre-absorbed with mammalian β-endorphin, the detection of intracellular (lane 3) or extracellular (lane 2) immunoreactivity was abolished, as was that of control β-endorphin (lane 1).

mimicking the effect of morphine and opioid peptides on this process (De Jesús and Renaud, 1989). As stated previously, *Stentor* also shows an opioid response: both morphine and β-endorphin will depress the probability that *Stentor coeruleus* will contract in response to mechanical stimulation, and this effect is naloxone-reversible, suggesting that it is receptor-mediated (Marino and Wood, 1993). Interestingly, a similar effect is caused by addition of *Tetrahymena* cellular extract to *Stentor* cultures: it reduces the response to mechanical stimulation by a naloxone-reversible mechanism (Fig. 2).

Phagocytosis by murine macrophages has also been reported to be inhibited by morphine and other opioids by a naloxone-reversible mechanism (Tomei and Renaud, 1997) and *Tetrahymena* extracts have a similar effect on these immune cells (Rodríguez, 1999). Therefore, the opioid-like activity present in these extracts can be recognized even by organisms that are evolutionarily very distant from ciliates. Nevertheless, the identification of these proteins as *bona fide* members of the opioid family awaits their purification and sequencing. Although classical members of the opioid family are characterized by the sequence YGGFM/L at the N-terminal end, there have been reports of other

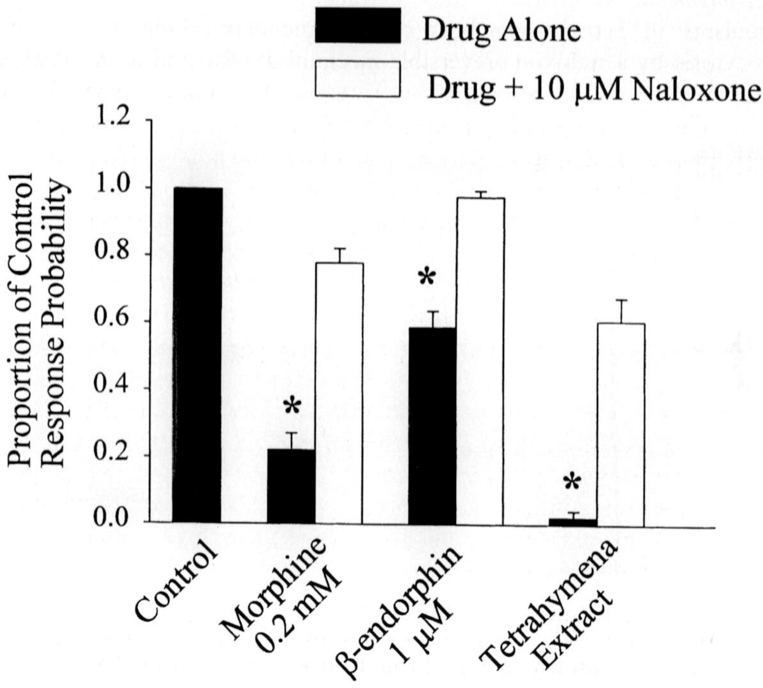

Figure 2. Opiate agonists and *Tetrahymena* extracts decrease mechanical sensitivity in *Stentor*. Cells were treated with the appropriate drug and tested for response to a mechanical stimulus 15 min later. Morphine, β-endorphin and the *Tetrahymena* intracellular extract produce a significant reduction in response probability ($*p < 0.05$), which is reversed by application of 10 μM naloxone. Each bar is mean ± SEM from 60–150 cells measured in at least three independent experiments.

opioids that show no obvious sequence similarities with classical members of the opioid family; deltorphins and casomorphin, for example (Erspamer *et al.*, 1989; Carr *et al.*, 1990), but yet are recognized as such by opioid receptors.

The molecular characterization of the opioid receptor in *Tetrahymena* has proved to be elusive so far. However, a pharmacological characterization by means of prototypic opioid agonists suggested that this receptor is mu-like, since morphine and β-endorphin are the most potent agonists (Chiesa, Silva and Renaud, 1993). In addition, binding of labelled β-endorphin to whole cells using a crosslinking agent detected a band of 110 kDa in *Tetrahymena*, rat and leech (Zipser *et al.*, 1988). This molecular mass is roughly double that reported for opioid receptors in mammalian cells, and perhaps reflects that the opioid receptor may associate to form dimers in the membrane, and this dimer is stabilized by the crosslinking agent. Nevertheless, there is no information as to whether this putative opioid receptor belongs to the seven transmembrane domains superfamily of receptors, as classical opioid receptors do (Minami and Satoh, 1995), or to a different category.

2.2. PHEROMONES IN *EUPLOTES* AND OTHER CILIATES

The existence of mating types in ciliates has been known for a long time, ever since they were discovered in *Paramecium* and *Euplotes* more than 60 years ago (Sonneborn, 1937; Kimball, 1939). However, the "mating type substances" that distinguish among different cell types and induce cells to unite with one another in mating (conjugal) pairs were identified much later. The most extensively studied ciliates, *Paramecium* and *Tetrahymena*, apparently carry these substances membrane-bound, particularly on ciliary membranes (Kitamura, 1988; Wolfe, Mpoke and Tirone, 1993), and contact between cells is required for mating recognition. On the other hand, other ciliates (*Blepharisma, Dileptus* and *Euplotes*) constitutively secrete these substances into the extracellular environment (see Mikaye, 1996, for a review). These soluble substances were first purified from *Blepharisma* and named "gamones", consistent with the view – repeatedly challenged (Nanney, 1980; Luporini and Miceli, 1986; Dini and Nyberg, 1993) – that they represent sexual factors, and ciliate mating types represent complementary sexes (Mikaye, 1981a). Two types of *Blepharisma* pheromones were described: one released by one cell type consisting of a basic glycoprotein of around 20 kDa, and another released by a different cell type consisting of a derivative of tryptophan (Mikaye, 1981b). A soluble pheromone has also been reported in *Dileptus anser*, consisting of a 3 kDa protein (Parfenova *et al.*, 1989). However, pheromones of *Euplotes* have been the best characterized structurally, either through direct chemical analysis of purified material, or through studies of their genes, which appear to be inherited as multiple alleles co-dominant at a single *mat* (or *mt*) locus (Luporini *et al.*, 1986; Heckmann and Kuhlmann, 1986). Proteins forming the pheromone family of *E. raikovi* are all 37–40 amino acid residues (Raffioni *et al.*, 1992), while those of the pheromone family of *E. octocarinatus* are formed by 85–109 residues

(Mollenbeck and Heckmann, 1999). Interestingly enough, there is not a great degree of similarity in amino acid sequence among different pheromones of the same family (Raffioni *et al.*, 1992). However, as illustrated in Fig. 3, these proteins do mimic each other in overall architecture as determined by comparative NMR studies in *E. raikovi* (Luginbühl *et al.*, 1994).

From an evolutionary point of view, it is of interest that pheromones of *E. raikovi* have some structural similarities with the cytokine IL-2 (Vallesi *et al.*, 1998). Pheromone E*r*-1 has been shown to be capable of binding to the α and β chains of the multimeric IL-2 receptor in mammalian cells, and IL-2 can bind the putative E*r*-1 receptor. These data strengthen the hypothesis of the

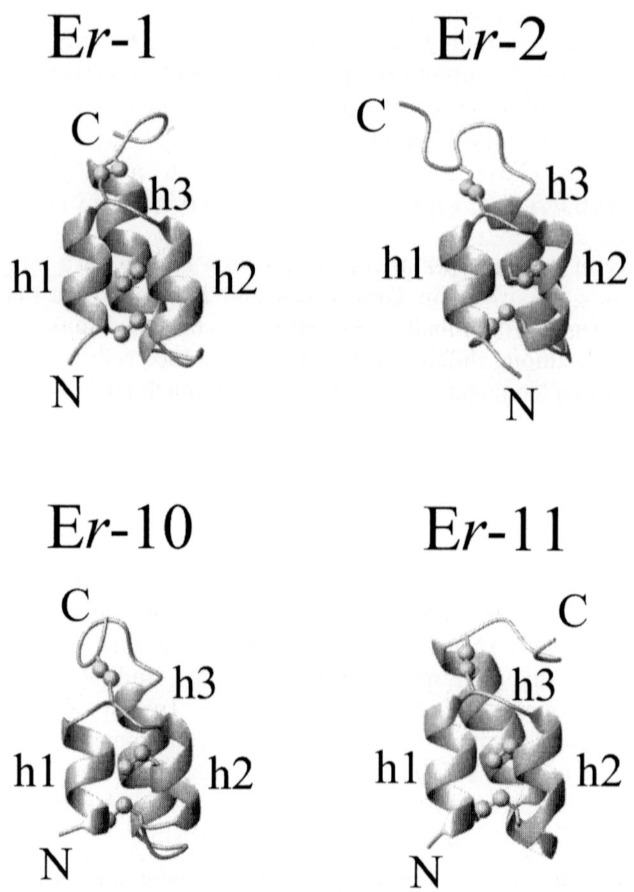

Figure 3. Ribbon representations (prepared by using the MOLMOL program) of the NMR solution structures of four pheromones of *Euplotes raikovi*, E*r*-1, E*r*-2, E*r*-10 and E*r*-11. The three helices that characterize every structure are denoted h1, h2 and h3, while N and C indicate the molecule amino- and carboxi-terminus, respectively. Spheres represent half-cystines involved in disulfide bonds. The general amino acid sequence of these molecules is $DX_{1-3}CX_6CX_4CX_{2-3}CX_{5-8}CX_{7-9}CX_{1-5}$ (Luginbühl *et al.*, 1994).

common origin in evolution of signalling pathways in protozoa and more complex organisms. Furthermore, it is clear that the role of these pheromones goes beyond that of stimulating mating. Sex in protozoa is certainly not indispensable, since many unicellular eukaryotes do not undergo sexual reproduction; therefore, the primordial function of pheromones was probably related to regulating growth – as discussed below – and acquired their role in mating later in evolution. The effect of the pheromone depends on whether it binds to receptors from the cells that secreted it, as part of an autocrine mechanism, or to receptors from other cells secreting other homologous pheromones, as part of a paracrine mechanism. Autocrine stimulation will result in increased cell vegetative (mitotic) proliferation, whereas paracrine stimulation temporarily arrests this proliferation and induces mating (Vallesi *et al.*, 1995). The pheromone receptors appear to be represented by non-secreted versions of the soluble pheromones (Ortenzi *et al.*, 2000) (Fig. 4). The extracellular domain of these membrane-bound pheromones is structurally equivalent to their soluble partner molecules, whereby these membrane proteins might also carry out ligand functions. This situation parallels the one reported for some animal growth factors and cytokines that can exist in both a free, soluble form and in a membrane-bound state (Massagué and Pandiella, 1993). Membrane localization of either cytokines or pheromones allows juxtacrine cell interactions between physically contacting cells. Therefore, binding of membrane-bound pheromones directly with one another across the extracellular space between

1
M G C T S D L C H M A S V L S K Y Q S L F L Y L C R S N N C

31
V G P L N S I N R F F S G T V H V I S F S I N F ***R*** **M N K** L A

61
I L A I I A M V L F S A N A F R F Q S R L R S N V E A K T G

91
D A C E Q A A I Q C V E S A C E S L C T E G E D R T G C Y M

121 130
Y I Y S N C P P Y V

Figure 4. Primary structure of pheromone receptors in type I cells of *Euplotes raikovi*. These receptors appear to be represented by membrane-anchored pheromone isoforms that are generated by a mechanism of alternate splicing of the primary transcripts of the pheromone genes. Of the 130 amino acids that form the complete sequence, 55 are printed in italics to distinguish the sequence region that is unique and represents the cytoplasmic domain of the molecule. The other region of 75 amino acids is equivalent to the pheromone precursor (i.e., pre-pro-pheromone) and represents the transmembrane and extracellular domains of the molecule. The two sequence segments that bear a significant similarity with the G-protein coupled receptor signature (Attwood, Eliopoulos and Findlay, 1991) are underlined; the segment that recalls the motif BBXXB (where B is a basic amino acid and X any other residue) indicative of a G protein activating domain (Pennington, 1995) is in bold (Miceli *et al.*, 1992; Ortenzi *et al.*, 2000).

facing cell membranes should promote transient intercellular communication and adhesion for mating pair formation. Important insights into the nature of these interactions between soluble and membrane-bound (receptor) pheromones have been derived from the crystal structure of pheromone E*r*-1 (Weiss *et al.*, 1995). These data suggest cooperative protein oligomerization by helix-helix interactions. Another remarkable observation is that the cytoplasmic domain of the membrane-bound pheromone has two adjacent segments that bear a significant similarity with a sequence generally recognized as a signature for seven membrane-spanning domains receptor superfamily (Ortenzi *et al.*, 2000). This has evolutionary implications that will be discussed at the end of this Chapter.

2.3. CHEMOTACTIC AGENTS AND THEIR RECEPTORS IN CILIATES

Chemosensory transduction is a feature of a wide range of organisms, from ciliates (Van Houten, 1998) to leukocytes and slime molds (Parent and Devreotes, 1999). This process is key to survival in unicellular organisms, because chemotactic agents serve to attract the organisms to food sources. Although a wide variety of substances have been reported to have chemotactic activity on protozoa, this process has been studied most extensively in *Paramecium* and *Tetrahymena*. In *Paramecium* chemotactic agents may be classified in three groups according to their mechanism of action (Van Houten, 1998): hyperpolarizing agents that do not affect levels of cAMP, such as acetate; those that do affect levels of cAMP, such as glutamate; and those that alter intracellular pH, such as ammonium ions. In *Tetrahymena*, there is evidence for two different receptors for the chemorepellents GTP and lysozyme, and both result in a depolarization (Kuruvilla, Kim and Hennessey, 1997).

2.3.1. *Chemical Sensing in Paramecium tetraurelia*

P. tetraurelia responds to chemical cues in its environment that probably serve to locate bacterial food sources. The cells respond to these signals by changing swimming patterns that ultimately lead to attraction (or repulsion) of populations of cells. In general, attractant stimuli cause smooth, fast swimming, characteristic of hyperpolarization, whereas repellents cause depolarization and the converse behavioural response (Van Houten, 1998). Manipulation of speed of swimming and frequency of turning is sufficient to cause a population of paramecia to accumulate near, or disperse from, the source of chemical cues. Electrophysiological measurements confirm that, indeed, attractant and repellent stimuli hyperpolarize and depolarize the membrane, respectively. A brief discussion of the main characteristics of these signalling pathways follows below.

2.3.1.1. *Pathway 1.* Folate, cAMP, acetate, lactate and biotin, among others, attract *Paramecium* probably through the same transduction pathway.

However, biochemical, binding kinetic studies and behavioural assays suggest separate receptors for each of these ligands (Van Houten, 1998). However, only the 48 kDa cyclic AMP receptor has been purified and partially sequenced (Van Houten *et al.*, 1991). Following binding, the cell hyperpolarizes, and this change in potential makes cells swim faster and smoother (Preston and Van Houten, 1987). There are many unknown steps between receptor binding and hyperpolarization. The latter appears to be initiated by K^+ conductances upon stimulation by acetate, folate or biotin. Hyperpolarization is sustained by a mechanism that is not due solely to Na^+ or K^+ channel activity, since no reversal or dependence on extracellular K^+ or Na^+ could be demonstrated (Van Houten, 1998). There is also indirect evidence that the plasma membrane calcium-ATPase pump may sustain this hyperpolarizing conductance. A calcium-ATPase activity is associated with the pellicle that has all the characteristics of a plasma membrane calcium pump, and its activity may be linked to chemoresponse (Elwess and Van Houten, 1997). This hypothesis is supported by the fact that antisense oligonucleotides against calmodulin mRNA inhibit the response to acetate (Yano *et al.*, 1996).

2.3.1.2. *Pathway 2.* There appear to be binding sites specific for glutamate on the *Paramecium* surface, some of which can also bind the chemorepellent IMP (Van Houten, 1998). The interesting fact is that, as opposed to pathway 1, glutamate induces a rapid increase in intracellular cAMP in these cells (Yang *et al.*, 1997). However, how the receptor couples to adenylyl cyclase is not clear at this time, although evidence will be presented in the next section that suggests that this is a G protein-mediated mechanism. Glutamate has also been reported to hyperpolarize cells, and this hyperpolarization probably is due to both channel activity and to activation of the calcium pump (Van Houten, 1998). Pump activity can be stimulated *in vitro* by heterologous PKA, suggesting that the glutamate-induced increase in cAMP may modulate the pump to sustain hyperpolarization. In terms of the *Paramecium* glutamate receptor, its pharmacology does not fit perfectly with that of glutamate receptors in mammalian cells, although synergy with GMP and glutamate-induced sustained hyperpolarization is also observed in taste cells (Van Houten, Yang and Bergeron, 2000). Confirmation of any homology with other glutamate receptors awaits purification of the *Paramecium* receptor.

2.3.1.3. *Pathway 3.* The third pathway has only one stimulus to date, NH_4Cl, and differs from the other two in that it does not appear to involve a receptor-mediated mechanism (Davis *et al.*, 1998). At pH 7, NH_4Cl is in equilibrium with NH_3, which is membrane-permeable and will increase intracellular pH. However, this alkalinization is reversible. It is suggested that there may be a pH-sensitive channel responsible for NH_4Cl-induced hyperpolarization.

2.3.2. *Chemorepellents and their Receptors in Tetrahymena thermophila*

T. thermophila, like *Paramecium*, responds to chemorepellents by repetitive bouts of backward swimming or "avoidance reactions". This behaviour is caused by a repellent-induced depolarization that triggers a ciliary, Ca^{2+}-based action potential (Kuruvilla, Kim and Hennessey, 1997). Two chemorepellents have been studied in detail in this organism: lysozyme and GTP. Interestingly enough, the avoidance reaction induced by these substances decreases with time, suggesting adaptation or desensitization. This decrease in response parallels a decrease in the number of binding sites for radioactive GTP and lysozyme, which suggests a time-dependent decrease in the number of putative receptors (Kuruvilla, Kim and Hennessey, 1997). Nevertheless, it is of interest that adaptation to lysozyme is independent of adaptation to GTP and the converse is also true, which suggests that their putative receptors are independently regulated. A recent exciting finding is the report that a chemorepellent receptor appears to have been purified from *Tetrahymena*: a 42 kDa protein isolated from *Tetrahymena* membranes binds lysozyme with a Kd of 200 nM, and antibodies to this protein will block the *in vivo* lysozyme chemoresponse (Kuruvilla and Hennessy, 1997). If confirmed by further research, this would constitute the first time that a chemorepellent receptor has been isolated from any unicellular eukaryote.

2.4. OTHER SIGNAL TRANSDUCTION MECHANISMS IN CILIATES

Various other transduction mechanisms have been reported in ciliates, and have been studied to varying degrees. Photosensory transduction has been reported in *Paramecium* (Shinozawa *et al.*, 1996), *Blepharisma* and *Stentor* (Fabczak *et al.*, 1993a). In the latter cases, this mechanism appears to take place by a change in intracellular pH. In *Blepharisma* the pink-coloured pigment (blepharismin) contained in the pigment granules is believed to be the photoreceptor responsible for the photophobic response (Matsuoka *et al.*, 1995). Curiously enough, as in the case of mammalian retinal rods and cones (Fung, Hurley and Stryer, 1981), the mechanism appears to be modulated by cGMP: a membrane-permeable analog of this nucleotide inhibits the photophobic response in *Blepharisma* (Fabczak *et al.*, 1993b,c).

3. Signalling Switches and Effector Molecules

External stimuli are transduced in multicellular organisms via one of five kinds of receptors: 7-transmembrane domains receptors (GR, for G protein-linked receptor), receptor tyrosine kinases, receptors that function as ion channels, soluble (non-membrane) nuclear receptors, and receptors with guanylyl cyclase activity (Wilson *et al.*, 1998). In lower eukaryotes, most of the progress has been made in delineating signalling pathways of the slime mold (*Dictyostelium*) and of yeast (*Saccharomyces*). However, receptors, G proteins and effectors remain poorly understood in protozoa.

3.1. HETEROTRIMERIC G PROTEINS AND THEIR RECEPTORS

There does not seem to be any conclusive evidence so far for the existence of G-protein-linked receptors in protozoa, although they are indeed present in *Dictyostelium* (Devreotes, 1994) and *Saccharomyces* (Kaziro *et al.*, 1988). The closest report of this type of receptor is the partial homology found in the cytoplasmic domain of pheromone receptors in *Euplotes* (Ortenzi *et al.*, 2000). Therefore, it seems likely that if G protein-mediated mechanisms do occur in ciliates, the nature of the receptor may be different from that of those present in multicellular organisms (Ulloa-Aguirre *et al.*, 1999).

The case for the existence of G proteins in protozoa is stronger. Heterotrimeric G proteins are ubiquitous signalling molecules which consist of three different subunits, designated α, β and γ (Kaziro *et al.*, 1991). The α-subunits (Gα) belong to a superfamily of GTP binding proteins, and are known to modulate effector molecules such as adenylyl cyclase and various ion channels (De Vivo and Iyengar, 1994). As integral components of intercellular signalling pathways, heterotrimeric G proteins play a crucial role in neuroendocrine and paracrine control in multicellular organisms. As has been discussed in previous sections, protozoa do indeed engage in intercellular communication for the purposes of mating (Luporini and Miceli, 1986), and they can also respond to environmental cues such as touch (Marino and Wood, 1993), food (Van Houten, 1998), light (Fabczak *et al.*, 1993a–c; Shinozawa *et al.*, 1996) and temperature (Nakaoka *et al.*, 1997). However, the role that heterotrimeric G proteins play in these processes is still under study. The only DNA sequence for a G protein gene reported in protozoa so far has been a partial sequence from *Stentor coeruleus* (Fabczak *et al.*, 1993c). Recently, one of the co-authors of this review (Michael J. Marino) cloned three Gi/o-like, and one Gs-like large sequences from *Stentor coeruleus* (Gene Bank AF 079839-42), that exhibit a 33–63% overall homology. The difficulty in cloning these proteins in protozoa may stem from inherent differences in G protein structure compared to higher taxa. Indeed, the Gαi homologues of *Saccharomyces* measure 50 and 54 kDa, significantly higher than the homologous Gαi subunits in mammals (Kaziro *et al.*, 1988). In addition, unusual genetic codes and an AT-rich genome among ciliates have made isolation and cloning of these genes more difficult. DNA sequences for β and γ subunits are also unknown, although a gene with β subunit homology has been identified in *Paramecium* (Forney and Rodkey, 1992).

Despite the lack of DNA sequences, there is mounting evidence that heterotrimeric G proteins play important roles in protozoa. Studies using the G protein modulators cholera toxin (CTX, maintains Gs activated) and pertussis toxin (PTX, inactivates Gi/o), as well as GTP and GDP analogues, have resulted in altered functions in various protozoa and suggest that G proteins play a role in backward swimming (Bernal, Kelsey and Ehrlich, 1991; Bernal and Ehrlich, 1993), chemoattraction (Symington *et al.*, unpublished work), photoreception (Shinozawa *et al.*, 1996), and cold transduction (Nakaoka *et al.*,

1997) in *Paramecium*, phagocytosis in *Tetrahymena* (Renaud *et al.*, 1995), proliferation in *Euplotes* (Vallesi *et al.*, 1995), mechanoresponse in *Stentor* (Marino and Wood, 1993), and development in *Trypanosoma* (Oz *et al.*, 1994; Coulter and Hide, 1995). More recently, efforts to detect protozoan G protein subunits by immunoblots have demonstrated the presence of protein bands in *Tetrahymena* using antibodies against conserved regions of Gα subunits (Renaud *et al.*, 1995); these protein bands (52 and 57 kDa) appear to be larger than most mammalian G proteins, a finding that is also true for *Euplotes* (Ortenzi *et al.*, 2000) and *Paramecium bursaria* (Shinozawa *et al.*, 1996). A 42 kDa protein is detected in the ciliary membranes of *Paramecium tetraurelia*; this protein is a substrate for PTX-mediated ADP-ribosylation and is thought to represent a Go-type α subunit (Symington *et al.*, unpublished work). Similar sized (38–43 kDa) protein bands have been detected in *Stentor* (Marino, unpublished work), *Euglena gracilis* (Torres-Marquez *et al.*, 1996), *Phytomonas* (Farber *et al.*, 1995), *Toxoplasma* (Halonen, Weidner and Siebenaller, 1996), *Entamoeba* (Paveto *et al.*, 1999), *Leishmania* (Cassel *et al.*, 1991) and *Trypanosoma* (Coulter and Hide, 1995; Oz *et al.*, 1994). We conclude from these observations that PTX-sensitive G proteins, similar to mammalian Go/Gi, and CTX-sensitive proteins, similar to mammalian Gs, exist in protozoa, and they affect similar cellular targets. In summary, both Gi-like and Gs-like proteins appear to be present in protozoa, although their definitive identification as members of the G protein family awaits their purification and sequencing.

Effector molecules in signal transduction events in metazoa include, among others, membrane proteins such as adenylate cyclase, ion channels, phospholipases and protein kinases. Among these, adenylate cyclase is one of the effector molecules that have been most extensively studied. This enzyme is present in protozoa, but it seems to be insensitive to ligands such as catecholamines (Kudo *et al.*, 1985). Furthermore, there is some evidence that it may not be ligand-modulated, but voltage-modulated (Schultz *et al.*, 1992), with the possible exception of the increase in cAMP that takes place as part of the mechanism of glutamate chemoattraction in *Paramecium* (Yang *et al.*, 1997). In the latter case it is not clear if AC stimulation is due to a direct ligand-mediated mechanism, or to hyperpolarization.

Another cyclic nucleotide-related protein that is important in some signal transduction mechanisms is guanylyl cyclase, which plays an important role in signal transduction in retinal rods and cones (Fung, Hurley and Stryer, 1981). It is of interest that this enzyme also seems to play a role in the photosensory mechanism present in *Blepharisma* (Fabczak *et al.*, 1993b,c): membrane-permeable analogs of cGMP cause a marked dose-dependent prolongation of the latency for the photophobic response; a similar effect was observed with a phosphodiesterase inhibitor, 3′-isobutylmethylxanthine, and with pertussis toxin. On the other hand, G protein activators that should lower the level of cGMP, significantly enhance the photoresponsiveness of these cells to light stimuli (Fabczak *et al.*, 1993c). These results suggest that, as in the retinal rods and cones, cGMP serves as a signal modulator in the photophobic response.

Increases in cAMP in multicellular organisms result in the activation of protein kinase A (PKA); activation of this enzyme involves cAMP binding by the regulatory subunit, which causes release of the activated catalytic subunit (Taylor *et al.*, 1993). A 44 kDa regulatory subunit of PKA has been purified from *Paramecium*, and its cDNA has a 31–38% positional identity with other cloned cAMP-dependent protein kinase regulatory subunits (Carlson and Nelson, 1996). However, its N-terminal lacks the dimerization domain found in most regulatory subunits of this family and also seems to code a unique autophosphorylation site. Furthermore, cilia apparently possess a unique version of this subunit (Hochstrasser *et al.*, 1994).

Concerning other effector proteins, there is a paucity of information. To the best of our knowledge, there have been no reports of receptors coupled to tyrosine kinase activity in protozoa. These receptors transduce their signal by initiating a cascade of activation of kinases (van der Geer, Hunter and Lindberg, 1994), including mitogen-activated kinases (MAPK). Nevertheless a stress-responsive, protein kinase-related enzyme has been described in *Tetrahymena*, and its predicted amino acid sequence resembles that of MAPK (Nakashima *et al.*, 1997). Therefore, there is the possibility that other elements of the tyrosine kinase transduction pathway are present in protozoa.

Another important signal transduction pathway in multicellular organisms is the phosphoinositide cascade, that is initiated by G protein-mediated activation of a membrane-bound phospholipase C. There are no clear reports of the existence of this pathway in protozoa, but there is at least some evidence for the presence of some of its components. Stomatin-induced differentiation in *Tetrahymena vorax* results in an increase in polyphosphoinositols and a concomitant decrease in phosphatidylinositol di- and triphosphate, suggesting activation of phospholipase C (Ryals, Bae and Patterson, 1999). On the other hand, in *T. pyriformis*, a PLC that preferentially hydrolyzes phosphatidylcholine has been described (Alam, Banno and Nozama, 1993).

4. Summary and Conclusions

The data discussed above suggest that the fundamental elements for G protein-mediated signal transduction are present in protozoa. However, many questions still remain concerning the particularities of receptors, signalling switches and effector molecules. So far there is no evidence for classical G protein-linked receptors (GRs) in protozoa, but the slight similarity between certain domains of the *Euplotes* pheromone receptors and some conserved domains in GRs suggests that perhaps seven membrane-spanning domains originated in evolution from simpler versions such as that of the pheromone receptor. The basic functions modulated by signalling molecules in protozoa are cell growth and reproduction and the acquisition of nutrients. Early in the evolution of signalling, the signalling molecules were indicators of the presence of food, such as amino acids, and did not necessitate the secretion of signalling molecules by the protozoan cell. Endogenous signalling molecules (pheromones) originated later in evolution to promote cell growth and reproduction. Therefore, the original function of pheromones was to be part of an autocrine loop to promote cell growth, and in this sense are primordial growth factors, the mating function being acquired only secondarily. A paracrine mechanism would be the next step in complexity, with cells secreting pheromones to signal other cells of the same species, including potential mates. With the advent of multicellularity, the increase in number and variety of responses required a further complexity in receptor structure in order to discriminate among the many ligands available. This gave rise to increased receptor complexity, resulting in the GR that we find in multicellular organisms today.

References

Alam, S., Banno, Y., and Nozawa, Y. (1993) Purification and characterization of phospholipase C preferentially hydrolysing phosphatidylcholine in *Tetrahymena* membranes. *Journal of Eukaryotic Microbiology* **40**, 775–781.

Attwood, T.K., Eliopoulos, E.E., and Findlay, B.C. (1991) Multiple sequence alignment of protein families showing low sequence homology: a methodological approach using database pattern-matching discriminators for G-protein-linked receptors. *Gene* **98**, 153–159.

Bernal, J., and Ehrlich, B.E. (1993) Guanine nucleotides modulate calcium currents in a marine *Paramecium. Journal of Experimental Biology* **176**, 117–133.

Bernal, J., Kelsey, A.M., and Ehrlich, B.E. (1991) GTP-γ-S increases duration of backward swimming behavior and the calcium ion potential in marine *Paramecium. Journal of Experimental Biology* **155**, 505–518.

Carlson, G.L., and Nelson, D.L. (1996) The 44-kDa regulatory subunit of the *Paramecium* cAMP-dependent protein kinase lacks a dimerization domain and may have a unique autophosphorylation sequence. *Journal of Eukaryotic Microbiology* **43**, 347–356.

Carr, R.I., Webster, D., Sadi, D., and Williams, H. (1990) Immunomodulation by opioids from dietary casein (exorphins). *Annals of the New York Academy of Sciences* **594**, 374–376.

Cassel, D., Shoubi, S., Glusman, G., Cukierman, E., Rotman, M., and Zilberstein, D. (1991) *Leishmania donovani*: characterization of a 38-kDa membrane protein that cross-reacts with the mammalian G-protein transducin. *Experimental Parasitology* **72**, 411–417.

Chiesa, R., Silva, W.I., and Renaud, F.L. (1993) Pharmacological characterization of an opioid receptor in *Tetrahymena. Journal of Eukaryotic Microbiogy* **40**, 800–804.

Coulter, L.J., and Hide, G. (1995) *Trypanosom brucei*: characterisation of a life cycle stage-specific G-protein. *Experimental Parasitology* **80**, 308–318.

Davis, D.P., Fiekers, J.F., and Van Houten, J.L. (1998) Intracellular pH and chemoresponse to NH_4^+ in *Paramecium. Cell Motility and the Cytoskeleton* **40**, 107–118.

De Jesús, S., and Renaud, F.L. (1989) Phagocytosis in *Tetrahymena thermophila*: naloxone-reversible inhibition by opiates. *Comparative Biochemistry and Physiology* **92C**, 139–142.

De Vivo, M., and Iyengar, R. (1994) G protein pathways: signal processing by effectors. *Molecular and Cellular Endocrinology* **100**, 65–70.

Devreotes, P.N. (1994) G protein-linked signaling pathways control the developmental program of *Dictyostelium. Neuron* **12**, 235–241.

Dini, F., and Nyberg, D. (1993) Sex in ciliates. In: J. Gwynfryn Jones (ed.) *Advances in Microbial Ecology, Vol. 13*. Plenum Press, New York, pp. 85–153.

Elwess, N.L., and Van Houten, J.L. (1997) Cloning and molecular analysis of the plasma membrane Ca^{2+}-ATPase gene in *Paramecium tetraurelia. Journal of Eukaryotic Microbiology* **44**, 250–257.

Erspamer, V., Melchiorri, P., Falconieri-Erspamer, G., Negri, L., Corsi, R., Severini, C., Barra, D., Simmaco, M., and Kreil, G. (1989) Deltorphins: a family of naturally occurring peptides with high affinity and selectivity for δ opioid binding sites. *Proceedings of the National Academy of Sciences, USA* **86**, 5188–5192.

Fabczak, H., Fabczak, S., Song, P.S., Checcucci, G., Ghetti, F., and Lenci, F. (1993a) Photosensory transduction in ciliates. Role of intracellular pH and comparison between *Stentor* and *Blepharisma japonicum. Journal of Photochemistry and Photobiology* **B21**, 47–52.

Fabczak, H., Tao, N., Fabczak, S., and Song, P.S. (1993b) Photosensory transduction in ciliates. 4. Modulation of the photomovement response of *Blepharisma japonicum* by cGMP. *Photochemistry and Photobiology* **57**, 889–892.

Fabczak, H., Tao, N., Fabczak, S., and Song, P.S. (1993c) Photosensory transduction in ciliates. 2. Possible role of G-protein and cGMP in *Stentor coerulus. Photochemistry and Photobiology* **57**, 702–706.

Farber, M.., Montagna, A.E., Paveto, C., Dollet, M., Sánchez-Moreno, M., Osuna, A., Torres, H.N., and Falviá, M.M. (1995) Adenylyl cyclase and G proteins in *Phytomonas. Journal of Eukaryotic Microbiology* **42**, 257–260.

Fenchel, T.C. (1987) *Ecology of Protozoa.* Springer-Verlag, Berlin.

Forney, J., and Rodkey, K. (1992) A repetitive DNA sequence in *Paramecium* macronuclei is related to the β subunit of G proteins. *Nucleic Acids Research* **20**, 5397–5402.

Fung, B.-K., Hurley, J.B., and Stryer, L. (1981) Flow of information in the light-triggered cyclic nucleotide cascade of vision. *Proceedings of the National Academy of Sciences, USA* **78**, 152–156.

Halonen, S.K., Weidner, E., and Siebenaller, J.F. (1996) Evidence for hetotrimeric GTP-binding proteins in *Toxoplasma gondii. Journal of Eukaryotic Microbiology* **43**, 187–193.

Heckmann, K., and Kulmann, H.W. (1986) Mating types and mating inducing substances in *Euplotes octocarinatus. Journal of Experimental Zoology* **237**, 87–96.

Hochstrasser, M., Carlson, G.L., Walczak, C.E., and Nelson, D.L. (1996) *Paramecium* has two regulatory subunits of cyclic AMP-dependent protein kinase, one unique to cilia. *Journal of Eukaryotic Microbiology* **43**, 356–362.

Josefsson, J.O., and Johansson, P. (1979) Naloxone-reversible effect of opioids on pinocytosis in *Amoeba proteus. Nature* **282**, 72–80.

Kaziro, Y., Itoh, H., Kozasa, T., Toyama, R., Tsukamoto, T., Matsuoka, M., Nakafuku, M., Obara, T., Takagi, T., and Hernández, R. (1988) Structures of the genes coding for G protein α subunits from mammalian and yeast cells. *Cold Spring Harbor Symposia on Quantitative Biology* **53**, 209–219.

Kaziro, Y., Itoh, H., Kozasa, T., Nakafuku, M., and Satoh, T. (1991) Structure and function of signal-transducing GTP-binding proteins. *Annual Review of Biochemistry* **60**, 349–400.

Kimball, R.F. (1939) Mating types in *Euplotes. American Naturalist* **73**, 451–456.

Kitamura, A. (1988) Mating-type substances. In: H.D. Gortz (ed.) *Paramecium.* Springer, Berlin, Heidelberg, New York, pp. 85–96.

Kudo, S., Muto, Y., and Nozawa, Y. (1985) Regulation by calcium of hormone insensitive adenylate cyclase and calmodulin dependent guanylate cyclase in *Tetrahymena* plasma membrane. *Comparative Biochemistry and Physiology* **80B**, 813–816.

Kuruvilla, H.G., and Hennessey, T.M. (1997) Purification and characterization of a novel chemorepellent receptor from *Tetrahymena thermophila. Journal of Membrane Biology* **162**, 51–57.

Kuruvilla, H.G., Kim, M.Y., and Hennessey, T.M. (1997) Chemosensory adaptation to lysozyme and GTP involves independently regulated receptors in *Tetrahymena thermophila. Journal of Eukaryotic Microbiology* **44**, 263–268.

Kyte, J., and Rodriguez, H. (1983) A discontinuous electrophoretic system for separating peptides on polyacrylamide gels. *Analytical Biochemistry* **133**, 515–522.

LeRoith, D., and Roth, J. (1984) Vertebrate hormones and neuropeptides in microbes: evolutionary origins of intercellular communication. *Frontiers in Neuroendocrinology* **8**, 1–25.

LeRoith, D., Shiloach, J., Roth, J., and Lesniak, M.A. (1980) Evolutionary origins of vertebrate hormones: substances similar to mammalian insulins are native to unicellular eukaryotes. *Proceedings of the National Academy of Sciences, USA* **77**, 6184–6188.

LeRoith, D., Liotta, S., Roth, J., Shiloach, J., Lewis, M.E., Pert, C.B., and Krieger, D.T. (1982) Corticotropin and β-endorphin-like materials are native to unicellular eukaryotes. *Proceedings of the National Academy of Sciences, USA* **79**, 2086–2090.

Luginbühl, P., Ottiger, M., Mronga, S., and Wüthrich, K. (1994) Structure comparison of the pheromones E*r*-1, E*r*-10 and E*r*-2 from *Euplotes raikovi. Protein Science* **3**, 1537–1547.

Luporini, P., and Miceli, C. (1986) Mating pheromones. In: J.G. Gall (ed.) *The Molecular Biology of Ciliated Protozoa.* Academic Press, New York, pp. 263–299.

Luporini, P., Raffioni, S., Concetti. A., and Miceli, C. (1986) The ciliate *Euplotes raikovi* heterozygous at the *mat* locus coreleases two individual species of the mating pheromone: genetic and biochemical evidence. *Proceedings of the National Academy of Sciences, USA* **83**, 2889–2893.

Luporini, P., Miceli, C., Vallesi, A., and Bradshaw, R.A. (1995) Chemical signaling in ciliates. *Journal of Eukaryotic Microbiology* **42**, 208–212.

Marino, M.J., and Wood, D.C. (1993) β-endorphin modulates a mechanoreceptor channel in the protozoan *Stentor. Journal of Comparative Physiology* **A173**, 233–240.

Massagué, J., and Pandiella, A. (1993) Membrane-anchored growth factors. *Annual Review of Biochemistry* **62**, 515–541.

Matsuoka, T., Watanabe, Y., Sagara, Y., Takayanagi, M., and Kato, Y. (1995) Additional evidence for blepharismin photoreceptor pigment mediating step-up photophobic response of unicellular organism, *Blepharisma. Photochemistry and Photobiology* **62**, 190–193.

Miceli, C., La Terza, A., Bradshaw, R.A., and Luporini, P. (1992) Identification and structural characterization of a cDNA clone encoding a membrane-bound form of the polypeptide pheromone E*r*-1 in the ciliate protozoan *Euplotes raikovi. Proceedings of the National Academy of Sciences, USA.* **89**, 1988–1992.

Mikaye, A. (1981a) Cell interactions by gamones in *Blepharisma.* In: D.H. O'Day and P. Horgen (eds.) *Sexual Interactions in Eukaryotic Microbes.* Academic Press, New York, pp. 95–129.

Mikaye, A. (1981b) Physiology and biochemistry of conjugation in ciliates. In: M. Levandowsky and S.H. Hutner (eds.) *Biochemistry and Physiology of Protozoa. Vol. 4.* Academic Press, New York, pp. 125–198.

Mikaye, A. (1996) Fertilization and sexuality in ciliates. In: K. Hausmann and P.C. Bradbury (eds.) *Ciliates: Cells as Organisms.* Springer-Verlag, Berlin, Heidelberg, New York, pp. 243–290.

Minami, M., and Satoh, M. (1995) Molecular biology of the opioid receptors: structures, functions and distributions. *Neuroscience Research* **23**, 121–145.

Mollenbeck, M., and Heckmann, K. (1999) Characterization of two genes encoding a fifth so far unknown pheromone of *Euplotes octocarinatus. European Journal of Protistology* **35**, 225–231.

Nakaoka, Y., Tanaka, T., Kurio, T., and Murata, T.A. (1997) Possible mediation of G-proteins in cold sensory transduction in *Paramecium multinucleatum. Journal of Experimental Biology* **200**, 1025–1030.

Nakashima, S., Wang, S., Hisamoto, N., Sakai, H., Andoh, M., Matsumoto, K., and Nozawa, Y. (1999) Molecular cloning and expression of stress-responsive mitogen-activated protein kinase-related kinase from *Tetrahymena. Journal of Biological Chemistry* **274**, 9976–9983.

Nanney, D.L. (1980) *Experimental Ciliatology.* Wiley, New York.

Ortenzi, C., Alimenti, C., Vallesi, A., Di Pretoro, B., La Terza, A., and Luporini, P. (2000) The autocrine mitogenic loop of the ciliate *Euplotes raikovi*: the pheromone membrane-bound forms are the cell binding sites and potential signaling receptors of soluble pheromones. *Molecular and Cellular Biology* **11**, 1445–1455.

Oz, S.H., Huang, H., Wittner, M., Tanowitz, H.B., Bilezikian, J.P., and Morris, S.A. (1994) Evidence for guanosine triphosphate-binding proteins in *Trypanosopma cruzi. American Journal of Tropical Medicine and Hygiene* **50**, 620–631.

Parent, C.A., and Devreotes, P. (1999) A cell's sense of direction. *Science* **284**, 765–770.

Parfenova, E.V., Afon'kin, S.Y., Yudin, A.L., and Etingof, R.N. (1989) Characterization and partial purification of mating pheromone excreted by mating type II cells of the ciliate *Dileptus anser. Acta Protozoologica* **28**, 11–21.

Paveto, C., Torres, H.N., Flawia, M.M., García-Espitia, M., Ortega, A., and Orozco, M. (1999) *Entamoeba histolytica*: signaling through G proteins. *Experimental Parasitology* **91**, 170–175.
Pennington, S.R. (1995) GTP-binding proteins. 1. Heterotrimeric G proteins. *Protein Profile* **2**, 167–315.
Preston, R.R., and Van Houten, J. (1987) Chemoreception in *Paramecium*: Acetate- and folate-induced membrane hyperpolarization. *Journal of Comparative Physiology* **A160**: 525–536.
Raffioni, S., Miceli, C., Vallesi, A., Chowdhury, S.K., Chait, B.T., Luporini, P., and Bradshaw, R.A. (1992) Primary structure of *Euplotes raikovi* pheromones: comparison of five sequences of pheromones with variable mating interactions. *Proceedings of the National Academy of Sciences, USA* **89**, 2071–2075.
Renaud, F.L., Chiesa, R., De Jesús, J., Lopez, A., Miranda, J., and Tomassini, N. (1991) Hormones and signal transduction mechanisms in protozoa. *Comparative Biochemistry and Physiology* **100A**, 41–45.
Renaud, F.L., Colón, I., Lebrón, J., Ortiz, N., Rodríguez, F., and Cadilla, C. (1995) A novel opioid mechanism seems to modulate phagocytosis in *Tetrahymena*. *Journal of Eukaryotic Microbiology* **42**, 205–207.
Rodriguez, E. Identification of a β-endorphin-like opioid in *Tetrahymena thermophila*. M.Sc. Thesis, University of Puerto Rico, Río Piedras Campus, 1999.
Ryals, P.E., Bae, S., and Patterson, C.E. (1999) Evidence for early signaling events in stomatin-induced differentiation of *Tetrahymena vorax*. *Journal of Eukaryotic Microbiology* **46**, 77–83.
Schultz, J.E., Klumpp, S., Benz, R., Schürhoff Goeters, W.J., and Schmid, A. (1992) Regulation of adenylyl cyclase from *Paramecium* by intrinsic potassium conductance. *Science* **255**, 600–603.
Shinozawa, T., Hasimoto, H., Fujita, J., and Nakaoka, Y. (1996) Participation of GTP-binding protein in the photo-transduction of *Paramecium bursaria*. *Cell Structure and Function* **21**, 461–474.
Sonneborn, T.M. (1937) Sex, sex inheritance and sex determination in *Paramecium aurelia*. *Proceedings of the National Academy of Sciences, USA* **23**, 378–385.
Sprang, S.S. (1997) G protein mechanisms: insights from structural analysis. *Annual Review of Biochemistry* **66**, 639–678.
Symington, S.B., de Ondarza, J., Clark, J.M., and Van Houten, J. G-protein modulators alter the swimming behavior and calcium efflux of *Paramecium tetraaurelia* (in preparation).
Taylor, S.S., Knighton, D.R., Zheng, J., Sowadski, J.M., Gibbs, C.S., and Zoller, M.J. (1993) A template for the protein kinase family. *Trends in Biochemical Sciences* **18**, 84–89.
Tomei, E.Z., and Renaud, F.L. (1997) Effect of morphine on Fc-mediated phagocytosis by murine macrophages *in vitro*. *Journal of Neuroimmunology* **74**, 111–116.
Torres-Márquez, M.E., Macías-Silva, M., and Vega Segura, A. (1996) Identification of a functional Gs protein in *Euglena gracilis*. *Comparative Biochemistry and Physiology* **115C**, 233–237.
Ulloa-Aguirre, A., Stanislaus, D., Japovick, J.A., and Conn, P.M. (1999) Structure activity relationships of G protein coupled receptors. *Archives of Medical Research* **30**, 420–435.
Vallesi, A., Giuli, G., Bradshaw, R.A., and Luporini, P. (1995) Autocrine mitogenic activity of pheromones produced by the protozoan ciliate *Euplotes raikovi*. *Nature* **376**, 522–524.
Vallesi, A., Giuli, G., Ghiara, P., Scapigliati, P., and Luporini, P. (1998) Structure-function relationships of pheromones of the ciliate *Euplotes raikovi* with mammalian growth factors: cross reactivity between E*r*-1 and interleukin-2 systems. *Experimental Cell Research* **241**, 253–259.
van der Geer, P., Hunter, T., and Lindberg, R.A. (1994) Receptor protein-tyrosine kinases and their signal transduction pathways. *Annual Review of Cell Biology* **10**, 251–337.
Van Houten, J. (1994) Chemoreception in microorganisms: Trends for neuroscience? *Trends in Neurosciences* **17**, 62–71.
Van Houten J. (1998) Chemosensory transduction in *Paramecium*. *European Journal of Protistology* **34**, 301–307.
Van Houten, J.L., Yang, W.Q., and Bergeron, A. (2000) Chemosensory signal transduction in *Paramecium*. *Journal of Nutrition* **130**, 946S–949S.
Van Houten, J.L., Cote, B.L., Zhang, J., Baez, J., and Gagnon, M.L. (1991) Studies of the cyclic

adenosine monophosphate chemoreceptor of *Paramecium. Journal of Membrane Biology* **119**, 15–24.

Weiss, M.S., Anderson, D.H., Raffioni, S., Bradshaw, R.A., Ortenzi, C., Luporini, P., and Eisenberg, D. (1995) A cooperative model for ligand recognition and cell adhesion: evidence from the molecular packing in the 1.6 Å crystal structure of the pheromone E*r*-1 from the ciliate protozoan *Euplotes raikovi. Proceedings of the National Academy of Sciences, USA* **92**, 10172–10176.

Wilson, J.D., Foster, D.W., Kronenberg, H.M., and Williams, R.H. (1998) *Williams Textbook of Endocrinology*, 9th edn. W.B. Saunders, Philadelphia.

Wolfe, J., Mpoke, S., and Tirone, S.F. (1993) Cilia, ciliary concanavalin A-binding proteins and mating recognition in *Tetrahymena thermophila. Experimental Cell Research* **209**, 342–349.

Yang, W.Q., Braun, C., Plattner, H., Purvee, J., and Van Houten, J.L. (1997) Cyclic nucleotides in glutamate chemosensory signal transduction of *Paramecium. Journal of Cell Science* **110**, 2567–2572.

Yano, J., Fraga, D., Hinrichsen, R., and Van Houten, J.L. (1996) Effects of calmodulin antisense oligonucleotides on chemoresponse in *Paramecium. Chemical Senses* **21**, 55–58.

Zipser, B., O'Neill, J.B., Ruff, M.R., Smith, C.C., Higgins, W.J., and Pert, C.B. (1988) The opiate receptor: a single 110 kDa recognition molecule appears to be conserved in *Tetrahymena*, leech and rat. *Brain Research* **463**, 296–304.

CHAPTER 4

SIGNALLING SYSTEMS IN CNIDARIA

WERNER MÜLLER

Department of Physiology and Developmental Biology, Zoological Institute, University, INF 230, D 69120 Heidelberg

Summary

The freshwater polyp, *Hydra* has long been used as a model for studying mechanisms controlling pattern formation and the axial polarity of the body column. It has been suggested that body patterning is dependent on the presence of internal morphogens. A variety of different candidate molecules, in *Hydra* and other cnidarians, have been proposed as putative morphogens. They are derived from different cellular sources and the evidence for their ability to act as morphogens is critically assessed. The actions of any morphogens will be mediated by signal transduction pathways and such pathways have been demonstrated in cnidarians. The phosphatidylinositol-protein kinase C system appears to be the major pathway involved in pattern formation. Cnidarians are able to respond to external cues which trigger key events in their life cycle, such as metamorphosis. Internal neurohormones, particularly members of the GLWamide family of neuropeptides, transmit the signals to all parts of the body and the signalling cascades involved in these events have been explored. Colonial forms possess the ability to recognise each other as self or non-self. Interactions between neighbouring colonies lead either to fusion or to the destruction of one of the colonies. The phenomenon, which may be linked to the competition for living space, is reminiscent of the allorecognition/allorejection responses observed in other invertebrates and vertebrates. The presence of these mechanisms in such simple organisms as cnidarians suggests that they arose early in evolution. To date, the signalling systems involved in the behaviours in cnidarians have not been studied in any depth.

1. Morphogens, Inducers and Signal Transduction Systems Involved in Pattern Formation

1.1. HYPOTHESES POSTULATING THE EXISTENCE OF MORPHOGENS

When the British mathematician and pioneer of computer-based computations Alan Turing (1952) coined the term morphogen and proposed his speculative "reaction-diffusion systems" of chemical pattern generation, he pointed to *Hydra* as an example, suggesting that the tentacle pattern in this animal might be created by prepatterns of such morphogens. Since those days many papers have been published reporting putative or alleged morphogens in *Hydra* or other Cnidarians.

Hydra indeed displays properties suggestive of the action of morphogens. Any isolated fragment of the body column can regenerate into a miniature

I. Fairweather (ed.), Cell Signalling in Prokaryotes and Lower Metazoa, 91–114.
© 2004 *Kluwer Academic Publishers. Printed in Great Britain.*

version of the animal and even aggregates of dissociated cells will, starting from almost chaotic initial conditions, reconstitute into viable polyps in a process of self-organization. Why could Turing-type reaction-diffusion systems help to create new patterns in processes of self-organization? In Turing's proposal, morphogens are hypothetical substances which diffuse through masses of cells, react with each other and form concentration patterns in space, such as chemical standing waves, *de novo*. The distribution of morphogens in space is thought to control the subsequent events leading to morphogenesis. Intending to develop computable models, Turing reduced living cells to geometrical points at which changes in local morphogen concentrations can be calculated using differential equations. An ultimately stable pattern of concentrations can be achieved if at least two substances, x and y, with different diffusion constants interact with each other. The interaction is based on auto- and cross-catalysis. In a modification of Turing's theory, Gierer and Meinhardt (1972) renamed x and y as activator and inhibitor. By applying the theory to pattern regulation in regenerating hydras, the morphogens "head activator" and "head inhibitor" were created. In extensions of this model, Meinhardt (1993) added a "foot activator" and "foot inhibitor", and a "tentacle activator" and "tentacle inhibitor".

It must be emphasized that the existence of all these morphogens was not deduced from biochemical work, but merely inferred from interpretations of transplantation studies and the apparent success of computer simulations (Webster, 1971; Shostak, 1972; Wolpert, Clarke and Hornbruch, 1972; Wolpert, Hornbruch and Clarke, 1974; MacWilliams and Kafatos, 1974; MacWilliams, 1983a,b). The transplantation studies showed that:

1. any transplant taken from a hydra has the intrinsic potential to organize the development of a head if it is transplanted down the body column into a region the positional value of which is considerably lower than the value of the transplant; and
2. a head inhibits the development of competing heads down the body column.

A re-examination of such transplantation studies showed that many phenomena are based on short- and long-range interactions that are probably not mediated by chemical standing waves. The apparent head inhibition was shown to coincide exactly with long-range foot promotion exerted by the head (Müller, 1995a,b, 1996a). A head assists in the down-regulation of positional value at the other end of the body column and this effect can be attributed to long-range competition for resources, such as precursor cells and hormonal factors that are distributed in the network of interstitial spaces just like hormones in vascular systems (Müller, 1995b, 1996a,b). Processes of head induction might occur when cells equipped with different cell adhesion molecules come into direct contact.

On the other hand, models of pattern formation based on morphogens can be modified to adopt new findings (Berking, 1997) and still have heuristic value. Irrespective of how studies on the biological level can be interpreted, the assumption that diffusable signalling substances exist was, and still is, a plausible hypothesis.

1.2. PUTATIVE AND ALLEGED MORPHOGENS IN *HYDRA*

1.2.1. *"The head activator HA"*

When the Gierer–Meinhardt model was developed, efforts were immediately begun to isolate the proposed morphogens. The attempts appeared to be successful. Four morphogens were presented: a head activator and inhibitor and a foot activator and inhibitor. All four morphogens were claimed to be neurohormones (Schaller, 1979; Schmidt, Grimmelikhuijzen and Schaller, 1980; Schaller and Bodenmüller, 1987).

The most attention has been attracted by a peptide called "head activator", or HA. The peptide has been presented as a neuropeptide, head inducer, morphogen and growth factor (Schaller and Bodenmüller, 1981; Schaller, Hoffmeister and Dubel, 1989). Previous studies had already reported the presence in extracts of *Hydra* of substances promoting head development at the basal end of excised gastric segments (Lentz, 1965; Lesh and Burnett, 1966), or speeding up apical head regeneration and bud development (Schaller and Gierer, 1973). The activities were attributed to neurosecretory granules. An extractable, presumably peptidergic activity causing head development at the basal end of polyps had also been demonstrated in the hydroid, *Hydractinia echinata* (Müller, 1969). Eventually, a peptide with the amino acid sequence pE-P-P-G-G-S-K-V-I-L-F (pGlu-Pro-Pro-Gly-Gly-Ser-Lys-Val-Ile-Leu-Phe) was presented as being the head activator of the traditional theories (Schaller and Bodenmüller, 1981).

Yet, in contrast to its suggestive denomination, the (synthetic) peptide with this sequence did not induce ectopic head formation, nor alter the final body pattern of regenerating hydras (Javois and Tombe, 1991 in *Hydra oligactis*; Müller, unpublished re-examinations in *Hydra vulgaris* and *Hydra magnipapillata*). The synthetic HA peptide speeds up the initial phase in regeneration and budding by shortening the G2 period in which most cycling hydra cells rest, unless a new round of cell division is started by (unknown) mitogenic signals. HA also causes a slight increase in nerve cell density (Hoffmeister and Schaller, 1987; Holstein, Schaller and David, 1986; Hobmayer, E., Holstein and David, 1990). Whether this augmentation of nerve cells is caused by recruitment and commitment of pluripotent stem cells, as assumed by the investigators, or by terminal differentiation of already committed neuroblasts is not definitively settled, as no markers for a reliable discrimination between uncommitted and committed precursors are available.

In cell cultures, terminal differentiation of committed, yet still cycling, cells can often be induced by removal of growth factors or by blocking receptors for growth factors. Therefore, in the interpretation of the cellular effects of HA as well, (non-specific) blockage of receptors should be taken into consideration.

The synthetic HA peptide and several derivatives of it bind to a membrane-associated receptor, the homologous mammalian counterpart of which binds lipoproteins circulating in the blood. The HA-binding receptor of the hydra was also found in a secreted form (Neubauer *et al.*, 1991; Hampe *et al.*, 1999). The cell types expressing the membrane-associated receptor have not been identified in detail.

By comparing HPLC profiles and performing RIAs or ELISAs using anti-HA antibodies, HA-like peptides were found in many extracts and fluids of biological origin. Based on such studies, it has been claimed that HA with identical amino acid sequence is present throughout the animal kingdom, including humans (Bodenmüller and Schaller, 1981; Schaller, Hoffmeister and Dubel, 1989). However, all working groups who tried to identify corresponding genes met with disappointment (personal communications). A gene encoding a polypeptide with the published amino acid sequence has not been found in *Hydra*, nor in any other organism.

1.2.2. *Heady, a Candidate for a Morphogen Involved in Axis Specification*

Head-inducing activities are expected to occur in head regeneration, budding and the formation of a new axis in aggregates. A new approach led to the identification of a novel peptide, the role of which is not congruent with traditional notions about the putative "head activator" (Lohmann and Bosch, 2000).

Using differential display PCR techniques, a particular mRNA has been identified that was transiently present during *de novo* axis formation in aggregates. The structure of the associated cDNA indicated that this mRNA coded for precursor proteins that can be processed to yield amidated peptides, as it displayed a leader sequence and putative sites for amidation. The amino acid sequence of one of the putative, processed peptides was deduced to be FHTMILLDTQSP-NH_2 (Phe-His-Thr-Met-Ile-Leu-Leu-Asp-Thr-Gln-Ser-Pro-amide). It was named HEADY.

Using RT-PCR, it was shown that the *Heady* gene is only weakly expressed in intact polyps, and is transiently up-regulated 6 hours after decapitation. No increase in *Heady* expression could be observed when the foot was removed. *In situ* hybridization rendered *Heady* expression visible exclusively in a small number of endodermal cells (and not in nerve cells!) at the regenerating tip of the body column, with highest levels in the apicalmost cells. No accumulation of *Heady* transcripts could be detected when the foot was removed. In the reg16 mutant, whose potential for head regeneration is strongly reduced (Sugiyama and Fujisawa, 1977), *Heady* was not induced upon decapitation. When the regeneration capacity of reg16 was restored by repeated wounding or treatment with the PKC activator TPA, *Heady* expression was restored to wild-type levels.

In budding (wild type) hydras, *Heady* is initially expressed evenly throughout the endoderm of the evaginating bud. During subsequent development of the bud, expression is progressively confined to the apical end. A few hours after the bud detached from the parental animal, *Heady* transcripts were no longer present in the head of the young polyp.

By means of a polyclonal antiserum, HEADY peptide precursors were detected in endodermal vesicles that were evenly distributed throughout the gastric region. HEADY seemed to be released in early stages of bud formation, since stained vesicles could not be detected in putative bud fields. Similarly, in head regenerating experiments immediately following decapitation, no staining could be detected at the apical end of the body column. It was concluded that HEADY plays a role in the induction but not in the maintenance of the polarity of the body column. Therefore, the putative role of HEADY is not congruent with the role of the "head activator" (HA), as proposed in the model of Gierer and Meinhardt (1972).

To determine whether HEADY is required for head regeneration, *Heady* activity was reduced by dsRNA-mediated interference. This procedure caused a depletion of *Heady* transcripts, and a reduction in the capacity to regenerate a head. Long-term exogenous application of synthetic HEADY caused an acceleration of bud development and an increase in positional value in the gastric region (an effect that is more drastically brought about by PKC activators; see section 1.4.1.).

1.2.3. *Another Candidate for Head-Organizing Activities: WNT*

The WNT signalling pathway is involved in many developmental processes of multicellular animals. *Wnt* genes encode a large family of secreted, cysteine-rich proteins that play key roles as intercellular signalling molecules in many developmental processes. The pathway has been found to be involved in axis specification in the embryos of *Caenorhabditis elegans*, sea urchins, ascidians and vertebrates. In *Xenopus*, the WNT signalling pathway is part of the primary induction system that leads to the establishment of the dorsoventral axis (review: Wodarz and Nusse, 1998).

In *Hydra*, a *Wnt* gene is expressed in the bud field. Eventually, expression is confined to a spot at the tip of a bud in the area of the putative head organizer. Expression continues in a few ectodermal and endodermal cells at the tip of the hypostome throughout the life of the polyp. Since WNT signals are received by transmembrane receptors that are closely linked to an intracellular signalling cascade, more information is given below in section 1.4.3.

1.2.4. *Eicosanoids: Arachidonic Acid and 12-R-HETE*

When *Hydra* is stimulated with diacylglycerol to form ectopic head structures (see section 1.4.1.) it releases arachidonic acid (AA) from membrane-associated lipids. Periodic stimulation of *Hydra magnipapillata* with external AA over 12 or more days can cause ectopic head formation (Müller *et al.*, 1993; Müller, 1995a). The effect has been attributed to the activating effect of AA on protein kinase C (PKC). In addition, AA is the source of a variety of metabolites, which together with AA are collectively called eicosanoids. Some of them support head or bud formation (Leitz *et al.*, 1994a,b): while low doses of 12-S-HETE stimulate budding, 12-R-HETE has a specific effect on the patterning of the tentacles. The number of tentacles within the tentacle whorl is increased, but the longitudinal body pattern is not affected and no additional heads are induced. This observation (Müller *et al.*, 1998) warns against taking the number of tentacles as sole criterion for head activation or head inhibition in bioassays of putative morphogens.

1.2.5. *Putative Head Inhibitor(s)*

In *Hydra*, the existing apical head suppresses head induction by transplants. Similarly, regenerating heads and emerging buds are mutually inhibitory (Berking, 1977; Müller 1995b). In attempts to isolate the putative inhibitory morphogen, a low-molecular-weight component that reversibly inhibited early events in head and bud formation was described (Berking, 1977, 1979). However, the chemical identity of this activity could not be elucidated. One of the effects ascribed to the supposed head inhibitor, namely, arrest of nerve cell development from stem cells, is (also) brought about by an identified hydra peptide, designated Hym-33H (Takahashi *et al.*, 1997).

A living cell contains something like 100,000 different substances. If only 1/1000 of them have a reversible inhibitory influence on developmental processes, we are left with 100 candidates for inhibitory morphogens. It is extremely difficult to unequivocally discriminate between specific and non-specific effects and to measure concentration patterns of low-molecular-weight substances within a living tissue. In spite of these difficulties, candidates for inhibitory morphogens have been presented in the context of periodic patterning in colonial marine hydroids (see below section 1.3.2.).

1.2.6. *Peptides Promoting Foot Formation*

The parallel development of models of pattern formation based on diffusable morphogens and the biochemical search for such substances converged in the belief that (neuro-)peptides are the dominant morphogenetic signalling molecules in *Hydra*. This belief has prompted a "hydra peptide project" (Takahashi *et al.*, 1997). More than 500 peptides have been isolated from *Hydra magnipapillata*, among them Hym-346 and Hym-330. The peptide Hym-346 was shown to lower positional value, thus promoting foot formation (Grens *et al.*, 1999). Independently, Hoffmeister (1996) isolated two similarly acting peptides, called pedin and pedibin, from *Hydra vulgaris*. Pedibin and Hym-346 are nearly identical, whereas pedin is different in size and amino acid composition and is nearly identical with Hym-330. The authors are confident that these peptides are part of the system activating foot formation. However, *in situ* hybridizations showed the presence of transcripts not in nerve cells, but in endodermal epithelial cells. Moreover, pedibin transcripts were found not only in the endoderm of the foot but also of the basal part of the tentacles (Hoffmeister-Ullerich and Herrmann, 1999).

The second peptide, called pedin or Hym-330, is expressed in the endodermal epithelial cells throughout the body column except in the hyopostome and tentacles (Hatta *et al.*, 1999). Previous papers had ascribed the foot-activating function to nerve cells and regarded the substances involved as neurohormones (Grimmelikhuijzen, 1979; Schaller, 1979; Schmidt, Grimmelikhuijzen and Schaller, 1980; Schaller and Bodenmüller, 1987).

1.3. MORPHOGENETIC SUBSTANCES FOUND IN OTHER CNIDARIA

1.3.1. *The Proportion-Altering Factor PAF-E.rac, a Very Potent Signalling Molecule*

From marine colonial hydroids, a low-molecular-weight substance has been extracted that alters the morphology of developing hydroids more drastically and effectively than all other endogenous substances extracted from any Cnidarian species so far. Only the factor SIF (see section 2.2.) has similar, though different, potencies. The main effect of PAF-E.rac is to promote head development and increase the dimensions of the head, or to enlarge the larval precursor region of the head, at the expense of the basal structures (stolons or their larval precursor region). Due to this apicalizing effect, the substance has been named proportion-altering factor PAF. It was recently renamed and specified as PAF-E.rac as it was extracted from *Eudendrium racemosum*. The factor has also been extracted from *Eirene viridula* and *Hydractinia echinata*. The starting material for chromatographic purification was 10 kg of *Eudendrium racemosum*. The factor was not found in *Hydra* and has little influence on body proportions in *Hydra* (Plickert, 1987).

Most bioassays were performed with developmental stages of the marine colonial hydroid *Hydractinia*. The development of this colonial hydroid proceeds through a sexual life cycle. Fertilised eggs develop into free-living, spindle-shaped planula larvae, which settle on a substrate and metamorphose into hydra-like primary polyps, the founders of new colonies. At their base, the primary polyps sprout tubular runners, called stolons, on which new polyps arise. In the development of *Hydractinia*, exogenous application of PAF-E.rac has the following effects (Plickert, 1989; Kroiher and Plickert, 1992):

- In developing larvae, the tail region is lengthened, and/or multiple tails develop. Lengthening of the tail is due to enhanced proliferation. The tail of the larva is the precursor of the head in the polyp.
- When the anterior part of the larvae was removed, in the presence of PAF the posterior fragment frequently regenerated a second, mirror-image tail instead of the lost anterior part.
- When PAF was applied during the metamorphosis, the primary polyps

emerged with an enlarged head, while the mass of the stolons was reduced or the development of stolons completely prevented.

- In the enlarged head region, the tentacle whorl was shifted to a lower position and the number of nerve cells containing neuropeptides of the RF-amide family was strongly increased (Plickert, 1989). An increase in the number of new nerve cells differentiating from cycling precursor cells was also found in PAF-treated *Hydra vulgaris* (Plickert, 1991).
- In the body column, the expression compartments of several genes were enlarged (LW-amides and RF-amides expressing nerve cells) and/or shifted to more basal regions (Gajewski and Plickert, 1999).

Striated muscle cells isolated from the medusa of *Podocoryne carnea* ordinarily transdifferentiate partly into smooth muscle cells and partly into RF-amide-positive nerve cells; however, in the presence of PAF they all entered the pathway towards neuronal development (Schmid and Plickert, 1990).

At present the chemical nature and the cellular source of PAF-E.rac have not yet been identified.

1.3.2. *The Stolon-Inducing Factor SIF*

Like PAF, SIF is produced by marine colonial hydrozoa. In particular, it was detected in *Hydractinia*, where it plays a role in the post-metamorphic development of the stolonal net which interconnects the individual members of the colony (Lange and Müller, 1991). The development of this network shows striking parallels to the development of the network of capillaries (angiogenesis) in vertebrates. Stolon tips are motile and serve as pathfinders when the stolons elongate and branch. The tips emit signals which attract other tips. If two tips make physical contact, they fuse to form an anastomosis. When the advancing tip of a growing stolon approaches the flank of another stolon, the signals emitted by the tip induce the formation of a lateral stolon tip in the flank of this stolon. This lateral tip will fuse with the approaching inducing tip, or it will give rise to a branch.

From the supernatant of several hundred thousands of primary polyps, a factor was isolated that has such a stolon-inducing effect if locally applied to the flank of stolons with a micropipette.

When entire metamorphosing primary polyps were bathed in SIF solutions, the SIF exerted effects opposite to those brought about by PAF. The positional level from which stolons sprout was shifted from the base to more apical positions. In the presence of high SIF concentrations, the animals completely transformed into huge stolons (Lange and Müller, 1991). The factor has been identified as a glycoprotein containing N-acetylglucosamine moieties (Müller, unpublished observations). The amino acid sequence has not yet been determined.

1.3.3. *Methyl Donors and Other Putative Inhibitory Morphogens Involved in Spacing of Periodic Patterns*

The putative morphogens which suppress bud formation in *Hydra* have also been implicated in the spatial distribution of buds along the body column. A new bud arises from the preceding bud at an angle of 180 degrees. By exerting lateral inhibition, an existing bud prevents the establishment of a new bud field in its close vicinity.

However, spacing of buds and control of periodic patterns in general have been better studied in colonial hydroids. Frequently polyps (also called hydranths) appear at regular distances along tubular stolons or hydrocauli (upright stolons). The stolons themselves may branch at regular intervals. Experimental studies revealed that branching patterns are also controlled by inhibitory fields which emanate from existing structures.

Spacing of polyps: In *Eirene viridula* and *Hydractinia echinata*, a certain minimal distance along the stolon is always maintained between existing polyps and new buds. The reason is that existing polyps are sources of inhibition, the intensity of which decreases with distance. Only when this lateral inhibition has fallen below a critical threshold can a new polyp arise (Plickert, Heringer and Hiller, 1987, in *Eirene viridula*). Candidate molecules which might convey such lateral inhibition are low-molecular-weight donors of methyl groups, in particular N-methylpicolinic acid (homarine), N-methylnicotinic acid (trigonelline) and N-trimethylglycine (betaine) (Berking, 1986b, 1987).

Spacing of stolon branches: Inhibitory morphogens have also been implicated in the control of branching patterns in the stolons (Müller and Plickert, 1982). Growing stolon tips suppress the formation of secondary tips, and hence of branches, in their vicinity. Although activities suppressing branch formation accumulate in conditioned medium (Müller, Hauch and Plickert, 1987), the causative substances have not yet been identified.

1.3.4. *A Role for Retinoic Acid?*

Retinoic acid (RA) and related derivatives of vitamin A are abundantly used to specify morphogenetic fields and to regionalize cell differentiation in vertebrates. Only very few reports point to a role for retinoids in controlling patterning in Cnidarians as well. In *Hydractinia*, retinoids influence pattern specification in a way that suggests interference with the generation and transmission of signals responsible for the dimensions and spacing of structures. Upon exposure to RA, metamorphosing primary polyps develop more densely distributed tentacles around the hypostome, increase the length of the stolons and the distances between stolon branches, but bud secondary hydranths at shorter temporal intervals and spatial distances (Müller, 1984). The presence of nuclear hormone receptors with remarkable homology to the vertebrate retinoic acid receptors RXR has been reported from the cubozoan, *Tripedalia cystophora* (Kostrouch *et al.*, 1998).

1.4. SIGNAL TRANSDUCTION SYSTEMS INVOLVED IN PATTERN FORMATION

1.4.1. *The Phosphatidylinositol – PKC System*

The stimulating action of growth factors and most hormonal signal molecules is mediated by signal transduction systems which convert the external signal into several intracellular signals, thereby amplifying the signal and directing the message into the various compartments of the cell. As many supposed signal molecules, in particular the often proposed factor(s) activating head formation, were unknown, attempts were made to imitate the reception of such external signals by directly stimulating key elements of signal transduction systems (Müller, 1985, 1989). A signal transduction system often involved in the control of developmental processes is the PI-PKC system (phosphatidyl-inositol – protein kinase C system).

The PI system operates with the following elements. Upon reception of an external signal the membrane-associated lipid PIP_2 is cleaved. Products of this cleavage are the hydrophilic second messenger, inositol trisphosphate (Ins $(1,4,5)P_3$), which frequently is transformed into several additional inositol phosphates, and lipids of the diacylglycerol class. The diacylglycerols (DAGs) activate protein kinases, in particular, kinases of the PKC family. Their activity initiates cascades of secondary events which result in cell proliferation or cell differentiation.

PKC-dependent intracellular pathways can artificially be activated by applying activators of PKC, for instance the tumor-promoting phorbol ester TPA or the synthetic DAG dioctanoylglycerol (diC8). These are compounds to which the cell is permeable and which can substitute for natural, DAG-type second messengers.

These PKC-activating agents induce multiple head formation in several hydroids. Gastric segments excised from polyps of *Hydractinia echinata* not only regenerate a head at their apical end, but often form a second head at their basal end, and sometimes additional heads in the middle of the body column (Müller, 1985). With pulses of TPA or diC8, given on successive days over a period of one to two weeks, *Hydra magnipapillata*, strain wt 105, can be caused to form ectopic heads in its gastric region (Müller, 1989, 1990). Ectopic head formation is preceded by an increase in positional value (Müller, 1990) and the appearance of head-specific nerve cells in the gastric region (Müller, 1991). Arachidonic acid, a known coactivator of PKC, strongly enhanced the effectiveness of diC8, whereas inhibitors of PKC such as chelerythrine caused buds to detach with heads reduced in size or absent (Müller *et al.*, 1993).

Whether PKC is present in *Hydra* was first checked at the level of transcripts. From cDNA clones prepared from *Hydra vulgaris*, the existence of at least three different PKC messages was deduced. Two transcripts encode members of the calcium-dependent cPKC family, named HvPKC1a and HvPKC1b. The mRNA of these PKC1a is found in all parts of the body except the foot. The message was detected in ectodermal and endodermal epithelial cells and in interstitial stem cells. The expression of HvPKC1a is up-regulated during head differentiation (Hassel, 1998). Polyclonal antibodies to expressed proteins assigned HvPKC1b to ganglionic nerve cells in the head region (Kleinholz, personal communication). A third PKC is a member of a calcium-independent nPKC family. Increased expression of this PKC is an early event during head and foot formation, with the rise being restricted to the endodermal cells underlying the regenerating ends (Hassel *et al.*, 1998). At present, it is unclear which of the PKCs contribute to the development of DAG-induced ectopic head structures.

In *Hydra vulgaris* pulses of DAG do not evoke ectopic heads but cause an enlargement of the existing head (Müller, 1995a). In this species pattern formation is more readily affected by lithium ions. However, the results presented to date cannot easily be interpreted. The stepwise lowering of the concentration from 4 to 1 mM LiCl evoked the development of ectopic head structures, whereas long-term treatment suppressed head and bud formation and caused the differentiation of ectopic feet. Ectopic head formation was preceded by an increase in the level of inositol phosphates, indicating an activation of the PI cycle (Hassel and Bieller, 1996).

Lithium ions are known to interfere with several steps of the PI cycle (for references see Hassel and Bieller, 1996). Lithium ions also interfere with WNT-type signal transduction cascades, and elements of this cascade have been identified in *Hydra* (see section 1.4.3.).

1.4.2. *The cAMP System*

Transcription factors of the CREB/CREM class and the corresponding cAMP-responsive elements (CRE) in the promoter region of cAMP-dependent genes are evolutionarily conserved and are present in hydra as well. Their biological function is subject to conflicting speculations. Both the cellular effects of the (synthetic) "head activator" HA and the promoting effect of pedin and pedibin on foot formation have been ascribed to activation of the cAMP-PKA pathway (review: Galliot, 1996).

1.4.3. *The WNT – β-Catenin – GSK3 System*

The WNT signalling pathway is involved in many developmental processes in eumetazoa ranging from the Cnidaria to the mammals. Important elements of this pathway are extracellular signals such as WINGLESS, the intracellular enzyme Glycogen-synthase kinase (GSK3), the regulator of GSK3 activity Dishevelled (Dsh), β-Catenin, and the transcription Factor TCF/LEF. β-Catenin is known as a building block of the cytoskeleton but also serves as a transcriptional cofactor in vertebrates. In this function, β-Catenin cooperates with the transcription factor, TCF which contains a DNA-binding HMG domain. With several molecular cloning techniques, homologous genes have been identified in hydra (Hobmayer, B. *et al.*, 2000).

Wnt is expressed in a few rows of cells at the apical tip of the head in intact polyps. *In situ* hybridization shows *Wnt* expression first in the protruding bud when a new body axis perpendicular to the parental body axis emerges. Later in development, when the bud forms, the head *Wnt* transcripts are found in a ring surrounding the mouth, and *Wnt* is eventually confined to a few ectodermal and entodermal cells at the very tip of the body axis in the lips of the mouth. The transcription domain of the transcription factor, TCF is broader, comprising the entire hypostome and exhibiting a gradient-shaped distribution down the body column.

Prior to *Wnt*, the genes for β-Catenin and TCF display enhanced expression in the bud field. Also, other genes are already being expressed in the bud field before evagination occurs. For instance, the gene *budhead*, a member of the *fork head/HNF-3* class of transcription factors (Martinez *et al.*, 1997), appears to be expressed earlier than *Wnt*. Therefore, the WNT signalling pathway may have a role in the actual axis formation rather than in the early specification and demarcation of the bud field. However, functional assays remain to be performed and the antibodies available at present did not show the expected translocation of β-Catenin from the cytoplasm into the nucleus of cells.

2. Signalling Molecules Involved in the Control of the Life Cycle

2.1. SYNCHRONISATION OF SPAWNING AND GAMONES

In hydrozoans, mature eggs are either sheltered in containers, often called gonangia, or they are released into the sea. The spawning of *Hydractinia* is triggered and synchronized by light (Ballard, 1942; Yoshida, 1959). Whether eggs are released into the sea or kept in a maternal gonangium, for sperm cells the probability of finding mature eggs by pure chance is extremely low. To increase the chance of encounters, eggs emit sperm-attracting gamones. The molecules responsible are polar, heat stable and have a molecular weight below 700 (Miller, 1972, 1980; Miller and O'Rand, 1975). Unfortunately, these pioneering studies have not been continued in more recent days.

2.2. RELEASE OF METAMORPHOSIS

2.2.1. *External Triggering Cues*

In most coelenterates, embryogenesis terminates in a planula, a ciliated, cone-shaped, mouthless larva. In the colonial hydroid, *Hydractinia echinata* which became the model organism to study cnidarian embryogenesis and metamorphosis, the mature planula does not undergo metamorphosis spontaneously. Instead, cell proliferation, differentiation, and morphogenesis cease in the larva when it matures and acquires the ability to metamorphose (Plickert, Kroiher and Munck, 1988). The mature larva uses its sensory cells at the anterior end to prospect for a habitat suited to its future sessile phase of life.

The metamorphosis of cnidarian larvae into sessile polyps is naturally triggered by external cues and not by a decrease or increase in the concentration of internal hormones, as in insects or vertebrates (reviews: Burke, 1983; Leitz, 1997, 1998a). Therefore, metamorphosis is dependent on the presence of an environmental key stimulus. For example, in the hydrozoan *Coryne uchidai*, the alga *Sargassum tortile* is the source of a chemical stimulus that causes the larvae to settle and to metamorphose on this alga. The active component was identified as α-tocotrienol (Kato *et al.*, 1975).

In *Hydractinia*, the inducing stimulus derives from bacteria of the genus *Alteromonas*. These bacteria cover the substratum on which the larvae settle (Müller, 1973; Leitz and Wagner, 1993). The larvae must make direct contact with the bacteria, which do not release an inducing substance. However, an inducing agent can be removed from the bacteria by applying an osmotic shock. The chemical nature of the agent has not yet been identified. In the laboratory, the bacterial agent can be replaced by several depolarizing agents (reviews: Leitz 1997, 1998a).

2.2.2. *Neurohormones as Internal Trigger Signals*

Posterior fragments of larvae of *Hydractinia* are unable to respond to the bacterial inducer, and depolarizing agents likewise act only on anterior fragments. However, if grafted onto stimulated anterior fragments, posterior fragments participate in metamorphosis (Schwoerer-Böhning, Kroiher and Müller, 1990). These observations led to the hypothesis that the anterior region is equipped with sensory cells capable of sensing the presence of the bacteria-derived inducer. Upon reception of this external signal, the anterior region would produce an internal hormonal factor that is transmitted to the posterior region, stimulating it to participate in metamorphosis.

In a pioneering attempt to identify the internal signal, Leitz, Morand and Mann (1994) first isolated a peptide, pEQPGLW-NH_2, capable of inducing metamorphosis of posterior fragments of *Hydractinia* planulae, from an abundantly available heterologous source, the sea anemone, *Anthopleura elegantissima*. The peptide, called metamorphosin A, turned out to be the prototype of a new family of neuropeptides, the GWL-amide-(Gly-Leu-Trp-NH_2) family. In the sea anemone, a preprohormone contains numerous copies of this neuropeptide (Leviev and Grimmelikhuijzen, 1995). However, with molecular cloning techniques, no identical sequence was found in *Hydractinia*. Instead, a cDNA was derived from *Hydractinia* larvae that coded for a precursor protein containing repeats of two different GLW-amides, pERPPGLW-NH_2 and KPPGLW-NH_2 (Gajewski *et al.*, 1996). All these different GLW-amides, those of the heterologous and those of the autologous source, were able to trigger metamorphosis in posterior fragments, as well as in complete *Hydractinia* larvae. Consensus sequence for induction of metamorphosis is the family attribute, the sequence Gly-Leu-Trp-NH_2 (Leitz 1997, 1998b).

In other cnidarian larvae, other neurotransmitters or neurohormonal signal molecules appear to be used as internal trigger signals. In *Phialidium gregarium*, metamorphosis can be induced with serotonin (McCauley, 1997), in *Halocordyle disticha* with catecholamines (Edwards *et al.*, 1987).

2.2.3. *Signal Transducing Systems and Downstream Events in the Initiation of Metamorphosis*

As outlined above (section 2.2.1.), external metamorphosis-triggering signals are received by the larvae of *Hydractinia* by means of neurosensory cells. Apparently, these cells use the PI-PKC system to transduce the message into the interior and thus prepare for the exocytosis of neuropeptides, because metamorphosis can artificially be triggered by activators of PKC such as tumour-promoting phorbol esters or dioctanoyl-glycerol (Müller, 1985; Leitz and Müller, 1987; Leitz and Klingmann, 1990). Moreover, antibodies to PKC stain neurosensory cells in the front region of the larva (Leitz, 1993). However, at present it is not clear whether only neurosensory cells make use of the PI-PKC system or also target cells which subsequently respond to the released hormonal neuropeptides. Shortly after induction of metamorphosis, the concentrations of inositol triphosphate and diacylglycerol rise in the larvae (review: Leitz, 1997). Considerations regarding the quantity of metabolites and the kinetics of their appearance make it likely that target cells also use the PI cycle to transduce the message across the cell membrane. In addition, in the translucent larvae of *Mitrocomella polydiademata*, calcium transients have been seen to travel through the epithelial layers upon induction of metamorphosis. The calcium waves were naturally made visible by endogenous photoproteins; the travelling calcium transients were proposed to activate PKC (Freeman and Ridgway, 1987, 1990).

Several downstream events have been traced: a protein kinase of the src-type is phosphorylated (Barnekow and Müller, 1986), lysophosphatidylcholine, arachidonic acid and several lipoxygenase products of arachidonic acid are generated (Leitz *et al.*, 1994a), and taurine, an endogenous, abundantly stored substance which stabilizes the larval state is released into the surrounding medium (Berking, 1988). Other internal low-molecular-weight metamorphosis-inhibiting compounds, such as homarine, trigonelline and betaine are probably used as donors of methyl groups in transmethylation processes. By transferring their methyl group to acceptor substrates, these compounds also lose their metamorphosis-inhibiting potential (Berking, 1986a,b, 1998). A multitude of further events, including resumption of gene activities in the previously quiescent larval animal, remain to be explored.

3. Recognition and Rejection of Allogeneic Competitors for Living Space

The ability to discriminate between self and non-self and to reject and destroy foreign tissue is not restricted to the vertebrate immune system, but is also known from a range of invertebrates, including the Cnidaria. A particular expression of this system is intolerance of a given genotype to transplanted allogeneic tissue. The biological significance of this phenomenon in vertebrates is enigmatic, as transplantations do not happen in nature. Studies with invertebrates, in particular the protochordate (tunicate) *Botryllus schlosseri* and the hydrozoan *Hydractinia*, have provided the first hints as to why such abilities may have evolved (reviews: Grosberg, 1988; Mokady, 1996; Rinkevich, 1996; Rinkevich and Weissmann, 1987).

3.1. ALLORECOGNITION AND ALLOREJECTION: THE PHENOMENON

3.1.1. *The Phenomenon Exemplified with Hydractinia as Model System*

In *Hydractinia*, each primary polyp, the founder of a new colony, derives from one fertilised egg. Therefore, different primary polyps are allogeneic (heterogeneic): their genomes contain different allelic combinations. Primary polyps give rise to new colonies by sprouting stolons, which bud new polyps. Often a substrate, for instance, the shell of a hermit crab, is colonised by several larvae (Yund, Cunningham and Buss, 1987). Therefore, frequently several different colonies develop and expand on a spatially restricted substrate. If the advancing tip of a growing stolon encounters a stolon of a neighbouring colony, two different outcomes can be observed (Buss, McFadden and Keene, 1984):

1. in rare cases the two stolons fuse in spite of their allogeneic origin. Step by step, the two neighbouring colonies grow together to form a uniform chimerical colony with a common gastrovascular system. The ability to fuse and to establish a common colony is strictly correlated with enduring histocompatibility observed upon surgical transplantation of tissue. Compatibility is rather frequent among siblings and close kin, but infrequent in encounters of remotely related specimens (Hauenschild, 1954, 1956). Genetic analyses revealed that fusibility is controlled by a single-locus gene system with codominantly expressed alleles. The alleles present in a population are highly polymorphic. Fusion is possible if the two encountering colonies share one allele on this histocompatibility locus (Mokady and Buss, 1996; Mokady, 1996);
2. if specimens that do not share a common allele come into contact with each other, stolons do not fuse and grafted tissue is rejected. In the stolonal net, a system of defence is activated: each colony tries to kill its neighbour. *Hydractinia* grows a particular type of nematocytes, large microbasic mastigophores, which are only found in the stolonal compartment of the colony. These nematocytes patrol along the stolons and accumulate at sites where two stolons come in contact with each other. Here, the nematocytes integrate into the ectodermal epithelium, focus their cnidocils toward the same point in the opposing stolon, and suddenly discharge almost synchronously, injecting toxins into the neighbour (Buss, McFadden and Keene, 1984; Lange, Plickert and Müller, 1989; Lange, Dick and Müller, 1992). One single attack results in only a local lesion, but not the death of the neighbour. Continued contact between the two colonies causes the formation of many hyperblastic stolon branches filled with increasing numbers of newly generated nematocytes (Lange, Plickert and Müller, 1989; Buss and Grosberg, 1990a,b).

Repetition of mutual aggressive attacks at many contact sites eventually results in the destruction of one of the two colonies. *Hydractinia* attacks not only conspecific colonies but also other species (Lange, Dick and Müller, 1992). In interspecific encounters, *Hydractinia* is not always the winner.

3.1.2. *Putative Signalling Systems Involved*

The aggressive behaviour of *Hydractinia* against allogeneic and xenogeneic competitors for substrate space is based on signal molecules at three levels:

1. a growing stolon that comes into contact with another stolon must recognize whether the encountered tissue is isogeneic (belonging to its own colony) or allogeneic/xenogeneic (belonging to a foreign colony). In conspecific encounters, it is likely that the epithelial cells expose molecules encoded by the histocompatibility locus which allow mutual recognition. These molecules may be related to homophilic cell adhesion molecules. However, in interspecific encounters *Hydractinia* brings its nematocytes into position before the advancing stolon tip actually has established physical contact with its neighbour (Lange, Dick and Müller, 1992). This suggests that some of the recognition molecules are released into the surroundings;
2. nematocytes migrate to contact sites. This suggests that upon contact with foreign tissue, the epithelial cells of the stolon emit attracting signals;
3. nematocytes discharge almost synchronously. This suggests mutual communication.

None of these presumed signal systems has yet been identified in terms of the molecules involved.

The use of nematocysts to destroy allogeneic neighbours is probably not an ancestral feature in Cnidaria, but was subsequently linked to histoincompatibility. When grafted together, not only heterogeneic stolons but also heterogeneic polyps of *Hydractinia* are capable of recognizing and rejecting one another, yet no nematocytes were seen accumulating at the contact sites in polyps. Other species such as *Eirene viridula* are able to kill neighbours, for instance, colonies of *Hydractinia*, without using nematocysts. Apparently, the stolons release toxins from other sources (Lange, Dick and Müller, 1992).

3.2. THE BIOLOGICAL SIGNIFICANCE OF ALLORECOGNITION AND ALLOREJECTION

Two complementary theories have been put forward with respect to the evolutionary pressure which led to the evolution of systems of allorecognition and allorejection:

1. The first theory considers the danger of invading foreign cells. Upon tissue contact and fusion, colonial invertebrates may be prone to germ cell parasitism. In chimeras, migratory germ cell precursors can pass from one partner into the other. Invading foreign germ cell precursors would enhance the

probability of the donor organism reproducing itself at the expense of the recipient (Müller, 1964; Buss, 1982; Rinkevich and Weismann, 1987; Grosberg, 1988; Rinkevich, 1996). The inability to fuse with foreign tissue prevents invasion of parasitic primordial germ cells. This hypothesis plausibly explains why the ability evolved to recognize allogeneic tissue and to prevent fusion with it. The observation that fusion with close kin may occur supports this theory. The degree of genetic similarity among siblings is high and primordial germ cells invading from siblings reduce the fitness of the infected recipient less than invasions from distantly related conspecifics. However, this hypothesis alone does not explain why the neighbour is eventually killed and foreign species are also attacked.

2. The second theory suggests an ecological role for allorejection. Sedentary organisms compete for living space. This competition favours mechanisms to expel or kill competitors. The use of such aggressive reactions necessitates precise identification of foreign tissues, because in colonial organisms the extending parts of the colony may encounter not only foreign tissue but also self tissue. The observation that close kin frequently is not expelled is not in conflict with either of the two theories proposed, as siblings share many alleles. The presence of siblings and even the formation of a chimerical colony can be tolerated without losing much of the original colony's fitness.

3.3. ALLORECOGNITION AND REJECTION IN OTHER CNIDARIA

Probably all sessile cnidarians engage in intra- and inter-specific competition for limited substrate space using a variety of aggressive structures and behaviours (reviews: Grosberg, 1988; Leddy and Greene, 1990). Allogeneic interactions have been described in the hydrocoral *Millepora dichotoma* (Frank and Rinkevich, 1994), several reef-building hexa- and octocorals (Frank *et al.*, 1995, 1997; Rinkevich *et al.*, 1994) and in the soft coral *Pareythropodium* (Frank, Bak and Rinkovich, 1996). In sedentary reef-building cnidarians, two responses to allogeneic contacts were described: retreat growth until a bare area a few mm wide separated the allogeneic neighbours, and unilateral overgrowth which results in dying of the overgrown colony.

The aggressive behaviour of sea anemones displays features which call to mind the behaviour of *Hydractinia*, and is exemplified here by the behaviour of *Anthopleura elegantissima*. Like other anemones, *Anthopleura* lives in clonal colonies and possesses a cellular recognition system of remarkable specificity. Members of its own clone are not attacked, but allogeneic individuals of the same species are, as well as members of other species. Attack is initiated by physical contact and involves the inflation of special tentacle-like structures known as acrorhagi, which contain numerous nematocytes. These stinging cells only discharge when the tip of the acrorhagus is in physical contact with a foreign anthozoan. Contact with isogeneic individuals, that is, with members of an individual's own clone, or organisms other than anthozoans, or with inanimate objects does not elicit discharge (Lubbock, 1980). Recognition of allogeneic tissue, which presumably is mediated by cell surface markers and receptors, is accompanied by particular local electrical activities. The latter may trigger the discharge of the stinging cell (Lubbock and Shelton, 1981). Injured neighbours respond by attacking in turn, or are caused to move away to a more distant location.

4. Bioluminescence

4.1. LIGHT-EMITTING SUBSTANCES

Many marine cnidarians are capable of emitting light. Analysis of substances involved in light-emitting reactions among bioluminescent coelenterates has revealed a remarkable uniformity in the structural features of initial reactants, the "luciferins", modifying enzymes, light-emitter products and light-modifying photo-proteins. The initial substrate (luciferin) is an imidazolopyrazine, called coelenterazine. Oxidation by oxygen, catalyzed by enzymes collectively called luciferases, sets free energy which is emitted in the form of photons. The photon-emitting product is structurally identical among different classes of coelenterates. It has been identified, for instance, in Hydrozoa, in particular in the medusa of *Aequorea victoria*, in Anthozoa, namely, in the sea cactus, *Cavernularia*, the sea pansy, *Renilla*, and the sea pen, *Leioptilus*, and is very likely present also in the Scyphozoa, for example, the jellyfish, *Pelagia* (Shimomura and Johnson, 1975; Hastings, 1996).

In *Aequorea*, which became the prototype of the research in this field, the primary photon-emitting luciferin-luciferase system is associated with a photo-protein which alters the quality of the light. The entire complex became known as aequorin. The complex is activated by free calcium transients. The photo-protein shifts the wavelength of the emitted light from UV into the range of blue. In several medusae, an additional green fluorescent protein, GFP, shifts the wavelength still further, into the green range. GFP is unusual as it does not contain, or cooperate with, a separate chromophore moiety. [Instead] GFP forms a chromophore itself by a post-translational modification of three of its amino acids. The sequence Ser-Tyr-Gly forms a ring-like chromophore by itself. (GFP has become a valuable tool in cell and molecular biology: see review by Stearns, 1995).

4.2. ON THE BIOLOGICAL SIGNIFICANCE OF BIOLUMINESCENCE

In the absence of well-founded knowledge, speculations on the biological function of bioluminescence are all that can be offered. As coelenterates do not permanently emit light, but do so only at night and on certain occasions, a signalling role is presumed. Sudden light flashes might disturb predators. This interpretation is suggested because light-emission is triggered by mechanical stimuli. Light flashes might also be used to synchronize common activities such as spawning or swimming manoeuvres.

5. Concluding and Supplementing Remarks

Cnidarians display a larger array of signalling systems than their apparently simple morphological organization suggests. In addition to the systems reviewed here, Cnidarians possess two systems of electrical signal transmission: epithelial conduction through gap junctions, similar to signal transmission in the heart of vertebrates (e.g. King and Spencer, 1979), and nervous transmission involving the use of peptidergic and other transmitters (see reviews by Anderson and Schwab, 1982; Grimmelikhuijzen and Westfall, 1995; Westfall, 1996). Neurotransmitters are beyond the scope of this review. Moreover, a multitude of hormonal signal molecules is likely to be propagated through the network of interstitial spaces and the gastrovascular systems. These signalling systems still await exploration.

References

Anderson, P.A., and Schwab, W.E. (1982) Recent advances and model systems in coelenterate neurobiology. *Progress in Neurobiology* **19**, 213–236.

Ballard, W.W. (1942) The mechanism for synchronous spawning in *Hydractinia* and *Pennaria. Biological Bulletin* **82**, 329–339.

Barnekow, A., and Müller, W.A. (1986) An *src*-related tyrosine kinase activity in the hydroid, *Hydractinia echinata. Differentiation* **33**, 29–33.

Berking, S. (1977) Analysis of early stages of budding in *Hydra* by means of an endogenous inhibitor. *Wilhelm Roux's Archives of Developmental Biology* **182**, 117–129.

Berking, S. (1979) Analysis of head and foot formation in *Hydra* by means of an endogenous inhibitor. *Wilhelm Roux's Archives of Developmental Biology* **186**, 189–210.

Berking,, S. (1986a) Transmethylation and control of pattern formation in Hydrozoa. *Differentiation* **32**, 10–16.

Berking, S. (1986b) Is homarine a morphogen in the marine hydroid *Hydractinia*? *Roux's Archives of Developmental Biology* **195**, 33–38.

Berking, S. (1987) Homarine (N-methylpicolinic acid) and trigonelline (N-methylnicotinic acid) appear to be involved in pattern control in a marine hydroid. *Development* **99**, 211–220.

Berking, S. (1988) Taurine found to stabilize the larval state is released upon induction of metamorphosis in the hydrozoan *Hydractinia. Roux's Archives of Developmental Biology* **197**, 321–327.

Berking, S. (1997) Pattern formation in Hydrozoa. *Naturwissenschaften* **84**, 381–388.

Berking, S. (1998) Hydrozoa metamorphosis and pattern formation. *Current Topics in Developmental Biology* **38**, 81–131.

Bodenmüller, H., and Schaller, H.C. (1981) Conserved amino acid sequence of a neuropeptide, the head activator, from coelenterates to humans. *Nature* **293**, 579–580.

Burke, R.D. (1983) The induction of metamorphosis in marine invertebrate larvae. *Canadian Journal of Zoology* **61**, 1701–1719.

Buss, L.W. (1982) Somatic cell parasitism and the evolution of somatic tissue compatibility. *Proceedings of the National Academy of Sciences, USA* **79**, 5337–5341.

Buss, L.W., and Grosberg, R.K. (1990a) Morphogenetic basis for phenotypic differences in hydroid competitive behaviour. *Nature* **343**, 60–63.

Buss, L.W., and Grosberg, R.K. (1990b) Morphological variation, morphogenetic potential, and the regulation of hydroid competitive ability. *Nature* **343**, 63–66.

Buss, L.W., McFadden, C., and Keene, D.R. (1984) Biology of hydractiniid hydroids. 2. Histocompatibility effector systems/competitive mechanisms mediated by nematocyst discharge. *Biological Bulletin* **167**, 139–158.

Edwards, N.C., Thomas, M.B., Long, B.A., and Amyotte, S.J. (1987) Catecholamines induce metamorphosis in the hydrozoan *Halochordyle disticha* but not in *Hydractinia echinata. Roux's Archives of Developmental Biology* **196**, 381–384.

Frank, U., and Rinkevich, B. (1994) Nontransitive patterns of historecognition phenomena in the Red Sea hydrocoral *Millepora dichotoma. Marine Biology* **118**, 723–729.

Frank, U., Bak, R.P.M., and Rinkevich, B. (1996) Allorecognition responses in the soft coral *Parerthropodium fulvum fulvum* from the Red Sea. *Journal of Experimental Marine Biology and Ecology* **197**, 191–201.

Frank, U., Brickner, I., Rinkevich, B., Loya, Y., Bak, R.P.M., Achituv, Y., and Ilan, M. (1995) Allogeneic and xenogeneic interactions in reef-building corals may induce tissue growth without calcification. *Marine Ecology – Progress Series* **124**, 181–188.

Frank, U., Oren, U., Loya, Y., and Rinkevich, B. (1997) Alloimmune maturation in the coral *Stylophora pistillata* is achieved through three distinctive stages, 4 months post-metamorphosis. *Proceedings of the Royal Society of London Series B – Biological Sciences* **264**, 99–104.

Freeman, G., and Ridgway, E.B. (1987) Endogenous photoproteins, calcium channels and calcium transients during metamorphosis in hydrozoans. *Roux's Archives of Developmental Biology* **196**, 30–50.

Freeman G., and Ridgway, E.B. (1990) Cellular and intracellular pathways mediating the metamorphic stimulus in hydrozoan planulae. *Roux's Archives of Developmental Biology* **199**, 63–79.

Gajewski, M., and Plickert, G. (1999) Pattern formation in the hydroid *Hydractinia echinata* as modulated by a putative morphogen – differential display of spatial expression patterns of selected genes. *13th Scientific Meeting of the Gesellschaft fur Entwicklungsbiologie*, abstract.

Gajewski, M., Leitz, T., Schlosser, J., and Plickert, G. (1996) Lwamides from Cnidarians constitute a novel family of neuropeptides with morphogenetic activity. *Roux's Archives of Developmental Biology* **205**, 232–242.

Galliot, B. (1996) Signaling molecules in regenerating hydra. *Bioessays* **19**, 37–46.

Gierer, A., and Meinhardt, H. (1972) A theory of biological pattern formation. *Kybernetik* **12**, 30–39.

Grens, A., Shimizu, H., Hoffmeister, S.A.H., Bode, H.R., and Fujisawa, T. (1999) The novel signal peptides, pedibin and Hym-346, lower positional value thereby enhancing foot formation in hydra. *Development* **126**, 517–524.

Grimmelikhuijzen, C.J.P. (1979) Properties of the foot activator from hydra. *Cell Differentiation* **8**, 267–273.

Grimmelikhuijzen, C.J., and Westfall, J.A. (1995) The nervous systems of Cnidarians. In: O. Breidbach and W. Kutsch (eds.) *The Nervous Systems of Invertebrates: an Evolutionary and Comparative Approach*. Birkhauser Verlag, Basel, pp. 7–24.

Grosberg, R.K. (1988) The evolution of allorecognition specificity in clonal invertebrates. *Quarterly Review of Biology* **63**, 377–412.

Hampe, W., Urny, J., Franke, I., Hoffmeister-Ullereich, S.A.H., Herrmann, D., Petersen, C.M., Lohmann, J., and Schaller, H.C. (1999) A head-activator binding protein is present in hydra in a soluble and a membrane-bound form. *Development* **126**, 4077–4086.

Hassel, M. (1998) Upregulation of *Hydra vulgaris* cPKC gene is tightly coupled to the differentiation of head structures. *Development Genes and Evolution* **207**, 489–501.

Hassel, M., and Bieller A. (1996) Stepwise transfer from high to low lithium concentrations increase the head-forming potential in *Hydra vulgaris* and possibly activates the PI cycle. *Developmental Biology* **177**, 439–448.

Hassel, M., Bridge, D.M., Stover, N.A., Kleinholz, H., and Steele, R.E. (1998) The level of expression of a protein kinase C gene may be an important component of the patterning process in *Hydra. Development Genes and Evolution* **207**, 502–514.

Hastings, J.W. (1996) Chemistry and colors of bioluminescent reactions: a review. *Gene* **173**, 5–11.

Hatta, M., Takahashi, T., Yum, S., Hoffmeister, S., and Fujisawa, T. (1999) Hym-330 enhances bud formation in *Hydra.* In: C.N. David and T.C.G. Bosch (eds.) *8th International Workshop on Hydroid Development*. Abstract, p. 109.

Hauenschild, C. (1954) Genetische und entwicklungsphysiologische Untersuchungen über

Intersexualitt und Gewebeverträglichkeit bei *Hydractinia echinata* Flem. *Wilhelm Roux' Archiv fur Entwicklungsmechanik der Organism* **147**, 1–41.
Hauenschild, C. (1956) Über die Vererbung einer Gewebeverträglichkeits-Eigenschaft bei dem Hydroidpolypen *Hydractinia echinata. Zeitschrift fur Naturforschung* **116**, 132–138.
Hobmayer, B., Rentzsch, F., Kuhn, K., Happel, C.M., von Laue, C.C., Snyder, P., Rothbacher, U., and Holstein, T.W. (2000) WNT signalling molecules act in axis formation in the diploblastic metazoan *Hydra. Nature* **407**, 186–189.
Hobmayer, E., Holstein, T.W., and David, C.N. (1990) Tentacle morphogenesis in *Hydra.* I. The role of head activator. *Development* **109**, 887–895.
Hoffmeister, S.A.H. (1996) Isolation and characterization of two new morphogenetically active peptides from *Hydra vulgaris. Development* **122**, 1941–1948.
Hoffmeister, S.A.H., and Schaller, H.C. (1987) Head activator and head inhibitor are signals for nerve cell differentiation in hydra. *Developmental Biology* **122**, 72–77.
Hoffmeister-Ullerich, S.A.H., and Herrmann, D. (1999) Cloning and expression pattern of the foot differentiation stimulating peptide pedibin in *Hydra vulgaris.* In: C.N. David and T.C.G. Bosch (eds.) *18th International Workshop on Hydroid Development.* Abstract, p. 103.
Holstein, T., Schaller, H.C., and David, C.N. (1986) Nerve cell differentiation in hydra requires two signals. *Developmental Biology* **115**, 9–17.
Javois, L.C., and Tombe, V.K. (1991) Head activator does not qualitatively alter head morphology in regenerates of *Hydra oligactis. Roux's Archives of Developmental Biology* **199**, 402–408.
Kato, T., Kumanireng, A. S., Ichinose, I., Kitahara, Y., Kakinuma, Y., Nishihira, M., and Kato, M. (1975) Active components of *Sargassum tortile* affecting the settlement of swimming larvae of *Coryne uchidai. Experientia* **31**, 433–434.
King, M.G., and Spencer, A.N. (1979) Gap and septate junctions in the excitable endoderm of *Polyorchis penicillatus* (Hydrozoa, Anthomedusae). *Journal of Cell Science* **36**, 391–400.
Kostrouch, Z., Kostrouchowa, M., Love, W., Jannini, E., Piatigorsky, J., and Rall, J. (1998) Retinoic acid X receptor in the diploblast, *Tripedalia cystophora. Proceedings of the National Academy of Sciences, USA* **95**, 13442–13447.
Kroiher, M., and Plickert, G. (1992) Analysis of pattern formation during embryonic development of *Hydractinia echinata. Roux's Archives of Developmental Biology* **201**, 95–104.
Lange, R.G., and Müller, W.A. (1991) SIF, a novel morphogenetic inducer in hydrozoa. *Developmental Biology* **147**, 121–132.
Lange, R.G., Plickert, G., and Müller, W.A. (1989) Histoincompatibility in *Hydractinia echinata*: analysis of the mechanisms of rejection. *Journal of Experimental Zoology* **249**, 284–292.
Lange, R.G., Dick, H.H., and Müller, W.A. (1992) Specificity and early ontogeny of historecognition in the hydroid *Hydractinia echinata. Journal of Experimental Zoology* **262**, 307–316.
Leddy, S.V., and Greene, D.R. (1990) Historecognition in the Cnidaria. In: G.W. Warr and N. Cohen (eds.) *Phylogenies of Immune Functions.* CRC Press, Boca Raton, pp. 103–116.
Leitz, T. (1993) Biochemical and cytological bases of metamorphosis in *Hydractinia echinata. Marine Biology* **116**, 559–564.
Leitz, T. (1997) Induction of settlement and metamorphosis of Cnidaria larvae: signals and signal transduction. *Invertebrate Reproduction and Development* **31**, 109–122.
Leitz, T. (1998a) Induction of metamorphosis of the marine hydrozoan *Hydractinia echinata* Fleming, 1828. *Biofouling* **12**, 173–187.
Leitz, T. (1998b) Metamorphosin A and related compounds: a novel family of neuropeptides with morphogenetic activity. *Annals of the New York Academy of Sciences* **839**, 105–110.
Leitz, T., and Müller, W.A. (1987) Evidence for the involvement of PI-signaling and diacylglycerol second messengers in the initiation of metamorphosis in the hydroid *Hydractinia echinata* Fleming. *Developmental Biology* **121**, 82–89.
Leitz, T., and Klingmann, G. (1990) Metamorphosis in *Hydractinia*: studies with activators and inhibitors aiming at protein kinase C and potassium channels. *Roux's Archives of Developmental Biology* **199**, 107–113.
Leitz, T., and Wagner, T. (1993) The marine bacterium *Alteromonas espejiana* induces metamorphosis of the hydroid *Hydractinia echinata. Marine Biology* **115**, 173–178.

Leitz, T., Morand, K., and Mann, M. (1994) Metamorphosin A, a novel peptide controlling development of the lower metazoan *Hydractinia echinata*. *Developmental Biology* **163**, 440–446.

Leitz, T., Beck, H., Stephan, M., Lehmann, W.D., DePetrocellis, L., and DiMarzo, V. (1994a) Possible involvement of arachidonic acid and eicosanoids in metamorphic events in *Hydractinia echinata* (Coelenterata; Hydrozoa). *Journal of Experimental Zoology* **269**, 422–431.

Leitz, T., Müller, W., DePetrocellis, L., and Di Marzo, V. (1994b) Enantiospecific synthesis of bioactive hydroxyeicosatetraenoic acids (HETEs) in *Hydra magnipapillata*. *Biochimica et Biophysica Acta* **1213**, 215–223.

Lentz, T. (1965) *Hydra*: Induction of supernumerary heads by isolated neurosecretory granules. *Science* **150**, 633–635.

Lesh, G.E., and Burnett, A.L. (1966) An analysis of the chemical control of polarized form in *Hydra*. *Journal of Experimental Zoology* **163**, 55–78.

Leviev, I., and Grimmelikhuijzen, C.J.P. (1995) Molecular cloning of a preprohormone from sea anemones containing numerous copies of a metamorphosis-inducing neuropeptide: A likely role for dipeptidyl aminopeptidase in neuropeptide precursor processing. *Proceedings of the National Academy of Sciencs, USA* **92**, 11647–11651.

Lohmann, J.U., and Bosch, T.C.G. (2000) The novel peptide HEAVY specifies apical fate in a simple radially symmetrical metazoan. *Genes and Development* **14** , 2771–2777.

Lubbock, R. (1980) Clone-specific cellular recognition in a sea anemone. *Proceedings of the National Academy of Sciences, USA* **77**, 6667–6669.

Lubbock, R., and Shelton, G.A. (1981) Electrical activity following cellular recognition of self and non-self in a sea anemone. *Nature* **289**, 59–60.

McCauley, D.W. (1997) Serotonin plays an early role in the metamorphosis of the hydrozoan *Phialidium gregarium*. *Developmental Biology* **190**, 229–240.

MacWilliams, H.K. (1983a) Hydra transplantation phenomena and the mechanism of hydra head regeneration. 1. Properties of the head inhibition. *Developmental Biology* **96**, 217–238.

MacWilliams, H.K. (1983b) Hydra transplantation phenomena and the mechanism of hydra head regeneration. 2. Properties of the head activation. *Developmental Biology* **96**, 239–257.

MacWilliams, H.K., and Kafatos, F.C. (1970) The basal inhibition in *Hydra* may be mediated by a diffusing substance. *American Zoologist* **14**, 633–645.

Martinez, D.E., Dirksen, M.-L., Bode, P.M., Jamrich, M., Steele, R.E., and Bode, H.R. (1997) *Budhead*, a *fork head/HNF-3* homologue, is expressed during axis formation and head specification in hydra. *Developmental Biology* **192**, 523–536.

Meinhardt, H. (1993) A model for pattern formation of hypostome, tentacles, and foot in hydra: How to form structures close to each other, how to form them at a distance. *Developmental Biology* **157**, 321–333.

Miller, R.L. (1972) Gel filtration of the sperm attractants of some marine hydrozoa. *Journal of Experimental Zoology* **82**, 281–298.

Miller, R.L. (1980) Species specificity of sperm chemotaxis in the hydromedusae. In: P. Tardent and R. Tardent (eds.) *Developmental and Cellular Biology of Coelenterates*. Elsevier/North-Holland Biomedical Press, Amsterdam, pp. 89–94.

Miller, R.L., and O'Rand, M.G. (1975) Utilization of chemical specificity during fertilization in the hydrozoa. In: B.A. Afzelius (ed.) *The Functional Anatomy of the Spermatozoon*. Pergamon Press, New York, pp. 15–26.

Mokady, O. (1996) Occam's razor, invertebrate allorecognition and IG superfamily evolution. *Research in Immunology* **147**, 45–50.

Mokady, O., and Buss, L.W. (1996) Transmission genetics of allorecognition in *Hydractinia symbiolongicarpus* (Cnidaria: Hydrozoa). *Genetics* **143**, 823–827.

Müller, W.A. (1964) Experimentelle Untersuchungen ber Stockentwicklung, Polypendifferenzierung und Sexualchimären bei *Hydractinia echinata*. *Wilhelm Roux' Archiv fur Entwicklungsmechanik der Organism* **155**, 181–268.

Müller, W.A. (1969) Die Steuerung des morphogenetischen Fliessgleichgewichts in den Polypen von *Hydractinia echinata*. II. Chemisch-analytische Untersuchungen. *Wilhelm Roux's Archives of Developmental Biology* **163**, 357–374.

Müller, W.A. (1973) Induction of metamorphosis by bacteria and ions in the planulae of *Hydractinia echinata*; an approach to the mode of action. *Publications. Seto Marine Biological Laboratory* **20**, 195–208.
Müller, W.A. (1984) Retinoids and pattern formation in a hydroid. *Journal of Embryology and Experimental Morphology* **81**, 253–271.
Müller, W.A. (1985) Tumor-promoting phorbol esters induce metamorphosis and multiple head formation in the hydroid *Hydractinia. Differentiation* **29**, 215–222.
Müller, W.A. (1989) Diacylglycerol-induced multihead formation in *Hydra. Development* **105**, 309–316.
Müller, W.A. (1990) Ectopic head and foot formation in *Hydra*: Diacylglycerol-induced increase in positional value and assistance of the head in foot formation. *Differentiation* **42**, 131–143.
Müller, W.A. (1991) Stimulation of head-specific nerve cell formation in *Hydra* by pulses of diacylglycerol. *Developmental Biology* **147**, 460–463.
Müller, W.A. (1995a) Competition for factors and cellular resources as a principle of pattern formation in *Hydra*. I. Increase of the potentials for head and bud formation and rescue of the regeneration-deficient mutant reg-16 by treatment with diacylglycerol and arachidonic acid. *Developmental Biology* **167**, 159–174.
Müller, W.A. (1995b) Competition for factors and cellular resources as a principle of pattern formation in *Hydra*. II. Assistance of foot formation by heads and buds and a new model of pattern control. *Developmental Biology* **167**, 175–189.
Müller, W.A. (1996a) Competition-based head versus foot decision in chimeric hydras. *International Journal of Developmental Biology* **40**, 1133–1139.
Müller, W.A. (1996b) Pattern formation in the immortal *Hydra. Trends in Genetics* **12**, 91–96.
Müller, W.A., and Plickert, G. (1982) Quantitative analysis of an inhibitory gradient field in the hydrozoan stolon. *Wilhelm Roux's Archives of Developmental Biology* **191**, 56–63.
Müller, W.A., Hauch, A., and Plickert, G. (1987) Morphogenetic factors in hydroids. I. Stolon tip activation and inhibition. *Journal of Experimental Zoology* **243**, 111–124.
Müller, W.A., Leitz, T., Stephan, M., and Lehmann, W.D. (1993) Arachidonic acid and the control of body pattern in *Hydra. Roux's Archives of Developmental Biology* **202**, 70–76.
Müller, W.A., Bartsch, C., Bartsch, H., Maidonis, I., and Bayer, E. (1998) Low-molecular-weight hormonal factors that affect head formation in *Hydra. International Journal of Developmental Biology* **42**, 825–828.
Neubauer, K.H., Christians, S., Hoffmeister, S.A.H., Kreger, S., and Schaller, H.C. (1991) Characterization of two types of head activator receptor on hydra cells. *Mechanisms of Development* **33**, 39–48.
Plickert, G. (1987) Low-molecular-weight factors from colonial hydroids affect pattern formation. *Roux's Archives of Developmental Biology* **196**, 248–256.
Plickert, G. (1989) Proportion-altering factor (PAF) stimulates nerve cell formation in *Hydractinia echinata. Cell Differentiation and Development* **26**, 19–28.
Plickert, G. (1991) A low-molecular weight factor from colonial hydroids affects body proportioning and cell differentiation. *Hydrobiologia* **216/217**, 83–89.
Plickert, G., Heringer, A., and Hiller, B. (1987) Analysis of spacing pattern in a periodic pattern. *Developmental Biology* **120**, 399–411.
Plickert, G., Kroiher, M., and Munck, A. (1988) Cell proliferation and early differentiation in *Hydractinia echinata. Development* **103**, 795–803.
Rinkevich, B. (1996) Links between alloimmune responses and their genetic background in colonial urochordates and cnidarians: evidence for the recognition of nonself as opposed to self. In: J.S. Stolen et al. (eds.) *Modulators of Immune Responses: the Evolutionary Trail.* SOS Publishers, Fair Haven, pp. 1–13.
Rinkevich, B., and Weissmann, I.L. (1987) Chimeras in colonial invertebrates: a synergistic symbiosis or somatic and germ cell parasitism? *Symbiosis* **4**, 117–134.
Rinkevich, B., Frank, U., Bak, R.P.M., and Muller, W.E.G. (1994) Alloimmune responses between *Acropora hemprichi* conspecifics: nontransitive patterns of overgrowth and delayed cytotoxicity. *Marine Biology* **118**, 731–737.

Schaller, H.C. (1979) Neuropeptides in hydra. *Trends in Neurosciences* **1**, 120–122.
Schaller, H.C., and Gierer, A. (1973) Distribution of the head-activating substance in hydra and its localization in membranous particles in nerve cells. *Journal of Embryology and Experimental Morphology* **29**, 39–52.
Schaller, H.C., and Bodenmuller, H. (1981) Isolation and amino acid sequence of a morphogenetic peptide from hydra. *Proceedings of the National Academy of Sciences, USA* **78**, 7000–7004.
Schaller, H.C., and Bodenmüller, H. (1987) Neurohormones from hydra. In: G. Adelman (ed.) *Encyclopaedia of Neuroscience*. Birkhäuser, Boston, pp. 505–506.
Schaller, H.C., Hoffmeister, S.A.H., and Dubel, S. (1989) Role of the neuropeptide head activator for growth and development in hydra and man. *Development* **107** (Supplement), 99–107.
Schmid, V., and Plickert, G. (1990) The proportion altering factor (PAF) and the *in vitro* transdifferentiation of isolated striated-muscle of jellyfish into nerve-cells. *Differentiation* **44**, 95–102.
Schmidt, T., Grimmelikhuijzen, C.J.P., and Schaller, H.C. (1980) Morphogenetic substances in hydra. In: P. Tardent and R. Tardent (eds.) *Development and Cellular Biology of Coelenterates*. Elsevier/North-Holland Biomedical Press, Amsterdam, pp. 395–399.
Schwoerer-Böhning, B., Kroiher, M., and Müller, W.A. (1990) Signal transmission and covert prepattern in the metamorphosis of *Hydractinia echinata* (Hydrozoa). *Roux's Archives of Developmental Biology* **198**, 245–251.
Shimomura, O., and Johnson, F.H. (1975) Chemical nature of bioluminescence systems in coelenterates. *Proceedings of the National Academy of Sciences, USA* **72**, 1546–1549.
Shostak, S. (1972) Inhibitory gradients of head and foot regeneration in *Hydra viridis*. *Developmental Biology* **28**, 620–633.
Stearns, T. (1995) Green fluorescent protein. The green revolution. *Current Biology* **5**, 362–364.
Sugiyama, T., and Fujisawa, T. (1977) Genetic analysis of developmental mechanisms in *Hydra*. I. Sexual reproduction of *Hydra magnipapillata* and isolation of mutants. *Development Growth and Differentiation* **19**, 187–200.
Takahashi, T., Muneoka, Y., Lohmann, J., deHaro, M.S.L., Solleder, G., Bosch, T.C.G., David, C.N., Bode, H.R., Koizumi, O., Shimizu, H., Hatta, M., Fujisawa, T., and Sugiyama, T. (1997) Systematic isolation of peptide signal molecules regulating development in hydra: LWamide and PW families. *Proceedings of the National Academy of Sciences, USA* **94**, 1241–1246.
Turing, A. (1952) The chemical basis of morphogenesis. *Philosophical Transactions of the Royal Society of London B* **237**, 37–72.
Webster, G. (1971) Morphogenesis and pattern formation in hydroids. *Biological Reviews* **46**, 1–46.
Westfall, I.A. (1996) Ultrastructure of synapses in the first-evolved nervous systems. *Journal of Neurocytology* **25**, 735–746.
Wodarz, A., and Nusse, R. (1998) Mechanisms of Wnt signaling in development. *Annual Review of Cell and Developmental Biology* **14**, 59–88.
Wolpert, L., Clarke, M.R.B., and Hornbruch, A. (1972) Positional signalling along hydra. *Nature, New Biology* **239**, 101–106.
Wolpert, L., Hornbruch, A., and Clarke, M.R.B. (1974) Positional information and positional signalling in hydra. *American Zoologist* **14**, 647–663.
Yoshida, M. (1959) Spawning in Coelenterates. *Experientia* **15**, 11.
Yund, P.O., Cunningham, C.W., and Buss, L.W. (1987) Recruitment and postrecruitment interactions in a colonial hydroid. *Ecology* **68**, 971–982.

CHAPTER 5

NEUROPEPTIDES IN CNIDARIANS

CORNELIS J.P. GRIMMELIKHUIJZEN, MICHAEL WILLIAMSON, and GEORG N. HANSEN

Department of Cell Biology, Zoological Institute, University of Copenhagen, Universitetsparken 15, DK-2100 Copenhagen Ø, Denmark

Summary

Cnidarians are real neuropeptide factories. From a single sea anemone species, 17 different neuropeptides have been isolated, and we believe that this is only the tip of the iceberg. A similar picture is now emerging from *Hydra*. Cnidarian neuropeptides can be neurotransmitters or neuromodulators involved in signal transduction, but also neurohormones that steer developmental processes such as metamorphosis. Cnidarians synthesize their neuropeptides as preprohormones of varying sizes that may contain up to 38 neuropeptide copies per precursor molecule. The cnidarian prohormones are processed by both known and unknown (novel) processing enzymes. We have cloned the enzymes that are responsible for neuropeptide C-terminal amidation, showing that at least two enzyme genes are involved in this two-step reaction (in contrast to one gene in mammals). By using neuropeptide immunocytochemistry, we found that the cnidarian nervous system is more sophisticated than we believed before, having neuronal concentrations in the form of ganglion-like structures, neuronal plexuses and nerve tracts. By using a whole-mount two-colour *in situ* hybridization technique and RNA probes coding for various *Hydra* preprohormones, we found that *Hydra* has at least six different populations of nerve cells, of which some coexpress two different preprohormone mRNAs. This is the first example of coexpression of two well-characterized preprohormones (yielding two well-characterized neurohormone families) in cnidarians.

1. Introduction

Cnidarians are primitive animals such as *Hydra*, sea anemones and corals. The phylum Cnidaria can be subdivided into four classes: Hydrozoa (such as *Hydra*, *Hydractinia*), Cubozoa (comb jellyfishes), Scyphozoa (real jellyfishes), and Anthozoa (animals such as sea anemones and corals). Most hydrozoans, cubozoans and scyphozoans have a life cycle that includes a mobile planula larva, a sessile polyp form, and a medusa. In anthozoans, the medusa form is lacking.

Cnidarians are interesting from an evolutionary point-of-view, because they are the lowest animal group having a nervous system. By studying the primitive nervous systems of cnidarians, therefore, one could hope to learn more about the evolutionary origin of nervous systems. In addition to these evolutionary aspects, cnidarians are interesting models for developmental biologists, because they are simple and have strong regenerative capacities (e.g., from small transverse slices of *Hydra*, or even from disintegrated and subsequently reaggregated *Hydra* cells, new animals develop within a few days: Gierer *et al.*, 1972;

I. Fairweather (ed.), Cell Signalling in Prokaryotes and Lower Metazoa, 115–139.
© 2004 *Kluwer Academic Publishers. Printed in Great Britain.*

Gierer 1977; see also Müller, this volume). Furthermore, it is possible to create *Hydra* specimens that only consist of one cell type, the epitheliomuscular cells (Campbell, 1976; Sugiyama and Fujisawa, 1978). One can transplant neuronal stem cells ("interstitial cells") into these "epithelial" *Hydra* and in this way observe the subsequent differentiation and development of a whole new nervous system in animals that were originally nerve-free (Minobe, Koizumi and Sugiyama, 1995).

The cnidarian nervous system consists basically of a nerve net that, in some regions, can condense to form nerve plexuses, nerve tracts, or nerve rings (Grimmelikhuijzen and Spencer, 1984; Grimmelikhuijzen, 1985; Grimmelikhuijzen, Spencer and Carre, 1986; Anderson *et al.*, 1992; Grimmelikhuijzen, Leviev and Carstensen, 1996). No real ganglia or a brain exist in cnidarians. Electronmicroscopy of *Hydra* showed that all *Hydra* neurons have a cilium (are sensory neurons), form synaptic contacts with epitheliomuscular cells (are motorneurons) or with other, adjacent neurons (are interneurons), and have neurosecretory vesicles at non-synaptic release sites (are neurosecretory) (Westfall, 1973; Westfall and Kinnamon, 1978). These data suggested that there was only one single type of primitive, multifunctional nerve cell in *Hydra* (Westfall and Kinnamon, 1978).

2. Neuropeptides

For a long time, it has been difficult to visualize the nervous systems of cnidarians at the light-microscopic level, which was mainly due to the absence of adequate histological stains. However, in the early 1980s, using various antibodies against mammalian and invertebrate neuropeptides, we discovered that *Hydra* and other cnidarians probably use neuropeptides as their neurotransmitters (Grimmelikhuijzen, Dockray and Schot, 1982). These findings gave us the possibility of staining the cnidarian neurons with neuropeptide antibodies. Over the following years, using various radioimmunoassays, we isolated numerous authentic neuropeptides from various cnidarians and elucidated their structures (Grimmelikhuijzen and Graff, 1986; Grimmelikhuijzen and Groeger, 1987; Graff and Grimmelikhuijzen, 1988a,b; Grimmelikhuijzen *et al.*, 1988, 1990; Nothacker, Rinehardt and Grimmelikhuijzen, 1991; Nothacker *et al.*, 1991; Carstensen *et al.*, 1992, 1993; Grimmelikhuijzen, Rinehardt and Spencer, 1992; Moosler, Rinehardt and Grimmelikhuijzen, 1996, 1997) (Tables 1 and 2). The structural knowledge on the cnidarian neuropeptides enabled us to develop better and more specific antisera and to visualize the cnidarian nervous system in an, at that time, unmatched quality and detail (Grimmelikhuijzen and Spencer, 1984; Grimmelikhuijzen, 1985; Grimmelikhuijzen, Spencer and Carre, 1986; Anderson, Moosler and Grimmelikhuijzen, 1992; Carstensen *et al.*, 1992; Grimmelikhuijzen *et al.*, 1992; Grimmelikhuijzen, Leviev and Carstensen, 1996) (Fig. 1). We also used these antisera for immunoelectronmicroscopy, where we could localize the peptides in neuronal dense-cored vesicles of *Hydra* and other

Table 1. Neuropeptide families isolated from anthozoans

Species	Structure	Name	Reference
Anthopleura elegantissima	L-3-phenyllactyl-Phe-Lys-Ala-NH_2	Antho-KAamide	[1]
Anthopleura elegantissima	L-3-phenyllactyl-Tyr-Arg-Ile-NH_2	Antho-RIamide I	[2]
Anthopleura elegantissima	Tyr-Arg-Ile-NH_2	Antho-RIamide II	[2]
Anthopleura elegantissima	L-3-phenyllactyl-Leu-Arg-Asn-NH_2	Antho-RNamide I	[3]
Anthopleura elegantissima	Leu-Arg-Asn-NH_2	Antho-RNamide II	[4]
Anthopleura elegantissima	<Glu-Ser-Leu-Arg-Trp-NH_2	Antho-RWamide I	[5]
Anthopleura elegantissima	<Glu-Gly-Leu-Arg-Trp-NH_2	Antho-RWamide II	[6]
Anthopleura elegantissima	<Glu-Asn-Phe-His-Leu-Arg-Pro-NH_2	Antho-RPamide II	[7]
Anthopleura elegantissima	Leu-Pro-Pro-Gly-Pro-Leu-Pro-Arg-Pro-NH_2	Antho-RPamide I	[8]
Anthopleura elegantissima	Gly-Pro-Hyp-Ser-Leu-Phe-Arg-Pro-NH_2	Antho-RPamide IV	[9]
Anthopleura elegantissima	<Glu-Val-Lys-Leu-Tyr-Arg-Pro-NH_2	Antho-RPamide III	[9]
Anthopleura elegantissima	Tyr-Arg-Pro-NH_2	Antho-RPamide V	[9]
Anthopleura elegantissima	<Glu-Gly-Arg-Phe-NH_2	Antho-RFamide	[10]
Renilla köllikeri	<Glu-Gly-Arg-Phe-NH_2	Antho-RFamide	[11]
Anthopleura elegantissima	<Glu-Gln-Pro-Gly-Leu-Trp-NH_2	Metamorphosin A (MMA or Antho-LWamide II)	[12]

The references are: [1], Nothacker, Rinehart and Grimmelikhuijzen, 1991; [2], Nothacker *et al.*, 1991; [3], Grimmelikhuijzen *et al.*, 1990; [4], Jacob and Grimmelikhuijzen, unpublished; [5], Graff and Grimmelikhuijzen, 1988a; [6], Graff and Grimmelikhuijzen, 1998b; [7], Carstensen *et al.*, 1993; [8], Carstensen *et al.*, 1992; [9], Carstensen and Grimmelikhuijzen, unpublished; [10], Grimmelikhuijzen and Graff, 1986; [11], Grimmelikhuijzen and Groeger, 1987; [12], Leitz, Morand and Mann, 1994.

Table 2. Neuropeptide families isolated from hydrozoans and scyphozoans

Species	Structure	Name		Reference
Cyanea lamarckii	Gly-Arg-Phe-NH$_2$	Cyanea-RFamide III		[1]
Cyanea lamarckii	<Glu-Pro-Leu-Trp-Ser-Gly-Arg-Phe-NH$_2$	Cyanea-RFamide II		[1]
Polyorchis penicillatus	<Glu-Leu-Leu-Gly-Gly-Arg-Phe-NH$_2$	Pol-RFamide I		[2]
Hydra magnipapillata	<Glu-Trp-Leu-Gly-Gly-Arg-Phe-NH$_2$	Hydra-RFamide I		[3]
Hydra magnipapillata	<Glu-Trp-Phe-Asn-Gly-Arg-Phe-NH$_2$	Hydra-RFamide II		[3]
Polyorchis penincillatus	<Glu-Trp-Leu-Lys-Gly-Arg-Phe-NH$_2$	Pol-RFamide II		[4]
Cyanea lamarckii	<Glu-Trp-Leu-Arg-Gly-Arg-Phe-NH$_2$	Cyanea-RFamide I		[1]
Hydra magnipapillata	Lys-Pro-His-Leu-Arg-Gly-Arg-Phe-NH$_2$	Hydra-RFamide III		[3]
Hydra magnipapillata	His-Leu-Arg-Gly-Arg-Phe-NH$_2$	Hydra-RFamide IV		[3]
Hydra magnipapillata	Glu-Pro-Leu-Pro-Ile-Gly-Leu-Trp-NH$_2$	Hydra-LWamide II	(Hym-331)	[5]
Hydra magnipapillata	Glu-Pro-Pro-Pro-Gly-Leu-Trp-NH$_2$	Hydra-LWamide I	(Hym-248)	[5]
Hydra magnipapillata	Glu-Pro-Hyp-Pro-Gly-Leu-Trp-NH$_2$		(Hym-38)	[5]
Hydra magnipapillata	Lys-Pro-Ile-Pro-Gly-Leu-Trp-NH$_2$	Hydra-LWamide III	(Hym-249)	[5]
Hydra magnipapillata	Asn-Pro-Tyr-Pro-Gly-Leu-Trp-NH$_2$	Hydra-RFamide IV	(Hym-53)	[5]
Hydra magnipapillata	Glu-Pro-Met-Thr-Gly-Leu-Trp-NH$_2$	Hydra-RFamide V	(Hym-54)	[5]
Hydra magnipapillata	Ala-Pro-Phe-Ile-Phe-Pro-Gly-Pro-Lys-Val-NH$_2$	Hydra-KVamide	(Hym-176)	[6]
Hydra magnipapillata	Phe-Pro-Gln-Ser-Phe-Leu-Pro-Arg-Gly-NH$_2$	Hydra-RGamide	(Hym-335)	[7]

The references are: [1], Moosler, Rinehart and Grimmelikhuijzen, 1997; [2], Grimmelikhuijzen *et al.*, 1998; [3], Moosler, Rinehart and Grimmelikhuijzen, 1996; [4], Grimmelikhuijzen, Rinehart and Spencer, 1992; [5], Takahashi *et al.*, 1997; [6], Yum *et al.*, 1998b; [7], Takahashi *et al.*, 2000.

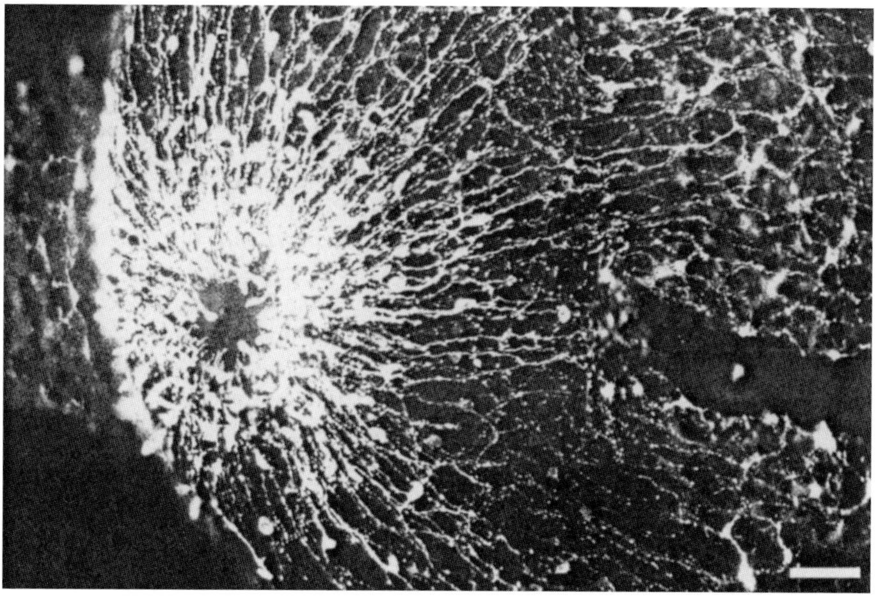

Figure 1. Whole-mount staining of *Hydra* with an antiserum to Arg-Phe-NH_2 (RFamide). This antiserum stains all neurons containing Hydra-RFamide peptides (see Table 2). The photograph shows a dense plexus of neurons in the hypostome (a head region around the mouth opening) and a nerve net in the tentacles of *Hydra*. The bar represents 50 μm.

cnidarians (Koizumi *et al.*, 1989; Westfall and Grimmelikhuijzen, 1993; Westfall *et al.*, 1995). The ultrastructural localization of the cnidarian neuropeptides suggested that they might be involved in neurotransmission. This was subsequently confirmed by physiological experiments, using intact sea anemones, sea anemone muscle preparations and isolated muscle cells, where we found that many of the sea anemone neuropeptides isolated by radioimmunoassays (Table 1) were indeed active, either as excitatory or inhibitory neurotransmitters or as neuromodulators (McFarlane, Graff and Grimmelikhuijzen, 1987; McFarlane and Grimmelikhuijzen, 1991; McFarlane, Anderson and Grimmelikhuijzen, 1991; Nothacker *et al.*, 1991; McFarlane, Reinscheid and Grimmelikhuijzen, 1992; McFarlane *et al.*, 1993).

Another way of finding cnidarian neuropeptides is by using a bioassay to monitor their purification. This approach was used by Leitz and coworkers, who found that extracts of the sea anemone, *Anthopleura elegantissima* induced morphogenesis in planula larvae of the marine hydroid, *Hydractinia echinata* (Leitz *et al.*, 1994). After purification, the bioactive component turned out to be a peptide, <Glu-Gln-Pro-Gly-Leu-Trp-NH_2 (metamorphosin A, or MMA; see Table 1).

A group of mainly Japanese researchers followed a third approach and isolated peptides from a *Hydra* extract without any monitoring system (neither radioimmunoassays, nor bioassays); they simply injected a crude *Hydra* extract into an HPLC system and purified each peak in subsequent HPLC runs, finally followed by amino acid sequencing (Takahashi *et al.*, 1997). In this way, the Japanese researchers and their coworkers obtained a large number of peptides, among them six peptides with the C-terminal sequence Gly-Leu-Trp-NH_2 (resembling the sea anemone neuropeptide MMA; these peptides were called Hym-38, -53, -54, -248, -249, -331 or Hydra-LWamides I-V; see Table 2), one peptide with the structure Ala-Pro-Phe-Ile-Phe-Pro-Gly-Pro-Lys-Val-NH_2 (called Hym-176 or Hydra-KVamide) and one peptide with the structure Phe-Pro-Gln-Ser-Phe-Leu-Pro-Arg-Gly-NH_2 (called Hym-355 or Hydra-RGamide) (Takahashi *et al.*, 1997, 2000; Yum *et al.*, 1998a; Table 2). By immunocytochemistry, it was subsequently established that the above-mentioned *Hydra* peptides were neuropeptides (Leitz and Lay, 1995; Yum *et al.*, 1998b; Takahashi *et al.*, 2000). Furthermore, it was demonstrated for both the five Hydra-LWamides and for Hydra-KVamide (Hym-176) that they caused contractions of the epitheliomuscular cells of *Hydra*, suggesting that they are involved in neuromuscular transmission (Takahashi *et al.*, 1997; Yum *et al.*, 1998a). For Hydra-RGamide (Hym-355), it was shown that it stimulated the interstitial cells to differentiate into nerve cells (Takahashi *et al.*, 2000). This, then, is the second example that a cnidarian neuropeptide is developmentally active.

3. Neuropeptide Biosynthesis

In mammals, neuropeptides are made as larger precursor proteins, the preprohormones, which consist of a hydrophobic, N-terminal signal sequence [for translocation across the rough endoplasmic reticulum (RER) membrane; the pre-part] and a prohormone part, containing one or more copies of the immature neuropeptide. After translocation across the RER membrane and removal of the pre-part, the prohormone is transported to the Golgi complex, sorted and packaged into neurosecretory, dense-cored vesicles together with a series of prohormone processing enzymes. The processing enzymes catalyze the following steps (see Fig. 2):

(1) First, there is an initial, endoproteolytic cleavage at the C-terminal side of paired or single basic residues. The enzymes responsible for this are called prohormone convertases and several of them have been cloned: PC1 (also called PC3) and PC2 (Smeekens and Steiner, 1990; Seidah *et al.*, 1990, 1991; Smeekens *et al.*, 1991);

(2) The endoproteolytic cleavage of the prohormone results in various small peptides with one or more C-terminal basic amino acid residues. These basic amino acid residues are removed by a carboxypeptidase B-like enzyme (step 2 in Fig. 2; Fricker *et al.*, 1989);

(3) If the final, biologically active peptide carries a C-terminal amide group, its immature structure is a peptidyl-glycine (product of step 2 in Fig. 2). This peptidyl-glycine is converted into a peptidyl-amide by the concerted action of two enzymes, peptidyl-glycine α-hydroxylating monooxygenase

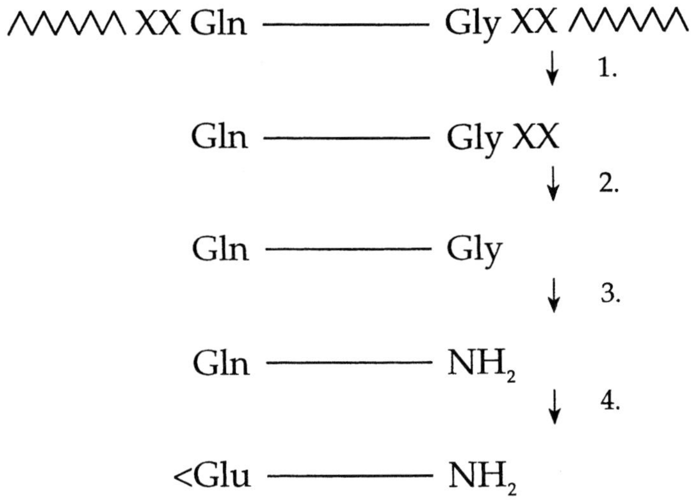

Figure 2. Prohormone processing in mammals. Top: The prohormone contains one or several copies of an immature neuropeptide sequence that is flanked by basic residues (marked by XX). The cleavage or processing steps are catalyzed by the following enzymes: 1. An endoproteinase cleaving at the C-terminal side of pairs or single basic residues (PC1/PC3 or PC2). 2. A carboxypeptidase B-like enzyme. 3. PHM and PAL (see text). 4. Glutaminyl cyclase. (From Grimmelikhuijzen, Leviev, and Carstensen, 1996)

(PHM) and peptidyl-α-hydroxyglycine α-amidating lyase (PAL). PHM hydroxylates peptidyl-glycine, thereby yielding peptidyl-hydroxyglycine, which is then cleaved in an N-C lyase reaction by PAL (Fig. 3). The two amidating enzymes PHM and PAL will be discussed in more detail below;

(4) Finally, if the mature neuropeptide carries an N-terminal <Glu group, the immature neuropeptide carries an N-terminal Gln residue, which is converted into a <Glu residue by the enzyme glutaminyl cyclase (Fischer and Spiess, 1987; Pohl *et al.*, 1991).

We have cloned most of the preprohormones for the isolated neuropeptides given in Tables 1 and 2 (Darmer *et al.*, 1991, 1998; Schmutzler *et al.*, 1992; Reinscheid and Grimmelikhuijzen, 1994; Schmutzler, Diekhoff and Grimmelikhuijzen, 1994; Leviev and Grimmelikhuijzen, 1995; Grimmelikhuijzen, Leviev and Carstensen, 1996; Leviev, Williamson and Grimmelikhuijzen, 1997), and the Japanese groups have subsequently cloned some of their *Hydra* neuropeptide precursors (Yum *et al.*, 1998b; Takahashi *et al.*, 2000). The cnidarian preprohormones vary in size and can be small (only containing one neuropeptide sequence) or large (containing up to 38 neuropeptide sequences). Fig. 4 shows an example of a large preprohormone, namely, that for metamorphosin A (MMA) from the sea anemone, *Anthopleura elegantissima* (Leviev and Grimmelikhuijzen, 1995). This preprohormone contains eight copies of authentic MMA (printed in boldface type and underlined in Fig. 4) and 29 copies of other, putative neuropeptides (underlined only in

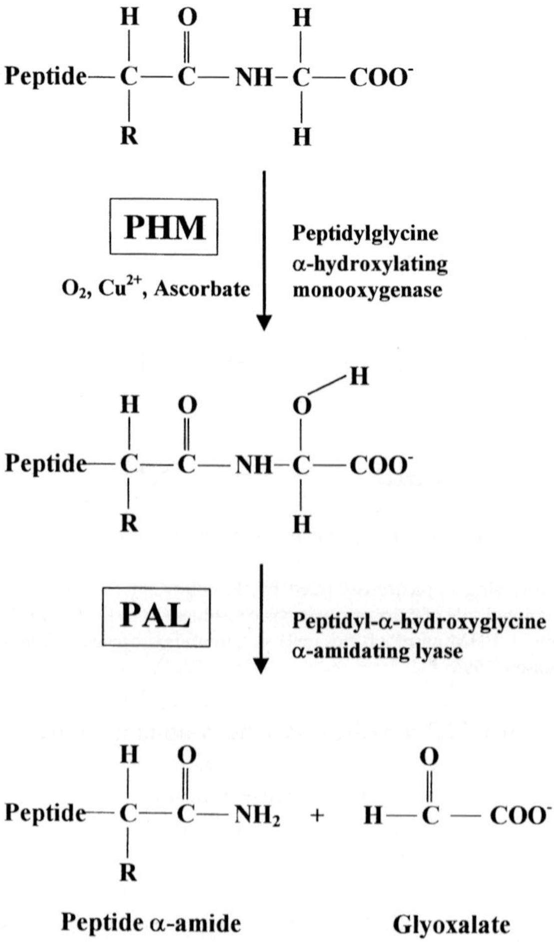

Figure 3. Conversion of a peptidyl-glycine into a peptidyl-amide (step 3 of Fig. 2). In a first step, the enzyme PHM (peptidyl α-hydroxylating monooxygenase) hydroxylates peptidyl-glycine (top). This reaction requires O_2, Cu^{2+}, and ascorbate. In a second step, the enzyme PAL (peptidyl-α-hydroxyglycine α-amidating enzyme) cleaves peptidylhydroxyglycine into a peptidyl-amide and glyoxylic acid (bottom). (From Grimmelikhuijzen, Leviev, and Carstensen, 1996)

Fig. 4). A Lys-Arg sequence follows all eight immature MMA sequences, suggesting that cnidarians must have a prohormone processing enzyme resembling PC1/PC3 or PC2. In fact, such an enzyme has been cloned from *Hydra* (Chan *et al.*, 1992), showing that prohormone convertases do indeed exist in cnidarians. Furthermore, a C-terminal Gly residue occurs in all immature MMA sequences, whereas in the mature sequence an amide group is found (Table 1). This points to the existence of PHM- and PAL-like enzymes in cnidarians. Again, both PHM- and PAL-like enzymes have recently been cloned from cnidarians (Hauser, Williamson and Grimmelikhuijzen, 1997; see also below).

```
CAATGAACTGAGTGGAACACAAGTAATACATATTCTTCACTTCGGTTGATA ATG GCC CTC AAG TGT CAT CTA GTT CTA CTG    81
                                                    Met Ala Leu Lys Cys His Leu Val Leu Leu    10

GCC ATT ACT TTA CTA TTA GCA CAG TGT TCA GGG TCA GTA GAC AAG AAG GAT AGT ACG ACG AAT CAC TTA   150
Ala Ile Thr Leu Leu Leu Ala Gln Cys Ser Gly Ser Val Asp Lys Lys Asp Ser Thr Thr Asn His Leu    33

GAT GAG AAG AAA ACA GAT TCC ACA GAA GCA CAT ATT GTA CAA GAA ACA GAC GCG TTA AAA GAA AAT TCT   219
Asp Glu Lys Lys Thr Asp Ser Thr Glu Ala His Ile Val Gln Glu Thr Asp Ala Leu Lys Glu Asn Ser    56

TAT CTT GGC GCC GAG GAG GAA TCT AAA GAA GAA GAC AAG AAG AGA TCC GCC GCT CCT CAG CAG CCT GGC   288
Tyr Leu Gly Ala Glu Glu Glu Ser Lys Glu Glu Asp Lys Lys Arg Ser Ala Ala Pro Gln Gln Pro Gly    79

CTC TGG GGG AAA CGC CAG AAA ATA GGA CTA TGG GGA AGA TCC GCT GAC GCA GGA CAG CCA GGC CTC TGG   357
Leu Trp Gly Lys Arg Gln Lys Ile Gly Leu Trp Gly Arg Ser Ala Asp Ala Gly Gln Pro Gly Leu Trp   102

GGC AAA CGA CAA AGT CCC GGA TTA TGG GGA AGA TCC GCT GAC GCA GGA CAG CCA GGC CTC TGG GGC AAA   426
Gly Lys Arg Gln Ser Pro Gly Leu Trp Gly Arg Ser Ala Asp Ala Gly Gln Pro Gly Leu Trp Gly Lys   125

CGT CAA AAT CCC GGA TTA TGG GGA AGA TCC GCT GAC GCA GGA CAG CCA GGC CTC TGG GGC AAA CGT CAA   495
Arg Gln Asn Pro Gly Leu Trp Gly Arg Ser Ala Asp Ala Gly Gln Pro Gly Leu Trp Gly Lys Arg Gln   148

AAT CCC GGA TTA TGG GGA AGA TCG GCT GAC GCA GGA CAG CCA GGC CTC TGG GGC AAA CGT CAA AAT CCC   564
Asn Pro Gly Leu Trp Gly Arg Ser Ala Asp Ala Gly Gln Pro Gly Leu Trp Gly Lys Arg Gln Asn Pro   171

GGA TTA TGG GGA AGG TCC GCT GAC GCA AGA CAA CCC GGA CTC TGG GGC AAA CGT GAA ATC TAC GCA TTA   633
Gly Leu Trp Gly Arg Ser Ala Asp Ala Arg Gln Pro Gly Leu Trp Gly Lys Arg Glu Ile Tyr Ala Leu   194

TGG GGA GGA AAA CGT CAA AAT CCC GGA CTT TGG GGA AGA TCC GCT GAT CCA GGA CAG CCC GGC CTC TGG   702
Trp Gly Gly Lys Arg Gln Asn Pro Gly Leu Trp Gly Arg Ser Ala Asp Pro Gly Gln Pro Gly Leu Trp   217

GGC AAA CGT GAA CTC GTC GGA TTA TGG GGG GGA AAA CGT CAA AAC CCC GGA TTG TGG GGA AGA TCG GCT   771
Gly Lys Arg Glu Leu Val Gly Leu Trp Gly Gly Lys Arg Gln Asn Pro Gly Leu Trp Gly Arg Ser Ala   240

GAA GCA GGA CAG CCA GGA CTT TGG GGA AAA CGC CAA AAA ATA GGA TTG TGG GGA CGT TCG GCT GAC CCA   840
Glu Ala Gly Gln Pro Gly Leu Trp Gly Lys Arg Gln Lys Ile Gly Leu Trp Gly Arg Ser Ala Asp Pro   263

CTT CAG CCT GGC CTC TGG GGC AAA CGT CAA AAT CCC GGA TTA TGG GGA AGA TCT GCT GAC CCG CAG CAG   909
Leu Gln Pro Gly Leu Trp Gly Lys Arg Gln Asn Pro Gly Leu Trp Gly Arg Ser Ala Asp Pro Gln Gln   286

CCT GGC CTC TGG GGC AAA CGT CAA AAT CCC GGA TTA TGG GGA AGA TCT GCT GAC CCG CAG CAG CCT GGC   978
Pro Gly Leu Trp Gly Lys Arg Gln Asn Pro Gly Leu Trp Gly Arg Ser Ala Asp Pro Gln Gln Pro Gly   309

CTC TGG GGC AAA CGT CAA AAT CCC GGA TTA TGG GGA AGA TCT GCT GAC CCG CAG CAG CCT GGC CTC TGG  1047
Leu Trp Gly Lys Arg Gln Asn Pro Gly Leu Trp Gly Arg Ser Ala Asp Pro Gln Gln Pro Gly Leu Trp   332

GGC AAA CGT CAA AAT CCC GGA TTA TGG GGA AGA TCT GCT GAC CCG CAG CAA CCT GGC CTC TGG GGC AAA  1116
Gly Lys Arg Gln Asn Pro Gly Leu Trp Gly Arg Ser Ala Asp Pro Gln Gln Pro Gly Leu Trp Gly Lys   355

AGC CCC GGT TTA TGG GGA CGA TCC GCT GAC CCA CAA CAG CCT GGA CTT TGG GGG AAA CGC CAA AAT CCC  1185
Ser Pro Gly Leu Trp Gly Arg Ser Ala Asp Pro Gln Gln Pro Gly Leu Trp Gly Lys Arg Gln Asn Pro   378

GGA TTT TGG GGA AGA TCT GCT GAC CCG CAG CAG CCT GGC CTC TGG GGC AAA CGT CAA AAT CCC GGA TTA  1254
Gly Phe Trp Gly Arg Ser Ala Asp Pro Gln Gln Pro Gly Leu Trp Gly Lys Arg Gln Asn Pro Gly Leu   401

TGG GGA AGA TCT GCT GAC CCG CAG CAA CCT GGC CTC TGG GGC AAA CGT CAA AAT CCC GGA TTA TGG GGA  1323
Trp Gly Arg Ser Ala Asp Pro Gln Gln Pro Gly Leu Trp Gly Lys Arg Gln Asn Pro Gly Leu Trp Gly   424

AGA TCT GCT GAC CCG CAG CAA CCT GGC CTC TGG GGC AAA CGT CAA AAC CCC GGT TTA TGG GGA CGA TCC  1392
Arg Ser Ala Asp Pro Gln Gln Pro Gly Leu Trp Gly Lys Arg Gln Asn Pro Gly Leu Trp Gly Arg Ser   447

GCT GAC CCA CAA CAG CCT GGA CTT TGG GGG AAA CGC CAA AAT CCA GGA CTA TGG GGA AGA AGT GCT GGC  1461
Ala Asp Pro Gln Gln Pro Gly Leu Trp Gly Lys Arg Gln Asn Pro Gly Leu Trp Gly Arg Ser Ala Gly   470

TCC GGT CAA CTC GGA CTT TGG GGT AAA AGG CAA TCA CGC ATT GGA TTA TGG GGA AGA TCT GCC GAG CCT  1530
Ser Gly Gln Leu Gly Leu Trp Gly Lys Arg Gln Ser Arg Ile Gly Leu Trp Gly Arg Ser Ala Glu Pro   493

CCA CAA TTT GAA GAT TTA GAA GAT TTA AAG AAA AAA TCA GCA ATT CCC CAA CCA AAA GGA CAA TGA TAA  1599
Pro Gln Phe Glu Asp Leu Glu Asp Leu Lys Lys Lys Ser Ala Ile Pro Gln Pro Lys Gly Gln Stop Stop 514

TATCCTAGGATCTTCAAAAGTTATCCCGATCATCAATCCCCGGACAAGAGATATTTTAATTTCTGCCGCACGATTGACAGTTCCATTCCAT  1690
TACGAAGAACAAAAAGCTACGTTTCTTTAAGATAATAAATCAAATTCAATATTGTTTGAAGCAATGCACTTCAGGTTTTCACACAAAACTA  1781
ATACAAAAGTTATAAACATAAATAATAAACAAAAAAGGGGTAAGAAACCTGGTTTTTCGTTTTAGACTTTTCAAATTGGTCCTGCATGTCA  1872
TTGTAAGAGGTCGGCAGTAGAAGCTCTTTGAAACACTGTAAGTAGTCATTGAATGCACATTTATGGCAATAAAGATCACGCAGTGGTCTGC  1963
GGCCGCGAA                                                                                    1972
```

Figure 4. cDNA and deduced amino acid sequence of the MMA (<Glu-Gln-Pro-Gly-Leu-Trp-NH_2) preprohormone from the sea anemone, *Anthopleura elegantissima*. Copies of authentic MMA sequences are underlined and printed in boldface type. Putative neuropeptide sequences are underlined only. (From Leviev and Grimmelikhuijzen, 1995)

The N-terminal amino acid residue of mature MMA is a <Glu group (Table 1). In the immature sequence this residue in Gln, showing again that, as in mammals, an N-terminal Gln group is converted into <Glu. Thus, the same 4–5 processing enzymes that we find in mammals (Fig. 2) probably have their

counterparts in cnidarians and the processing enzymes, therefore, must have a long evolutionary history.

In addition to all the similarities that occur between mammalian and cnidarian neuropeptide biosynthesis, there are also clear differences. One major difference lies in the amino acid sequences preceding the immature neuropeptide sequence. In mammals, as mentioned earlier, these sequences consist of one or more basic amino acid residues. In cnidarians, the sequences can include many types of residues, including basic residues. In the MMA precursor, for example, the immature MMA sequences are preceded by X-Ala or X-Pro sequences (Fig. 4). This points to the involvement of a dipeptidyl aminopeptidase in MMA precursor processing (Leviev and Grimmelikhuijzen, 1995). From earlier work, dipeptidyl aminopeptidase was known to be responsible for the final processing of N-terminal extended yeast α-mating factor, honey bee mellitin (a venome peptide), moth cecropin (an antibacterial peptide) and caerulein and xenopsin, two peptides from the frog skin (Kreil, 1990). The MMA prohormone in sea anemones, however, is the first example, where dipeptidyl aminopeptidase would play a role in neurohormone processing.

In addition to X-Ala or X-Pro sequences, other amino acid residues could precede the immature neuropeptide sequences in cnidarian prohormones. These could be basic residues (Arg, Lys), acidic residues (Asp, Glu), Asn, aliphatic or aromatic residues (Tyr, Phe, Leu, Val), Thr, Ser, or Gly (Grimmelikhuijzen, Leviev and Carstensen, 1996). For removal of some of these residues an unspecific aminopeptidase could be responsible, whereas for the acidic residues some arguments exist that this could be a more specific endoproteinase (Grimmelikhuijzen, Leviev and Carstensen, 1996). An example of a cnidarian prohormone, where N-terminal processing occurs at flanking acidic residues, is shown in Fig. 5. This Figure shows the preprohormone for Antho-RFamide (<Glu-Gly-Arg-Phe-NH_2) from the sea pansy, *Renilla köllikeri* (Anthozoa). The immature Antho-RFamide sequences (36 copies *plus* two putative neuropeptide sequences!) are organized in a very regular way and are nearly exclusively preceded by Glu residues.

Thus, in addition to the 4–5 processing enzymes that we know from mammals, cnidarians use various other enzymes that all process at the N terminus of the immature neuropeptide sequence. This makes neuropeptide biosynthesis in cnidarians more complex than in higher organisms.

4. Neuropeptide Amidation

We have chosen to investigate the peptide amidation enzymes (step 3 in Fig. 2) in cnidarians in more detail, because C-terminal peptide amidation is very essential for the biological actions of the cnidarian neuropeptides: all neuropeptides isolated so far from cnidarians are C-terminally amidated (Tables 1 and 2) and without this amidation, these peptides are not active (cf., Yum *et al.*, 1998a). Before discussing the cnidarian amidating enzymes, we would like to give a

```
                                                        TTTAGTATTGCGTCGTGTGTGTGTGTGTGTGTGTG   35
TTTGGCTCATTTTCTACGATCAACGTTGCTCTACACATTTGTATGGAGAAAACCTGTCAGTCTATAACGTATATACTTC  114
AAAATAATGGACTTAGTAATATAAGTATCAACTGCATTATTGTTGTCAATGTCCA ATG GTT AGT CTG GGT TTT  187
                                                        Met Val Ser Leu Gly Phe    6

TTC GTT CGT GAT GTT ACT CCA ACA TTC ATT GTA GAT CAT ATG TTC TAC ATG CTA TTC CCG  247
Phe Val Arg Asp Val Thr Pro Thr Phe Ile Val Asp His Met Phe Tyr Met Leu Phe Pro   26

ATG GAT TTT ACG TGT TAC GTT GCC GGG CTG TTG TTG ATA TTA AAT ACT TAC AGT TTG GCC  307
Met Asp Phe Thr Cys Tyr Val Ala Gly Leu Leu Leu Ile Leu Asn Thr Tyr Ser Leu Ala   46

GGG CCC TCA ACT AGC GAA GGA CTA AAC GAA CGG AAC TTG CTG GAT AAA ACA GAG TTG TCG  367
Gly Pro Ser Thr Ser Glu Gly Leu Asn Glu Arg Asn Leu Leu Asp Lys Thr Asp Leu Ser   66

ATA AAT GAC GAG ATA TTT AGC GAA GAT GAT GAT ATG CTG GCA AGG GAC GCT GAA GAC AAA  427
Ile Asn Asp Glu Ile Phe Ser Glu Asp Asp Asp Met Leu Ala Arg Asp Ala Glu Asp Lys   86

CAA GGA CGA TTT AAT CGT AAA TTA AAT AAC AAG TTG AAT GAA GCG GTA CAA GGA CGC TTC  487
Gln Gly Arg Phe Asn Arg Lys Leu Asn Asn Lys Leu Asn Glu Ala Val Gln Gly Arg Phe  106

GGG AGA AAT GAG AGA AAA GAA TCG GAA GAA GAA CAA GGA AGG TTC GGG CGA GAA AAT GAA  547
Gly Arg Asn Glu Arg Lys Glu Ser Glu Glu Glu Gln Gly Arg Phe Gly Arg Glu Asn Glu  126

AAA CAA GGA AGA TTT GGA AGG GAA AGC GAG GAG CAA GGG AGA TTT GGA CGA GAA AAC AAA  607
Lys Gln Gly Arg Phe Gly Arg Glu Ser Glu Glu Gln Gly Arg Phe Gly Arg Glu Asn Lys  146

GAA CAA GGA AGA TTT GGA AGG GAA AAC AAA GAA CAA GGA AGG TTT GGA CGA GAA AAC GAG  667
Glu Gln Gly Arg Phe Gly Arg Glu Asn Lys Glu Gln Gly Arg Phe Gly Arg Glu Asn Glu  166

GAA CAA GGA AGA TTT GGA AGG GAA AGC GAG GAG CAA GGG AGA TTT GGA AGA GAA AAC GAA  727
Glu Gln Gly Arg Phe Gly Arg Glu Ser Glu Glu Gln Gly Arg Phe Gly Arg Glu Asn Glu  186

GAA CAA GGA AGA TTT GGA CGA GAA AAC GAA GAA CAA GGA AGA TTT GGA CGA GAA AAC GAA  787
Glu Gln Gly Arg Phe Gly Arg Glu Asn Glu Glu Gln Gly Arg Phe Gly Arg Glu Asn Glu  206

GAA CAA GGA AGG TTT GGA CGA GAA AAC GAG GAG CAA GGG AGA TTT GGA AGG GAG AAC GAA  847
Glu Gln Gly Arg Phe Gly Arg Glu Asn Glu Glu Gln Gly Arg Phe Gly Arg Glu Asn Glu  226

GTT CAA GGA AGA TTT GGA CGA GAG AAC GAG GAG CAA GGA AGA TTT GGA CGA GAA AAC GAA  907
Val Gln Gly Arg Phe Gly Arg Glu Asn Glu Glu Gln Gly Arg Phe Gly Arg Glu Asn Glu  246

GAA CAA GGA AGG TTT GGA AGA GAA AAC GAA GAA CAA GGA CGA TTT GGA CGA GAA AAC GAG  967
Glu Gln Gly Arg Phe Gly Arg Glu Asn Glu Glu Gln Gly Arg Phe Gly Arg Glu Asn Glu  266

GAG CAA GGA AGA TTT GGA CGA GAA AAT GAA GAA CAA GGA AGA TTT GGA CGA GAA AAC GAG 1027
Glu Gln Gly Arg Phe Gly Arg Glu Asn Glu Glu Gln Gly Arg Phe Gly Arg Glu Asn Glu  286

GAG CAA GGA AGA TTT GGA CGA GAA AAC GAA AAA CAA GGA CGA TTT GGA CGA GAA AAC GAA 1087
Glu Gln Gly Arg Phe Gly Arg Glu Asn Glu Lys Gln Gly Arg Phe Gly Arg Glu Asn Glu  306

GAA CAA GGA AGA TTT GGA CGA GGA AAC GAG GAG CAA GGG AGA TTT GGA AGG GAA AAC GAA 1147
Glu Gln Gly Arg Phe Gly Arg Gly Asn Glu Glu Gln Gly Arg Phe Gly Arg Glu Asn Glu  326

GAA CAA GGA AGA TTT GGA CGA GAA AAC GAA GAA CAA GGA AGA TTT GGA CGA GAA AAC GAA 1207
Glu Gln Gly Arg Phe Gly Arg Glu Asn Glu Glu Gln Gly Arg Phe Gly Arg Glu Asn Glu  346

GAA CAA GGA CGA TTT GGG CGA GAA AAC GAA GAA CAA GGA AGG TTT GGG CGA GAA AAC GAA 1267
Glu Gln Gly Arg Phe Gly Arg Glu Asn Glu Glu Gln Gly Arg Phe Gly Arg Glu Asn Glu  366

GAA CAA GGA AGG TTT GGA AGG GAG AAC GAG AAG CAA GGG AGA TTT GGA AGA GGG GAC GAA 1327
Glu Gln Gly Arg Phe Gly Arg Glu Asn Glu Lys Gln Gly Arg Phe Gly Arg Gly Asp Glu  386

GAA CAA GGA AGG TTT GGA AGG GAG AAC GAG GAG CAA GGG AGA TTT GGA AGA GGG GAC GAA 1387
Glu Gln Gly Arg Phe Gly Arg Glu Asn Glu Glu Gln Gly Arg Phe Gly Arg Gly Asp Glu  406

GAA CAA GGA AGA TTT GGA CGA GAA AAC GAA GAG CAA GGG AGA TTT GGA AGA GAA AAC AAA 1447
Glu Gln Gly Arg Phe Gly Arg Glu Asn Glu Glu Gln Gly Arg Phe Gly Arg Glu Asn Lys  426

GAA CAA GGA AGA TTT GGA AGA GAA AAC GAA GAA CAA GGA AGA TTT GGA AGG GGG AAC AAA 1507
Glu Gln Gly Arg Phe Gly Arg Glu Asn Glu Glu Gln Gly Arg Phe Gly Arg Gly Asn Lys  446

GAA CAA GGA AGA TTT GGA CGA GAA AAC GAA CAA GGA AGA TTT GGA AGA GAA AAC GAA GTT 1567
Glu Gln Gly Arg Phe Gly Arg Glu Asn Glu Gln Gly Arg Phe Gly Arg Glu Asn Glu Val  466

CAA GGA AGG TTT GGA AGA TTC AGT CGG GAG TTG GCG AAA GGT TTA AAG ATT GAC GAT GTT 1627
Gln Gly Arg Phe Gly Arg Phe Ser Arg Glu Leu Ala Lys Gly Leu Lys Ile Asp Asp Val  486

CTC TGA  CAATGAACTAATTACGTGAATTACTAGAAACAAACTTAGAGAAAGTTGTTTATGAATCTATCAATAGTAT 1703
Leu  *                                                                             487

TTAAAAAGCGTTTCCAAATATTTAGTGTGAAATGATATTTTAAAAAAATATACACCGAAAAAAAAAAAAAGGCCGATCG 1782
GGCCG(A)32                                                                        1819
```

Figure 5. cDNA and deduced amino acid sequence of the Antho-RFamide (<Glu-Gly-Arg-Phe-NH_2) preprohormone from the sea pansy, *Rennilla köllikeri*. Neuropeptide sequences are underlined. (From Reinscheid and Grimmelikhuijzen, 1994)

short review on the occurrence of these enzymes in higher animals. In mammals, PHM and PAL (Fig. 3) are part of one bifunctional enzyme PAM (peptidyl α-amidating monooxygenase) that is coded for by a single gene. In this gene, the PHM coding exons are located at the 5′ end, whereas those coding for PAL are at the 3′ end (Kato *et al.*, 1990; Perkins, Husten and Eipper, 1990; Eipper, Stoffers and Mains, 1992; Quafik *et al.*, 1992). In the frog, *Xenopus laevis* there are two closely related genes for PAM (Mizuno *et al.*, 1987; Ohsuye *et al.*, 1988). In the fruitfly, *Drosophila melanogaster*, however, a single PHM occurs, and no bifunctional PAM. On the genomic level, one gene for PHM could be cloned, but PAL appeared to be coded for by a separate and yet unidentified gene (Kolhekar *et al.*, 1997a). In the pond snail, *Lymnaea stagnalis*, finally, there exists a single PAM gene that codes for a multifunctional enzyme containing one PAL copy and four different copies of a PHM (Spijker *et al.*, 1999).

The situation in cnidarians is another variant of the above-mentioned possibilities and appears to be the hitherto most complex one. In the sea anemone, *Calliactis parasitica* there are at least two rather different genes for α-amidating enzymes. Gene-1 (CP1) codes only for PHM and not for PAL (Fig. 6; Hauser, Williamson and Grimmelikhuijzen, 1997). This gene expresses two splicing variants, A and B, that differ from each other by the alternative use of two exons (exons 8 and 9 in Fig. 6). The two splicing variants have 95% of their amino acid sequence in common. Fig. 7 shows the amino acid sequence of splicing variant B (Fig. 6B). This PHM is 364 amino acid residues long and has 42% sequence identity with rat PHM (the PHM part of rat PAM). Six residues in the rat PHM (five His and one Met residue, marked with an asterisk in Fig. 7) are known to form a complex with two Cu^{2+} ions, which are essential for the catalytic activity of PHM (Eipper *et al.*, 1995; Kolhekar *et al.*, 1997b; Prigge *et al.*, 1997). These six residues are conserved between the sea anemone PHM and the rat, *Drosophila* and *Xenopus* enzymes. Furthermore, the three-dimensional structure of rat PHM is assumed to be fixed with the help of five Cys-Cys bridges (Kolhekar *et al.*, 1997b). Eight of the ten Cys residues are conserved between the sea anemone, rat, *Drosophila* and *Xenopus* PHMs (marked with a filled circle in Fig. 7). In all important features, therefore, the sea anemone PHM resembles the PHMs (or PHM parts of PAM) from higher organisms.

After screening of *Calliactis parasitica* cDNA and genomic libraries with the cDNA coding for the first sea anemone PHM, we found a second sea anemone PHM gene that also coded for a sea anemone PAL (Williamson, Hauser and Grimmelikhuijzen, 2000). Fig. 8 gives a schematic representation of this gene (CP2) that contains 22 exons and 21 introns and that is more than 7 kb in size. As in mammals, the PHM part is located at the 5′ end and the PAL part at the 3′ end and the whole organization of the gene looks like mammalian PAM, suggesting that the CP2 gene codes for a bifunctional enzyme. The PHM part of CP2 has about 40% sequence identity with both CP1 and the PHM

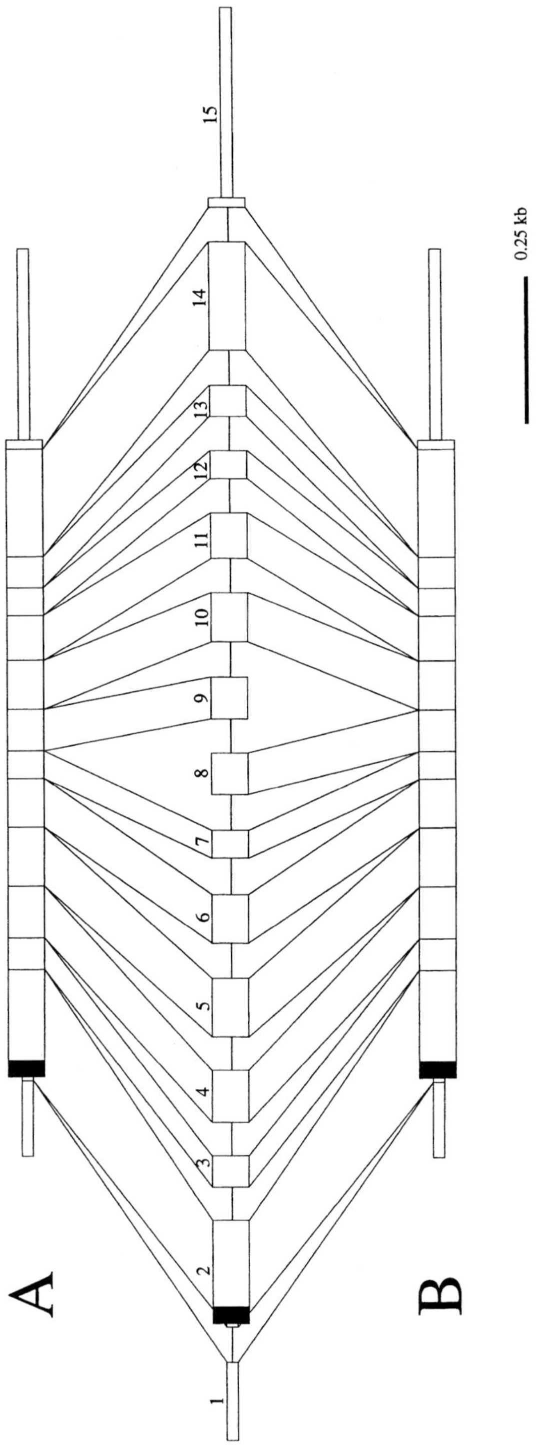

Figure 6. Schematic representation of the first PHM gene (CP1) from the sea anemone, *Calliactis parasitica* (middle) and its two splicing variants (A) and (B). Introns are given as lines and not drawn to scale. The exons are given as bars, being broad for coding regions and small for noncoding regions and are marked 1–15. The region coding for the signal sequence is highlighted in black. The two splicing variants differ by the alternative use of exons 8 (B) and 9 (A).

Figure 7. Alignments of the PHMs from *Calliactis parasitica* (CP1), *Drosophila melanogaster* (DRO), *Xenopus laevis* (XL1) and rat (RAT). The CP1 sequence given here corresponds to the PHM splicing variant containing exon 8 (Fig. 6B). Amino acid residues that are identical between the sea anemone enzyme and at least one of the other PHMs are boxed. The five His and one Met residues that form complexes with two Cu^{2+} atoms are indicated by asterisks and are printed in boldface type. The eight conserved Cys residues that form four putative Cys-Cys bridges, are marked with closed circles and are printed in boldface type as well. The data from *Drosophila*, *Xenopus* and rat are from Ohsuye *et al.* (1988), Kato *et al.* (1990), Quafik *et al.* (1992), and Kolhekar *et al.* (1997a). (From Hauser, Williamson, and Grimmelikhuijzen, 1997)

part of mammalian PAM. Its structural organization (with respect to the positions of essential His, Met and Cys residues) resembles that of CP1 (Fig. 7).

A Southern blot, using CP2 cDNA, strongly suggests that there is another gene in sea anemones resembling CP2 (Williamson, unpublished). Therefore the situation could be even more complex than depicted here. It is clear, however, that more enzymes are involved in cnidarian than in mammalian peptide α-amidation. This is a situation that resembles peptide N-terminal

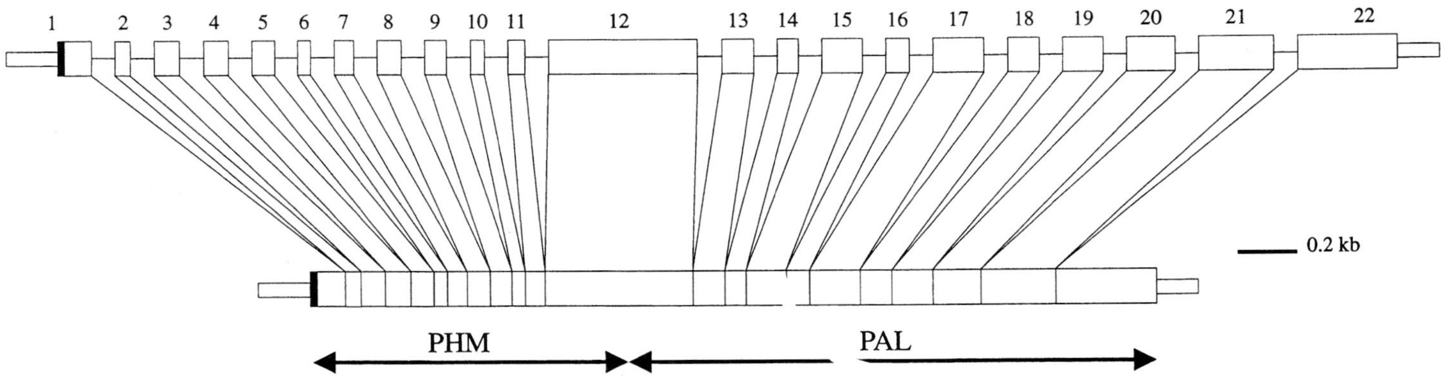

Figure 8. Schematic representation of the second PHM gene (CP2) from the sea anemone, *Calliactis parasitica* that also codes for a cnidarian PAL. This gene, therefore, resembles the organization of the mammalian PAM gene. The introns are given as lines and are not drawn to scale. The exons are given as bars, being broad for coding regions and small for noncoding regions and are numbered 1–22. The region coding for the signal sequence is highlighted in black. The gene is given at the top, the cDNA at the bottom.

processing (see above) and appears to be a paradox, considering the evolutionary status of cnidarians.

5. Coexpression of Preprohormones

The twelve neuropeptides that we and other groups have isolated from *Hydra* (Table 2) are synthesized by six different preprohormones, of which five are schematically represented in Fig. 9. Preprohormone A contains one copy of each Hydra-RFamide I, Hydra-RFamide II, Hydra-RFamide III/IV and two putative neuropeptide sequences. Preprohormone B contains one copy of

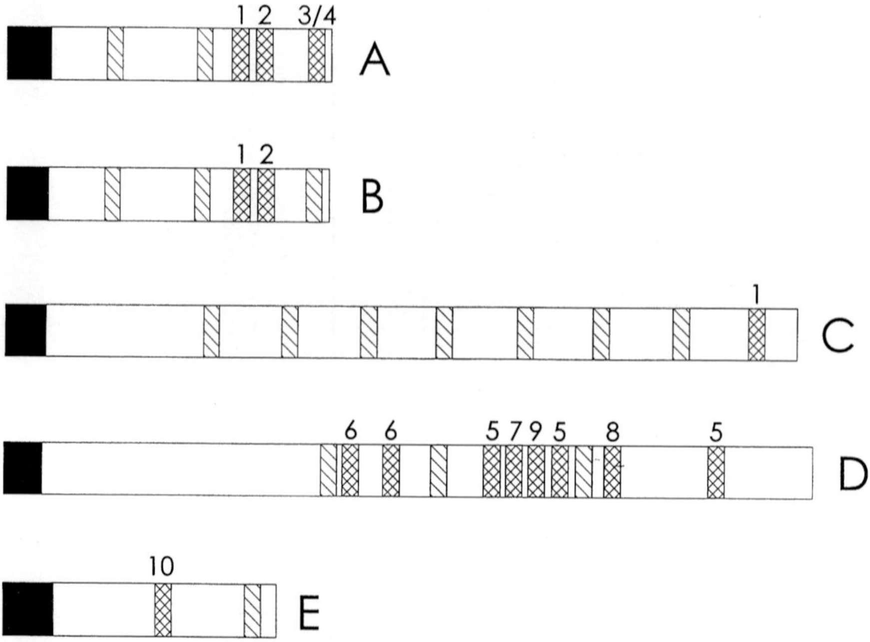

Figure 9. Schematic representation of five preprohormones in *Hydra*. (A) Preprohormone-A, containing a signal sequence (black box), one unprocessed sequence of Hydra-RFamide I (hatched and marked by 1), Hydra-RFamide II (2), Hydra-RFamide III/IV (3, 4) and several putative neuropeptide sequences (hatched, without numbers). (B) Preprohormone-B, containing a signal sequence (black) and one unprocessed sequence of Hydra-RFamide I (1), Hydra-RFamide II (2) and several putative neuropeptide sequences (hatched, without numbers). (C) Preprohormone-C, containing a signal sequence (black), one copy of unprocessed Hydra-RFamide I (1) and seven putative neuropeptide sequences (hatched, without numbers). (D) The Hydra-LWamide precursor, containing a signal sequence (black), unprocessed sequences of the isolated peptides Hydra-LWamide I-V (hatched and marked by 5–9) and three putative neuropeptide sequences (hatched, without numbers). (E) The Hydra-KVamide precursor, containing a signal sequence (black), one copy of unprocessed Hydra-KVamide (hatched and marked by 10) and another copy of a putative neuropeptide (hatched without number). The structures of the isolated *Hydra* peptides are given in Table 2. (From Hansen, Williamson, and Grimmelikhuijzen, 2000)

each Hydra-RFamide I and II and three putative neuropeptide sequences. Preprohormone C contains one copy of Hydra-RFamide I and seven putative neuropeptide sequences (Darmer *et al.*, 1998). Prepro-Hydra-LWamide (Fig. 9D) contains eight copies of the isolated Hydra-LWamides and three putative neuropeptide sequences (Leviev, Williamson and Grimmelikhuijzen, 1997). Prepro-Hydra-KVamide (Fig. 9E) contains one copy of Hydra-KVamide (Hym-176) and one putative neuropeptide (Yum *et al.*, 1998b), whereas prepro-Hydra-RGamide (not shown) contains one copy of Hydra-RGamide (Hym-355) and one other putative neuropeptide sequence (Takahashi *et al.*, 2000).

One interesting question, then, is the location of these six *Hydra* preprohormones. Are they localized in the same neurons, that is, is there, neurochemically speaking, only one nerve cell type? Or are there different populations of neurons that are neurochemically different and, thereby, also functionally different? We have investigated these questions with the help of whole-mount *in situ* hybridizations. Using cRNA probes specific for preprohormone-A, -B, -C and -Hydra-LWamide mRNAs, we found that these preprohormones were produced by different populations of neurons (Darmer *et al.*, 1998; Mitgutsch, Hauser and Grimmelikhuijzen, 1999; Hansen, Williamson and Grimmelikhuijzen, 2000). Preprohormone A is produced by a population of neurons occurring in the tentacles, hypostome (a region around the mouth opening), upper gastric region (just below the tentacles) and peduncle (a region just above the basal disk) (see also Fig. 10B). Preprohormone-B is produced by neurons of the hypostome and upper gastric region, and does not occur in other body parts, whereas preprohormone-C only occurs in neurons of the tentacles. Prepro-Hydra-LWamide mRNA is produced by neurons of the basal disk and tentacles in some animals, whereas in other animals it is produced by neurons occurring in all body regions (the reasons for this are unclear; Mitgutsch, Hauser and Grimmelikhuijzen, 1999; Hansen *et al.*, 2000). By using a two-colour whole-mount double-labelling *in situ* hybridization technique, we observed that different populations of neurons in the tentacles produce preprohormone-A, -C and -Hydra-LWamide mRNAs and that different neuronal populations in the foot produce preprohormone-A and -Hydra-LWamide (Mitgutsch, Hauser and Grimmelikhuijzen, 1999; Hansen *et al.*, 2000). Because of technical reasons (a cross-reaction between the preprohormone-A probe and preprohormone-B mRNA) we could not decide whether the same neurons in the hypostome and upper gastric region contained both preprohormone-A and -B or whether these were different populations (Mitgutsch, Hauser and Grimmelikhuijzen, 1999).

Using single-labelling *in situ* hybridization, Yum and coworkers (1998b) already determined that prepro-Hydra-KVamide was produced by a population of neurons in the peduncle and (more weakly) by some neurons in the rest of the *Hydra* body. We confirmed the presence of this preprohormone mRNA in neurons of the peduncle (Fig. 10C), but observed weakly expressing neurons only in the upper gastric region (Hansen *et al.*, 2000). Using two-colour double-labelling *in situ* hybridization, we subsequently determined that the neurons in the peduncle express both prepro-Hydra-KVamide and preprohormone-A

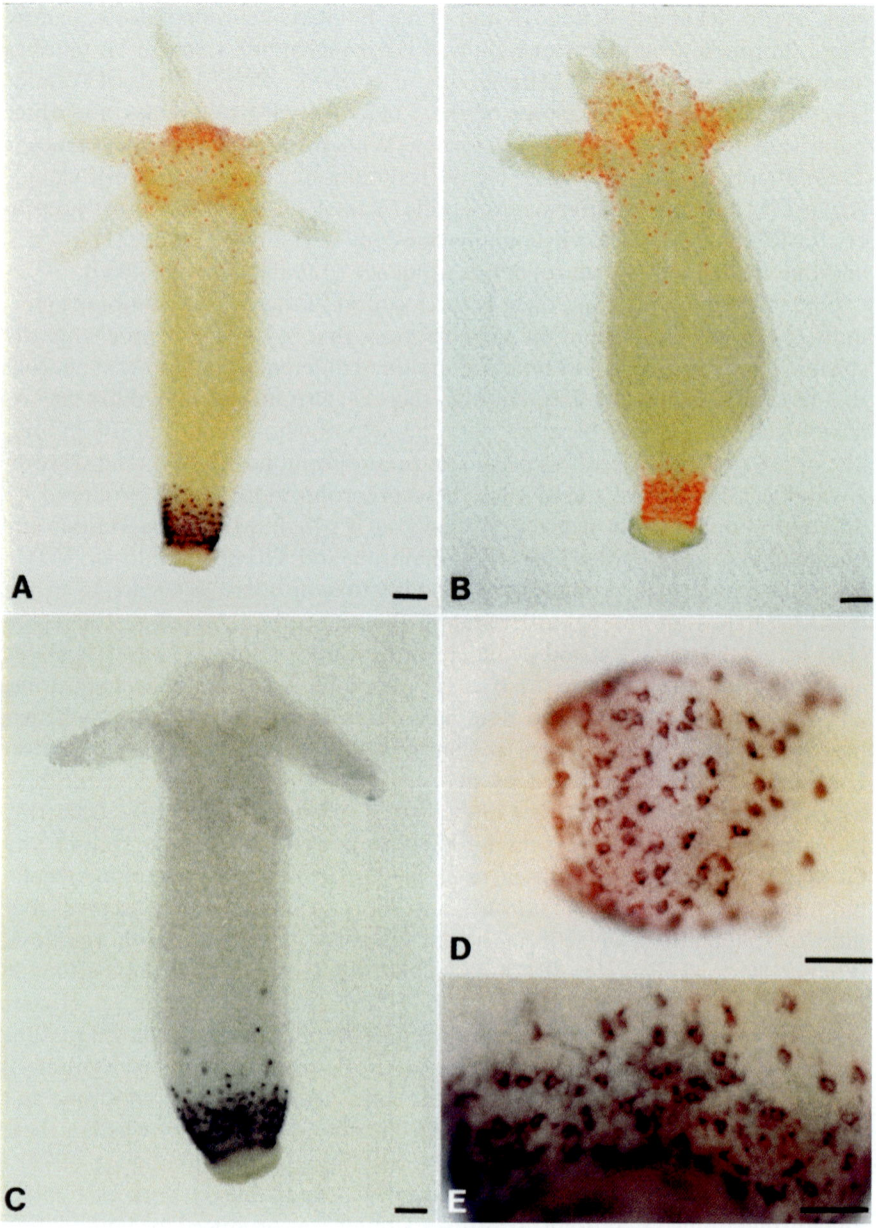

Figure 10. Whole-mount *in situ* hybridization of *Hydra* with cRNA probes coding for preprohormone-A and prepro-Hydra-KVamide. (A) Double-labelling *in situ* hybridization with specific cRNA probes coding for preprohormone-A (stained with Fast Red) and prepro-KVamide (stained with BM purple). Note that coexpression of the two mRNAs occurs in neurons from the peduncle, which now have obtained a brilliant violet colour. This colour is clearly different from the one shown in Figs. B and C. (B) *In situ* hybridization with the two specific prepro-KVamide and preprohormone-A cRNA probes. Staining, however, was only carried out for preprohormone-A

mRNAs (Fig. 10A, D, E), whereas the neurons in the upper gastric region express either one of the two mRNAs and so belong to different neuronal populations (Hansen *et al.*, 2000). This is the first example of the neuronal coexpression of two well-characterized preprohormones (yielding two classes of well-characterized neuropeptides) in cnidarians. Whether prepro-Hydra-RGamide (Takahashi *et al.*, 2000) is expressed by a separate population of neurons or is coexpressed by one of the neuronal populations described above, is presently unclear, but will be investigated. Clear at this moment is, that there are at least six neurochemically different classes of neurons in *Hydra*.

Westfall and coworkers concluded that *Hydra* has a single type of multifunctional neuron, the so-called "granule-containing sensory-motor-interneurons" (Westfall, 1973; Westfall and Kinnamon, 1978). In the light of our present experiments, this picture of the *Hydra* nervous system has to be revised somewhat. The neurons in *Hydra* might indeed be multifunctional, but they are not the same, and their different neurohormone contents suggest that they might have different functions.

6. Neurohormone Receptors

Because there are so many neurohormones in cnidarians (Tables 1 and 2), one would expect that there would be many neurohormone receptors as well. In an attempt to find these receptors, we applied PCR using cDNA from *Anthopleura elegantissima* as a template and primers coding for conserved sequences of mammalian G protein-coupled receptors. This approach yielded a receptor that clearly was not a receptor for one of the small sea anemone neuropeptides of Table 1 but, instead, showed a striking structural identity with members of the mammalian glycoprotein hormone (TSH, FSH, LH/CG) receptor family (Nothacker and Grimmelikhuijzen, 1993). This was an exciting

---

mRNA (Fast Red), showing that the colour nuance (red) in the penducle is clearly different from the one shown in A (violet). The *Hydra* specimen shown in this Figure is taken from a batch of animals that finally also yielded the animal shown in A, the only difference being that BM purple was not added. (C) *In situ* hybridization with two specific cRNA probes coding for either preprohormone-A or prepro-Hydra-KVamide mRNA. Staining was only carried out for prepro-Hydra-KVamide mRNA (BM purple) and the Fast Red staining (preprohormone-A mRNA) was omitted. Note that the colour nuance (dark blue) in the neurons from the peduncle is clearly different from that in A (violet). (D) Double-labelling *in situ* hybridization of the *Hydra* peduncle with cRNA probes specific for either preprohormone-A mRNA (stained with Fast Red) or prepro-Hydra-KVamide (stained with BM purple). Note the violet staining of the peduncle neurons, indicating coexistence of the two preprohomone mRNAs. The colour nuance in the neurons is about the same as in Fig. E, where the staining for the two preprohormone mRNAs was reverse, indicating that the two mRNAs are present in roughly comparable amounts. (E) Double-labelling *in situ* hybridization of *Hydra* peduncle neurons with cRNA probes specific for either preprohormone-A (stained with BM purple) or prepro-Hydra-KVamide (stained with Fast Red). Note that the colour nuance in the neurons (*violet*) is about the same in Fig. D. The bars in (A–C) represent 100 μm, in (D, E) 50 μm. (Adapted from Hansen, Williamson, and Grimmelikhuijzen, 2000)

finding, since glycoprotein hormone receptors had, in 1993, only been cloned from mammals and not from other vertebrates or invertebrates. We have recently also cloned the sea anemone glycoprotein hormone receptor gene (Vibede *et al.*, 1998), which confirmed the evolutionary relationship between the sea anemone and the mammalian receptors. Furthermore, we have cloned two glycoprotein hormone receptors from *Drosophila*, and other workers have cloned similar receptors from snails and nematodes, showing that these types of receptors are generally occurring in invertebrates (Tensen *et al.*, 1994; Hauser, Nothacker and Grimmelikhuijzen, 1997; Eriksen *et al.*, 2000; Kudo *et al.*, 2000). Consequently, the results show that the endocrinology of sea anemones must already be quite complex, involving at least 17 neuropeptides and, in addition, a new class of presumed glycoprotein hormones.

7. Conclusion

The primitive nervous system of cnidarians is strongly peptidergic. From a single sea anemone species, *Anthopleura elegantissima*, 17 neuropeptides have been isolated so far (14 of them are shown in Table 1). When we clone the preprohormones for these peptides (e.g., for MMA, see Fig. 4), we find several extra putative neuropeptides (e.g., eight putative peptides on the MMA preprohormone). From this, one can estimate that the total number of neuropeptides in a sea anemone species is 50–100. A similar number will probably be present in *Hydra* and other cnidarians. In addition to neuropeptides, other hormonal systems might be present in cnidarians, such as glycoprotein hormones and their receptors. This, together with the biosynthetic and colocalization data of the neuropeptides (headings 3–5), shows that the endocrinological systems in cnidarians are more complex and sophisticated than we previously might have assumed.

Acknowledgement

We want to thank Lotte Steffensen for typing the manuscript and the Danish National Science Research Council and Carlsberg Foundation for financial support.

References

Anderson, P.A.V., Moosler, A., and Grimmelikhuijzen, C.J.P. (1992) The presence and distribution of Antho-RFamide-like material in scyphomedusae. *Cell and Tissue Research* **267**, 67–74.

Campbell, R.D. (1976) Elimination of *Hydra* interstitial and nerve cells by means of colchicine. *Journal of Cell Science* **21**, 1–13.

Carstensen, K., Rinehart, K.L., McFarlane, I.D., and Grimmelikhuijzen, C.J.P. (1992) Isolation of Leu-Pro-Pro-Gly-Pro-Leu-Pro-Arg-Pro-NH_2 (Antho-RPamide), an N-terminally protected, biologically active neuropeptide from sea anemones. *Peptides* **13**, 851–857.

Carstensen, K., McFarlane, I.D, Rinehart, K.L, Hudman, D., Sun, F., and Grimmelikhuijzen, C.J.P. (1993) Isolation of <Glu-Asn-Phe-His-Leu-Arg-Pro-NH_2 (Antho-RPamide II), a novel, biologically active neuropeptide from sea anemones. *Peptides* **14**, 131–135.

Chan, S.J., Oliva, A.A., Lamendola, J., Grenns, A., Bode, H., and Steiner, D.F. (1992) Conservation of the prohormone convertase gene family in metazoa: Analysis of cDNAs encoding a PC3-like protein from hydra. *Proceedings of the National Academy of Sciences, USA* **89**, 6678–6682.

Darmer, D., Schmutzler, C., Diekhoff, D., and Grimmelikhuijzen, C.J.P. (1991) Primary structure of the precursor for the sea anemone neuropeptide AnthoRFamide (<GluGlyArgPheNH_2). *Proceedings of the National Academy of Sciences, USA* **88**, 2555–2559.

Darmer, D., Hauser, F., Nothacker, H.-P., Bosch, T.C.G., Williamson, M., and Grimmelikhuijzen, C.J.P. (1998) Three different prohormones yield a variety of Hydra-RFamide (Arg-Phe-NH_2) neuropeptides in *Hydra magnipapillata*. *Biochemical Journal* **332**, 403–412.

Eipper, B.A., Stoffers, D.A., and Mains, R.E. (1992) The biosynthesis of neuropeptides: Peptide α-amidation. *Annual Review of Neurosciences* **15**, 57–85.

Eipper, B.A., Quon, A.S.W., Mains, R.E., Boswell, J.S., and Blackburn, N.J. (1995) The catalytic core of peptidylglycine α-hydroxylating monooxygenase-investigation by site-directed mutagenesis, Cu X-ray absorption spectroscopy and electron-paramagenetic resonance. *Biochemistry* **34**, 2857–2865.

Eriksen, K.K., Hauser, F., Schiøtt, M., Pedersen, K.-M., Søndergaard, L., and Grimmelikhuijzen, C.J.P. (2000) Molecular cloning, genomic organization, developmental regulation, and a knock-out mutant of a novel Leu-rich repeats-containing G protein-coupled receptor (DLGR-2) from *Drosophila melanogaster*. *Genome Research* **10**, 924–938.

Fischer, W.H., and Spiess, J. (1987) Identification of a mammalian glutaminyl cyclase converting glutaminyl into pyroglutaminyl peptides. *Proceedings of the National Academy of Sciences, USA* **84**, 3628–3632.

Fricker, L.D., Adelman, J.P., Douglass, J., Thomson, R.C., von Strandmann, R.P., and Hutton, J. (1989) Isolation and sequence analysis of cDNA for rat carboxypeptidase E [EC 3.4.17.10], a neuropeptide processing enzyme. *Molecular Endocrinology* **3**, 666–673.

Gierer, A. (1977) Biological features and physical concepts of pattern formation exemplified by *Hydra*. *Current Topics in Developmental Biology* **11**, 17–59.

Gierer, A., Berking, S., Bode, H., David, C.N., Flick, K., Hansmann, G., Schaller, H., and Trenckner, E. (1972) Regeneration of *Hydra* from reaggregated cells. *Nature, New Biology* **239**, 98–101.

Graff, D., and Grimmelikhuijzen, C.J.P. (1988a) Isolation of <Glu-Ser-Leu-Arg-Trp-NH_2, a novel neuropeptide from sea anemones. *Brain Research* **442**, 354–358.

Graff, D., and Grimmelikhuijzen, C.J.P. (1988b) Isolation of <Glu-Gly-Leu-Arg-Trp-NH_2 (Antho-RWamide II), a novel neuropeptide from sea anemones. *FEBS Letters* **239**, 137–140.

Grimmelikhuijzen, C.J.P. (1985) Antisera to the sequence Arg-Phe-amide visualize neuronal centralization in hydroid polyps. *Cell and Tissue Research* **241**, 171–182.

Grimmelikhuijzen, C.J.P., and Spencer, A.N. (1984) FMRFamide immunoreactivity in the nervous system of the medusa *Polyorchis penicillatus*. *Journal of Comparative Neurology* **230**, 361–371.

Grimmelikhuijzen, C.J.P., and Graff, D. (1986) Isolation of <Glu-Gly-Arg-Phe-NH_2 (Antho-RFamide), a neuropeptide from sea anemones. *Proceedings of the National Academy of Sciences, USA* **83**, 9817–9821.

Grimmelikhuijzen, C.J.P., and Groeger, A. (1987) Isolation of the neuropeptide pGlu-Gly-Arg-Phe-amide from the pennatulid *Renilla köllikeri*. *FEBS Letters* **211**, 105–108.

Grimmelikhuijzen, C.J.P., Dockray, G.J., and Schot, L.P.C. (1982) FMRFamide-like immunoreactivity in the nervous system of *Hydra*. *Histochemistry* **73**, 499–508.

Grimmelikhuijzen, C.J.P., Spencer, A.N., and Carré, D. (1986) Organization of the nervous system of physonectid siphonophores. *Cell and Tissue Research* **246**, 463–479.

Grimmelikhuijzen, C.J.P., Rinehart, K.L., and Spencer, A.N. (1992) Isolation of the neuropeptide <Glu-Trp-Leu-Lys-Gly-Arg-Phe-NH_2 (Pol-RFamide II) from the hydromedusa *Polyorchis penicillatus*. *Biochemical and Biophysical Research Communications* **183**, 375–382.

Grimmelikhuijzen, C.J..P., Leviev, I., and Carstensen, K. (1996) Peptides in the nervous systems of cnidarians: structure, function and biosynthesis. *International Review of Cytology* **167**, 37–89.

Grimmelikhuijzen, C.J.P., Hahn, M., Rinehart, K.L., and Spencer, A.N. (1988) Isolation of <Glu-Leu-Leu-Gly-Gly-Arg-Phe-NH_2 (Pol-RFamide), a novel neuropeptide from hydromedusae. *Brain Research* **475**, 198–203.

Grimmelikhuijzen, C.J.P., Rinehart, K.L., Jacob, E., Graff, D., Reinscheid, R.K., Nothacker, H.-P., and Staley, A.L. (1990) Isolation of L-3-phenyllactyl-Leu-Arg-Asn-NH_2 (Antho-RNamide), a sea anemone neuropeptide containing an unusual amino-terminal blocking group. *Proceedings of the National Academy of Sciences, USA* **87**, 5410–5414.

Grimmelikhuijzen, C.J.P., Carstensen, K., Darmer, D., Moosler, A., Nothacker, H.P., Reinscheid, R.K., Schmutzler, C., Vollert, H., McFarlane, I.D., and Rinehart, K.L. (1992b) Coelenterate neuropeptides: structure, action and biosynthesis. *American Zoologist* **32**, 112.

Hansen, G.N., Williamson, M., and Grimmelikhuijzen, C.J.P. (2000) Two-color double labeling *in situ* hybridization of whole mount *Hydra* using RNA probes for five different *Hydra* neuropeptide preprohormones: evidence for colocalisation. *Cell and Tissue Research* **301**, 245–253.

Hauser, F., Nothacker, H.-P., and Grimmelikhuijzen, C.J.P. (1997) Molecular cloning, genomic organization and developmental regulation of a novel receptor from *Drosophila melanogaster* structurally related to members of the thyroid-stimulating hormone, follicle-stimulating hormone, luteinizing hormone/choriogonadotropin receptor family from mammals. *Journal of Biological Chemistry* **272**, 1002–1010.

Hauser, F., Williamson, M., and Grimmelikhuijzen, C.J.P. (1997) Molecular cloning of a peptidyl-glycine α-hydroxylating monooxygenase from sea anemones. *Biochemical and Biophysical Research Communications* **241**, 509–512.

Kato, I., Yonekura, H., Tajima, M., Yanagi, M., Yamamoto, H., and Okamoto, H. (1990) Two enzymes concerned in peptide hormone a-amidation are synthesized from a single mRNA. *Biochemical and Biophysical Research Communications* **172**, 197–203.

Koizumi, O., Wilson, J.D., Grimmelikhuijzen, C.J.P., and Westfall, J.A. (1989) Ultrastructural localization of RFamide-like peptides in neuronal dense-cored vesicles in the peduncle of *Hydra*. *Journal of Experimental Zoology* **249**, 17–22.

Kolhekar, A.S., Keutmann, H.T., Mains, R.E., Quon, A.S.W., and Eipper, B.A. (1997a) Peptidylglycine a-Hydroxylating Monooxygenase: Active Site Residues, Disulfide Linkages, and a Two-Domain Model of the Catalytic Core. *Biochemistry* **36**, 10901–10909.

Kolhekar, A.S., Roberts, M.S., Jiang, N., Johnson, R.C., Mains, R.E., Eipper, B.A., and Taghert, P.H. (1997b) Neuropeptide amidation in *Drosophila*: separate genes encode the two enzymes catalyzing amidation. *Journal of Neuroscience* **17**, 1363–1376.

Kreil, G. (1990) Processing of precursors by dipeptidylaminopeptidases: A case of molecular ticketing. *Trends in Biochemical Sciences* **15**, 23–26.

Kudo, M., Chen, T., Nakabayashi, K., Hsu, S.Y., and Hsueh, A.J.W. (2000) The nematode leucine-rich repeat-containing, G protein-coupled receptor (LGR) protein homologous to vertebrate gonadotropin and thyrotropin receptors is constitutively activated in mammalian cells. *Molecular Endocrinology* **14**, 272–284.

Leitz, T., and Lay, M. (1995) Metamorphosin A is a neuropeptide. *Roux's Archives of Developmental Biology* **204**, 276–279.

Leitz, T., Morand, K., and Mann, M. (1994) Metamorphosin A: A novel peptide controlling development of the lower metazoan *Hydractinia echinata* (Coelenterata, Hydrozoa). *Developmental Biology* **163**, 440–446.

Leviev, I., and Grimmelikhuijzen, C.J.P. (1995) Molecular cloning of a preprohormone from sea anemones containing numerous copies of a metamorphosis inducing neuropeptide: A likely role for dipeptidyl aminopeptidase in neuropeptide precursor processing. *Proceedings of the National Academy of Sciences, USA* **92**, 11647–11651.

Leviev, I., Williamson, M., and Grimmelikhuijzen, C.J.P. (1997) Molecular cloning of a preprohormone from *Hydra magnipapillata* containing multiple copies of Hydra-LWamide (Leu-Trp-NH_2) neuropeptides: Evidence for processing at Ser and Asn residues. *Journal of Neurochemistry* **68**, 1319–1325.

McFarlane, I.D., and Grimmelikhuijzen, C.J.P. (1991) Three anthozoan neuropeptides, Antho-RFamide and Antho-RWamides I and II, modulate spontaneous tentacle contractions in sea anemones. *Journal of Experimental Biology* **155**, 669–673.

McFarlane, I.D., Graff, D., and Grimmelikhuijzen, C.J.P. (1987) Exitatory actions of Antho-RFamide, an anthozoan neuropeptide, on muscles and conducting systems in the sea anemone *Calliactis parasitica*. *Journal of Experimental Biology* **133**, 157–168.

McFarlane, I.D., Anderson, P.A.V., and Grimmelikhuijzen, C.J.P. (1991) Effects of three anthozoan neuropeptides, Antho-RWamide I, Antho-RWamide II and Antho-RFamide, on slow muscles from sea anemones. *Journal of Experimental Biology* **156**, 419–431.

McFarlane, I.D., Reinscheid, R.K., and Grimmelikhuijzen, C.J.P. (1992) Opposite actions of the anthozoan neuropeptide Antho-RNamide on antagonistic muscle groups in sea anemones. *Journal of Experimental Biology* **164**, 295–299.

McFarlane, I.D., Hudman, D., Nothacker, H.-P., and Grimmelikhuijzen, C.J.P. (1993) The expansion behaviour of sea anemones may be coordinated by two inhibitory neuropeptides, Antho-KAamide and Antho-RIamide. *Proceedings of the Royal Society of London, Series B* **253**, 183–188.

Minobe, S., Koizumi, O., and Sugiyama, T. (1995) Nerve cell differentiation in epithelial *Hydra* from precursor cells introduced by grafting. I. Tentacles and hypostome. *Developmental Biology* **172**, 170–181.

Mitgutsch, C., Hauser, F., and Grimmelikhuijzen, C.J.P. (1999) Expression and developmental regulation of the Hydra-RFamide and Hydra-LWamide preprohormone genes in *Hydra*: evidence for transient phases of head formation. *Developmental Biology* **207**, 189–203.

Mizuno, K., Ohsuye, K., Wada, Y., Fuchimura, K., Tanaka, S., and Matsuo, H. (1987) Cloning and sequencing of a cDNA encoding a peptide C-terminal α-amidating enzyme from *Xenopus laevis*. *Biochemical and Biophysical Research Communications* **148**, 546–552.

Moosler, A., Rinehart, K.L., and Grimmelikhuijzen, C.J.P. (1996) Isolation of four novel neuropeptides, the Hydra-RFamides I-IV, from *Hydra magnipapillata*. *Biochemical and Biophysical Research Communications* **229**, 596–602.

Moosler, A., Rinehart, K.L., and Grimmelikhuijzen, C.J.P. (1997) Isolation of three novel neuropeptides, the Cyanea-RFamides I-III, from scyphomedusae. *Biochemical and Biophysical Research Communications* **236**, 743–749.

Muller, W.A. (2003) Signalling systems in cnidaria. In: I. Fairweather (ed.) *Cell Signalling in Prokaryotes and Lower Metazoa*. Kluwer Academic Publishers, Dordrecht, pp. 91–114.

Nothacker, H.-P., and Grimmelikhuijzen, C.J.P. (1993) Molecular cloning of a novel, putative G protein-coupled receptor from sea anemones structurally related to members of the FSH, TSH, LH/CG receptor family from mammals. *Biochemical and Biophysical Research Communications* **197**, 1062–1069.

Nothacker, H.-P., Rinehart, K.L., and Grimmelikhuijzen, C.J.P. (1991) Isolation of L-3-phenyllactyl-Phe-Lys-Ala-NH_2 (Antho-KAamide), a novel neuropeptide from sea anemones. *Biochemical and Biophysical Research Communications* **179**, 1205–1211.

Nothacker, H.-P., Rinehart, K.L., McFarlane, I.D., and Grimmelikhuijzen, C.J.P. (1991) Isolation of two novel neuropeptides from sea anemones: The unusual, biologically active L-3-phenyllactyl-Tyr-Arg-Ile-NH_2 and its des-phenyllactyl fragment Tyr-Arg-Ile-NH_2. *Peptides* **12**, 1165–1173.

Ohsuye, K., Kitano, K., Wada, Y., Fuchimura, K., Tanaka, S., Mizuno, K., and Matsuo, H. (1988) Cloning of cDNA encoding a new peptide C-terminal α-amidating enzyme having a putative membrane-spanning domain from *Xenopus laevis* skin. *Biochemical and Biophysical Research Communications* **150**, 1275–1281.

Perkins, S.N., Husten, E.J., and Eipper, B.A. (1990) The 108 kDA peptidylglycine α-amidating monooxygenase precursor contains two separable enzymatic activities involved in peptide amidation. *Biochemical and Biophysical Research Communications* **171**, 926–932.

Pohl, T., Zimmer, M., Mugele, K., and Spiess, J. (1991) Primary structure and functional expression of glutaminyl cyclase. *Proceedings of the National Academy of Sciences, USA* **88**, 10059–10063.

Prigge, S.T., Kolhekar, A.S., Eipper, B.A., Mains, R.E., and Amzel, L.M. (1997) Amidation of bioactive peptides: the structure of α-hydroxylating monooxygenase. *Science* **278**, 1300–1305.

Quafik, L., Stoffers, D.A., Campbell, T.A., Johnson, R.C., Bloomquist, B.T., Mains, R.E., and Eipper,

B.A. (1992) The multifunctional peptidylglycine α-amidating monooxygenase gene: exon/intron organization of catalytic, processing, and routing domains. *Molecular Endocrinology* **6**, 1571–1584.

Reinscheid, R.K., and Grimmelikhuijzen, C.J.P. (1994) Primary structure of the precursor for the anthozoan neuropeptide Antho-RFamide from *Renilla köllikeri*: Evidence for unusual processing enzymes. *Journal of Neurochemistry* **62**, 1214–1222.

Schmutzler, C., Diekhoff, D., and Grimmelikhuijzen, C.J.P. (1994) The primary structure of the Pol-RFamide neuropeptide precursor protein from the hydromedusa *Polyorchis penicillatus* indicates a novel processing proteinase activity. *Biochemical Journal* **299**, 431–436.

Schmutzler, C., Darmer, D., Diekhoff, D., and Grimmelikhuijzen, C.J.P. (1992) Identification of a novel type of processing sites in the precursor for the sea anemone neuropeptide AnthoRFamide (<GluGlyArgPheNH$_2$) from *Anthopleura elegantissima*. *Journal of Biological Chemistry* **267**, 22534–22541.

Seidah, N.G., Gaspar, L., Mion, P., Marcinkiewicz, M., Mbikay, M., and Chrétien, M. (1990) cDNA sequence of two distinct pituitary proteins homologous to Kex2 and furin gene products: Tissue-specific mRNAs encoding candidates for pro-hormone processing proteinases. *DNA and Cell Biology* **9**, 415–424.

Seidah, N.G., Marcinkiewicz, M., Benjannet, S., Gaspar, L., Beaubien, G., Mattei, M.G., Lazure, C., Mbikay, M., and Chrétien, M. (1991) Cloning and primary sequence of a mouse candidate prohormone convertase PC1 homologous to PC2, furin and Kex2: Distinct chromosomal localization and messenger RNA distribution in brain and pituitary compared to PC2. *Molecular Endocrinology* **5**, 111–122.

Smeekens, S.P., and Steiner, D.F. (1990) Identification of a human insulinoma cDNA encoding a novel mammalian protein structurally related to the yeast dibasic processing protease Kex2. *Journal of Biological Chemistry* **265**, 2997–3000.

Smeekens, S.P., Avruch, A.S., LaMendola, J., Chan, S.J., and Steiner, D.F. (1991) Identification of a cDNA encoding a second putative prohormone convertase related to PC2 in AtT20 cells and islets of Langerhans. *Proceedings of the National Academy of Sciences, USA* **88**, 340–344.

Spijker, S., Smit, A.S., Eipper, B.A., Malik, A., Mains, R.E., and Geraerts, W.P.M. (1999) A molluscan peptide α-amidating enzyme precursor that generates five distinct enzymes. *FASEB Journal* **13**, 735–748.

Sugiyama, T., and Fujisawa, T. (1978) Genetic analysis of developmental mechanisms in *Hydra*. II. Isolation and characterization of an interstitial cell deficient strain. *Journal of Cell Science* **29**, 35–52.

Takahashi, T., Muneoka, Y., Lohmann, J., Lopez de Haro, M.S., Solleder, G., Bosch, T.C.G., David, C.N., Bode, H.R., Koizumi, O., Shimizu, H., Hatta, M., Fujisawa, T., and Sugiyama, T. (1997) Systematic isolation of peptide signal molecules regulating development in *Hydra*; LWamide and PW families. *Proceedings of the National Academy of Sciences, USA* **94**, 1241–1246.

Takahashi, T., Koizumi, O., Ariura, Y., Romanovitch, A., Bosch, T.C.G., Kobayakawa, Y., Mohri, S., Bode, H.R., Yum, S., Hatta, M., and Fujisawa, T. (2000) A novel neuropeptide, Hym-355, positively regulates neuron differentiation in *Hydra*. *Development* **127**, 997–1005.

Tensen, C.P., van Kesteren, E.R., Planta, R.J., Cox, K.J.A., Burke, J.F., van Heerikhuizen, H., and Vreugdenhil, E. (1994) A G protein-coupled receptor with low density lipoprotein-binding motifs suggests a role for lipoproteins in G-linked signal transduction. *Proceedings of the National Academy of Sciences, USA* **91**, 4816–4820.

Vibede, N., Hauser, F., Williamson, M., and Grimmelikhuijzen, C.J.P. (1998) Genomic organization of a receptor from sea anemones, structurally and evolutionarily related to glycoprotein hormone receptors from mammals. *Biochemical and Biophysical Research Communications* **252**, 497–501.

Westfall, J.A. (1973) Ultrastructural evidence for a granule-containing sensory-motor interneuron in *Hydra littoralis*. *Journal of Ultrastructure Research* **42**, 268–282.

Westfall, J.A., and Kinnamon, J.C. (1978) A second sensory-motor-interneuron with neurosecretory granules in *Hydra*. *Journal of Neurocytology* **7**, 365–379.

Westfall, J.A., and Grimmelikhuijzen, C.J.P. (1993) Antho-RFamide immunoreactivity in neuronal synaptic and nonsynaptic vesicles of sea anemones. *Biological Bulletin* **185**, 109–114.

Westfall, J.A., Sayyar, K.L., Elliott, C.F., and Grimmelikhuijzen, C.J.P. (1995) Ultrastructural localization of Antho-RWamide I and II at neuromuscular synapses in the gastrodermis and oral sphincter muscle of the sea anemone *Calliactis parasitica*. *Biological Bulletin* **189**, 280–287.

Williamson, M., Hauser, F., and Grimmelikhuijzen, C.J.P. (2000) Genomic organization and splicing variants of a peptidylglycine alpha-hydroxylating monooxygenase from sea anemones. *Biochemical and Biophysical Research Communications* **277**, 7–12.

Yum, S., Takahashi, T., Hatta, M., and Fujisawa, T. (1998a) The structure and expression of a preprohormone of a neuropeptide, Hym-176 in *Hydra magnipapillata*.. *FEBS Letters* **439**, 31–34.

Yum, S., Takahashi, T., Koizumi, O., Ariura, Y., Kobayakawa, Y., Mohri, S., and Fujisawa, T. (1998b) A novel neuropeptide, Hym-176, induces contraction of the ectodermal muscle in *Hydra*. *Biochemical and Biophysical Research Communications* **248**, 584–590.

CHAPTER 6

SIGNALLING MECHANISMS IN PLATYHELMINTHS

IAN FAIRWEATHER

School of Biology and Biochemistry, The Queen's University of Belfast, Belfast BT9 7BL, Northern Ireland

Summary

Platyhelminths, or flatworms, include both free-living forms such as the freshwater planarians and parasitic forms – the flukes and tapeworms. Parasitic species infect both humans and livestock and a number represent important zoonotic infections of humans. The flatworms are the first metazoans to show centralization of the nervous system, with the formation of a "brain". They possess a variety of potential transmitter substances and this chapter will summarise what is known about their biological activity. Evidence is accumulating for the presence of individual components of signalling pathways that are known to exist in higher organisms. This evidence will be reviewed, as will the extent to which functional signalling pathways have been unravelled in flatworms. Manipulation of signalling pathways for therapeutic intervention will also be considered. Parasitic species utilize a multiplicity of environmental and host signals for different phases of their life cycle, and they have evolved mechanisms to mimic and/or manipulate host signals and responses to favour their establishment and development; these mechanisms will be discussed.

1. Introduction

The platyhelminths, or flatworms, occupy a strategic position at the base of the metazoan tree, being the first animals to display bilateral symmetry and undergo cephalization, with the development of a "brain". They are triploblastic acoelomate animals, the spaces between the organ systems being filled with a form of packing or connective tissue, termed parenchyma. The flatworms are typically dorsoventrally flattened, varying in shape from the leaf-like flukes to more elongate worms, the extreme of which is represented by the tapeworms, whose long, ribbon-like body is divided into a series of segments, or proglottides, each of which contains a set of reproductive organs. Flatworms lack a circulatory system and the gut (where present) has only one opening, so a definitive anus is absent. They are typically hermaphrodite, with complex reproductive systems, and the parasitic forms in particular are specialised for egg production (for egg output figures, see Fairweather and Halton, 1992).

As a group, the platyhelminths include the turbellarians, which are predominantly free-living forms as typified by the freshwater planarians. The trematodes (flukes) and cestodes (tapeworms) are exclusively parasitic, as ecto- or endoparasites, and their life cycles can be complex, involving a number of invertebrate and vertebrate hosts. A number of the parasites are of considerable

I. Fairweather (ed.), Cell Signalling in Prokaryotes and Lower Metazoa, 141–194.
© 2004 *Kluwer Academic Publishers. Printed in Great Britain.*

economic importance, either as parasites of man or livestock. For example, the human blood fluke, *Schistosoma spp.* infects some 260 million people in 74 countries, with 500–600 million people at risk of infection (Savioli *et al.*, 1997). The liver fluke, *Fasciola hepatica* infects 300 million cattle and 250 million sheep worldwide, causing economic losses of US $3 billion per annum (Boray, 1994). *F. hepatica* also infects man, with estimates ranging from 2.4 million (Crompton, 1999) to 17 million (Hopkins, 1992). The larval form of the dog tapeworm, *Echinococcus granulosus* infects man, causing the disease hydatidosis, although its normal intermediate host is sheep and other livestock. The infection level in man is 2–3 million, compared with more than 200 million sheep (Craig, Rogan and Allan, 1996). More familiar human tapeworms include the so-called beef tapeworm, *Taenia saginata* and the pork tapeworm, *Taenia solium*, with infection levels of 77 and 10 million, respectively (Crompton, 1999). In comparison, more than 50 million pigs are infected with *T. solium* (Craig, Rogan and Allan, 1996). Larval infections of man with these tapeworms are also important, resulting in the disease cysticercosis. Treatment of the neural form of this disease (neurocysticercosis) in the USA, Mexico and Brazil cost over US $180 million in 1992 (Craig, Rogan and Allan, 1996).

The nervous system of platyhelminths represents an important intermediate stage in the evolution of nervous systems. Part remains in the form of diffuse, peripheral nerve nets typical of the more simple cnidarians. The remainder has sunk to a submuscular position and become concentrated into longitudinal nerve cords, with two main ganglia in the head region forming the "brain". In the absence of true endocrine glands and a circulatory system, the nervous system of flatworms assumes the added role of endocrine system, as it is the only system capable of producing and transporting substances that can act in a hormone-like fashion. Historically, this capacity was linked to its possession of glandular nerve cells, the neurosecretory cells. The use of the term neurosecretion has been superceded by studies on neuropeptides, although monoamines can also serve as neurosecretory substances. Recent studies on the peptidergic component of the nervous system have helped to re-kindle interest in the flatworm nervous system which, for parasites in particular, had been largely ignored through a (false) assumption that it had become reduced or degenerate with the adoption of a parasitic lifestyle. Flatworms represent potentially important experimental models for studying processes under the control of hormone-type substances. For example, a typical adult tapeworm has been described as "essentially made up of a string of embryos of increasing degree of maturity, so that all developmental stages can be studied in one organism" (Smyth and, McManus, 1989). The pseudophyllidean tapeworm, *Diphyllobothrium dendriticum* displays a never-ending pattern of growth. Certain tapeworm larvae undergo interesting forms of asexual reproduction, involving longitudinal fission, and external or internal budding, making them good models for developmental studies, too. Freshwater planarians have been used extensively for studies on regeneration, their remarkable capacity for replacing missing parts underpinning the fissioning process that occurs in asexual strains.

Neglect of the flatworm nervous system has also stemmed from some of the problems associated with working with the nervous tissue. The nervous system has proved to be fairly inaccessible for electrophysiological studies because it cannot be dissected free of the surrounding parenchymal tissue and the small size of the nerve cells makes it difficult to penetrate them with microelectrodes. Many species are small and this, combined with the low concentrations of neuropeptides present, has made isolation of peptides difficult, with the result that very few have been identified in comparison with other groups.

The initial part of this Chapter will provide an overview of the range of potential signalling molecules present in flatworms, together with their functions. The chapter will then consider what evidence exists for the presence of individual components of signalling (second messenger) pathways in flatworms and discuss functional signalling pathways linked to individual neurotransmitters. Finally, it will consider how knowledge of signalling pathways could be exploited therapeutically for the treatment of parasitic infections.

2. Range of Transmitter Substances in Flatworms

2.1. FREE-LIVING FLATWORMS

Free-living flatworms possess a wide range of potential classical and peptidergic neurotransmitters. For example, the monoamines 5-hydroxytryptamine (5-HT, or serotonin), adrenaline, noradrenaline, dopamine and histamine have been localised within the nervous system by fluorescence and immunocytochemical techniques (see reviews by Martelly and Franquinet, 1984; Reuter and Gustafsson, 1995). Few physiological studies have been carried out to determine the neuromuscular roles of these substances, although dopaminergic agonists can induce screw-like, hyperkinetic-type patterns of behaviour (Venturini *et al.*, 1989). Gamma-aminobutyric acid (GABA) immunoreactivity has been localised in the planarian, *Dugesia tigrina* (Eriksson and Panula, 1994) and GABA has been shown to depress electrically evoked activity recorded from the longitudinal nerve cord of the polyclad flatworm, *Notoplana acticola* (Keenan, Koopowitz and Bernardo, 1979). Nitric oxide has been demonstrated in *D. tigrina* (Eriksson, 1996). As will be discussed later, monoamines play a significant role in regeneration (section 5.5.).

The presence of peptidergic nerve cells in turbellarians has been known for some time as a result of the demonstration of neurosecretory cells (see reviews by Lender, 1974; Webb, 1988a). Cycles of neurosecretory activity have been linked to sexual reproductive cycles in planarians, suggesting a role for neurosecretions in stimulating gametogenesis. Sexualisation of asexual strains has been achieved by feeding them neurosecretory-rich fractions from sexual strains (see Fairweather, 1997 for references). A cycle of neurosecretory activity is associated with the scissiparity (fissioning) cycle during asexual reproduction. Fission occurs *after* the release of neurosecretions, the latter presumably being responsible for initiating the cycle. A cycle of RNA synthesis parallels the neurosecretory cycle, although it is slightly displaced. This suggests that the neurosecretory material may activate the totipotent neoblast cells during the second half of the scissiparity cycle, inducing them to synthesise RNA and so prepare the worm for the regeneration that follows the next fission (Lender, 1974, 1980; Webb, 1988a). A more substantial role for neurosecretions in planarian regeneration was provided by a classic set of experiments by Webb and Friedel (Webb and Friedel, 1978, 1979; Friedel and Webb, 1979). In these experiments, a peptidergic fraction was isolated from *Dugesia tigrina* and was shown to be localised in neurosecretory cells in the brain and ventral nerve cords. Addition of this fraction to post-pharyngeal pieces of worm stimulated protein and RNA synthesis, also mitotic activity. Thus, the experiments provided evidence for a morphogenetic role of an uncharacterised peptide in planarians.

More recent immunocytochemical studies using antisera to vertebrate and invertebrate peptides have shown that homologues to these peptides are present in the nervous system of turbellarians (see reviews by Fairweather and Skuce, 1995; Reuter and Gustafsson, 1995). Three peptides endogenous to turbellarians have been isolated: neuropeptide F (NPF), a homologue of the vertebrate neuropeptide Y superfamily, from the predatory flatworm, *Artioposthia triangulata* (Curry *et al.*, 1992); and the FMRFamide-related peptides (or FaRPs), GYIRFamide from *D. tigrina* and RYIRFamide from *A. triangulata* (Maule *et al.*, 1994; Johnston *et al.*, 1995). With regard to the potential physiological roles of peptides in turbellarians, met-enkephalin has been shown to decrease spontaneous activity in *Dugesia gonocephala* (Venturini *et al.*, 1981) and it has been demonstrated that a number of vertebrate peptides (substance P, neurokinin A, neuropeptide K, bradykinin, growth-hormone releasing factor, met-enkephalin and somatostatin) are capable of acting as mitogens (for references see Baguna *et al.*, 1990; Fairweather, 1997; Hori, 1997). Epidermal growth factor (EGF) and fibroblast growth factor (FGF) have also been shown to be capable of acting as mitogens in *D. tigrina*. Immunoreactivity for EGF has been localised in the nervous system of the worm and a faint reaction observed on the membrane of the neoblast cells with EGF-receptor (EGF-r) antibodies (Baguna *et al.*, 1990). A third growth factor, nerve growth factor, has been shown to stimulate neuronal hypertrophy in *D. gonocephala* (Palladini *et al.*, 1988). Immunostaining for EGF, EGF-r and basic fibroblast growth factor (bFGF) has been demonstrated in the nervous system of two microturbellarian species which reproduce asexually by the formation of chains of zooids. Immunoreactive nerve cells were particularly concentrated in the growth cones of developing zooids, suggesting that the proliferation of nerve cells may be under the control of growth factor-like substances and that the outgrowing nerve fibres subsequently promote the further development of the zooid. Furthermore, antibodies to the growth factors were able to block asexual reproduction (Reuter and Kuuisto, 1992).

2.2. PARASITIC FLATWORMS

In parasitic flatworms, acetylcholine, monoamines, amino acids, nitric oxide, peptides and growth factors have been localised in the nervous system. The presence of acetylcholine and the mapping of the cholinergic component of the nervous system has been based largely on the histochemical demonstration of acetylcholinesterase activity (see review by Davis and Stretton, 1995). Acetylcholine-immunoreactivity has been demonstrated in the rat tapeworm, *Hymenolepis diminuta* (Samii and Webb, 1990). Pharmacological studies indicate that acetylcholine exerts an inhibitory effect on neuromuscular activity. The responses of neuromuscular preparations to vertebrate acetylcholine agonists and antagonists vary from species to species; this has led to the suggestion that the acetylcholine receptor(s) possesses a mixture of classical nicotinic and muscarinic properties and is distinct from the equivalent receptor in the vertebrate host (see reviews by Davis and Stretton, 1995; Pax *et al*., 1996; Walker, Brooks and Holden-Dye, 1996).

A number of monoamines have been demonstrated in parasitic platyhelminths by biochemical and histochemical means. They are: 5-HT, adrenaline, noradrenaline, dopamine, octopamine and histamine. Far more is known about 5-HT than the other putative transmitters, in terms of its receptors; its synthesis, metabolism and inactivation; and its physiological roles. These aspects will be discussed later (sections 5.2.–5.4.). Noradrenaline appears to have an inhibitory effect on neuromuscular activity, whereas the responses to the other monoamines vary from species to species. Pharmacological data, albeit limited, highlights differences between parasite receptors and their vertebrate (host) counterparts (Davis and Stretton, 1995).

Glutamate-immunoreactivity has been demonstrated in the cestodes, *H. diminuta* and *Mesocestoides corti* and the liver fluke, *F. hepatica* (Webb and Eklove, 1989; Eklove and Webb, 1990; Brownlee and Fairweather, 1996). It exerts a stimulatory effect on neuromuscular activity (see section 5.5.). It has also been shown that glutamate causes excitation of neurons in the longitudinal nerve cords of the cesode, *Gyrocotyle fimbriata* (Keenan and Koopowitz, 1982). There has only been one report of GABA-immunoreactivity in parasitic flatworms: very limited immunostaining was observed in the peripheral nervous system of *F. hepatica* and the sheep tapeworm, *Moniezia expansa* (Eriksson *et al*., 1995). This finding remains to be confirmed.

Nitric oxide has been demonstrated by histochemical, chromatographic and radiometric means in a number of parasitic flatworms, including the cestodes *H. diminuta*, *D. dendriticum* and *Mesocestoides vogae* (Gustafsson *et al*., 1996; Lindholm, Reuter and Gustafsson, 1998; Terenina, Reuter and Gustafsson, 1999; Terinina *et al*., 2000). Little is known about its physiological role(s), although it has been suggested that it is involved in the control of sucker activity (Terenina, Reuter and Gustafsson, 1999).

Many immunocytochemical studies have been carried out on parasitic flatworms, using antisera to a range of invertebrate and vertebrate peptides (the data is covered by the reviews by Davis and Stretton, 1995; Fairweather and Skuce, 1995; Reuter and Gustafsson, 1995; Fairweather, 1997). Collectively, the results indicate that the peptidergic component of the nervous system is likely to contain a large number of potential transmitter molecules, a number that far outstrips the combined number of other transmitter candidates. In other words, the peptidergic nervous system predominates over all other components. Two endogenous neuropeptides have been isolated from the tapeworm, *M. expansa*: NPF and the FaRP, GNFFRFamide (Maule *et al.*, 1991, 1993). It has also been demonstrated that the human blood fluke, *S. mansoni* possesses a proopiomelanocortin (POMC)-related gene at all stages in its life cycle. Moreover, it synthesises and releases a number of peptides derived from the POMC precursor, namely, ACTH, β-endorphin, α-MSH and met-enkephalin; the worm also possesses a δ-like opioid receptor. The release of these peptides has been linked to a downregulation of the immune response in both the snail and mammalian hosts, thereby enhancing the chances of parasite survival (Duvaux-Miret *et al.*, 1992, 1993; Leung *et al.*, 1995). Another example of the secretion of hormone-like molecules by parasites concerns the plerocercoid (sparganum) larva of the tapeworm, *Spirometra mansonoides*. It produces a plerocercoid factor (PGF) which induces accelerated growth in its rodent host. PGF is synthesized in the tegumental cells, released from the surface of the parasite, enters the bloodstream of the host and binds to growth hormone receptors. It possesses many of the biological, immunological and molecular characteristics of mammalian growth hormone. Interestingly, PGF is both a growth hormone-like factor and a cysteine proteinase active at neutral pH. Collagen is a substrate for the proteinase, indicating a likely role in tissue penetration for the enzyme (Phares, 1992, 1996; Phares and Kubik, 1996). The proteolytic activity of PGF may also contribute to the ability of plerocercoids to evade the host's immune response since it has the capacity to cleave IgG (Kong *et al.*, 1994).

In parasitic flatworms, neurosecretions have been implicated in the control of sexual reproduction in certain trematode species and morphogenesis in the cestode, *D. dendriticum* (see Fairweather and Skuce, 1995). A number of vertebrate peptides exert either stimulatory (CCK, motilin, somatostatin) or inhibitory (caerulein) effects on the motility of *F. hepatica* and *H. diminuta in vitro* (Sukhdeo *et al.*, 1984; Sukhdeo and Sukhdeo, 1989). Intravenous administration of secretin and CCK induced an anterior migration of *H. diminuta* within the rat gut, whereas insulin produced a posterior migration (Mettrick and Podesta, 1982; Cho, 1984). The results suggest that the worm possesses receptors that interact with gut hormones and so help it to determine its position within the gut during its diurnal migration. The turbellarian peptides GYIRFamide and RYamide exert a stimulatory effect on isolated muscle strips from the liver fluke, *F. hepatica*, although the two cestode peptides NPF and GNFFRFamide had no effect on motility. Excitatory responses were also observed with the molluscan peptides, FMRFamide and FLRFamide and with the nematode peptides AF-1. AF-2, AF-8 and PF-4 (Graham, Fairweather and McGeown,

1997). It has been shown that a number of peptides are capable of affecting protein synthesis in parasites and so may act as mitogens. The actions are either stimulatory (FMRFamide), inhibitory (NPF) or without effect (GNFFRFamide) (see Fairweather, 1997). Additional roles for neuropeptides have been suggested on the basis of the distribution of peptidergic nerve elements. They include gametogenesis, egg production and sensory perception (Fairweather, 1997). Finally, immunostaining for the growth factors, FGF and EGF has been localised in the nervous system of the tapeworm, *D. dendriticum* (Gustafsson and Eriksson, 1992; Gustafsson, Eriksson and Hyden, 1995). An increase in the number of EGF-immunoreactive nerve cells during development of the larval stage to the adult has been linked to a role for the growth factor in the development of the parasite (Gustafsson *et al.*, 1995). A gene encoding a homologue of the EGF receptor has been isolated from the blood fluke, *S. mansoni* (Shoemaker *et al.*, 1992). This will be discussed further in section 4.2.1.

3. Overview of Signalling Pathways

3.1. THE PARASITE PERSPECTIVE

Communication *within* a multicellular organism such as a flatworm will be brought about by the use of signalling molecules, produced largely by the nervous system. Parasitic members of the group, especially endoparasites that live within the body of another animal, are exposed to a variety of external, host-derived signals that may influence their behaviour and development. The life cycle of the human blood fluke, *S. mansoni* will illustrate this point (Fig. 1). The adult worms reside in the mesenteric veins that drain the intestine. The life cycle in the mammalian host begins with penetration of the cercarial larva into the skin, where it becomes transformed into the schistosomule. After spending a few days in the skin, the schistosomule enters the venous circulation and migrates to the lungs. It stays there for several days, with no change in size or evidence of mitosis. The schistosomule migrates again to the portal vessels of the liver where it undergoes a period of rapid growth and sexual maturation. The male and female worms mate and move back along the hepatic portal veins to their final niche in the mesenteric veins. Here, eggs are laid which reach the lumen of the gut and are passed out in the faeces. Within the blood circulation, the worms are surrounded by host molecules such as antibodies, hormones, cytokines and growth factors. They must have the ability to receive and respond to these host signals for site-finding behaviour in order to complete the complex migration just described. They must also be able to receive and respond to signals that will regulate cell proliferation, differentiation and development. Even after maturation, adult worms receive signals from the host that regulate development: for example, TNFα stimulates egg laying. Evidence for the possession of receptors and pathways to deal with these signals will be discussed in sections 4.2. and 4.3. Interactions also occur between the male and female worms, because the male is necessary for the initiation and

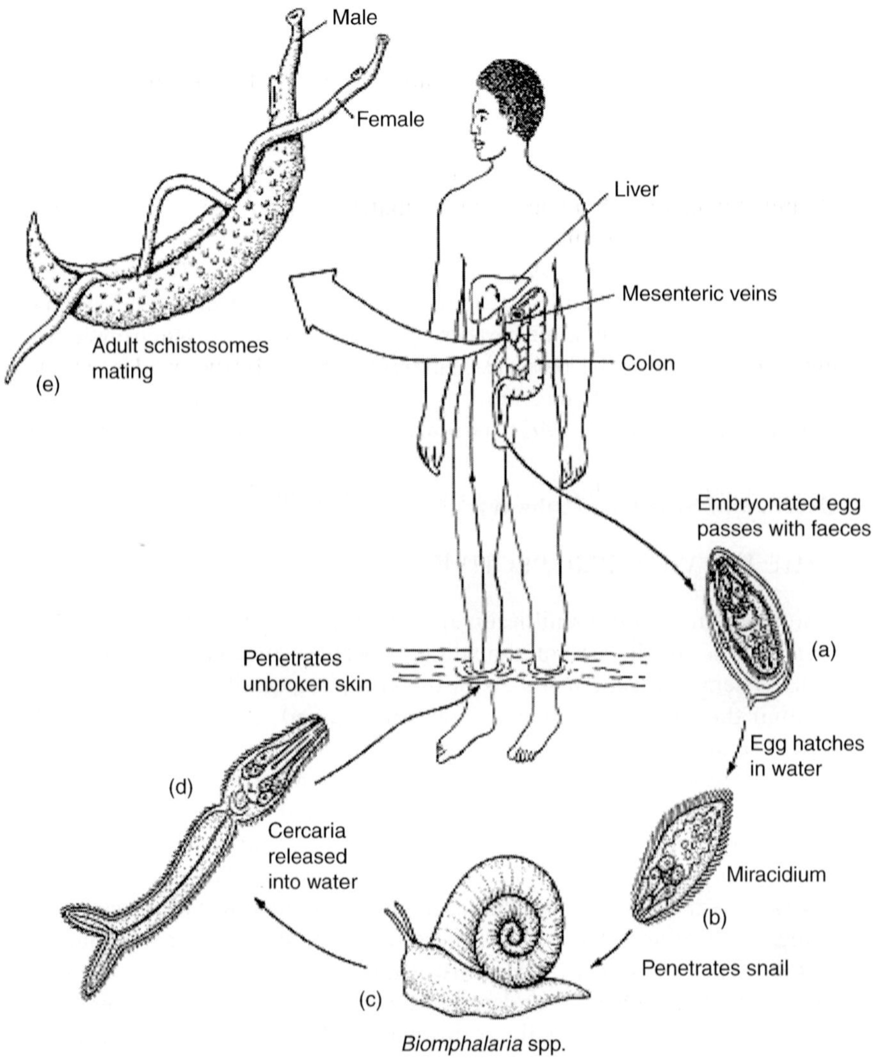

Figure 1. The life cycle of *Schistosoma mansoni* (from Roberts and Janovy, 2000).

maintenance of sexual maturity in the female and for egg production; the male also regulates gene expression in the female (see Fairweather, 1997; also section 4.2.1.).

The rest of the life cycle of *S. mansoni* involves a snail intermediate host and a number of free-living and parasitic stages. Following their release in the faeces, the eggs hatch in water to release a miracidium larva. This is a free-swimming larva which locates a suitable snail host belonging to the genus *Biomphalaria*. The miracidium penetrates the snail, typically in the foot, and develops into the sporocyst larva, which migrates to the digestive gland where it develops into the cercaria larva. Considerable multiplication of larval numbers occurs within the snail host. Eventually, the cercaria migrates to the mantle collar, from where it emerges, again into water. The cercaria is another free-swimming stage and will locate a human host, penetrating the skin and commencing the mammalian phase of the life cycle once again. As can be seen, then, the life cycle of a parasite such as *S. mansoni* involves a number of hosts and a number of morphological transformations through a variety of larval stages that inhabit very different environments. It also involves a series of behaviour patterns that culminate in each successive stage reaching its correct site for development into the next stage of the life cycle. For useful reviews on parasite behaviours, including host finding and migration, the reader is referred to Sukhdeo and Mettrick (1987), Haas (1994), Kalbe, Haberl and Haas (2000) and Sukhdeo and Sukhdeo (2002). The parasite must be highly sensitive to different stimuli and be able to respond to them in the most appropriate way. The release of the right behaviour pattern at the right time (and in the correct sequence) requires a well-developed signalling system that, in terms of nerve cells, is dependent on a relatively small population, perhaps a few hundred cells at most. Free-living flatworms also need to be able to recognize and respond to signals in their environment.

3.2. GENERAL ORGANISATION OF SIGNALLING PATHWAYS

In order to recognize and react to extracellular signals, cells possess receptors, typically located on the cell surface. In other cases, small hydrophobic signalling molecules can diffuse directly across the plasma membrane of target cells and bind to receptors that are located inside the cells. Such molecules include steroid hormones, thyroid hormones and retinoids. Following binding to their receptors, the signal molecules activate the receptors which bind to DNA and regulate the transcription of specific genes, using mRNA as second messenger. The receptors bind to specific DNA sequences adjacent to the genes the ligand regulates.While some receptors reside in the cytoplasm and enter the nucleus after ligand binding, others (including thyroid and retinoid receptors) are bound to DNA in the nucleus even without the ligand. The transcription of particular genes and the production of many mRNA molecules provides the amplification of the initial signal. The subsequent synthesis of new proteins will alter the activity of the cell in some way, depending on whether the protein is a structural protein or enzyme, for example, and this change represents the response of the cell to the original stimulus.

Most cell surface receptors fall into three groups: ion-channel-linked, G protein-linked and enzyme-linked receptors. Ion-channel-linked receptors are typically involved in neurotransmission. G protein-linked receptors indirectly activate or inactivate plasma membrane-bound enzymes or ion channels via GTP-binding proteins, or G proteins; they form the largest group of cell-surface receptors. Enzyme-linked receptors, when activated, either act directly as enzymes or are associated with enzymes that they activate. In these signalling cascades, the ligand remains on the outside of the cell, so the signal or message it carries must be taken into the cell and converted into an intracellular one. The intracellular signal will ultimately serve to alter a particular target protein within the cell and this will trigger changes in the cell that, again, constitute the response of that cell to that particular signalling molecule. Activated G protein-linked and enzyme-linked receptors relay the signal into the cell by activating a series of intracellular signalling proteins that transduce, amplify or spread the signal as they relay it.

3.3. G PROTEIN-LINKED RECEPTORS

The binding of a signal to its receptor causes a conformational change in the receptor that allows it to activate G proteins. In turn, the activated G proteins trigger the production of second messenger molecules that relay the signal into the interior of the cell along particular pathways (the "first messengers" being the extracellular signals). Second messenger molecules include cAMP, Ca^{++}, Ins(1,4,5)P_3 (or IP_3) and diacylglycerol. A brief outline of each of these pathways will be given below.

3.1.1. *cAMP Pathway*

cAMP is synthesized from ATP by adenylyl (or adenylate) cyclase, a plasma membrane-bound enzyme that is activated by G proteins and has its catalytic domain on the cytosolic side of the membrane. The production of cAMP via G proteins and adenylyl cyclase represents both a transduction and amplification of the original extracellular signal. cAMP activates cAMP-dependent protein kinase A (PKA), or protein kinase A. In turn, PKA regulates the activity of target proteins, by phosphorylating them, thus altering the activity of the cell. cAMP can be inactivated by a number of mechanisms: by the inactivation of G proteins via their intrinsic GTPase activity; by the inhibition of adenylyl cyclase activity by inhibitory G proteins; by the inactivation of PKAs by phosphatase enzymes; by the hydrolysis of cAMP to adenosine 5′-monophosphate (5′AMP) by cyclic AMP phosphodiesterase enzymes.

3.1.2. *Inositol Phospholipid Pathway*

In this pathway, binding of a signal molecule to its receptor is linked via a G protein to phospholipase C, a plasma membrane-bound enzyme analogous to adenylyl cyclase. Phospholipase C serves to cleave a lipid component of the plasma membrane itself, phosphatidylinositol 4,5-bisphosphate into two products, each of which can act as a second messenger. The two are: inositol 1,4,5-trisphosphate (Ins(1,4,5)P_3, or IP_3) and diacylglycerol. The production of IP_3, like that of cAMP, represents a transduction and amplification step. IP_3 leaves the plasma membrane and rapidly diffuses into the cytoplasm, where it triggers the release of Ca^{++} from stores such as the endoplasmic reticulum (ER), by binding to and opening IP_3-gated Ca^{++}-release channels in the ER membrane. The concentration of Ca^{++} in the cytosol rises. Some of the cellular actions of Ca^{++} are mediated by calmodulin, a Ca^{++}-binding protein. Ca^{++} binds to calmodulin and activates it: in this state, Ca^{++}/calmodulin can bind directly to target proteins and alter their activity. More typically, however, the effects of Ca^{++} are indirect and are mediated by Ca^{++}/calmodulin-dependent protein kinases (CaM-kinases). These enzymes act to phosphorylate target proteins, thus enabling the cell's response. Termination of the IP_3-mediated response is brought about in a number of ways: IP_3 can be dephosphorylated by a phosphatase enzyme, or phosphorylated by a kinase enzyme; Ca^{++} can be pumped out of the cytoplasm by Ca^{++} pumps in the plasma membrane and in the ER and mitochondrial membranes; and phosphorylated proteins can be dephosphorylated by phosphatases.

Diacylglycerol remains in the membrane and can activate a Ca^{++}-dependent protein kinase enzyme, protein kinase C (PKC). PKC acts to phosphorylate target proteins, to bring about the cell's response to the initial stimulus. Alternatively, diacylglycerol can be cleaved to release/form arachidonic acid, which can act as a messenger in its own right or be used as a precursor in the synthesis of eicanosoids such as prostaglandins.

3.2. ENZYME-LINKED RECEPTORS

As with G protein-linked receptors, enzyme-linked receptors have their ligand-binding domain on the outer surface of the plasma membrane. However, the cytosolic domain does not interact with a G protein, but has intrinsic enzyme activity or associates directly with a protein. Enzyme-linked receptors belong to a number of classes: e.g. receptor tyrosine kinase; tyrosine kinase-associated receptors; receptor serine/threonine kinase; receptor guanylyl cyclases; histidine kinase-associated receptors. In flatworms, there is only evidence for the first three classes, so the foregoing discussion will be restricted to them; for information on the other types of receptors, see Alberts *et al.* (2002).

3.2.1. *Receptor Tyrosine Kinase*

Receptor tyrosine kinases are involved in the response to various hormones and growth factors, including epidermal growth factor (EGF), insulin and nerve growth factor (NGF). Binding of the signal protein to the ligand-binding domain on the outside of the cell activates the intracellular tyrosine kinase domain via an autophosphorylation process which also serves to create high-affinity docking sites for the binding of a number of intracellular signalling proteins in the target cell. The formation of this complex functions to relay the signal along various signalling pathways within the cell. Binding proteins typically possess phosphotyrosine domains such as SH2 or proline-binding domains such as SH3. Some of the docked proteins act as adaptors to couple the receptor to the important downstream signalling protein Ras. Ras belongs to a group of small monomeric GTPases that also includes Rho and Rab proteins. Activated Ras acts as a transducer, changing the nature of the signal and passing it downstream along several pathways, by activating other signalling proteins. One of the pathways is a serine/threonine phosphorylation cascade that converges on a mitogen-activated protein kinase (MAP-kinase). The latter relays the signal to the nucleus by phosphorylating gene regulatory proteins there. The Ras-MAP-kinase pathway, therefore, serves to alter the pattern of gene expression in significant ways, for example, those genes involved in cell proliferation, differentiation and growth. Ras exists in activated and inactivated states; GTPase-activating proteins (GAPs) induce activated Ras to inactivate itself by hydrolyzing its bound GTP to GDP. MAP-kinases are inactivated by dephosphorylation and phosphatases will reverse the phosphorylations in the phosphorylation cascade.

3.2.2. *Tyrosine Kinase-Associated Receptors*

These receptors lack a tyrosine kinase domain, but act via cytoplasmic tyrosine kinases which are associated with the receptors and phosphorylate various target proteins following binding of the ligand to the receptor. Cytoplasmic tyrosine kinases belong to the Src family of protein kinases that includes Src and Fyn. The enzymes are located on the cytoplasmic side of the plasma membrane and contain SH2 and SH3 domains.

The cytokine receptor family constitutes the largest family of receptors in this class. When stimulated by ligand binding, the receptors activate Janus kinase (JaK) cytoplasmic tyrosine kinases, which phosphorylate STATS (Signal Transducers and Activators of Transcription). The STATS then dimerize, migrate to the nucleus and activate the transcription of specific genes.

3.2.3. *Receptor Serine/Threonine Kinases*

These kinases are activated by signalling proteins of the transforming growth factor-beta (TGF-β) superfamily. There are two classes of receptor serine/threonine kinases: type I and type II. Once activated, the receptor complex relays the signal to the nucleus. The type-I receptor binds and phosphoylates a latent gene regulatory protein of the Smad family. A Smad complex is formed which moves into the nucleus and associates with other gene regulatory proteins, binds to specific sites in DNA and activates a particular set of target genes that may be involved in cell growth, differentiation or proliferation, for example. The Smad pathway can be blocked by the action of inhibitory Smads: the latter bind to activated type-I receptors and prevent other Smads from binding there. Since active Smad complexes cannot be formed, this effectively shuts off the response to the TGF-β family ligand. Alternatively, extracellular inhibitory proteins can bind to the signal molecule and stop them from activating their receptors on target cells.

4. Evidence for Elements of Signalling Systems in Platyhelminths

4.1. G PROTEIN-MEDIATED PATHWAYS

4.1.1. *G Proteins*

A number of G proteins have been demonstrated in the liver fluke, *F. hepatica* and the human blood fluke, *S. mansoni*. They include a 42.5 kDa protein in *F. hepatica* and a 41 kDa protein in *S. mansoni*, both of which are ADP-ribosylated by cholera toxin, and are the homologues of $G_{s\alpha}$ in mammals. Both parasites possess a 41 kDa protein that is ADP-ribosylated with pertussis toxin and is the equivalent of the $G_{o\alpha}$ subunit. A 36 kDa G_{β} subunit has also been identified in the two parasites and is immunologically similar to bovine transducin. Additional proteins in *F. hepatica* and *S. mansoni* react with antisera to G protein subunits, a result which indicates the presence of further G proteins, although their molecular weights and ADP-ribosylation profiles do not equate with the properties of known mammalian G proteins (Mansour and Mansour, 1986, 1989; Estey and Mansour, 1988). A 42 kDa putative G protein α-subunit has been isolated from a tegumental brush border plasma membrane preparation from the rat tapeworm, *Hymenolepis diminuta* (Walker and Barrett, 1993). The protein is susceptible to ADP ribosylation by both cholera and pertussis toxins, but it shows no resemblance to any of the four major mammalian G protein classes in terms of its Mr and toxin susceptibility. The results suggest that a G protein-linked signal transduction system operates within the tegumental syncytium of *H. diminuta*.

A cDNA clone (designated SG12) encoding the α-subunit of a G protein has been isolated from *S. mansoni* (Iltzsch *et al*., 1992). The protein encoded by the clone shows a marked sequence homology with mammalian $G_{s\alpha}$: a 65% identical match and a 78% match when conserved amino acid substitutions are taken into account. Comparison of Sg12 with other G_{α} proteins indicates that all the major structural features of a $G_{s\alpha}$ protein are present. Thus, the schistosome clone contains an ADP-ribosylation site for cholera toxin; the amino acids are completely conserved at the GTP binding sites and the sequence equivalent to the adenylate cyclase activation site has 67% identity and 78% similarity to the same region of bovine $G_{s\alpha}$ (Iltzsch *et al*., 1992). The expression of the gene appears to be developmentally regulated, being absent in the miracidial and cercarial larval stages. The gene transcript is detectable (albeit at a low level) in 1-day-old schistosomules, but does not reach full expression until the adult stage (Iltzsch *et al*., 1992).

A gene (designated *smrab*) encoding a rab-related low-molecular weight GTP-binding protein has been cloned from *S. mansoni* (Loeffler and Bennett, 1996). The gene is expressed to a much greater extent in male than female worms. Given that the female lies within the gynaecophoric canal of the male, which affords her protection against the environment, and the known roles of rab proteins in secretory mechanisms, it has been suggested that the smrab protein plays a role in the maintenance and turnover of the tegument in the male worm, which is part of the parasite's immune evasion mechanism. Sexual maturity in the female is dependent on physical contact with the male, so smrab may have an additional, indirect effect on female fecundity.

4.1.2. *Adenylate Cyclase*

Adenylate cyclase activity has been demonstrated in a number of freshwater planarians. The enyme is stimulated by 5-HT and 5-HT agonists (Franquinet, Le Moigne and Hanoune, 1978; Venturini *et al.*, 1986; Creti *et al.*, 1992) and is linked to a 5-HT receptor of the 5-HT 1a type (Creti *et al.*, 1992). It is insensitive to the catecholamines epinephrine, norepinephrine and octopamine at low concentrations (1×10^{-5} M, 1×10^{-6} M) (Franquinet, Stengel and Hanoune, 1976; Franquinet, Le Moigne and Hanoune, 1978; Creti *et al.*, 1992), although norepinephrine at the higher concentration of 1×10^{-4} M does stimulate activity in *Polycelis tenuis* (Franquinet, Stengel and Hanoune, 1976). A dopamine-sensitive adenylate cyclase has been demonstrated in *Dugesia gonocephala*, dopamine and dopaminergic agonists stimulating activity, whereas dopaminergic antagonists inhibit it (Venturini *et al.*, 1986; Venturini, 1993). Adenylate cyclase activity varies during regeneration, with peaks at 3 and 15 hours following transection (Franquinet and Le Moigne, 1979; Martelly, Franquinet and Le Moigne, 1981). Its role during regeneration will be discussed in section 5.5.

Cytochemical studies have shown that adenylate cyclase activity is localised on the plasma membrane of mucous gland cells, rhabdite-forming cells, pharyngeal gland cells and the ductules and ducts of the protonephridial system (Pascolini *et al.*, 1979, 1982; Moraczewski and Duma, 1981). Enzyme activity is also localised on the plasma membrane and the membrane of the sarcoplasmic reticulum of muscle cells and nerve endings associated with ciliated sensory receptors in the epidermis (Pascolini *et al.*, 1979; Moraczewski and Duma, 1981). In view of the regenerative capacity of planarians, it is significant that there is an increase in synthesis and surface expression of adenylate cyclase activity in the undifferentiated neoblast cells during the early stages of regeneration (Moraczewski and Duma, 1978). It is the neoblast cells that serve as stem cells for the differentiation of other cell types for the replacement of missing tissues during regeneration.

Adenylate cyclase activity has been demonsrated in the trematode parasites, *F. hepatica* and *S. mansoni*, the enzyme in these two parasites being amongst the most active adenylate cyclases known in eukaryotes (Northup and Mansour, 1978a,b; Northup *et al.*, 1979; Kasschau and Mansour, 1982a). Adenylate cyclase activity is stimulated by 5-HT (25–30-fold in *F. hepatica*; 50–100 times in *S. mansoni*), but the stimulation is dependent on GTP (Northup and Mansour, 1978b; Kasschau and Mansour, 1982a). In *F. hepatica*, the enzyme is not activated by other putative transmitters (dopamine, epinephrine, histamine, norepinephrine, octopamine), nor by carbachol (Abrahams, Northup and Mansour, 1976; Northup and Mansour, 1978a); epinephrine and norepinephrine are similarly without effect on adenylate cyclase in *S. mansoni*, but histamine is stimulatory (Higashi *et al.*, 1973). Pharmacological data suggests that there is a single class of adenylate cyclase which is regulated by only one type of 5-HT receptor (Northup and Mansour, 1978a). The fluke enzyme differs from its mammalian counterpart in terms of the kinetics of nucleotide activation and detergent binding (Northup and Mansour, 1978b; Northup *et al.*, 1979).

Developmental changes in 5-HT-stimulated adenylate cyclase activity have been described during the life cycle of *S. mansoni*. Activity in the cercarial stage is low, gradually increasing through the schistosomule stage to reach a maximum level in the adult (Kasschau and Mansour, 1982a,b). The increase in activity is not due simply to an increase in enzyme activity *per se* or a change in the level of 5-HT receptors, but to increases in both these factors and the G protein component of the 5-HT signalling pathway (Estey and Mansour, 1987, 1988). Sex-linked differences in adenylate cyclase activity have also been described in this dioecious parasite. Adult females display only 40% as much sodium fluoride-stimulated adenylate cyclase activity as males and 5-HT stimulation of adenylate cyclase in females is only 20% as much as in males (Kasschau and Mansour, 1982a). Interestingly, paired worms respond to activators in a manner similar to that in males, suggesting activation of the female enzyme by the male (Kasschau and Mansour, 1982a).

Data on the localisation of adenylate cyclase activity in parasitic flatworms is only available for the tapeworm, *Hymenolepis microstoma*. The enzyme is located in the crypts between the microtriches on the tegumental surface of mature proglottides (or segments). It was suggested that the presence of adenylate cyclase in the surface membranes of the parasite is indicative of signallling systems linked to metabolite uptake from the host gut lumen, ion regulation or response to host hormones (Conway-Jones and Rothman, 1978). Adenylate cyclase activity in the related parasite, *H. diminuta* is stimulated by glutamate and glutamate agonists and is linked to a glutamate receptor that possesses some characteristics of the NMDA-type receptor (Eklove and Webb, 1991). Adenylate cyclase activity in this parasite is also stimulated by 5-HT and 5-HT agonists and is linked to a 5-HT receptor of the 5-HT 1 type (Ribeiro and Webb, 1987). Glutamate eliminates the 5-HT-stimulated adenylate cyclase response, suggesting that glutamate may act as a neuromodulator of 5-HT responses (Eklove and Webb, 1991). Adenylate cyclase activity in *H. diminuta* is inhibited by epinephrine, but not affected by either dopamine or octopamine (Sangster and Mettrick, 1987).

4.1.3. *Cyclic AMP*

cAMP has been demonstrated in the freshwater planarian, *P. tenuis* (Franquinet, Stengel and Hanoune, 1976; Franquinet and Le Moigne, 1979; Martelly, Franquinet and Le Moigne, 1981). Treatment of regenerating fragments of *D. dorotocephala* with cAMP and the cAMP analogue dibutyryl cAMP stimulated the rate of mitotic activity, suggesting a role for cAMP in regeneration (Weinstein and Gavurin, 1977). It is believed that cAMP is involved in DNA and RNA synthesis during regeneration in response to 5-HT and dopamine, respectively; cAMP levels reach a peak at 2 hours after sectioning (Franquinet and Le Moigne, 1979) and are increased by dopaminergic agonists (Venturini *et al.*, 1986). The role of cAMP during regeneration will be discussed in more detail in section 5.5.

cAMP has been demonstrated in three parasites: *F. hepatica*, *S. mansoni* and *H. diminuta*. Production of cAMP is stimulated by 5-HT in each of the parasites, but is not affected by epinephrine, norepinephrine or histamine in *F. hepatica* (Mansour *et al.*, 1960; Higashi *et al.*, 1973; Abrahams, Northup and Mansour, 1976; Ribeiro and Webb, 1987). In *H. diminuta*, cAMP formation is increased by 5-HT agonists and reduced by 5-HT antagonists (Ribeiro and Webb, 1987). cAMP production in *F. hepatica* is stimulated by forskolin, an activator of adenylate cyclase activity (McNall and Mansour, 1985). cAMP stimulates protein kinase A activity in *F. hepatica* and *S. mansoni* (Higashi *et al.*, 1973; Gentleman, Abrahams and Mansour, 1976).

5-HT has been shown to stimulate the phosphorylation of proteins in the cestode, *H. diminuta*; the response was also triggered by 5-HT agonists, but inhibited by 5-HT antagonists (Ribeiro and Webb, 1991). cAMP exerts a similar stimulatory action, suggesting that it mediates the response to 5-HT, presumably via the action of protein kinase A enzymes involved in phosphorylation reactions (Ribeiro and Webb, 1991). cAMP is also believed to act as second messenger for 5-HT in the regulation of motility and carbohydrate metabolism in a number of parasites: these roles will be discussed in more detail in sections 5.2. and 5.3.

4.1.4. *Protein Kinase A*

cAMP-dependent protein kinase (PKA) activity has been demonstrated in the freshwater planarian, *P. tenuis* (Moraczewski and Franquinet, 1980; Moraczewski, Franquinet and Le Moigne, 1983). A separate, cAMP-independent, calcium-dependent protein kinase also occurs in the worm (Moraczewski, Franquinet and Le Moigne, 1983). The levels of both enzymes vary during regeneration, with peaks of PKA, for example, at 3 and 18 hours following transection (Moraczewski, Franquinet and Le Moigne, 1983). Their roles in regeneration will be dealt with in section 5.5.

cAMP-dependent protein kinase (PKA) activity has been demonstrated in the trematode parasites, *F. hepatica* and *S. mansoni*. Activity is stimulated by cAMP in both parasites and by 5-HT in *F. hepatica* alone (Higashi *et al.*, 1973; Gentleman, Abrahams and Mansour, 1976). Three PKAs (designated A1, A2 and B) have been isolated from *F. hepatica* (Iltzsch and Mansour, 1987). Their elution profiles suggest that A2 and B are analogous to type II mammalian PKAs, whilst A1 falls between the type I and type II classification for the mammalian enzyme. The estimated molecular mass of B (187 kDa) is consistent with a tetrameric form of the enzyme (i.e. two regulatory, two catalytic subunits). The other two enzymes have lower molecular masses (98 kDa) which suggests that they are dimers of one regulatory and one catalytic subunit. The general properties of the fluke catalytic subunit are similar to those of the mammalian enzyme although the kinetic effects of Mg^{++} are different. Reconstitution experiments show that there is sufficient homology between the fluke and mammalian (bovine heart) PKAs for the catalytic subunit from one species to associate with the regulatory subunit of the other species to form an inactive holoenzyme. The latter can redissociate in the presence of cAMP to release the active catalytic subunit. Homology is also indicated by the fact that mammalian protein kinase inhibitors can inhibit the activity of the fluke catalytic subunit (Iltzsch and Mansour, 1987). The combined data, then, suggest that the PKA enzyme has been well conserved during evolution. A possible role for PKA in the regulation of ion transport and membrane permeability has been suggested for *F. hepatica* (Gentleman *et al.*, 1976).

4.1.5. *Phosphodiesterase*

Phosphodiesterase activity has been localised in regenerating and non-regenerating planarians (Duma, 1980). The enzyme was associated with the plasma membrane of different cell types (epithelial, muscle, nerve and neoblast cells) and in vesicles below the plasma membrane of epithelial cells. An increase in enzyme activity was observed in the muscle and nerve cells of regenerating animals. Neoblast cells in non-regenerating planarians display no phosphodiesterase activity; enzyme activity is only present in neoblast cells in the immediate vicinity of the blastema in regenerating animals (Duma, 1980). These observations further underline the idea that cAMP plays a role in the regeneration process.

Phosphodiesterase activity has been demonstrated in the trematode parasites, *F. hepatica* and *S. mansoni*; in both species, enzyme activity is sensitive to inhibition by mammalian phosphodieserase inhibitors (Higashi *et al.*, 1973; Mansour and Mansour, 1977). Enzyme kinetic data for *F. hepatica* phosphodiesterase suggests that it is a low km enzyme with low affinity for cAMP (Mansour and Mansour, 1977).

4.1.6. *Phospholipase C*

Phospholipase C activity has been demonstrated in detergent (i.e. tegumental) extracts of the liver fluke, *F. hepatica* (Hawn and Strand, 1993). The enzyme is cation-independent, inhibited by sulphydryl modifying agents and is similar in size to protozoal phospholipase C. In the same study, cercarial and adult extracts of the human blood fluke, *S. mansoni* were shown to contain a mixture of phospholipase C and D activities. A number of roles for phospholipase C in this parasite have been put forward, including one in immune evasion via shedding of antigen-antibody complexes from the parasite surface by the cleavage of surface GPI-anchored proteins; also, a role in signal transduction by production of diacylglycerol and the subsequent activation of protein kinase C (Hawn and Strand, 1993).

It has been shown that alkaline phosphatase and acetylcholinesterase are attached to the surface (tegumental) membrane of *S. mansoni* via phosphatidylinositol anchors (Espinoza *et al.*, 1988). Treatment of worms with phosphatidylinositol-specific phospholipase C induces release of acetylcholinesterase via cleavage of the anchor. The release is accompanied by an increase in *de novo* synthesis of the total acetylcholinesterase activity of the parasite (Espinoza *et al.*, 1988, 1991). Synthesis of acetylcholinesterase is also stimulated by diacylglycerol mimetics and a protein kinase C activator (the phorbol ester, phorbol-12-myristate-13-acetate, TPA), but inhibited by the protein kinase inhibitor, polymyxin B (Espinoza *et al.*, 1991). The results suggest that the cleavage of the anchor by phospholipase C will liberate diacylglycerol molecules within the membrane which will act as second messengers to activate a signalling pathway via protein kinase C that will trigger the synthesis of acetylcholinesterase and so replenish the acetylcholinesterase lost from the surface of the worm. The tegument-associated acetylcholinesterase may be involved in interactions with the host immune system, as distinct from its more typical role in cholinergic neurotransmission within the parasite's nervous system (Espinoza *et al.*, 1988, 1991).

4.1.7. *Inositol 1,4,5-trisphosphate*

It has been demonstrated that the tapeworm, *Hymenolepis diminuta* is capable of synthesising phosphoinositides (Webb and Mettrick, 1971). Production of inositol 1,4,5-trisphosphate (Ins(1,4,5)P_3, or IP_3) is stimulated by glutamate and acetylcholine, but not by 5-HT. A role for G protein(s) in the response to glutamate is supported by a number of pieces of evidence: sodium fluoride stimulates IP_3 production; GTP and GTP-γ-S potentiate the response, whereas GDP-β-S inhibits the response (Samii and Webb, 1996). On the basis of sensitivity to toxins, the G protein(s) involved in IP_3 production is(are) different from G_s and G_i proteins in other systems and such a difference(s) represents a potential target for therapeutic exploitation (Samii and Webb, 1996). It is presumed that the IP_3 response requires the intermediary action of PLC following activation of G-protein(s). In turn, the elevation of IP_3 levels is likely to lead to the release of calcium from intracellular stores in muscle cells in the worm, resulting in muscle contraction; glutamate is known to stimulate muscle contraction in *H. diminuta* (Thompson and Mettrick, 1989). However, not all of the glutamate-stimulated increase in IP_3 levels necessarily occurs via a G-protein and metabotropic receptors; rather, the primary response may be on post-synaptic ionotropic receptors, with the glutamate response being dependent on external calcium (Webb, 1997). Pharmacological characterisation of the ionotropic receptor by use of agonists and antagonists suggests that it is of the NMDA type (Webb, 1997). However, there may be some co-operativity between metabotropic and ionotropic receptors, the influx of calcium resulting from binding of glutamate to ionotropic rceptors leading to activation of PLC and elevation of IP_3 levels, with some of the PLC activation being mediated by via a G-protein (Webb, 1997).

4.1.8. *Diacylglycerol and Arachidonic Acid*

Functional arachidonic metabolising pathways leading to the production of prostaglandins and other eicosanoids (or arachidonic acid metabolites) have been shown to be present in the cercarial larval stage of the blood fluke, *S. mansoni* (Fusco, Salafsky and Kevin, 1985). It has been suggested that eicosanoid production is necessary for successful skin penetration of the mammalian host by cercariae because the pathways are stimulated by free fatty acids that stimulate skin penetration behaviour and the response is blocked by eicosanoid inhibitors (Fusco, Salafsky and Kevin, 1985; Fusco, Salafsky and Delbrook, 1986). Similar eicosanoids are produced by the cercariae of the related fluke, *Trichobilharzia ocellata*, a parasite of ducks and these compounds may have similar functions, namely, skin penetration (via vasodilation and secretion of the contents of the acetabular glands), transformation of the tegument and immune evasion (via disruption of neutrophil activation) (Nevhutalu *et al*., 1993). The schistosomule and adult of *S. mansoni* are also capable of synthesising and secreting a wide range of eicosanoids (Salafsky and Fusco, 1987). In view of the known immunoregulatory properties of eicosanoids, the ability of the parasite to secrete these compounds has been linked to a possible role in immune suppression as part of the parasite's mechanism for evasion of the host's immune response (Salafsky and Fusco, 1987).

The metabolism of arachidonic acid to produce prostaglandins has also been demonstrated in the larvae of two cestodes, namely, the plerocercoid larva (sparganum) of *Spirometra erinaceieuropaei* (Fukushima *et al*., 1993; Gao *et al*., 1998) and the strobilocercus larva of *Taenia taeniaeformis* (Leid and McConnell, 1983). The release of prostaglandin PGE2 into the incubation medium has been demonstrated *in vitro* and again linked to a possible role *in vivo* in immune suppression during tissue penetration and migration (Leid and McConnell, 1983; Fukushima *et al*., 1993; Gao *et al*., 1998).

Prostaglandins have also been demonstrated in the tapeworms, *Hymenolepis nana* and *Taenia saginata* and in the liver fluke, *F. hepatica* (Grzywacz and Szkudlinski, 1989; Ali, Joachim and Daugschies, 1999).The physico-chemical properties of the prostaglandins are dissimilar to those of their mammalian counterparts. Differences in the properties of the enzymes involved in their synthesis in *S. mansoni* have also been described (Salafsky and Fusco, 1985). The genes coding for the enzymes in *S. mansoni* display a different organisation from those in mammals, and they show developmental changes in their expression (Abdel Baset, O'Neill and Ford-Hutchinson, 1995).

Prostaglandins may have a role in carbohydrate metabolism in *F. hepatica*, because PGE1 and PGF2α stimulate glycogen phosphorylase activity, thus promoting glycogenolysis (Simonic, Sartorelli and Locatelli, 1983). Adenylate cyclase activity in *S. mansoni* is stimulated by prostaglandins and this, too, has been linked to glycogen utilisation (Higashi *et al*., 1973). The role of diacylglycerol in the synthesis of acetylcholinesterase has been discussed in the previous section (section 4.1.6.) (Espinoza *et al*., 1991). In addition to functions in parasite metabolism and physiology, prostaglandins play a variety of roles in the host-parasite relationship. For example, they have been linked to pathological aspects of infection (e.g. granuloma formation), to parasite invasion and to immune suppression, actions that will modulate haemostasis and serve to facilitate parasite establishment in the host: these roles are reviewed in more detail by Daugschies and Joachim (2000). In the *Taenia crassiceps*/mouse relationship, treatment of female mice with PGE2 prior to infection with *T. crassiceps* lead to enhanced parasite growth, low levels of IL-2 and IFN-γ (Th1 cytokines) and high levels of the Th2 cytokines, IL-6 and IL-10 (Terrazas *et al.*, 1999b). These effects were reversed by indomethacin, a PGE2 synthesis inhibitor. The results suggest that PGE2 plays a role in the susceptibility of mice to infection and may act synergistically with cytokines to inactivate the protective Th1-type response and promote the shift to the more permissive Th2-type response that is seen in *T. crassiceps* infections (Terrazas *et al.*, 1999b) (see section 4.2.2.). It is not known whether the parasite induces PGE2 production by the host, but the ability of parasites such as *T. taeniaeformis* (a close relative of *T. crassiceps*) to secrete PGE2, as described above, suggests that prostaglandins of parasite origin may be important in this response.

4.1.9. *Protein Kinase C*

Protein kinase C activity has been demonstrated in the freshwater planarian, *D. gonocephala*, in the human blood fluke, *S. mansoni* and the rat tapeworm, *H. diminuta* (Martelly *et al.*, 1987; Moraczewski *et al.*, 1987; Blair, Bennett and Pax, 1988; Wiest *et al.*, 1992; Gordon and Webb, 1996). Enzyme activity is stimulated by diacyglycerol and diacylglycerol agonists (Martelly *et al.*, 1987; Wiest *et al.*, 1992; Gordon and Webb, 1996). It is also stimulated by phorbol esters, which are known activators of protein kinase C (and diacylglycerol receptor agonists), and inhibited by inhibitors of mammalian protein kinase C activity, such as H-7, melittin and staurosporine (Martelly *et al.*, 1987; Wiest *et al.*, 1992; Gordon and Webb, 1996). The enzyme in *D. gonocephala* is calcium-independent (Martelly *et al.*, 1987), whereas that in *H. diminuta* is calcium-dependent (Gordon and Webb, 1996). The data on *S. mansoni* is contradictory (see Blair, Bennett and Pax, 1988 and Wiest *et al.*, 1992).

There is an increase in protein kinase C activity during the early stages of regeneration in *D. gonocephala* and this is accompanied by a translocation of the enzyme from the cytosol to the membrane: this is typical of enzyme activation (Moraczewski *et al.*, 1987). The role of protein kinase C in planarian regeneration will be discussed more fully in section 5.5. In contrast to the previous study, which indicates a positive role for protein kinase C in regeneration, a separate study on *D. lugubris* has shown that the phorbol ester TPA (phorbol-12-myristate-13-acetate, or PMA) inhibits regeneration (Pascolini *et al.*, 1989). TPA appears to activate the fixed parenchymal cells and stimulate their migration to the wound area, where they phagocytose damaged cells and induce cell lysis. The fixed parenchymal cells have an inhibitory effect on the migration of the neoblast cells that serve as stem cells for the differentiation of different cell types during regeneration (Pascolini *et al.*, 1989).

In the blood fluke, *S. mansoni* a 9-fold increase in protein kinase C activity has been observed during the transformation of the larval cercarial stage, through the schistosomule, to the adult stage (Wiest *et al.*, 1992). The result suggests that the expression of protein kinase C activity is developmentally regulated and that the enzyme plays a role in the regulation of parasite development or metabolism. The transformation of the cercaria to the schistosomule was associated with a translocation of protein kinase C activity from the cytosol to the membrane; as indicated above, this phenomenon is typical of enzyme activation.

Treatment of schistosomes with phorbol esters and a synthetic analogue of diacylglycerol induced vesiculation and blebbing of the surface membrane and internal vacuolization of the tegument (Wiest, Kunz and Miller, 1994). The effect was more pronouced in the male than the female worm. The results suggest that protein kinase C has a role in maintaining the structural integrity of the surface tegument, which is essential to parasite survival within the mammalian host. However, the mechanism by which this is brought about is unknown (Wiest, Kunz and Miller, 1994). The role of protein kinase C in the synthesis of tegument-associated acetylcholinesterase activity has been discussed in section 4.1.6. (Espinoza *et al.*, 1991).

Phorbol esters also induce an increase in muscle tone in schistosomes, indicating a role for protein kinase in the modification of muscle activity (Blair, Bennett and Pax, 1988). However, the identity of the neurotransmitter responsible for activation of protein kinase C is unknown. The phorbol ester-induced muscle contraction was not accounted for by changes in the release or postsynaptic sensitivity to putative transmitters such as acetylcholine, 5-HT or glutamate, nor by the peptides, FMRFamide and substance P (Blair, Bennett and Pax, 1994). In contrast, it has been shown that protein kinase C mediates the stimulation of motility of the liver fluke, *Fasciola hepatica* in response to the turbellarian peptide, GYRIFamide (Graham, Fairweather and McGeown, 2000). GYRIFamide is a FaRP and this group of peptides represent putative transmitters in this parasite. The GYRIFamide response was significantly reduced by neomycin sulphate, a drug which inhibits both phospholipase C and phospholipase D activity, enzymes which normally catalyse the production of diacylglycerol from membrane phospholipids; diacylglycerol subsequently acts to stimulate the activity of protein kinase C activity. The phorbol ester, PMA significantly increased the frequency of contractions in muscle strips from *F. hepatica*, whereas chelerythrine chloride, a potent and selective inhibitor of mammalian protein kinase C activity, inhibited the response to GYRIFamide. The reults from this study indicate that, following G protein activation, the response to GYRIFamide appears to involve activation of the phospholipase C, diacylglycerol, protein kinase C pathway (Graham, Fairweather and McGeown, 2000).

4.1.10. *Calmodulin*

Calmodulin has been demonstrated in the freshwater planarians, *D. lugubris* and *P. tenuis* (Martelly *et al.*, 1983). In these species and *D. gonocephala*, DNA and RNA synthesis have been shown to be calcium-dependent processes (Martelly, 1983; Martelly *et al.*, 1983). Calmodulin potentiates the effect of calcium; the latter can be blocked by the calmodulin antagonist, trifluoperazine (TFP, or Stellazine), but the inhibition can be overcome by calmodulin (Martelly, 1983; Martelly *et al.*, 1983). TFP has also been shown to inhibit the 5-HT stimulation of DNA synthesis and the dopamine stimulation of RNA synthesis (Martelly, 1984). During the regeneration of freshwater planarians, calmodulin levels decline over the first 48 hours, although there is a transient peak at 12 hours, which coincides with the first peak of DNA synthesis, suggesting that calmodulin may be involved in inducing the first peak of DNA synthesis (Martelly *et al.*, 1983). The possible role of calmodulin in regeneration is discussed in more detail in section 5.5.

A number of calcium binding proteins with homology to calmodulin have been isolated from the human blood fluke, *S. mansoni*. The first was purified and characterised from adult worms by classical biochemical techniques and has an apparent molecular mass of 19 kDa (Thompson *et al.*, 1986). Subsequently, several calcium binding proteins were purified by hydrophobic affinity chromatography and characterized by SDS-PAGE followed by fluorography: one of the proteins was identified as calmodulin via immunoblot analysis (Siddiqui, Podesta and Clarke, 1991). cDNA clones encoding 8 and 20 kDa proteins with homology to calcium binding proteins have been isolated and identified. The 8 kDa protein is only expressed in the cercarial stage and not in the sporocyst (the stage immediately preceding it) or in the adult stage which follows it (Ram *et al.*, 1989). In contrast, the 20 kDa protein is expressed in the schisosomule and adult stages, but not in the eggs. This protein appears to be associated with the inner face of the tegumental membrane (Havercroft *et al.*, 1990).

A 16.9 kDa protein with homology to calmodulin has also been isolated from the rat tapeworm, *H. diminuta* (Branford White, Hipkiss and Peters, 1984). An antibody to mammalian calmodulin has been used to determine the distribution of calcium binding proteins in the tapeworm, by means of electron immunocytochemistry. Immunostaining was evident in muscle cells, flame cells and spermatozoa; also in the tegument (in association with the microtriches bordering the syncytium and in the distal cytoplasm) and in the cytoplasmic connections between the tegument and the underlying tegumental cell bodies (Eastlake, Branford-White and Whish, 1994).

Trifluoperazine (TFP) has been used to shed light on possible roles for calmodulin in these two parasites. In *H. diminuta*, treatment with TFP leads to a decrease in glucose uptake and an inhibition of a number of enzymes in isolated brush border membrane fractions (Hipkiss and Branford White, 1986; Hipkiss *et al.*, 1986; Hipkiss, Branford White and Peters, 1987). In *S. mansoni*,

TFP induces a biphasic contraction of longitudinal muscle and a depolarisation of tegumental potential (Thompson *et al.*, 1986). Trifluoperazine causes severe disruption of the tegument in both species, but the effect may be secondary to its inhibitory actions on tegumental enzymes such as Ca^{++}-ATPase, because the latter effects are evident at lower concentrations than those which cause morphological changes (Thompson *et al.*, 1986; Hipkiss, Branford White and Peters, 1987; Hipkiss, Skinner and Branford White, 1987). The results suggest a possible role for calmodulin in osmotic/ionic regulation across the tegumental membranes.

4.1.11. *Rhodopsin*

Rhodopsin is a seven-pass transmembrane molecule analogous to other G protein-linked receptors. It operates through a G protein, but in this case the signal is a photon of light. Rhodopsin plays an important role in visual transduction responses in the vertebrate retina, the light signal being converted into an electrical one. A cDNA clone encoding a rhodopsin homologue has been isolated from *S. mansoni*. The homologue, SmRHO (*S. mansoni* rhodopsin) is most similar to rhodopsins in water-dwelling organisms (bivalve and cephalopod molluscs) and is expressed at much higher levels in the free-living miracidial and cercarial stages than in the adult worms (Hoffmann *et al.*, 2001). Immunostaining for SmRHO in the cercaria is particularly strong in sub-tegumental cells towards the anterior end of the larva. It has been suggested that the schistosome homologue is involved in light reception responses and may play a role in host location behaviours, involving both the intermediate, snail host and the definitive, vertebrate host (Hoffmann *et al.*, 2001).

4.2. ENZYME-LINKED RECEPTOR PATHWAYS

4.2.1. *Receptor Tyrosine Kinase*

A gene encoding an EGF receptor homologue has been isolated from *S. mansoni*. The encoded protein shares substantial sequence homology with other members of the EGF receptor family and has been designated SER, or *Schistosoma mansoni* epidermal growth factor homologue (Shoemaker *et al.*, 1992; Ramachandran, Skelly and Shoemaker, 1996). It is present in the cercarial larvae and adult schistosomes (Shoemaker *et al.*, 1992). The schistosome receptor is a functional receptor, having been shown to be capable of tyrosine autophosphorylation. Immunostaining for SER is localized predominantly in the sub-tegumental muscle layers, pointing to a possible role in the regulation of muscle contractile activity or a more mitogenic role in the development or turnover of schistosome muscle (Ramachandran, Skelly and Shoemaker, 1996).

cDNA clones encoding a Ras homologue have been isolated from *S. mansoni*. Analysis of the amino acid sequence of the encoded protein revealed strong homology to Ras proteins from other organisms (Kampkotter *et al.*, 1999; Osman, Niles and Lo Verde, 1999). Developmental expression of the Ras homologue was studied by reverse transcriptase polymerase chain reaction (RT-PCR) and it was shown to be present in paired and unpaired males and females; in all stages of development from the schistosomule to the adult worm; and in the miracidia and cercariae as well (Kampkotter *et al.*, 1999; Osman, Niles and Lo Verde, 1999). Ras is over-expressed in female worms compared to male worms and this was attributed to a role in vitelline cell maturation and/or egg laying (Osman, Niles and Lo Verde, 1999). Using an antibody raised against the *S. mansoni* Ras homologue, the protein was observed in the sub-tegumental tissues and not exposed on the tegumental surface (Osman, Niles and Lo Verde, 1999).

Ras, GAP and MAP kinases have also been identified in *S. mansoni* by means of immunoblotting, using antibodies raised against conserved epitopes of these signal transducing molecules (Schussler, Grevelding and Kunz, 1997). Moreover, the molecules were shown to be developmentally regulated. For example, Ras expression was much higher in mature/paired females than in immature/unpaired females, whereas it was equally expressed in mature and immature males. Interestingly, the expression of GAP in females depends on pairing with the male. Thus, the level of GAP decreased following separation of females from female/male pairs, disappearing altogether after 6 days, then was re-expressed in females that had been re-mated with males (Schussler, Grevelding and Kunz, 1997). MAP kinases (MAP1 and MAP2 kinases) are expressed more in females than males (Schussler, Grevelding and Kunz, 1997). The combined data suggest that the Ras pathway is involved in the male-directed maturation of the female worm.

A novel receptor tyrosine kinase receptor molecule from *S. mansoni* has been cloned and sequenced: it is designated SmRTK-1 (*S. mansoni* receptor tyrosine kinase-1) (Vicogne *et al.*, 2003). It contains a conserved intracellular catalytic domain characteristic of members of the insulin receptor family. However, its extracellular ligand-binding domain is unusual in that it possesses a structure similar to the Venus Flytrap (VFT) module of bacterial periplasmic binding proteins (Vicogne *et al.*, 2003). The receptor is associated with membranes and is expressed in the miracidium, sporocyst and cercaria larvae and in adult worms from bisexual infections. Its location in possible germinal cells of the larvae and ovocytes in mature females suggests a role in growth and differentiation. Another role in male-female communication and maturation of the female worm has been proposed (Vicogne *et al.*, 2003).

4.2.2. *Tyrosine Kinase-Associated Receptors*

A gene encoding a tyrosine kinase enzyme from *S. mansoni* has been cloned and sequenced. The protein (designated TK5) exhibits strong homology to Src-like tyrosine kinases, with the highest homology to members of the Fyn sub-family. TK5 contains a number of hallmark domains: SH4, SH3, SH2 and the catalytic tyrosine kinase domain. It is expressed in the miracidial and cercarial larval stages and in adult male and female worms (Kapp *et al.*, 2001).

Cytokines may play a role in parasite growth and reproduction. For example, egg production by *S. mansoni* is dependent on tumour necrosis factor α (TNFα). There is a shift in cytokine production during schistosome infection from a more host-protective T-helper cell 1 (Th1)-type profile to a more permissive, less host-protective, Th2-type profile (Grzych *et al.*, 1991; Amiri *et al.*, 1992; Leptak and McKerrow, 1997; McKerrow, 1997). Synthesis of TNFα (a Th2-type cytokine) coincides with migration of worms to the liver and appears to be induced by the parasite itself (McKerrow, 1997). Interleukin-7 (IL-7) may also be involved in schistosome development (Salzet, Capron and Stefano, 2000).

A number of experiments have been carried out to evaluate the effects of various cytokines on parasite infections. Treatment of *T. crassiceps*-infected mice with interferon-γ (INF-γ) and IL-2 (Th1-type cytokines) decreased the worm burden, an effect reversed by treatment with an anti-IFN-γ antibody. In contrast, treatment with IL-10 (a Th2-type cytokine) increased the susceptibility of mice to infection, whereas treatment with an anti-IL-10 antibody reduced the worm burden (Terrazas *et al.*, 1999a). These experiments showed that a Th1-type response protects against infection, whereas a Th2-type response favours parasite establishment. Administration of IFN-γ *in vivo* has also been shown to reduce the worm burden in *M. corti* infections (Jenkins *et al.*, 1991). *In vitro*, INF-γ stimulated the macrophage-mediated killing of the larvae (Jenkins *et al.*, 1990); this action was antagonized by TNFα (Jenkins *et al.*, 1992).

It has been demonstrated that excretory/secretory (E/S) products released by parasites can influence cytokine production by the host. For example, *Spirometra erinaceieuropaei* secretes a product that suppresses TNFα synthesis by stimulated murine peritoneal macrophages (Miura *et al.*, 2000). *T. solium* cysts produce a metacestode factor (MF) that inhibits production of cytokines such as IL-2, IL-4 and IFN-γ (by stimulated spleen cells) and TNFα (by stimulated macrophages) (Arechavaleta, Molinari and Tato, 1998). *T. crassiceps* larvae release factor(s) which induce anergy on T cells located near the cysticerci, causing a downregulation of Th1 cytokines (e.g. IL-2) and upregulation of Th2 cytokines (e.g. IL-4). The immunosuppression produced appears to be a localized phenomenon around the parasite (Villa and Kuhn, 1996). One of the E/S products, designated p66, released by *T. crassiceps* has been shown to possess immunological and functional characteristics of mouse IFN-γ (Spolski *et al.*, 2002).

4.2.3. *Receptor Serine/Threonine Kinases*

A member of the TGF-β receptor family has been isolated from *S. mansoni* and designated SmRK-1 (*Schistosoma mansoni* receptor kinase-1) (Davies, Shoemaker and Pearce, 1998). The amino acid sequence of SmRK-1 shows a high degree of similarity to receptor serine/threonine kinases of the type I subfamily (Davies, Shoemaker and Pearce, 1998). It has been shown that the extracellular domain of SmRk-1 can interact with TGF-β and activate the signalling pathway involving Smads (see below) (Beall and Pearce, 2001). Using antibodies to SmRK-1, the receptor has been localized to the tubercles on the dorsal surface of the male worms, not elsewhere in the male or anywhere in the female (Davies, Shoemaker and Pearce, 1998). The receptor can be isolated in phosphorylated form, suggesting that it can function in transducing signals across the tegument, in response to host-derived ligands (Davies, Shoemaker and Pearce, 1998). Originally, the expression of SmRK-1 was said to occur only after 3 weeks infection in the mammalian host (Davies, Shoemaker and Pearce, 1998), but subsequently it was shown to occur as early as 5 days post-infection, when the parasites are preparing to leave the lungs and migrate to the liver (Beall, McGonigle and Pearce, 2000; McGonigle *et al.*, 2001).

Other downstream elements of the TGF-β-triggered pathway have been identified. For example, two *Smad* genes from *S. mansoni* have been cloned. The encoded proteins show strong sequence homology to Smad proteins from other organisms and have been designated SmSmad1 and SmSmad2 (*S. mansoni* Smads1 and 2) (Beall, McGonigle and Pearce, 2000; Osman, Niles and Lo Verde, 2001). Immunoreactivity to SmSmad2 has been observed in the tubercles of the tegument; in parenchymal cells of both males and females; in the ovary and vitellaria of the female; in developing embryos; and in the testis of the male. The localization was confirmed by *in situ* hybridization studies (Osman, Niles and Lo Verde, 2001). Interestingly, the expression of the SmSmads coincides with the expression of SmRK-1 and the SmSmads are capable of interacting with the intracellular domain of SmRK-1 *in vitro* (Beall, McGonigle and Pearce, 2000; Osman, Niles and Lo Verde, 2001). Both SmSmads show nuclear translocation in mammalian cells in response to TGF-β (Beall, McGonigle and Pearce, 2000; Osman, Niles and Lo Verde, 2001). Moreover, both proteins have been shown to be capable of increasing the transcription of a luciferase reporter gene in an assay involving mammalian RIB cells (Beall, McGonigle and Pearce, 2000; Beall and Pearce, 2001). The widespread tissue distribution of SmSmad immunoreactivity and the developmentally-regulated expression of the SmSmads suggest that they play multiple roles in the development of the schistosome and in the sexual maturation of the male and female worms in particular.

A SmRK-1-interacting protein (SIP) has been isolated from *S. mansoni* and may function as an adaptor protein, linking SmRK-1 activation to downstream signalling pathways (McGonigle *et al.*, 2001). The expression of this protein parallels that of SmRK-1, that is, it is not present in the cercaria or schistosomule, but is present in the lung stage (at 5 days post-infection) and in adult male and female worms (McGonigle *et al.*, 2001).

4.3. STEROID HORMONES, THYROID HORMONES AND RETINOIDS

During their development within mammalian hosts, flatworm parasites are exposed to a variety of host molecules, including hormones. Hormonal signals have an important influence on parasite establishment and survival, homing (that is, migration to their preferred site in the body), growth, sexual maturation and egg production. The communication is two-way, as parasites can have an effect on host fecundity, as exemplified by the feminization of mice with the larvae of *T. crassiceps*: see below. The strobilocercus larva of *Taenia taeniaeformis* releases an excretory/secreory product which, when injected into rats, decreases human chorionic gonadotropin (HCG)-stimulated serum and testicular testosterone concentrations. The product also decreases testosterone production by isolated Leydig cells *in vitro* (Rikihisa, Lin and Fukaya, 1985).

The sex of the mammalian host is an important determinant for parasite infection. The male is more susceptible to infection with *M. corti*, *E. granulosus*, *Cysticercus fasciolaris*, *F. hepatica* and *S. mansoni* (in the rat). In contrast, it is the female that is more susceptible to infection with *S. mansoni* (in the mouse), *T. crassiceps*, *Echinococcus multilocularis*, *Taenia multiceps* and *T. taeniaeformis* (for references, see Fairweather, 1997; de Mendonca *et al.*, 2000). Numerous *in vivo* hormonal manipulation experiments have been carried out and the results lend support to these generalisations. For example, in *S. mansoni* infections in mice, the treatment of female mice with testosterone before infection lead to a reduction in worm burden, while castration of male mice had the opposite effect (Nakazawa *et al.*, 1997). Female mice can be partially protected against infection by early treatment with dihydroepiandrosterone (de Mendonca *et al.*, 2000). Testosterone exerts its effects only early in infection, during the development of immature schistosomes, because female mice treated with testosterone 10 or more days after infection have worm burdens similar to female mice given no testosterone (Nakazawa *et al.*, 1997).

Female mice are more susceptible to infection with *T. crassiceps* than male mice (Sciutto *et al.*, 1991). Gonadectomy renders both sexes equally susceptible, but the effect is only seen in mice with intact immune systems, suggesting that the gonadal effect on mouse susceptibility is indirect and mediated via the immune system (Huerta *et al.*, 1992). In chronic infections, the sex-related difference in susceptibility disappears and males show a feminization process with elevated estradiol levels and decreased testosterone levels; infected female mice also display a slight feminization (Larralde *et al.*, 1995). There is a concomitant shift from a Th1-type to a more permissive Th2-type immune response, with a decrease of Th1-type cytokines IL-2 and IFN-γ and an increase in the Th2-type cytokines IL-4, IL-6 and IL-10 (Terrazas *et al.*, 1998). Treatment of castrated male and female mice with testosterone before infection led to reduced worm burdens, whereas treatment with oestradiol increased parasite loads. These changes were compared with cytokine responses. IL-2 and IFN-γ production was depressed in infected-castrated mice of both sexes, but the effect was reversed with testosterone treatment. Testosterone had no effect on production of Th2-type cytokines, but estradiol treatment stimulated production of IL-6 and IL-10. The results suggest that androgens influence (that is, depress) host susceptibility by stimulating cellular immunity, acting on the Th1, not Th2 cell population (Morales-Montor *et al.*, 2002). On the other hand, estradiol promotes cysticercus multiplication by interfering with thymus-dependent cellular immune mechanisms that obstruct parasite growth (Terrazas *et al.*, 1994).

While the ability of parasites to synthesise steroid hormones has been questioned, ecdysteroids (moulting hormones) have been demonstrated in a number of trematode and cestode parasites, including *S. mansoni*, *F. hepatica*, *E. granulosus* and *H. diminuta* (see review by Barker, Chitwood and Rees, 1990; Foster, Mercer and Rees, 1992). Ecdysone immunoreactivity has also been localized in the cercarial and adult stages of *S. mansoni* (Basch, 1986). Ecdysterone has been shown to stimulate an increase in asexual reproduction of *M. corti* larvae *in vivo* and to stimulate an increase in larval growth and asexual reproduction *in vitro* (Kowalski and Thorson, 1976).

Steroid hormones modulate gene expression via specific nuclear receptors. The nuclear receptors bind to the regulatory regions, called hormone response elements (HREs), of target genes. HRE-like elements have been found in some schistosome genes. For example, the F-10 gene, which encodes the eggshell protein F-10 of *S. mansoni*, is transcriptionally active only in adult females. It contains a hexanucleotide motif in the downstream, non-translated region that is identical to the HRE present in genes regulated by steroids (Rumjanek, Braga and Kelly, 1989). The presence of nuclear proteins binding this motif has been demonstrated by gel-shift assays (Rumjanek, Braga and Kelly, 1989; Engelender, Giannini and Rumjanek, 1993; Giannini *et al.*, 1995). Binding of female nuclear proteins was altered by tamoxifen, an oestrogen antagonist, suggesting that the regulation of transcription of the F-10 gene involves steroid receptors (Giannini *et al.*, 1995).

Five nuclear receptors for which no ligand is known have been identified in *S. mansoni*: one retinoid-X receptor (RXR) homologue; one member of the orphan receptor testicular receptor 2/4 (TR2/4) group; two members of the chicken ovalbumin upstream promoter-transcription factor (COUP-TF) family; and one member of the *fushi tarazu* factor 1 (FTZ-F1) family (Escriva *et al.*, 1997). A cDNA encoding a homologue of the RXR has been identified in *S. mansoni* and has been designated SmRXR (Freebern *et al.*, 1999). It binds to *cis* elements of the *p14* gene and activates transcription. The *p14* gene encodes an eggshell precursor expressed only in vitelline cells of sexually mature females (Chen, Rekosh and Lo Verde, 1992). This suggests a role for SmRXR in the regulation of female-specific gene expression. However, the expression of SmRXR is not confined to sexually mature females, as it has been identified in various developmental stages post-infection of the mammalian host, including immature females, mature males and males from single sex infections (Freebern *et al.*, 1999). Consequently, the *SmRXR* gene may play multiple roles during the life cycle of the parasite.

The thyroid status of the host can influence helminth infections. Thus, treatment with thyroxine, to induce a hyperthyroid state has been shown to promote the growth and egg production of a number of parasites, whereas a hypothyroid state reverses these effects, although exceptions to this rule occur (see review by Fairweather, 1997). Taking the blood fluke, *S. mansoni* as an example, worms from hyperthyroid mice were larger, matured earlier and produced more eggs than those from hypothyroid mice (Abdel Wahab, Warren and Levy, 1971). Similar results were obtained with the closely related species, *Schistosomatium douthitti* (Cornford, 1974). The results with the rat, a non-permissive laboratory host for *S. mansoni*, were rather different, however. Thyroidectomy changed the status of the rat to a permissive one and the worms were able to complete their development and produce viable eggs. Thyroxine treatment of thyroidectomized rats prior to infection partially restored their non-permissive status (Knopf and Linden, 1985). It is not clear whether thyroid hormones exert a direct effect on the parasite or not. In this context, it is worth noting that the effects of thyroid hormone in mice are mirrored by those of interleukin 7 (IL-7). Thus, fewer, smaller worms were recovered from IL-7-knockout mice and normal parasite growth was restored by injections of IL-7 (de Mendonca *et al.*, 2000). Therefore, it may be that factors such as thyroid hormone and IL-7 serve to create an environment that is favourable for schistosome development in mice and their effects are indirect.

4.4. ION CHANNELS

Little is known about the electrophysiological properties of nerve cells in flatworms, probably due to the problems (alluded to previously) of working with nervous tissue *in situ* and of isolating nerve cells for intracellular recordings. Among the turbellarians, recordings have been made from nerve cells *in situ* in the polyclad, *Notoplana acticola* and from nerve cells isolated from the triclad, *Bdelloura candida*, an ectoparasite of the horseshoe crab, *Limulus polyphemus*. The nerve cells from these two species possess a variety of Na^+, Ca^{++} and K^+ currents (Keenan and Koopowitz, 1981, 1984; Blair and Anderson, 1993). In contrast to the study on *N. acticola*, all neuronal Na^+ currents recorded from *B. candida* were sensitive to tetrodotoxin (TTX). Flatworms are the lowest animals to possess TTX-sensitive Na^+ currents and therefore represent an important stage in the evolution of Na^+ channels. It seems likely that the TTX receptor evolved prior to the evolution of the platyhelminths, but after the separation of the Cnidaria from the invertebrate lineage. Electrophysiological recordings have also been made *in situ* from the nerve cells of another polyclad, *Alloeoplana californica* and from the longitudinal nerve cords of *Gyrocotyle fimbriata*, a cestodarian parasite from the gut of the chimerid fish, *Hydrolagus colliei* (Solon and Koopowitz, 1982; Keenan, Koopowitz and Solon, 1984). A hyperpolarisation-activated cation current has been characterized in unipolar nerve cells isolated from the liver fluke, *Fasciola hepatica* (Kim *et al.*, 2002).

A few genes encoding ion channel proteins have been isolated from flatworms. They are: a Na^+ channel a subunit (BdNa1) from *B. candida* (Jeziorski, Greenberg and Anderson, 1997); a *Shaker*-like K^+ channel (Skv1.1) from the blood fluke, *S. mansoni* (Kim *et al.*, 1995); and three Ca^{++} channel α_1 subunits, also from *S. mansoni* (Kohn *et al.*, 2001).

5. Functional Signalling Pathways Linked to Individual Neurotransmitters

5.1. RECEPTORS

A serotonin receptor of the 5-HT_{1A} type, linked to adenylate cyclase, has been identified in the freshwater planarians, *Dugesia gonocephala* and *D. lugubris* (Creti *et al.*, 1992). A similar receptor has been identified in the related species, *D. japonica* by means of a radioligand binding study involving 5-HT receptor antagonists (Saitoh, Yuruzume and Nakata, 1996). Using a molecular approach, four G protein-coupled receptors have been identified in this species. They all showed significant homology to the human 5-HT_{1A} serotonin receptor; consequently, they have been designated as serotonin receptor-like planarian receptors 1–4, or 5HTLpla1–5HTLpla4 (Saitoh *et al.*, 1997). 5HTLpla1 and 5HTLpla4 have been sequenced and show a significant homology to the 5HTdrol serotonin receptor of the insect, *Drosophila.* 5HTLpla4 is the most abundantly expressed of the four receptors and the expression of 5HTLpla4 mRNA increases during regeneration, suggesting a role for the receptor in the regeneration mechanism (Saitoh *et al.*, 1997).

5-HT receptors coupled to adenylate cyclase have been demonstrated in *F. hepatica* using an adenylate cyclase assay and a radioligand binding assay (McNall and Mansour, 1984). The binding properties of the receptor were affected by 5-HT agonists and antagonists, but were not affected by other putative transmitters such as dopamine, epinephrine and norepinephrine. The pharmacological profile of the receptors appears to be different from that in mammals, suggesting that the differences could be exploited therapeutically, to disrupt 5-HT-regulated functions (McNall and Mansour, 1984). The stimulatory motility response to 5-HT in *S. mansoni* has been tested using a variety of 5-HT agonists and antagonists and 5-HT uptake inhibitors, in an attempt to pharmacologically characterise the 5-HT receptors in the fluke. The results indicated that the receptors are unlike those in their mammalian host, again highlighting receptors as potential targets for drug attack (Willcockson and Hillman, 1984). A putative 5-HT receptor has been identified in the tapeworm, *H. diminuta* that does not conform to any known class of 5-HT recognition site. Binding is affected by 5-HT agonists and antagonists, but not by 5-HT uptake inhibitors or by dopamine or adrenergic drugs (Ribeiro and Webb, 1986). A separate 5-HT receptor coupled to adenylate cyclase via a G protein has been characterised in the worm, the receptor having the pharmacological properties of a 5-HT 1 type of receptor (Ribeiro and Webb, 1987).

A novel G protein-coupled receptor cDNA has been cloned and isolated from *S. mansoni.* The receptor protein (designated SmGPCR) shares sequence homology with monoamine G protein-coupled receptors, but has no clear homology with any particular type of receptor (Hamdan *et al.*, 2002). It is expressed at the cell surface, with the N-terminus extracellular and the C-terminus intracellular. Among the monoamines tested in functional assays, only histamine was able to activate the receptor significantly, causing a dose-dependent increase in cellular calcium. No effect was observed in a cAMP-responsive reporter assay, indicating that SmGPCR signalling does not involve cAMP (Hamdan *et al.*, 2002). Instead, the receptor may be coupled to the IP_3 pathway. With regard to the potential function of the histamine-activated pathway in *S. mansoni*, two suggestions have been put forward: either the modulation of neuromuscular activity within the parasite, or a role in the evasion of the host's immune response via the reception of histamine produced by the host's immune cells (Hamdan *et al.*, 2002).

5.2. 5-HT AND MOTILITY

5-HT exerts a stimulatory effect on the motility of the liver fluke, *F. hepatica*, indicating a potential role as an excitatory transmitter in the fluke (Holmes and Fairweather, 1984; Tembe *et al.*, 1993). It has been localised within the nervous system of the fluke (Fairweather *et al.*, 1987) and the fluke has been shown to be capable of synthesising 5-HT from 5-hydroxytryptophan, but not from tryptophan (Mansour, Lago and Hawkins, 1957; Mansour and Stone, 1970). There is some evidence that the effect of 5-HT on muscle activity is mediated by a cAMP-dependent pathway. Thus, 5-HT receptors are linked to adenylate cyclase (Northup and Mansour, 1978a) and 5-HT has been shown to stimulate adenylate cyclase activity (Northup and Mansour, 1978a,b; Northup *et al.*, 1979), cAMP production (Mansour *et al.*, 1960; Abrahams, Northup and Mansour, 1976) and protein kinase A activity (Gentleman, Abrahams and Mansour, 1976). Forskolin, an activator of adenylate cyclase activity stimulates motility and potentiates the activation of adenylate cyclase by 5-HT (McNall and Mansour, 1985; Graham, McGeown and Fairweather, 1999). The cAMP analogue, 8-bromo-cAMP significantly increases spontaneous contractions (Graham, McGeown and Fairweather, 1999). Phosphodiesterase inhibitors such as IBMX increase the levels of cAMP and stimulate motility (Mansour and Mansour, 1977; McNall and Mansour, 1985; Graham, McGeown and Fairweather, 1999).

5-HT has been demonstrated in the nervous system of adult *S. mansoni* (Dei-Cas *et al.*, 1979; Gustaffson, 1987), although the worm is unable to synthesise 5-HT *de novo* from tryptophan (Bennett and Bueding, 1973; Catto and Ottesen, 1979). However, the flukes are unable to take up 5-HT *in vitro* (Chou, Bennett and Bueding, 1972; Bennett and Bueding, 1973; Bueding *et al.*, 1974; Senft *et al.*, 1976; Wood and Mansour, 1986). There is a marked impairment of tryptophan metabolism in *S. mansoni*-infected mice, indicating that the flukes take up tryptophan from the host's circulation (Niagi and Bender, 1990). Consequently, the parasite appears to be dependent on its host for a supply of 5-HT or its intermediate, 5-hydroxytryptophan. This requirement is met when the schistosomule arrives in the peripheral bloodstream, but more particularly when the worm reaches the mesenteric-portal venous system, where the highest levels of 5-HT (synthesised by the enterochromaffin cells of the gut lining) occur. 5-HT exerts a stimulatory effect on the motility of adult worms (Hillman and Senft, 1973; Tomosky, Bennett and Bueding, 1974; Fetterer, Pax and Bennett, 1977). It also induces the contraction of isolated muscle fibres of *S. mansoni* (Day, Bennett and Pax, 1994). This response is mimicked by forskolin, an adenylate cyclase activator, but is blocked by the PKA inhibitor, H89, suggesting that the response may be mediated by a cAMP-dependent pathway (Day, Bennett and Pax, 1994).

In tapeworms, studies on 5-HT mechanisms have focussed on the rat tapeworm, *H. diminuta.* The worm is capable of synthesising 5-HT from both tryptophan and 5-hydroxytryptophan (Ribeiro and Webb, 1983, 1984), although it probably derives 5-HT from its host as well (Cyr, Gruner and Mettrick, 1983; Ribeiro and Webb, 1984; Osloobi and Webb, 1999). Release of 5-HT from nervous tissue occurs by a calcium-dependent process (Gordon and Webb, 1989). 5-HT has been localised in the nervous system of *H. diminuta* (Webb and Mizukawa, 1985; Fairweather *et al.*, 1988; McKay *et al.*, 1991) and exerts a stimulatory effect on neuromuscular preparations (Sukhdeo *et al.*, 1984; Thompson and Mettrick, 1989). Taken together, then, there is substantial evidence that 5-HT functions as a neurotransmitter in this species. It seems likely that the motility response is mediated by a cAMP-dependent pathway. This idea is supported by the following observations: *H. diminuta* possesses a 5-HT receptor of the 5-HT 1 type which is coupled to adenylate cyclase via a G protein (Ribeiro and Webb, 1987); and 5-HT has been shown to stimulate adenylate cyclase activity and cAMP production (Ribeiro and Webb, 1987). However, it has been suggested that 5-HT does not act directly on muscles and, rather than functioning as a transmitter, 5-HT serves as a neuromodulator of another excitatory transmitter (most likely glutamate). In this connection, it is worth noting that 5-HT reduces the stimulatory effect of glutamate on muscle contraction (Thompson and Mettrick, 1989) and stimulates the uptake of glutamate (Webb, 1986).

5.3. 5-HT AND CARBOHYDRATE METABOLISM

In the liver fluke, *F. hepatica*, 5-HT stimulates glucose uptake, glycogen breakdown and lactic acid production (Mansour, 1959). The 5-HT-induced increase in glycolysis occurs via stimulation of phosphofructokinase (PFK), the rate-limiting enzyme in this pathway (Mansour, 1962; Mansour and Mansour, 1962) and the increase in glycogenolysis is believed to involve the action of glycogen phosphorylase (Mansour *et al.*, 1960; Gentleman, Abrahams and Mansour, 1976). Activation of PFK by 5-HT is indirect, being mediated by a cAMP-dependent pathway. Evidence for this is two-fold. On the one hand, 5-HT is known to stimulate adenylate cyclase activity (Northup and Mansour, 1978b), raise cAMP levels (Mansour *et al.*, 1960; Abrahams, Northup and Mansour, 1976) and increase PKA activity (Gentleman, Abrahams and Mansour, 1976).

On the other, cAMP is known to stimulate PFK activity (Mansour, 1962; Mansour and Mansour, 1962; Stone and Mansour, 1967). The phosphorylation and subsequent activation of PFK by the catalytic subunit of PKA has been demonstrated *in vitro* (Kamemoto and Mansour, 1986; Kamemoto *et al.*, 1987). Activation of PFK in response to 5-HT has also been shown to occur *in vivo* (Kamemoto, Lan and Mansour, 1989). The phosphorylation site in the PFK has been isolated, the amino acid sequence determined and the phosphorylated residue identified as threonine. The cAMP-dependent protein kinase recognition, or "target", sequence in PFK is Arg/Lys-Arg-Ser-Thr(P), where an initial basic residue is presumed from trypsin cleavage. This sequence is unusual in that threonine is the phosphorylated residue (Mahrenholz, Hefta and Mansour, 1991).

5-HT also stimulates carbohydrate utilisation and PFK activity in the blood fluke, *S. mansoni* (Mansour and Mansour, 1962; Bueding and Fisher,1982). However, the AC-cAMP-PKA-PDE pathway demonstrated in *S. mansoni* has been linked more closely to glycogen metabolism (breakdown) than to glycolysis (Higashi *et al.*, 1973). Nevertheless, these roles remain to be substantiated. In the rat tapeworm, *H. diminuta*, 5-HT stimulates glucose uptake and glycogen utilisation (Mettrick, Rahman and Podesta, 1981; Rahman, Mettrick and Podesta, 1983). 5-HT appears to have little direct stimulatory effect on PFK activity, whereas cAMP does (Rahman, Mettrick and Podesta, 1983; Sangster and Mettrick, 1987). 5-HT increases adenylate cyclase activity, which presumably in turn leads to activation of cAMP and binding of cAMP to PFK, resulting in deinhibition of the enzyme (Sangster and Mettrick, 1987). In contrast to its action in *H. diminuta*, 5-HT only induces depletion of glycogen reserves in the related tapeworm, *H. microstoma*, possibly via stimulation of glycogen phosphorylase activity. It has no effect on glycolytic enzymes such as PFK, although the activation of the latter is stimulated by cAMP (Rahman, Mettrick and Podesta, 1982).

5.4. MONOAMINES AND REGENERATION IN TURBELLARIANS

Planarians have remarkable powers of regeneration, every part of the body (except the pharynx) being able to regenerate all the other parts. This ability is dependent on the neoblast cells which are undifferentiated stem cells that are able to divide and differentiate into all other cell types according to their position in the body. The capacity to regenerate underpins the method of asexual reproduction displayed by these organisms, which takes place by a process of fissioning, the head and tail ends tearing apart and each fragment regenerating the appropriate missing section. There is a considerable amount of data to indicate an endocrine-like role for monoamines in regeneration, particularly in the early stages (over the first 24 hours) involving cellular activation prior to neurosecretory control. Indeed, monoamines may be required to stimulate the release of peptidergic neurosecretions (Ouahsine and Sauzin-Monnot, 1981). Following sectioning, there is an initial peak of

DNA synthesis at 12 hours. This is followed by increases in the rates of RNA sythesis (with a peak at 18 hours) and protein synthesis (Fig. 2a) (Martelly and Le Moigne, 1980; Martelly, Rey and Le Moigne, 1981). The mitotic index increases towards the end of the first day (Fig. 2a) (Gabriel, 1969). The levels of monoamines vary during the first 24 hours of regeneration: 5-hydroxytryptamine (or 5-HT) and norepinephrine (or noradrenaline) rise from 2 hours to 15 hours, with an initial peak for 5-HT at approximately 3 hours and a later one at 15 hours. After 15 hours, the levels of both return to control levels (Fig. 2b). Peaks of dopamine occur at 15 hours (Fig. 2b). Treatment of planarians with 5-HT and dopamine promotes regeneration by accelerating (i.e. shortening) the process, whilst antagonists of all three monoamines delay it (Franquinet, 1979; Franquinet and Le Moigne, 1979). 5-HT (and possibly norepinephrine as well) stimulates DNA synthesis and the mitotic index rises (Franquinet, 1981), whereas dopamine stimulates RNA synthesis (Franquinet and Martelly, 1981). The peaks of dopamine and RNA synthesis – at 15 and 18 hours, respectively – coincide well. The 5-HT-induced stimulation of DNA synthesis occurs at the expense of RNA synthesis, which is inhibited (Franquinet and Martelly, 1981).

The signalling mechanisms underlying these actions have been investigated in a series of experiments monitoring variations in concentrations of adenylate cyclase, calcium, cAMP and protein kinases and comparing them with monoamine and nucleic acid levels. The levels of adenylate cyclase activity peak at 3 hours and 15 hours, prior to the rises of DNA and RNA synthesis, respectively (Fig. 2c) (Franquinet and Le Moigne, 1979; Martelly, Franquinet and Le Moigne, 1981). Changes in the levels of adenylate cyclase activity are paralleled by those of 5-HT (Franquinet and Le Moigne, 1979) and 5-HT has been shown to stimulate adenylate cylase activity (Franquinet, Le Moigne and Hanoune, 1978; Venturini *et al.*, 1986; Creti *et al.*, 1992). Adenylate cyclase activity is also stimulated by dopamine (Venturini *et al.*, 1986; Venturini, 1993) and a high concentration of norepinephrine (Franquinet, Stengel and Hanoune, 1976). cAMP levels show a marked increase at 2 hours, returning to normal by 12 hours (Fig. 2c) (Franquinet and Le Moigne, 1979). The close link between adenylate cyclase activity and 5-HT levels and the known role of adenylate cyclase in stimulating cAMP synthesis suggest that 5-HT may stimulate DNA synthesis via a cAMP-dependent pathway. However, although cAMP can stimulate DNA synthesis (Martelly, 1984), the effect appears to be indirect and is dependent on calcium, since DNA and RNA synthesis have been shown to be calcium-dependent (Martelly, 1983; Martelly *et al.*, 1983; Martelly, 1984). The threshold level of calcium for triggering RNA synthesis is lower than that for DNA synthesis (Martelly, 1984). During the early stages of regeneration, calcium levels rise from 6 hours to reach a peak at 12 hours before falling to resting levels by 18 hours, then gradually rise to 30 hours (Fig. 2c) (Martelly *et al.*, 1983). The peak at 12 hours coincides with the peak of DNA synthesis. cAMP increases calcium uptake by planarian cells (Martelly, 1984). cAMP has

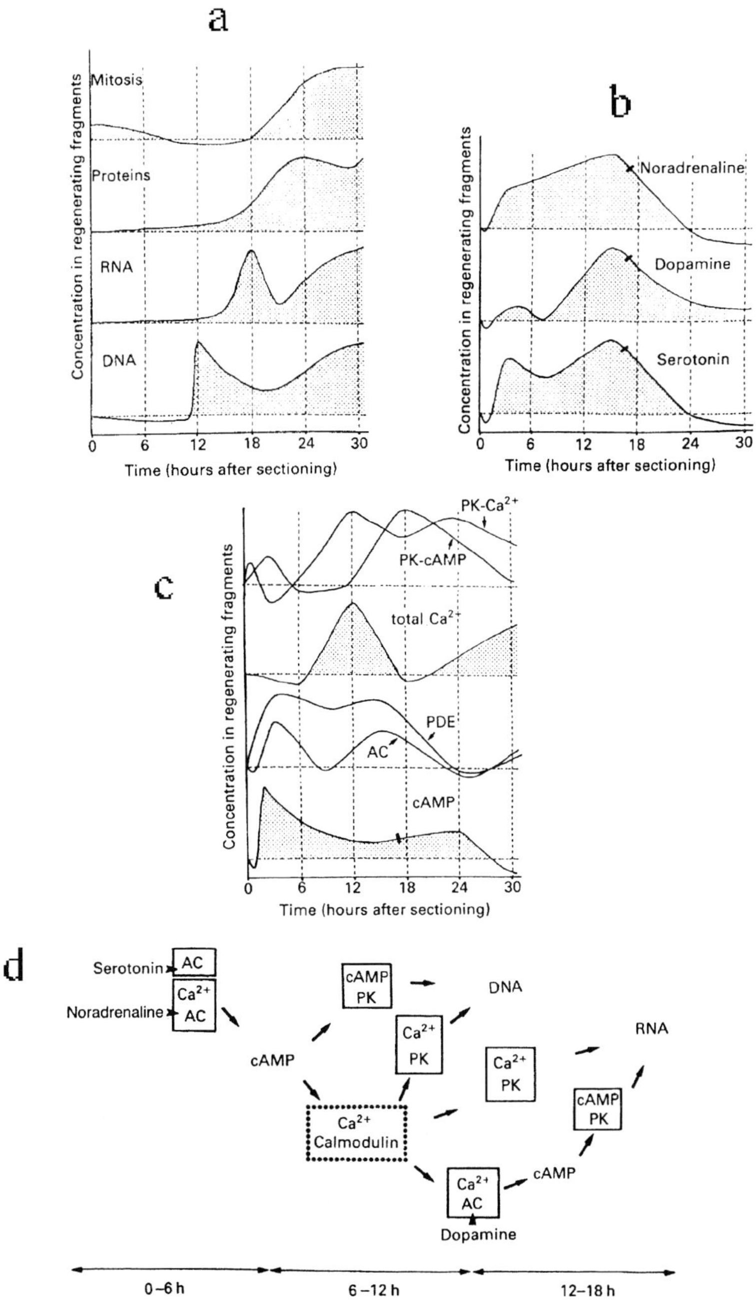

Figure 2. Temporal relationships between protein synthesis, monoamines and signalling molecules during the early stages of regeneration in planarians (after Martelly and Franquinet, 1984). (a) Temporal relationships between the rates of synthesis of DNA, RNA and protein and rate of mitosis. (b) Changes in concentration of noradrenaline, dopamine and 5-HT (serotonin). (c) Changes in concentration of cAMP and total calcium and activities of adenylate cyclase, phosphodiesterase, cAMP-dependent protein kinase and calcium-dependent protein kinase. (d) Summary of main events during cellular activation involved in planarian regeneration.

also been shown to stimulate the mitotic rate in regenerating planarians (Weinstein and Gavurin, 1977).

The calcium-dependent synthesis of DNA and RNA is probably mediated by the calcium binding protein calmodulin. The evidence for this is 4-fold. Thus, calmodulin potentiates the calcium stimulation of DNA synthesis (Martelly, 1983; Martelly *et al.*, 1983). The calmodulin antagonist TFP blocks the calcium effect (Martelly, 1983; Martelly *et al.*, 1983), but this inhibition can be overcome by calmodulin (Martelly *et al.*, 1983). TFP also inhibits the stimulation of DNA synthesis by 5-HT and of RNA synthesis by dopamine (Martelly, 1984). Finally, although the level of calmodulin declines over the first 48 hours of regeneration, there is a transient peak at 12 hours which coincides with the first peak of DNA synthesis (Martelly *et al.*, 1983).

Protein kinase enzymes are also involved in regeneration. Peaks of cAMP-dependent protein kinase (PKA) activity occur at 3 and 18 hours and the pattern of changes in PKA levels follows closely that of adenylate cyclase activity (Fig. 2c) (Moraczewski, Franquinet and Le Moigne, 1983). A separate, calcium-dependent protein kinase enzyme shows peaks of activity at 12 and 24 hours (Fig. 2c) (Moraczewski, Franquinet and Le Moigne, 1983). The first peak parallels the rise in calcium content. The roles for these two enzymes have not been clarified, but it has been suggested that they are concerned with the phosphorylation and activation of enzymes associated with nucleic acid synthesis and/or proteins involved in initiating mitosis and cell differentiation (Moraczewski, Franquinet and Le Moigne, 1983). Evidence in support of this idea has come from studies on protein phospborylation in regenerating planarians during the first 24 hours of regeneration. There are early peaks at 1 hour and 3 hours; the level then drops to a minimum at 6 hours, before rising to a third peak at 18 hours (Franquinet, Moraczewski and Le Moigne, 1985; Moraczewski, Martelly and Franquinet, 1986). The process of protein phosphorylation has been shown to be dependent on calcium. It has been suggested that the first peak at 1 hour is associated with a rise in calcium-dependent protein kinase (PK-Ca^{++}) activity; the latter may also be responsible for the rise in phosphorylation between 6 and 12 hours. In contrast, the peaks at 3 and 18 hours are believed to result from increases in PKA activity (Franquinet, Moraczewski and Le Moigne, 1985; Moraczewski, Martelly and Franquinet, 1986). The increases in phosphorylation at 9–12 and 15–18 hours correlate well with the peaks of DNA (at 12 hours) and RNA (at 18 hours) syntheses. As mentioned previously, DNA synthesis is a calcium-dependent process, changes in levels of calcium, calcium uptake and calmodulin correlating well with the peak of DNA synthesis at 12 hours and with PK-Ca^{++} activity. DNA synthesis is also very much dependent on the phosphorylation of histone proteins. Changes in a number of them have been demonstrated over the first 24 hours of regeneration, the patterns of their changes showing strong correlation with different phases in the cell cycle and with the initiation of nucleic acid synthesis during the early stages of regeneration (Franquinet, Moraczewski and Le Moigne, 1985; Moraczewski, Martelly and Franquinet, 1986). Protein kinase C (PKC) activity shows a rapid (< 15 min) increase and translocation (activation) after sectioning, then drops by 3 hours to control levels (Moraczewski *et al.*, 1987). This rise precedes the increase in protein phosphorylation and may be associated with 5-HT secretion. A second increase in PKC activity was observed at 12 hours, which coincides with the onset of DNA synthesis. This peak may be associated with the phosphorylation of enzymes and proteins required for DNA synthesis (Moraczewski *et al.*, 1987).

From the preceding discussion, it is clear that strong temporal relationships exist between the different elements involved in regeneration. They have led to the following hypotheses regarding the overall control of the process (summarized in Fig. 2d). The trauma of sectioning induces a loss of approximately 60% of the total calcium content and this activates 5-HT synthesis. 5-HT activates a calcium-independent adenylate cyclase and the cAMP level rises. This promotes the uptake of calcium, leading to a stimulation of calcium-dependent protein kinase activity, which in turn causes the phosphorylation of essential proteins for the initiation of DNA synthesis 12 hours after sectioning and subsequent stimulation of mitosis. Calcium levels start to fall and this allows the synthesis of RNA to begin, a process which requires a lower calcium threshold than DNA synthesis. RNA synthesis is stimulated by dopamine via an adenylate cyclase, cAMP, PKA pathway, the protein kinase enzymes phosphorylating the proteins involved in RNA synthesis, which peaks at 18 hours.The peaks of dopamine and RNA correlate well. The overall process is summarised in Fig. 2d.

5.5. GLUTAMATE AND MOTILITY

Glutamate immunoreactivity has been localised in the nervous system of the tapeworm, *H. diminuta*, with the most intense immunostaining being confined to the longitudinal nerve cords (Webb and Eklove, 1989; Eklove and Webb, 1990). It induces powerful contractions of the longitudinal muscles of the worm in isolated body wall strips (Thompson and Mettrick, 1989). High affinity uptake mechanisms for glutamate have been demonstrated in tissue slices of *H. diminuta* (Webb, 1986) and the uptake sites localised in the synaptic terminal region (Webb, 1995). The release of glutamate occurs by a calcium-dependent process (Webb, 1988b, 1995). Thus, there is strong evidence for glutamate acting as an excitatory neurotransmitter candidate in *H. diminuta*. Pharmacological evidence suggests that glutamate binds to ionotropic receptors of the NMDA type (Samii and Webb, 1996; Webb, 1997). However, metabotropic receptors are also present, the elevation of IP_3 production in *H. diminuta* by glutamate being effected via a G protein- and PLC-coupled pathway (Samii and Webb, 1996; Webb, 1997). The response is calcium-dependent, suggesting a functional cooperativity between metabotropic and ionotropic receptors. Following this idea, the initial binding may be to ionotropic receptors, causing an influx of calcium. In turn, this leads to activation of PLC, followed by elevation of IP_3 levels, where at least some of the activation of PLC is mediated by a G protein (Webb, 1997).

Glutamate has also been shown to stimulate adenylate cyclase activity in *H. diminuta*, the response operating via a glutamate receptor related to the NMDA type which is linked to a G protein (Eklove and Webb, 1991). However, the functional significance of this putative pathway is not known. It may be separate from the IP_3-mediated pathway described above, be linked to a metabolic pool of glutamate rather than the glutamate acting as a neurotransmitter and be involved in the modulation of glycolytic activity (Eklove and Webb, 1991). Glutamate eliminates the stimulation of adenylate cyclase activity by 5-HT, lending further support to the modulatory interactions between these two transmitters (Eklove and Webb, 1991).

5.6. PEPTIDES AND MOTILITY

Immunoreactivities to a large number of vertebrate and invertebrate peptides have been demonstrated in both the central and peripheral nervous systems of flatworms (Fairweather and Skuce, 1995), although only a small number of endogenous neuropeptides have been isolated to date. They include the FMRFamide-related peptides, GYIRFamide and RYIRFamide isolated from the turbellarians, *Artioposthia triangulata* and *Dugesia tigrina* (Maule *et al.*, 1994; Johnston *et al.*, 1995). These two peptides have been shown to exert a stimulatory effect on isolated muscle strips from the liver fluke, *F. hepatica*, with GYIRFamide increasing both the frequency and amplitude of contractions at concentrations as low as 50 nM (Graham, Fairweather and McGeown, 1997). Physiological data suggests that the response is mediated through a G protein, leading to activation of PLC activity, elevation of diacylglycerol formation and stimulation of PKC activity. Thus, the GYIRFamide response was significantly reduced by neomycin sulphate, an inhibitor of PLC activity; the phorbol ester and PKC activator, phorbol-12-myristate-13-acetate (PMA) increased contraction frequency and amplitude; and chelerythrine chloride, a potent and selective inhibitor of mammalian PKC, inhibited the GYIRFamide response in *F. hepatica* (Graham, Fairweather and McGeown, 2000).

6. Therapeutic Exploitation of Signalling Pathways in Flatworms

A well co-ordinated neuromuscular system is vital for a parasite's survival, and not just for those parasites that are attached to their host by means of hooks, suckers or other attachment organs. Paralysis or severe disruption of their neuromusculature would cause ectoparasites to become detached and internal parasites either to be eliminated from the host via the gut, for example, or (in the case of blood flukes) to be swept by the blood to the lungs where they are destroyed. Paralysis places an additional stress on the parasite because it ceases to feed properly and enters a state of starvation, drawing on its energy reserves for survival and may eventually succumb. Internally, too, the parasite is dependent on neuromuscular co-ordination for the movements of gametes and for the synchronisation of mechanisms resulting in egg formation, to underpin the often prodigious egg production capacity (for figures on egg production, see Fairweather and Halton, 1992).

Because of its significance to many aspects of parasite biology, the neuromuscular system represents an important target for anthelmintic drugs and a number are considered to act in this way. For example, metrifonate inhibits acetylcholinesterase activity in schistosomes, leading to a flaccid paralysis of the worms (Bueding, Liu and Rogers, 1972; Cioli, Pica-Mattoccia and Archer, 1995). Hycanthone has been alternatively described as a cholinergic antagonist or cholinesterase inhibitor in schistosomes, although its prime action is against DNA synthesis (Hillman and Senft, 1975; Hillman, Senft and Gibler, 1978; Archer *et al.*, 1990; Cioli, Pica-Mattoccia and Archer, 1995). Praziquantel perturbs calcium regulation of muscle contractility in schistosomes and tapeworms (Cioli, Pica-Mattoccia and Archer, 1995; Redman *et al.*, 1996). Other anthelmintics induce a rapid paralysis of flatworms, but their actions have not been characterised.

The essential feature of an anthelmintic drug is that it exploits a difference between host and parasite, the parasite target being far more sensitive to drug action than the same or comparable target in the host. In this review of data on individual components of signalling systems, differences between them and equivalent molecules in their mammalian hosts have been highlighted. Differences *are* apparent in relation to the properties of 5-HT receptors (Mansour, 1984; Estey and Mansour, 1987), G proteins (Mansour and Mansour, 1989; Samii and Webb, 1996), adenylate cyclase (Northup and Mansour, 1978b) and protein kinase A (Iltzsch and Mansour, 1987). Differences, too, have been described for the properties of the prostaglandins and the enzymes involved in their synthesis compared with their mammalian counterparts (Salafsky and Fusco, 1985; Grzywacz and Szkudlinski, 1989). Consequently, they represent potential targets for the rational design of new anthelmintics. This idea is a realistic one because a number of existing anthelmintics are known to act in this fashion, namely, levamisole, suramin, amoscanate and diethylcarbamazine.

The principal action of levamisole, for example, is as an acetylcholine agonist in nematodes, inducing a rapid spastic paralysis of the worms (Martin, Robertson and Bjorn, 1997). However, it also disrupts glycogen metabolism, inhibiting the mobilisation of glycogen required for muscle contraction. It achieves this by inhibiting adenylate cyclase, thus blocking cAMP-mediated glycogenolysis, which leads to the flaccid paralysis stage following the initial contraction of muscle that is characteristic of levamisole action (Donahue, Masaracchia and Harris, 1983). In nematodes, suramin inhibits cAMP-independent protein kinase activity (Walter and Schulz-Key, 1980; Ossikovski and Walter, 1984). It has been suggested that the protein kinase is involved in the regulation of activity of pyruvate dehydrogenase, which plays a major role in carbohydrate metabolism and energy production (Ossikovski and Walter, 1984). Amoscanate is active against the human blood fluke, *S. mansoni* and the nematode, *Onchocerca volvulus*, the causative agent of river blindness. It inhibits phosphodiesterase activity in the two parasites, causing accumulation of cAMP which then activates the glycogenolytic cascade causing depletion of glycogen (Walter and Albiez, 1984). The anti-filarial compound, diethylcarbamazine disrupts the metabolism of arachidonic acid in both microfilariae and host endothelial cells, although it is the latter effect that activates an innate immune responsem by the host and is responsible for destruction of the parasite, rather than any direct disruption of signalling pathways in the parasite itself (Martin, Robertson and Bjorn, 1997).

A role of eicosanoids in skin penetration behaviour by the cercaria larvae of the human blood fluke, *S. mansoni* has been described previously (section 4.1.8.). The possibility of exploiting this phenomenon in order to inhibit skin penetration and therefore prevent infection by the parasite has been suggested by Salafsky *et al.* (1989). The idea is to use dietary supplementation with lipids as oral anti-penetration agents for chemoprophylaxis. Treatment with castor oil led to >90% protection; ricinoleic acid was less effective (Salafsky *et al.*, 1989). An alternative to oral prophylaxis is the topical application of agents that prevent cercarial penetration. Among the agents tested, cedarwood oil has been shown to be effective (Campbell and Cuckler, 1961). More recently, the insect repellent DEET (N,N-diethyl-m-toluamide) and more especially in its long-acting form of LIPODEET, have been shown to confer a high level of protection against cercarial entry, although the action of the compound may not be via disruption of eicosanoid pathways (Salafsky *et al.*, 1998, 1999).

6. Conclusions

Flatworms possess a variety of signalling molecules that are similar to those used by higher organisms. The molecules are involved in neurotransmission, neuromodulation and neuromuscular transmission. They also serve more hormone-type roles in other areas of physiology, including carbohydrate metabolism, reproduction and regeneration. Individual elements of G protein-mediated and enzyme-linked receptor pathways have been shown to be present and genes for some of them have been isolated. However, while the individual pieces of the signalling jigsaw are present, there is little information at present as to how the pieces might fit together to form functional pathways linking individual molecules to their physiological actions.

Flatworms are of importance in terms of the evolution of signalling systems as they exhibit the first signs of centralization of the nervous system, yet predate the evolution of the endocrine system and the separation of neural and endocrine activities. Certain species of flatworms have found use as models to study aspects of neurophysiology, regeneration, development and reproduction.

Their potential as experimental models has yet to be fully realized and a greater understanding of signalling mechanisms in flatworms may help to shed greater insight into the evolution of signalling systems generally.

Flatworms have an added significance as members of the phylum are major parasites of humans and livestock. During the evolution of the host-parasite relationship, parasites have become adapted to recognize and respond to, as well as utilize, host molecules for the regulation of their behaviours and development. They have also evolved to produce and release molecules that will either mimic those of the host or will affect the hormonal or immune status of the host; the actions of the molecules serve to alter the host environment for the apparent benefit of the parasite. Despite neuromuscular co-ordination being an obvious and significant target for anthelmintic action, few anthelmintics have been characterized as acting in this way. This review has highlighted differences between the parasite and its host with regard to the properties of their respective receptors and signalling components; such differences offer considerable scope for future therapeutic exploitation.

References

Abdel Wahab, M.F., Warren, K.S., and Levy, R.P. (1971) Function of the thyroid and the host-parasite relationship in murine *schistosomiasis mansoni*. *Journal of Infectious Diseases* **124**, 161–171.

Abrahams, S.L., Northup, J.K., and Mansour, T.E. (1976) Adenosine cyclic 3′,5′-monophosphate in the liver fluke, *Fasciola hepatica*. I. Activation of adenylate cyclase by 5-hydroxytryptamine. *Molecular Pharmacology* **12**, 49–58.

Abdel Baset, H., O'Neill, G.P., and Ford-Hutchinson, A.W. (1995) Characterization of arachidonic-acid-metabolizing enzymes in adult *Schistosoma mansoni*. *Molecular and Biochemical Parasitology* **73**, 31–41.

Alberts, B., Johnson, A., Lewis, J., Raff, M., Roberts, K., and Walter, P. (2002) *Molecular Biology of the Cell* (4th edition). Garland Science, New York.

Ali, S.F., Joachim, A., and Daugschies, A. (1999) Eicosanoid production by adult *Fasciola hepatica* and plasma eicosanoid patterns during fasciolosis in sheep. *International Journal for Parasitology* **29**, 743–748.

Amiri, P., Kocksley, R.M., Parslow, T.G., Sadick, M., Rector, E., Ritter, D., and McKerrow, J.H. (1992) Tumour necrosis factor α restores granulomas and induces parasite egg-laying in schistosome-infected SCID mice. *Nature* **356**, 604–607.

Archer, S., El-Hamouly, W., Seyed-Mozaffari, A., Butler, R., Pica-Mattoccia, L., and Cioli, D. (1990) Mode of action of the schistosomicide hycanthone: site of DNA alkylation. *Molecular and Biochemical Parasitology* **46**, 89–96.

Arechavaleta, F., Molinari, J.L., and Tato, P. (1998) A *Taenia solium* metacestode factor nonspecifically inhibits cytokine production. *Parasitology Research* **84**, 117–122.

Baguna, J., Romero, R., Salo, E., Collet, J., Auladell, C., Ribas, M., Riutort, M., Garcia-Fernandez, J., Burgaya, F., and Bueno, D. (1990) Growth, degrowth and regeneration as developmental phenomena in adult freshwater planarians. In: H.-J. Marthy (ed.) *Experimental Embryology in Aquatic Plants and Animals*. Plenum Press, New York, pp. 129–162.

Barker, G.C., Chitwood, D.J., and Ress, H.H. (1990) Ecdysteroids in helminths and annelids. *Invertebrate Reproduction and Development* **18**, 1–11.

Basch, P.F. (1986) Immunocytochemical localization of ecdysteroids in the life history stages of *Schistosoma mansoni*. *Comparative Biochemistry and Physiology* **83A**, 199–202.

Beall, M.J., and Pearce, E.J. (2001) Human transforming growth factor-β activates a receptor serine/threonine kinase from the intravascular parasite *Schistosoma mansoni*. *Journal of Biological Chemistry* **276**, 31613–31619.

Beall, M.J., McGonigle, S., and Pearce, E.J. (2000) Functional conservation of *Schistosoma mansoni* Smads in TGF-β signaling. *Molecular and Biochemical Parasitology* **111**, 131–142.

Bennett, J.L., and Bueding, E. (1973) Uptake of 5-hydroxytryptamine by *Schistosoma mansoni*. *Molecular Pharmacology* **9**, 311–319.

Blair, K.L., and Anderson, P.A.V. (1993) Properties of voltage-activated ionic currents in cells from the brains of the triclad flatworm *Bdelloura candida*. *Journal of Experimental Biology* **185**, 267–286.

Blair, K.L., Bennett, J.L., and Pax, R.A. (1988) *Schistosoma mansoni*: evidence for protein kinase-C-like modulation of muscle activity. *Experimental Parasitology* **66**, 243–252.

Blair, K.L., Bennett, J.L., and Pax, R.A. (1994) *Schistosoma mansoni*: myogenic characteristics of phorbol ester-induced muscle contraction. *Experimental Parasitology* **78**, 302–316.

Boray, J.C. (1994) *Diseases of Domestic Animals Caused by Flukes*. Food and Agricultural Organisation of the United Nations, Rome, 49 pp.

Branford White, C.J., Hipkiss, J.B., and Peters, T.J. (1984) Evidence for a Ca^{2+}-dependent activator protein in the rat tapeworm *Hymenolepis diminuta*. *Molecular and Biochemical Parasitology* **13**, 201–211.

Brownlee, D.J.A., and Fairweather, I. (1996) Immunocytochemical localisation of glutamate-like immunoreactivity within the nervous system of the cestode *Mesocestoides corti* and the trematode *Fasciola hepatica*. *Parasitology Research* **82**, 423–427.

Bueding, E., and Fisher, J. (1982) Metabolic requirements of schistosomes. *Journal of Parasitology* **68**, 208–212.

Bueding, E., Liu, C.L., and Rogers, S.H. (1972) Inhibition by metrifonate and dichlorvos of cholinesterase in schistosomes. *British Journal of Pharmacology* **46**, 480–487.

Bueding, E., Bennett, J.L., Chou, T., Pert, C., and Tomosky, T. (1974) Effect of antischistosomal drugs on the uptake of 5-hydroxytryptamine by *Schistosoma mansoni*. *Proceedings of the Third International Congress on Parasitology* **3**, 1443–1444.

Campbell, W.C., and Cuckler, A.C. (1961) The prophylactic effect of topically applied cedarwood oil on infection with *S. mansoni* in mice. *American Journal of Tropical Medicine and Hygiene* **10**, 712–715.

Catto, B.A., and Ottesen, E.A. (1979) Serotonin uptake in schistosomules of *Schistosoma mansoni*. *Comparative Biochemistry and Physiology* **63C**, 235–242.

Chen, L., Rekosh, D.M., and Lo Verde, P.T. (1992) *Schistosoma mansoni* p48 eggshell protein gene: characterization, developmentally regulated expression and comparison to the p14 eggshell protein gene. *Molecular and Biochemical Parasitology* **52**, 39–52.

Cho, C.H. (1984) Study of the effects of insulin on the migration of *Hymenolepis diminuta* in rats. *Journal of Helminthology* **58**, 291–293.

Chou, T.-C.T., Bennett, J.L., and Bueding, E. (1972) Occurrence and concentrations of biogenic amines in trematodes. *Journal of Parasitology* **58**, 1098–1102.

Cioli, D., Pica-Mattoccia, L., and Archer, S. (1995) Antischistosomal drugs: Past, present ... and future? *Pharmacology and Therapeutics* **68**, 35–85.

Conway-Jones, P.B., and Rothman, A.H. (1978) *Hymenolepis microstoma*: ultrastructural localization of adenyl cyclase in the tegument. *Experimental Parasitology* **46**, 152–156.

Cornford, E.M. (1974) *Schistosomatium douthitti*: effects of thyroxine. *Experimental Parasitology* **36**, 210–221.

Creti, P., Capasso, A., Grasso, M., and Parisi, E. (1992) Identificaton of a 5-HT_{1A} receptor positively coupled to planarian adenylate cyclase. *Cell Biology International Reports* **16**, 427–432.

Crompton, D.W.T. (1999) How much human helminthiasis is there in the world? *Journal of Parasitology* **85**, 397–403.

Craig, P.S., Rogan, M.T., and Allan, J.C. (1996) Detection, screening and community epidemiology of taeniid cestode zoonoses: cystic echinococcosis, alveolar echinococcosis and neurocysticercosis. *Advances in Parasitology* **38**, 170–250.

Curry, W.J., Shaw, C., Johnston, C.F., Thim, L., and Buchanan, K.D. (1992) Neuropeptide F: primary structure from the turbellarian, *Artioposthia triangulata. Comparative Biochemistry and Physiology* **101C**, 269–274.

Cyr, D., Gruner, S., and Mettrick, D.F. (1983) *Hymenolepis diminuta*: uptake of 5-hydroxytryptamine (serotonin), glucose, and changes in worm glycogen levels. *Canadian Journal of Zoology* **61**, 1469–1474.

Daugschies, A., and Joachim, A. (2000) Eicosanoids in parasites and parasitic infections. *Advances in Parasitology* **46**, 181–240.

Davies, S.J., Shoemaker, C.B., and Pearce, E.J. (1998) A divergent member of the transforming growth factor β receptor family from *Schistosoma mansoni* is expressed on the parasite surface membrane. *Journal of Biological Chemistry* **273**, 11234–11240.

Davis, R.A., and Stretton, A.O.W. (1995) Neurotransmitters of helminths. In: J.J. Marr and M. Muller (eds.) *Biochemistry and Molecular Biology of Parasites*. Academic Press, New York, pp. 257–287.

Day, T.A., Bennett, J.L., and Pax, R.A. (1994) Serotonin and its requirement for maintenance of contractility in muscle fibres isolated from *Schistosoma mansoni. Parasitology* **108**, 425–432.

Dei-Cas, E., Dhainaut-Coutois, N., Dhainaut, A., and Vernes, A. (1979) Ultrastructural localization of tritiated 5-HT in adult *Schistosoma mansoni*: a preliminary report. *Biologie Cellulaire* **35**, 321–324.

Donahue, M.J., Masaracchia, R.A., and Harris, B.G. (1983) The role of cyclic AMP-mediated regulation of glycogen metabolism in levamisole-perfused *Ascaris suum* muscle. *Molecular Pharmacology* **23**, 378–383.

Duma, A. (1980) Activity in cAMP phosphodiesterase in the early regeneration stage of planarian *Dugesia lugubris* (O. Schmidt). Ultracytochemical studies. *Acta Medica Polona* **21**, 317–318.

Duvaux-Miret, O., Stefano, G.B., Smith, E.M., Dissous, C., and Capron, A. (1992) Immunosuppression in the definitive and intermediate hosts of the human parasite *Schistosoma mansoni* by release of immunoreactive neuropeptides. *Proceedings of the National Academy of Sciences, USA* **89**, 778–781.

Duvaux-Miret, O., Leung, M.K., Capron, A., and Stefano, G.B. (1993) *Schistosoma mansoni*: an enkephalinergic system that may participate in internal and host-parasite signaling. *Experimental Parasitology* **76**, 76–84.

Eastlake, J.L., Branford-White, C.J., and Whish, W.J.D. (1994) The distribution of calmodulin/calmodulin binding proteins in the rat tapeworm, *Hymenolepis diminuta. Comparative Biochemistry and Physiology* **108B**, 487–500.

Eklove, H., and Webb, R.A. (1990) Glutamate-like immunoreactivity in the cestode *Hymenolepis diminuta. Canadian Journal of Zoology* **68**, 2417–2423.

Eklove, H., and Webb, R.A. (1991) The effect of L-glutamate and related agents on adenylate cyclase in the cestode *Hymenolepis diminuta. Canadian Journal of Physiology and Pharmacology* **69**, 28–36.

Engelender, S., Giannini, A.L., and Rumjanek, F.D. (1993) Protein interactions with a gender-specific gene of *Schistosoma mansoni*: characterization by DNase I footprinting, band shift and UV cross-linking. *Molecular and Cellular Biochemistry* **124**, 159–168.

Eriksson, K.S. (1996) Nitric oxide synthase in the pharynx of the planarian *Dugesia tigrina. Cell and Tissue Research* **286**, 407–410.

Eriksson, K.S., and Panula, P. (1994) Gamma-aminobutyric acid in the nervous system of a planarian. *Journal of Comparative Neurology* **345**, 528–536.

Eriksson, K.S., Maule. A.G.M., Halton, D.W., Panula, P.A.J., and Shaw, C. (1995) GABA in the nervous system of parasitic flatworms. *Parasitology* **110**, 339–346.

Escriva, H., Safi, R., Hanni, C., Langlois, M.-C., Sauimouti-Laprade, P., Stehelin, D., Capron, A., Pierce, R., and Laudet, V. (1997) Ligand binding was acquired during evolution of nuclear receptors. *Proceedings of the National Academy of Sciences, USA* **94**, 6803–6808.

Espinoza, B., Tarrab-Hazdai, R., Silman, I., and Arnon, R. (1988) Acetylcholinesterase in *Schistosoma mansoni* is anchored to the membrane via covalently attached phosphatidylinositol. *Molecular and Biochemical Parasitology* **29**, 171–179.

Espinoza, B., Silman, I., Arnon, R., and Tarrab-Hadzai, R. (1991) Phosphatidylinositol-specific phospholipase C induces biosynthesis of acetylcholinesterase via diacylglycerol in *Schistosoma mansoni*. *European Journal of Biochemistry* **195**, 863–870.

Estey, S.J., and Mansour, T.E. (1987) Nature of serotonin-activated adenylate cyclase during development of *Schistosoma mansoni*. *Molecular and Biochemical Parasitology* **26**, 47–60.

Estey, S.J., and Mansour, T.E. (1988) GTP binding regulatory proteins of adenylate cyclase in *Schistosoma mansoni* at different stages of development. *Molecular and Biochemical Parasitology* **30**, 67–76.

Fairweather, I. (1997) Platyhelminthes. In: T.S. Adams (ed.) *Reproductive Biology of Invertebrates. Volume. VIII. Progress in Reproductive Endocrinology*. Oxford and IBH Publishing Co. PVT. Ltd., New Delhi, pp. 1–31.

Fairweather, I., and Halton, D.W. (1992) Regulatory peptide involvement in the reproductive biology of flatworm parasites. *Invertebrate Reproduction and Development* **22** , 117–125.

Fairweather, I., and Skuce, P.J. (1995) Flatworm neuropeptides – present status, future directions. *Hydrobiologia* **305**, 309–316.

Fairweather, I., Maule, A.G., Mitchell, S.H., Johnston, C.F., and Halton, D.W. (1987) Immunocytochemical demonstration of 5-hydroxytryptamine (serotonin) in the nervous system of the liver fluke, *Fasciola hepatica*. *Parasitology Research* **73**, 255–258.

Fairweather, I., Macartney, G.A., Johnston, C.F., Halton, D.W., and Buchanan, K.D. (1988) Immunocytochemical demonstration of 5-hydroxytryptamine (serotonin) and vertebrate neuropeptides in the nervous system of excysted cysticercoid larvae of the rat tapeworm, *Hymenolepis diminuta* (Cestoda, Cyclophyllidea). *Parasitology Research* **74**. 371–379.

Fetterer, R.H., Pax, R.A., and Bennett, J.L. (1977) *Schistosoma mansoni*: direct method for simultaneous recording of electrical and motor activity. *Experimental Parasitology* **43**, 286–294.

Foster, J.M., Mercer, J.G., and Rees, H.H. (1992) Analysis of ecdysteroids in the trematodes Schistosoma mansoni and Fasciola hepatica. *Tropical and Medical Parasitology* **43**, 239–244.

Franquinet, R. (1979) Role de la serotonine et des catecholamines dans la regeneration de la planaire *Polycelis tenuis*. *Journal of Embryology and Experimental Morphology* **51**, 85–95.

Franquinet, R. (1981) Synthese d'ADN dans les cellules de planaires cultivees *in vitro*. Role de la serotonine. *Biologie Cellulaire* **40**, 41–45.

Franquinet, R., and Le Moigne, A. (1979) Relation entre les variations des taux de serotonine et d'AMP cyclique au cours de la regeneration d'une planaire. *Biologie Cellulaire* **34**, 71–76.

Franquinet, R., and Martelly, I. (1981) Effects of serotonin and catecholamines on RNA synthesis in planarians: *in vitro* and *in vivo* studies. *Cell Differentiation* **10**, 201–209.

Franquinet, R., Stengel, D., and Hanoune, J. (1976) The adenylate cyclase system in a freshwater planarian (*Polycelis tenuis* Iijima). *Comparative Biochemistry and Physiology* **53B**, 329–333.

Franquinet, R., Le Moigne, A., and Hanoune, J. (1978) The adenylate cyclase system of planaria *Polycelis tenuis*. Activation by serotonin and guanine nucleotides. *Biochimica et Biophysica Acta* **539**, 88–97.

Franquinet, R., Moraczewski, J., and Le Moigne, A. (1985) Phosphorylation of endogenous proteins, including histones, during initiation of planarian regeneration. *Comparative Biochemistry and Physiology* **80B**, 661–669.

Freebern, W.J., Osman, A., Niles, E.G., Christen, L., and Lo Verde, P.T. (1999) Identification of a cDNA encoding a retinoid X receptor homologue from *Schistosoma mansoni*. Evidence for a role in female-specific gene expression. *Journal of Biological Chemistry* **274**, 4577–4585.

Friedel, T., and Webb, R.A. (1979) Stimulation of mitosis in *Dugesia tigrina* by a neurosecretory fraction. *Canadian Journal of Zoology* **57**, 1818–1819.

Fukushima, T., Isobe, A., Hojo, N., Shiwaku, K., Yamane, Y., and Torii, M. (1993) The metabolism of arachidonic acid to prostaglandin E2 in plerocercoids of *Spirometra erinacei*. *Parasitology Research* **79**, 634–638.

Fusco, A.C., Salafsky, B., and Kevin, M.B. (1985) *Schistosoma mansoni*: eicanosoid production by cercariae. *Experimental Parasitology* **59**, 44–50.

Fusco, A.C., Salafsky, B., and Delbrook, K. (1986) *Schistosoma mansoni*: production of cercarial eicanosoids as correlates of penetration and transformation. *Journal of Parasitology* **72**, 397–404.

Gabriel, A. (1969) Etude morphologique et evolution biochimique des neoblastes au cours des

premieres phases de la regeneration des planaires d'eau douce. *Annales d'Embryologie et de Morphogenese* **3**, 49–69.
Gao, T., Fukushima, T., Isobe, A., Hojo, N., Shiwaku, K., and Yamane, Y. (1998) Arachidonic acid metabolism to prostaglandin E2, D2, F2α, and I2 in the plerocercoid of *Spirometra erinaceieuropaei. Journal of Parasitology* **84**, 1107–1111.
Gentleman, S., Abrahams, S.L., and Mansour, T.E. (1976) Adenylate cyclic 3′,5′-monophosphate in the liver fluke, *Fasciola hepatica*. II. Activation of protein kinase by 5-hydroxytryptamine. *Molecular Pharmacology* **12**, 59–68.
Giannini, A.L.M., Caride, E.C., Braga, V.M.M., and Rumjanek, F.D. (1995) F-10 nuclear binding proteins of *Schistosoma mansoni*: structural and functional features. *Parasitology* **110**, 155–161.
Gordon, B., and Webb, R.A. (1989) Release of exogenously-supplied and endogenous serotonin from tissue slices of *Hymenolepis diminuta. Brain Research* **486**, 376–380.
Gordon, D.B.R., and Webb, R.A. (1996) Demonstration and partial characterization of protein kinase C in crude extracts of the cestode *Hymenolepis diminuta. Comparative Biochemistry and Physiology* **115B**, 355–361.
Graham, M.K., Fairweather, I., and McGeown, J.G. (1997) The effects of FaRPs on the motility of isolated muscle strips from the liver fluke, *Fasciola hepatica. Parasitology* **114**, 455–465.
Graham, M.K., McGeown, J.G., and Fairweather, I. (1999) Ionic mechanisms underlying spontaneous muscle contractions in the liver fluke, *Fasciola hepatica. American Journal of Physiology* **277** (*Regulatory, Integrative and Comparative Physiology* 46), R374–R383.
Graham, M.K., Fairweather, I., and McGeown, J.G. (2000) Second messengers mediating mechanical responses to the FaRP, GYIRFamide in the fluke, *Fasciola hepatica. American Journal of Physiology* **279** (*Regulatory, Integrative and Comparative Physiology*), R2089–R2094.
Grzych, J.M., Pearce, E., Cheever, A., Caulada, Z.A., Caspar, P., Heiny, S., Lewis, F., and Sher, A. (1991) Egg deposition is the major stimulus for the production of Th2 cytokines in murine schistosomiasis mansoni. *Journal of Immunology* **146**, 1322–1327.
Grzywacz, M., and Szkudlinski, J. (1989) Occurrence of prostaglandins in helminths. *Acta Parasitologica Polonica* **34**, 67–75.
Gustafsson, M.K.S. (1987) Immunocytochemical demonstration of neuropeptides and serotonin in the nervous system of adult *Schistosoma mansoni. Parasitology Research* **74**, 168–174.
Gustafsson, M.K.S., and Eriksson, K. (1992) Never ending growth and a growth factor. I. Immunocytochemical evidence for the presence of basic fibroblast growth factor in a tapeworm. *Growth Factors* **7**, 327–334.
Gustafsson, M.K.S., Eriksson, K., and Hyden, A. (1995) Never ending growth and a growth factor. II. Immunocytochemical evidence for the presence of epidermal growth factor in a tapeworm. *Hydrobiologia* **305**, 229–233.
Gustafsson, M.K.S., Lindholm, A.M., Terenina, N.B., and Reuter, M. (1996) NO nerves in a tapeworm. NADPH-diaphorase histochemistry in adult *Hymenolepis diminuta. Parasitology* **113**, 559–565.
Haas, W. (1994) Physiological analyses of host-finding behaviour in trematode cercariae: adaptations for transmission success. *Parasitology* **109**, S15–S29.
Hamdan, F.F., Abramovitz, M., Mousa, A., Xie, J., Durocher, Y., and Ribeiro, P. (2002) A novel *Schistosoma mansoni* G protein-coupled receptor is responsive to histamine. *Molecular and Biochemical Parasitology* **119**, 75–86.
Havercroft, J.C., Huggins, M.C., Dunne, D.W., and Taylor, D.W. (1990) Characterisation of Sm20, a 20-kilodalton calcium-binding protein of *Schistosoma mansoni. Molecular and Biochemical Parasitology* **38**, 211–220.
Hawn, T.R., and Strand, M. (1993) Detection and partial characterization of glycosylphosphatidylinositol-specific phospholipase activities from *Fasciola hepatica* and *Schistosoma mansoni. Molecular and Biochemical Parasitology* **59**, 73–82.
Higashi, G.I., Kreiner, P.W., Keirns, J.J., and Bitensky, M.W. (1973) Adenosine 3′,5′-cyclic monophosphate in *Schistosoma mansoni. Life Sciences* **13**, 1211–1220.
Hillman, G.R., and Senft, A.W. (1973) Schistosome motility measurements: responses to drugs. *Journal of Pharmacology and Experimental Therapeutics* **185**, 177–184.

Hillman, G.R., and Senft, A.W. (1975) Anticholinergic properties of the antischistomal drug hycanthone. *American Journal of Tropical Medicine and Hygiene* **24**, 827–834.

Hillman, G.R., Senft, A.W., and Gibler, W.B. (1978) The mode of action of hycanthone revisited. *Journal of Parasitology* **64**, 754–756.

Hipkiss, J.B., and Branford White, C.J. (1986) The influence of trifluoperazine (stellazine) on *Hymenolepis diminuta in vivo. Zeitschrift fur Parasitenkunde* **72**, 271–272.

Hipkiss, J.B., Branford White, C.J., and Peters, T.J. (1987) Effect of phenothiazines on *Hymenolepis diminuta* with special reference to the brush border Ca^{2+}-dependent ATPase. *Molecular and Biochemical Parasitology* **22**, 55–63.

Hipkiss, J.B., Skinner, A. and Branford White, C.J. (1987) Biochemical and ultrastructural investigation of the effect of Stelazine (trifluoperazine) on *Hymenolepis diminuta* (Cestoda). *Parasitology* **94**, 135–149.

Hipkiss, J.B., Branford White, C.J., Peters, T.J., and Skinner, A. (1986) Trifluoperazine as an anthelmintic. *Biochemical Society Transactions* **14**, 1035–1036.

Hoffmann, K.F., Davis, E.M., Fischer, E.R., and Wynn, T.A. (2001) The guanine protein coupled receptor rhodopsin is developmentally regulated in the free-living stages of *Schistosoma mansoni. Molecular and Biochemical Parasitology* **112**, 113–123.

Holmes, S.D., and Fairweather, I. (1984) *Fasciola hepatica*: the effects of neuropharmacological agents upon *in vitro* motility. *Experimental Parasitology* **58**, 194–208.

Hopkins, D.R. (1992) Homing in on helminths. *American Journal of Tropical Medicine and Hygiene* **46**, 626–634.

Hori, I. (1997) Cytological approach to morphogenesis in the planarian blastema. II. The effect of neuropeptides. *Journal of Submicroscopic Cytology and Pathology* **29**, 91–97.

Huerta, L., Terrazas, L.I., Sciutto, E., and Larralde, C. (1992) Immunological mediation of gonadal effects on experimental murine cysticercosis caused by *Taenia crassiceps* metacestodes. *Journal of Parasitology* **78**, 471–476.

Iltzsch, M.H., and Mansour, T.E. (1987) Partial purification and characterisation of cAMP-dependent protein kinase from *Fasciola hepatica. Molecular and Biochemical Parasitology* **23**, 265–274.

Iltzsch, M.H., Bieber, D., Vijayasarathy, S., Webster, P., Zurita, M., Ding, J., and Mansour, T.E. (1992) Cloning and characterization of a cDNA coding for the α-subunit of a stimulatory G protein from *Schistosoma mansoni. Journal of Biological Chemistry* **267**, 14504–14508.

Jenkins, P., Dixon, J.B., Rakha, N.K., and Carter, S.D. (1990) Regulation of macrophage-mediated larvicidal activity in *Echinococcus granulosus* and *Mesocestoides corti* (Cestoda) infection in mice. *Parasitology* **100**, 309–315.

Jenkins, P., Dixon, J.B., Haywood, S., Rakha, N.K., and Carter, S.D. (1991) Differential regulation of murine *Mesocestoides corti* infection by bacterial lipopolysaccharide and interferon-γ. *Parasitology* **102**, 125–132.

Jenkins, P., Spiers, S., Dixon, J.B., Carter, S.D., and May, S. (1992) The effects of tumour necrosis factor on host-parasite relations in murine *Mesocestoides corti* (Cestoda) infection. *Parasitology* **105**, 453–459.

Jeziorski, M.C., Greenberg, R.M., and Anderson, P.A.V. (1997) Cloning of a putative voltage-gated sodium channel from the turbellarian flatworm *Bdelloura candida. Parasitology* **115**, 289–296.

Johnston, R.N., Shaw, C., Halton, D.W., Verhaert, P., and Baguna, J. (1995) GYIRFamide: a novel FMRFamide-related peptide (FaRP) from the triclad turbellarian, *Dugesia tigrina. Biochemical and Biophysical Research Communications* **209**, 689–697.

Kalbe, M., Haberl, B., and Haas, W. (2000) Snail host finding by *Fasciola hepatica* and *Trichobilharzia ocellata*: compound analysis of "miracidia-attracting glycoproteins". *Experimental Parasitology* **96**, 231–242.

Kamemoto, E.S., and Mansour, T.E. (1986) Phosphofructokinase in the liver fluke *Fasciola hepatica.* Purification and kinetic changes by phosphorylation. *Journal of Biological Chemistry* **261**, 4346–4351.

Kamemoto, E.S., Lan, L., and Mansour, T.E. (1989) *In vivo* regulation of phosphorylation level and activity of phosphofructokinase by serotonin in *Fasciola hepatica. Archives of Biochemistry and Biophysics* **271**, 553–559.

Kamemoto, E.S., Iltzsch, M.H., Lan, L., and Mansour, T.E. (1987) Phosphofructokinase from *Fasciola hepatica*: activation by phosphorylation and other regulatory properties distinct from the mammalian enzyme. *Archives of Biochemistry and Biophysics* **258**, 101–111.

Kampkotter, A., Ridgers, I., Johnston, D.A., Rollinson, D., Kunz, W., and Grevelding, C.G. (1999) *Schistosoma mansoni*: cloning and characterization of the Ras homologue. *Experimental Parasitology* **91**, 280–283.

Kapp, K., Schussler, P., Kunz, W., and Grevelding, C.G. (2001) Identification, isolation and characterization of a Fyn-like tyrosine kinase from *Schistosoma mansoni. Parasitology* **122**, 317–327.

Kasschau, M.R., and Mansour, T.E. (1982a) Adenylate cyclase in adults and cercariae of *Schistosoma mansoni. Molecular and Biochemical Parasitology* **5**, 107–116.

Kasschau, M.R., and Mansour, T.E. (1982b) Serotonin-activated adenylate cyclase during development of *Schistosoma mansoni. Nature* **296**, 66–68.

Keenan, L., and Koopowitz, H. (1981) Tetrodotoxin-sensitive action potentials from the brain of the polyclad flatworm, *Notoplana acticola. Journal of Experimental Zoology* **215**, 209–213.

Keenan, L., and Koopowitz, H. (1982) Physiology and *in situ* identification of putative aminergic neurotransmitters in the nervous system of *Gyrocotyle fimbriata*, a parasitic flatworm. *Journal of Neurobiology* **13**, 9–21.

Keenan, L., and Koopowitz, H. (1984) Ionic bases of action potentials in identified flatworm neurons. *Journal of Comparative Physiology A* **155**, 197–208.

Keenan. L., Koopowitz, H., and Bernardo, K. (1979) Primitive nervous systems: action of aminergic drugs and blocking agents on activity in the ventral nerve cord of the flatworm *Notoplana acticola. Journal of Neurobiology* **10**, 397–407.

Keenan, L., Koopowitz, H. and Solon, M.H. (1984) Primitive nervous systems: electrical activity in the nerve cords of the parasitic flatworm, *Gyrocotyle fimbriata. Journal of Parasitology* **70**, 131–138.

Kim, E., Day, T.A., Bennett, J.L., and Pax, R.A. (1995) Cloning and functional expression of a *Shaker*-related voltage-gated potassium channel gene from *Schistosoma mansoni* (Trematoda: Digenea). *Parasitology* **110**, 171–180.

Kim, H.S., Kam, K.Y., Ryu, P.D., Hong, S.J., Jeon, J.S., Jeon, B.H., Kim, K.J., and Park, J.B. (2002) A gadolinium and pH-sensitive hyperpolarization-activated cation current in acutely isolated single neurons from *Fasciola hepatica. Parasitology* **125**, 423–430.

Knopf, P.M., and Linden, T. (1985) Completion of *Schistosoma mansoni* life cycle in thyroidectomized rats and effects of thyroid hormone replacement therapy. *Journal of Parasitology* **71**, 422–426.

Kohn, A.B., Lea, J.M., Roberts-Misterly, J.M., Anderson, P.A.V., and Greenberg, R.M. (2001) Structure of three high voltage-activated calcium channel α_1 subunits from *Schistosoma mansoni. Parasitology* **123**, 489–497.

Kong, Y., Chung, Y.-B., Cho, S.-Y., and Kang, S.-Y. (1994) Cleavage of immunoglobin G by excretory-secretory cathepsin S-like protease of *Spirometra mansoni* plerocercoid. *Parasitology* **109**, 611–621.

Kowalski, J.C., and Thorson, R.E. (1976) Effects of certain lipid compounds on growth and asexual multiplication of *Mesocestoides corti* (Cestoda) tetrathyridia. *International Journal for Parasitology* **6**, 327–331.

Larralde, C., Morales, J., Terrazas, I., Govezensky, T., and Romano, M.C. (1995) Sex hormone changes induced by the parasite lead to feminization of the male host in murine *Taenia crassiceps* cysticercosis. *Journal of Steroid Biochemistry and Molecular Biology* **52**, 575–580.

Leid, R.W., and McConnell, L.A. (1983) PGE2 generation and release by the larval stage of the cestode, *Taenia taeniaeformis. Prostaglandins Leukotrienes and Medicine* **11** , 317–323.

Lender, T. (1974) The role of neurosecretion in freshwater planarians. In: N.W. Riser and Morse, M.P. (eds.) *Biology of the Turbellaria.* McGraw-Hill, New York, pp. 460–475.

Lender, T. (1980) Endocrinologie des planaires. *Bulletin de la Societe Zoologique de France* **105**, 173–191.

Leptak, C.L, and McKerrow, J.H. (1997) Schistosome egg granulomas and hepatic expression of TNFα are dependent on immune priming during parasite maturation. *Journal of Immunology* **158**, 301–307.

Leung, M.K., Dissous, C., Capron, A., Woldegaber, H., Duvaux-Miret, O., Pryor, S., and Stefano, G.B. (1995) *Schistosoma mansoni*: the presence and potential use of opiate-like substances. *Experimental Parasitology* **81**, 208–215.

Lindholm, A.M., Reuter, M., and Gustafsson, M.K.S. (1998) The NADPH-diaphorase staining reaction in relation to the aminergic and peptidergic nervous system and the musculature of adult *Diphyllobothrium dendriticum. Parasitology* **117**, 283–292.

Loeffler, I.K., and Bennett, J.L. (1996) A rab-related GTP-binding protein in *Schistosoma mansoni. Molecular and Biochemical Parasitology* **77**, 31–40.

McGonigle, S., Feeney, E.L., Beall, M.J., and Pearce, E.J. (2001) SIP, a novel SH3 domain-containing protein, interacts with *Schistosoma mansoni* receptor kinase 1. *Molecular and Biochemical Parasitology* **114**, 119–123.

McKay, D.M., Fairweather, I., Johnston, C.F., Shaw, C., and Halton, D.W. (1991) Immunocytochemical and radioimmunometrical demonstration of serotonin- and neuropeptide-immunoreactivities in the adult rat tapeworm, *Hymenolepis diminuta* (Cestoda, Cyclophyllidea). *Parasitology* **103**, 275–289.

McKerrow, J.H. (1997) Cytokine induction and exploitation in schistosome infections. *Parasitology* **115**, S107–S112.

McNall, S.J., and Mansour, T.E. (1984) Novel serotonin receptors in *Fasciola*. Characterisation by studies on adenylate cyclase activation and [^{3}H]LSD binding. *Biochemical Pharmacology* **33**, 2789–2797.

McNall, S.J., and Mansour, T.E. (1985) Forskolin activation of serotonin-stimulated adenylate cyclase in the liver fluke *Fasciola hepatica. Biochemical Pharmacology* **34**, 1683–1688.

Mahrenholz, A.M., Hefta, S.A, and Mansour, T.E. (1991) Phosphofructokinase from *Fasciola hepatica*: sequence of the cAMP-dependent protein kinase phosphorylation site. *Archives of Biochemistry and Biophysics* **288**, 463–467.

Mansour, J.M., and Mansour, T.E. (1986) GTP-binding proteins associated with serotonin-activated adenylate cyclase in *Fasciola hepatica. Molecular and Biochemical Parasitology* **21**, 139–149.

Mansour, J.M., and Mansour, T.E. (1989) Identification of GTP-binding proteins in *Fasciola hepatica and Schistosoma mansoni* by immunoblotting. *Molecular and Biochemical Parasitology* **36**, 11–18.

Mansour, T.E. (1959) The effect of serotonin and related compounds on the carbohydrate metabolism of the liver fluke, *Fasciola hepatica. Journal of Pharmacology and Experimental Therapeutics* **126**, 212–216.

Mansour, T.E. (1962) Effect of serotonin on glycolysis in homogenates from the liver fluke, *Fasciola hepatica. Journal of Pharmacology and Experimental Therapeutics* **135**, 94–101.

Mansour, T.E. (1984) Serotonin receptors in parasitic worms. *Advances in Parasitology* **23**, 1–36.

Mansour, T.E., and Mansour, J.M. (1962) Effects of serotonin (5-hydroxytryptamine) and adenosine 3′,5′-phosphate on phosphofructokinase from the liver fluke *Fasciola hepatica. Journal of Biological Chemistry* **237**, 629–634.

Mansour, T.E., and Stone, D.B. (1970) Biochemical effects of lysergic acid diethylamide on the liver fluke *Fasciola hepatica. Biochemical Pharmacology* **19**, 1137–1146.

Mansour, T.E., and Mansour, J.M. (1977) Phosphodiesterase in the liver fluke, *Fasciola hepatica. Biochemical Pharmacology* **26**, 2325–2330.

Mansour, T.E., Lago, A.D., and Hawkins, J.L. (1957) Occurrence and possible role of serotonin in *Fasciola hepatica. Federation Proceedings* **16**, 319.

Mansour, T.E., Sutherland, E.W., Rall, T.W., and Bueding, E. (1960) The effect of serotonin (5-hydroxytryptamine) on the formation of adenosine 3′,5′-phosphate by tissue particles from the liver fluke, *Fasciola hepatica. Journal of Biological Chemistry* **235**, 466–470.

Martelly, I. (1983) Effets du calcium sur les synthese d'ARN et d'ADN de cellules de planaires cultivees *in vitro. Comptes Rendus des Seances de l'Academie des Sciences, Paris* **297**, 9–12.

Martelly, I. (1984) Calcium thresholds in the activation of DNA and RNA synthesis in cultured planarian cells: relationship with hormonal and DB cAMP effects. *Cell Differentiation* **15**, 25–36.
Martelly, I., and Franquinet, R. (1984) Planarian regeneration as a model for cellular activation studies. *Trends in Biochemical Sciences* **9**, 468–471.
Martelly, I., and Le Moigne, A. (1980) Ribonucleic acid metabolism during planarian regeneration. *Reproduction Nutrition Developpement* **20**, 1527–1537.
Martelly, I., Franquinet, R., and Le Moigne, A. (1981) Relationship between variations of cAMP, neuromediators and the stimulation of nucleic acid synthesis during planarian (*Polycelis tenuis*) regeneration. *Hydrobiologia* **84**, 195–201.
Martelly, I., Rey, C., and Le Moigne, A. (1981) Planarian regeneration: DNA metabolism in adults. *International Journal of Invertebrate Reproduction* **4**, 107–121.
Martelly, I., Molla, A., Thomasset, M., and Le Moigne, A (1983) Planarian regeneration: *in vivo* and *in vitro* effects of calcium and calmodulin on DNA synthesis. *Cell Differentiation* **13**, 25–34.
Martelly, I., Moraczewski, J., Franquinet, R., and Castagna, M. (1987) Protein kinase C activity in a freshwater planarian (*Dugesia gonocephala*). *Comparative Biochemistry and Physiology* **86B**, 405–409.
Martin, R.J., Robertson, A.P., and Bjorn, H. (1997) Target sites of anthelmintics. *Parasitology* **114**, S111–S124.
Maule, A.G.M., Shaw, C., Halton, D. W., Thim, L., Johnston, C.F., Fairweather, I., and Buchanan, K.D. (1991) Neuropeptide F: a novel parasitic flatworm regulatory peptide from *Moniezia expansa* (Cestoda: Cyclophyllidea). *Parasitology* **102**, 309–316.
Maule, A.G.M., Shaw, C., Halton, D.W., and Thim, L. (1993) GNFFRFamide: a novel FMRFamide-immunoreactive peptide isolated from the sheep tapeworm, *Moniezia expansa*. *Biochemical and Biophysical Research Communications* **193**, 1054–1060.
Maule, A.G.M., Shaw, C., Halton, D.W., Curry, W.J., and Thim, L. (1994) RYIRFamide: a turbellarian FMRFamide-related peptide (FaRP). *Regulatory Peptides* **50**, 37–43.
Mendonca, R.L. de, Escriva, H., Bouton, D., Laudet, V., and Pierce, R.J. (2000) Hormones and nuclear receptors in schistosome development. *Parasitology Today* **16**, 233–240.
Mettrick, D.F., and Podesta, R.B. (1982) Effect of gastrointestinal hormones and amines on intestinal motility and the migration of *Hymenolepis diminuta* in the rat small intestine. *International Journal for Parasitology* **12**, 151–154.
Mettrick, D.F., Rahman, M.S., and Podesta, R.B. (1981) Effect of 5-hydroxytryptamine (5-HT; serotonin) on *in vitro* glucose uptake and glycogen reserves in *Hymenolepis diminuta*. *Molecular and Biochemical Parasitology* **4**, 217–223.
Miura, K., Fukumoto, S., Dirgahayu, P., Hirai, K. (2000) Excretory/secretory products from plerocercoids of *Spirometra erinaceieuropaei* suppress gene expression and production of tumor necrosis factor-α in murine macrophages stimulated with lipopolysaccharide or lipoteichoic acid. *International Journal for Parasitology* **31**, 39–47.
Moraczewski, J., and Duma, A. (1978) Ultrastructural localization of adenyl cyclase activity in neoblasts of turbellarian *Dugesia lugubris* (O. Schmidt). *Journal of Experimental Zoology* **203**, 491–496.
Moraczewski, J., and Franquinet, R. (1980) Mise en evidence et activite des proteins kinases AMP cyclique dependantes et AMP cyclique independantes chez la planaire *Polycelis tenuis*. *Comptes Rendus des Seances de l'Academie des Sciences, Paris, Serie D* **290**, 1567–1570.
Moraczewski, J., and Duma, A. (1981) Localisation of adenylate cyclase activity in the tissues of an intact planarian *Dugesia lugubris* (O. Schmidt). *Histochemistry* **71**, 301–311.
Moraczewski, J., Franquinet, R., and Le Moigne, A. (1983) Characterisation and activity during planarian regeneration of two types of protein kinases: cyclic AMP-dependent and calcium-dependent, cyclic AMP-independent. *Biologe Cellulaire* **47**, 171–178.
Moraczewski, J., Martelly, I., and Franquinet, R. (1986) Protein phosphorylation and the role of Ca^{2+} in planarian turbellarian regeneration. *Hydrobiologia* **132**, 223–227.
Moraczewski, J., Martelly, I., Franquinet, R., and Castagna, M. (1987) Protein kinase C activity during planarian regeneration. *Comparative Biochemistry and Physiology* **87B**, 703–707.

Morales-Montor, J., Baig, S., Hallal-Calleros, C., and Damian, R.T. (2002) *Taenia crassiceps*: androgen reconstitution of the host leads to protection during cysticercosis. *Experimental Parasitology* **100**, 209–216.

Nakazawa, M., Fantappie, M.R., Freeman, G.L. Jr., Eloi-Santos, S., Olsen, N.J., Kovacs, W.J., Secor, W.E., and Colley, D.G. (1997) *Schistosoma mansoni*: susceptibility differences between male and female mice can be mediated by testosterone during early infection. *Experimental Parasitology* **85**, 233–240.

Nevhutalu, P.A., Salafsky, B., Haas, W., and Conway, T. (1993) *Schistosoma mansoni* and *Trichobilharzia ocellata*: comparison of secreted cercarial eicanosoids. *Journal of Parasitology* **79**, 130–133.

Niagi, E.N., and Bender, D.A. (1990) *Schistosoma mansoni*: effects on tryptophan metabolism in mice. *Experimental Parasitology* **70**, 43–54.

Northup, J.K., and Mansour, T.E. (1978a) Adenylate cyclase from *Fasciola hepatica*. 1. Ligand specificity of adenylate cyclase-coupled serotonin receptors. *Molecular Pharmacology* **14**, 804–819.

Northup, J.K., and Mansour, T.E. (1978b) Adenylate cyclase from *Fasciola hepatica*. 2. Role of guanine nucleotides in coupling adenylate cyclase and serotonin receptors. *Molecular Pharmacology* **14**, 820–833.

Northup, J.K., Renart, M.F., Grove, J.R., and Mansour, T.E. (1979) Serotonin-activated adenylate cyclase from *Fasciola hepatica*. Properties of the solubilised enzyme. *Journal of Biological Chemistry* **254**, 11861–11867.

Osloobi, N., and Webb, R.A. (1999) Localization of a sodium-dependent high-affinity serotonin transporter and recruitment of exogenous serotonin by the cestode *Hymenolepis diminuta*: an autoradiographic and immunohistochemical study. *Canadian Journal of Zoology* **77**, 1265–1277.

Osman, A., Niles, E.G., and Lo Verde, P.T. (1999) Characterization of the Ras homologue of *Schistosoma mansoni*. *Molecular and Biochemical Parasitology* **100**, 27–41.

Osman, A., Niles, E.G., and Lo Verde, P.T. (2001) Identification and characterization of a Smad2 homologue from *Schistosoma mansoni*, a transforming growth factor-β signal transducer. *Journal of Biological Chemistry* **276**, 10072–10082.

Ossikovski, E., and Walter, R.D. (1984) Cyclic AMP-dependent and independent protein kinases in *Ascaridia galli*. *Molecular and Biochemical Parasitology* **12**, 299–306.

Ouahsine, H., and Sauzin-Monnot, M.-J. (1981) Effets de la reserpine sur les secretions nerveuses au cours de la regeneration de la planaire *Dugesia gonocephala* (Turbellarie-Triclade). *Comptes Rendus des Sciences de l'Academie des Sciences, Paris* **293**, 527–530.

Palladini, G., Medolago-Albani, L., Gallo, V.P., Lauro, G.M., Diana, G., Scorsini, D., Alfei, L., and Margotta, V. (1988) The answer of the planarian, *Dugesia gonocephala*, neurons to nerve growth factor. *Cellular and Molecular Biology* **34**, 53–63.

Pascolini, R., Gargiulo, A.M., Spreca, A., and Orlacchio, A. (1979) Studies on the cytochemical localization of adenylate-cyclase activity in *Dugesia lugubris* s.l. *Experientia* **35**, 1315–1317.

Pascolini, R., Gargiulo, A.M., Spreca, A., and Ceccarelli, P. (1982) Cytochemical localization of adenylate cyclase activity in planarian tissues. *Journal of Experimental Zoology* **223**, 193–196.

Pascoloini, R., Panara, F., Di Rosa, I., and Lorvik, S. (1989) The phorbol ester TPA dramatically inhibits planarian regeneration. *Tissue and Cell* **21**, 395–401.

Pax, R.A., Day, T.A., Miller, C.L., and Bennett, J.L. (1996) Neuromuscular physiology and pharmacology of parasitic flatworms. *Parasitology* **113**, S83–S96.

Phares, C.K. (1992) Biological characteristics of the growth hormone-like factor from plerocercoids of the tapeworm *Spirometra mansonoides*. *Advances in Neuroimmunology* **2**, 235–247.

Phares, K. (1996) An unusual host-parasite relationship: the growth hormone-like factor from plerocercoids of spirometrid tapeworms. *International Journal for Parasitology* **26**, 575–588.

Phares, K., and Kubik, J. (1996) The growth factor from plerocercoids of *Spirometra mansonoides* is both a growth hormone agonist and a cysteine proteinase. *Journal of Parasitology* **82**, 210–215.

Rahman, M.S., Mettrick, D.F., and Podesta, R.B. (1982) 5-Hydroxytryptamine, glucose uptake, glycogen utilization and carbon dioxide fixation in *Hymenolepis microstoma* (Cestoda). *Comparative Biochemistry and Physiology* **73B**, 901–906.

Rahman, M.S., Mettrick, D.F., and Podesta, R.B. (1983) Effects of 5-hydroxytryptamine on carbohydrate metabolism in *Hymenolepis diminuta* (Cestoda). *Canadian Journal of Physiology and Pharmacology* **61**, 137–143.

Ram, D., Grossman, Z., Markovics, A., Avivi, A., Ziv, E., Lantner, F., and Schechter, I. (1989) Rapid changes in the expression of a gene encoding a calcium-binding ptotein in *Schistosoma mansoni*. *Molecular and Biochemical Parasitology* **34**, 167–176.

Ramachandran, H., Skelly, P.J., and Shoemaker, C.B. (1996) The *Schistosoma mansoni* epidermal growth factor receptor homologue, SER, has tyrosine kinase activity and is localized in adult muscle. *Molecular and Biochemical Parasitology* **83**, 1–10.

Redman, C.A., Robertson, A., Fallon, P.G., Modha, J., Kusel, J.R., Doenhoff, M.J., and Martin, R.J. (1996) Praziquantel: an urgent and exciting challenge. *Parasitology Today* **12**, 14–20.

Reuter, M., and Kuuisto, A. (1992) Growth factors in asexually reproducing Catenulida and Macrostomida (Platyhelminthes)? A confocal, immunocytochemical and experimental study. *Zoomorphology* **112**, 155–166.

Reuter M., and Gustafsson, M.K.S. (1995) The flatworm nervous system: pattern and phylogeny. In: O. Breidbach and W. Kutsch (eds.) *The Nervous Systems of Invertebrates: an Evolutionary and Comparative Approach*. Birkhauser Verlag, Basel, pp. 25–59.

Ribeiro, P., and Webb, R.A. (1983) The synthesis of 5-hydroxytryptamine from tryptophan and 5-hydroxytryptophan in the cestode *Hymenolepis diminuta*. *International Journal for Parasitology* **13**, 101–106.

Ribeiro, P., and Webb, R.A. (1984) The occurrence, synthesis and metabolism of 5-hydroxytryptamine and 5-hydroxytryptophan in the cestode *Hymenolepis diminuta*: a high performance liquid chromatographic study. *Comparative Biochemistry and Physiology* **79C**, 159–164.

Ribeiro, P., and Webb, R.A. (1986) Demonstration of specific high-affinity binding sites for [^{3}H]5-hydroxytryptamine in the cestode *Hymenolepis diminuta*. *Comparative Biochemistry and Physiology* **84C**, 353–358.

Ribeiro, P., and Webb, R.A. (1987) Characterisation of a serotonin transporter and an adenylate cyclase-linked serotonin receptor in the cestode *Hymenolepis diminuta*. *Life Sciences* **40**, 755–768.

Ribeiro, P., and Webb, R.A. (1991) Serotonin stimulates protein phosphorylation in the cestode *Hymenolepis diminuta*. *Comparative Biochemistry and Physiology* **100C**, 483–489.

Rikihisa, Y., Lin, Y.C., and Fukaya, T. (1985) *Taenia taeniaeformis*: inhibition of rat testosterone production by excretory-secretory product of the cultured metacestode. *Experimental Parasitology* **59**, 390–397.

Roberts, L., and Janovy, J. Jnr. (2000) *Foundations of Parasitology*. McGraw-Hill, Boston.

Rumjanek, F.D., Braga, V.M.M., and Kelly, C. (1989) DNA binding proteins of *Schistosoma mansoni* recognizing a hexanucleotide motif occurring in genes regulated by steroids. *Compararative Biochemistry and Physiology* **94B**, 807–812.

Saitoh, O., Yuruzume, E., and Nakata, H. (1996) Identification of planarian serotonin receptor by ligand binding and PCR studies. *NeuroReport* **8**, 173–178.

Saitoh, O., Yuruzume, E., Watanabe, K., and Nakata, H. (1997) Molecular identification of a G protein-coupled receptor family which is expressed in planarians. *Gene* **195**, 55–61.

Salafsky, B., and Fusco, A.C. (1985) *Schistosoma mansoni*: cercarial eicosanoid production and penetration response inhibited by esculetin and ibuprofen. *Experimental Parasitology* **60**, 73–81.

Salafsky, B., and Fusco, A.C. (1987) *Schistosoma mansoni*: a comparison of secreted vs nonsecreted eicanosoids in developing schistosomulae and adults. *Experimental Parasitology* **64**, 361–367.

Salafsky, B., Fusco, A.C., Li, L.H., Mueller, J., and Ellenberger, B. (1989) *Schistosoma mansoni*: experimental chemoprophylaxis in mice using oral anti-penetration agents. *Experimental Parasitology* **69**, 263–271.

Salafsky, B., Ramaswamy, K., He, Y.-X., Anderson, G.L., Nowicki, D.K., and Shibuya, T. (1998) Evaluation of N,N-diethyl-*m*-toluamide (DEET) as a topical agent for preventing skin penetration by cercariae of *Schistosoma mansoni*. *American Journal of Tropical Mediicine and Hygiene* **58**, 828–834.

Salafsky, B., Ramaswamy, K., He, Y.-X., Li, J., and Shibuya, T. (1999) Development and evaluation of LIPODEET, a new long-acting formulation of N,N-diethyl-*m*-toluamide (DEET) for the prevention of schistosomiasis. *American Journal of Tropical Medicine and Hygiene* **61**, 743–750.

Salzet, M., Capron, A., and Stefano, G.B. (2000) Molecular crosstalk in host-parasite relationships: schistosome- and leech-host interactions. *Parasitology Today* **16**, 536–540.

Samii, S.I., and Webb, R.A. (1990) Acetylcholine-like immunoreactivity in the cestode *Hymenolepis diminuta*. *Brain Research* **513**, 161–165.

Samii, S.I., and Webb, R.A. (1996) The stimulatory effect of L-glutamate and related agents on inositol 1,4,5-trisphosphate production in the cestode *Hymenolepis diminuta*. *Comparative Biochemistry and Physiology* **113C**, 409–420.

Sangster, N.C., and Mettrick, D.F. (1987) Effects of 5-hydroxytryptamine, cyclic AMP, AMP, and fructose 2,6-bisphosphate on phosphofructokinase activity in *Hymenolepis diminuta*. *Comparative Biochemistry and Physiology* **88B**, 317–321.

Savioli, L., Renganathan, E., Montresor, A., Davis, A., and Behbehani, K. (1997) Control of schistosomiasis – a global picture. *Parasitology Today* **13** , 444–448.

Schussler, P., Grevelding, C.G., and Kunz, W. (1997) Identification of Ras, MAP kinases, and a GAP protein in *Schistosoma mansoni* by immunoblotting and their putative involvement in male-female interaction. *Parasitology* **115**, 629–634.

Sciutto, E., Fragoso, G., Diaz, M.L., Valdez, F., Montoya, R.M., Govezensky, T., Lomeli, C., and Larralde, C. (1991) Murine *Taenia crassiceps* cysticercosis: H-2 complex and sex influence on susceptibility. *Parasitology Research* **77**, 243–246.

Senft, A.W., Senft, G.R., Hillman, D.P., and Kryger, S. (1976) Influence of hycanthone on morphology and serotonin uptake of *Schistosoma mansoni*. *American Journal of Tropical Medicine and Hygiene* **25**, 832–840.

Shoemaker, C.B., Ramachandran, H., Landa, A., dos Reis, M.G., and Stein, L.D. (1992) Alternative splicing of the *Schistosoma mansoni* gene encoding a homologue of epidermal growth factor receptor. *Molecular and Biochemical Parasitology* **53**, 17–32.

Siddiqui, A.A., Podesta, R.B., and Clarke, M.W. (1991) *Schistosoma mansoni*: characterization and identification of calcium-binding proteins associated with the apical plasma membrane and envelope. *Experimental Parasitology* **72**, 63–68.

Simonic, T., Sartorelli, P., and Locatelli, A. (1983) *Fasciola hepatica*: increase of glycogen phosphorylase activity due to prostaglandins. *Experimental Parasitology* **56**, 89–92.

Smyth, J.D., and McManus, D.P. (1989) *The Physiology and Biochemistry of Cestodes*. Cambridge University Press, Cambridge.

Solon, M.H., and Koopowitz, H. (1982) Multimodal interneurones in the polyclad flatworm, *Alloeoplana californica*. *Journal of Comparative Physiology A* **147**, 171–178.

Spolski, R.J., Thomas, P.G., See, E.J., Mooney, K.A., and Kuhn, R.E. (2002) Larval *Taenia crassiceps* secretes a protein with characteristics of murine interferon-γ. *Parasitology Research* **88**, 431–438.

Stone, D.B., and Mansour, T.E. (1967) Phosphofructokinase from the liver fluke *Fasciola hepatica*. I. Activation by adenosine 3′,5′-phosphate and by serotonin. *Molecular Pharmacology* **3**, 161–176.

Sukhdeo, M.V.K., and Mettrick, D.F. (1987) Parasite behaviour: understanding platyhelminth responses. *Advances in Parasitology* **26**, 73–144.

Sukhdeo, M.V.K., and Sukhdeo, S.C. (1989) Gastrointestinal hormones: environmental cues for *Fasciola hepatica*? *Parasitology* **98**, 239–243.

Sukhdeo, M.V.K., and Sukhdeo, S.C. (2002) Fixed behaviours and migration in parasitic flatworms. *International Journal for Parasitology* **32**, 329–342.

Sukhdeo, M.V.K., Hsu, S.C., Thompson, C.S., and Mettrick, D.F. (1984) *Hymenolepis diminuta*: behavioral effects of 5-hydroxytryptamine, acetylcholine, histamine and somatostatin. *Journal of Parasitology* **70**, 682–688.

Tembe, E.A., Holden-Dye, L., Smith, S.W.G., Jacques, P.A.M., and Walker, R.J. (1993) Pharmacological profile of the 5-hydroxytryptamine receptor of *Fsaciola hepatica* body wall muscle. *Parasitology* **106**, 67–73.

Terenina, N.B., Reuter, M., and Gustafsson, M.K.S. (1999) An experimental, NADPH-diaphorase histochemical and immunocytochemical study of *Mesocestoides vogae* tetrathyridia. *International Journal for Parasitology* **29**, 787–793.

Terinina, N.B., Onufriev, M.V., Gulyaeva, N.V., Lindholm, A.M., and Gustafsson, M.K.S. (2000) A radiometric analysis of nitric oxide synthase activity in *Hymenolepis diminuta. Parasitology* **120**, 91–95.

Terrazas, L.I., Bojalil, R., Govezensky, T., and Larralde, C. (1994) A role for 17-β-estradiol in immunoendocrine regulation of murine cysticercosis (*Taenia crassiceps*). *Journal of Parasitology* **80**, 563–568.

Terrazas, L.I., Bojalil, R., Govezensky, T., and Larralde, C. (1998) Shift from an early protective Th1-type immune response to a late permissive Th2-type response in murine cysticercosis (*Taenia crassiceps*). *Journal of Parasitology* **84**, 74–81.

Terrazas, L.I., Cruz, M., Rodrigez-Sosa, M., Bojalil, R., Garcia-Tamayo, F., and Larralde, C. (1999a) Th1-type cytokines improve resistance to murine cysticercosis caused by *Taenia crassiceps. Parasitology Research* **85**, 135–141.

Terrazas, L.I., Bojalil, R., Rodriguez-Sosa, M., Govezensky, T., and Larralde, C. (1999b) *Taenia crassiceps* cysticercosis: a role for prostaglandin E2 in susceptibility. *Parasitology Research* **85**, 1025–1031.

Thompson, C.S., and Mettrick, D.F. (1989) The effects of 5-hydroxytryptamine and glutamate on muscle contraction in *Hymenolepis diminuta* (Cestoda). *Canadian Journal of Zoology* **67**, 1257–1262.

Thompson, D.P., Chen, G-Z., Sample, A.K., Semeyn, D.R., and Bennett, J.L. (1986) Calmodulin: biochemical, physiological, and morphological effects on *Schistosoma mansoni. American Journal of Physiology* **251** (*Regulatory, Integrative and Comparative Physiology* 20), R1051–R1058.

Tomosky, T.K., Bennett, J.L., and Bueding, E. (1974) Tryptaminergic and dopaminergic responses of *Schistosoma mansoni. Journal of Pharmacology and Experimental Therapeutics* **190**, 260–271.

Venturini, G. (1993) Dopamine sensitive adenylate cyclase in planaria *Dugesia gonocephala. Comparative Biochemistry and Physiology* **105C**, 297–301.

Venturini, G., Carolei, A., Palladini, G., Margotta, V., and Cerbo, R. (1981) Naloxone enhances cAMP levels in planaria. *Comparative Biochemistry and Physiology* **69C**, 105–108.

Venturini, G., Carolei, A., Palladini, G., and Margotta, V. (1986) Peptide-monoamine interactions in planaria and hydra. In: G.B. Stefano (ed.) *CRC Handbook of Comparative Opioid and Related Neuropeptide Mechanisms.* CRC Press, Inc., Boca Raton, pp. 245–254.

Venturini, G., Stocchi, F., Margotta, V., Ruggieri, S., Bravi, D., Bellantuono, P., and Palladini, G. (1989) A pharmacological study of dopaminergic receptors in planaria. *Neuropharmacology* **28**, 1377–1382.

Vicogne, J., Pin, J.-P., Lardans, V., Capron, M., Noel, C., and Dissous, C. (2003) An unusual receptor tyrosine kinase of *Schistosoma mansoni* contains a Venus Flytrap module. *Molecular and Biochemical Parasitology* **126**, 51–62.

Villa, O.F., and Kuhn, R.E. (1996) Mice infected with the larvae of *Taenia crassiceps* exhibit a Th2-like immune response with concomitant anergy and downregulation of Th1-associated phenomena. *Parasitology* **112**, 561–570.

Walker, J., and Barrett, J. (1993) Evidence for a G protein system in the tegumental brush border plasma membrane of *Hymenolepis diminuta. International Journal for Parasitology* **23**, 281–284.

Walker, R.J., Brooks, H.L., and Holden-Dye, L. (1996) Evolution and overview of classical transmitter molecules and their receptors. *Parasitology* **113**, S3–S33.

Walter, R.D., and Schulz-Key, H. (1980) Interaction of suramin with protein kinase I from *Onchocerca volvulus.* In: H. Van den Bossche (ed.) *The Host-Invader Interplay.* Elsevier/North-Holland Biomedical Press, Amsterdam, pp. 709–712.

Walter, R.D., and Albiez, E.J. (1984) Interaction of amoscanate with the cyclic AMP-phosphodiesterase from *Schistosoma mansoni, Onchocerca volvulus* and bovine heart. *Tropenmedizin und Parasitologie* **35**, 78–80.

Webb, R.A. (1986) The uptake and metabolism of L-glutamate by tissue slices of the cestode *Hymenolepis diminuta. Comparative Biochemistry and Physiology* **85C**, 151–162.

Webb, R.A. (1988a) Endocrinology of Acoelomates. In: R.G.H. Downer and H. Laufer (eds.) *Invertebrate Endocrinology. Volume 2. Endocrinology of Selected Invertebrate Types.* Alan R. Liss, New York, pp. 31–62.

Webb, R.A. (1988b) Release of exogenously supplied [^{3}H]glutamate and endogenous glutamate from tissue slices of the cestode *Hymenolepis diminuta. Canadian Journal of Physiology and Pharmacology* **66** , 889–894.

Webb, R.A. (1995) Electrical field-stimulated release of L-[G-^{3}H]-glutamate from tissue slices of the cestode *Hymenolepis diminuta. Parasitology Research* **81**, 173–174.

Webb, R.A. (1997) Glutamatergic inositol 1,4,5-triphosphate production in tissue slices of the cestode *Hymenolepis diminuta. Comparative Biochemistry and Physiology* **118C**, 19–26.

Webb, R.A., and Mettrick, D.F. (1971) Pattern of incorporation of ^{32}P into the phosholipids of the rat tapeworm *Hymenolepis diminuta. Canadian Journal of Biochemistry* **49**, 1209–1212.

Webb, R.A., and Friedel, T. (1978) Localization of a putative regeneration hormone from *Dugesia* (Turbellaria). An immunoenzyme immunocytochemical study. *Bulletin of the Canadian Microscopical Society* **6**, 21–22.

Webb, R.A., and Friedel, T. (1979) Isolation of a neurosecretory substance which stimulates RNA synthesis in regenerating planarians. *Experientia* **35**, 657–658.

Webb, R.A., and Mizukawa, K. (1985) Serotoninlike immunoreactivity in the cestode *Hymenolepis diminuta. Journal of Comparative Neurology* **234**, 431–440.

Webb, R.A., and Eklove, H. (1989) Demonstration of intense glutamate-like immunoreactivity in the longitudinal nerve cords of the cestode *Hymenolepis diminuta. Parasitology Research* **75**, 545–548.

Weinstein, S., and Gavurin, L. (1977) The effect of adenosine-3′,5′-cyclic monophosphate on the mitotic rate in regenerating *Dugesia dorotocephala. Cell Differentiation* **5**, 311–322.

Wiest, P.M., Kunz, S.S., and Miller, K.R. (1994) Activation of protein kinase C by phorbol esters disrupts the tegument of *Schistosoma mansoni. Parasitology* **109**, 461–468.

Wiest, P.M., Burnham, D.C., Olds, G.R., and Bowen, W.D. (1992) Developmental expression of protein kinase C activity in *Schistosoma mansoni. American Journal of Tropical Medicine and Hygiene* **46**, 358–365.

Willcockson, W.S., and Hillman, G.R. (1984) Drug effects on the 5-HT response of *Schistosoma mansoni. Comparative Biochemistry and Physiology* **77C**, 199–203.

Wood, P.J., and Mansour, T.E. (1986) *Schistosoma mansoni*: serotonin uptake and its drug inhibition. *Experimental Parasitology* **62**, 114–119.

CHAPTER 7

CONTROL OF *CAENORHABDITIS ELEGANS* BEHAVIOUR AND DEVELOPMENT BY G PROTEINS BIG AND SMALL

CAROL A. BASTIANI, MELVIN I. SIMON, and PAUL W. STERNBERG

Division of Biology, California Institute of Technology, Pasadena, California 91125, USA

Summary

Caenorhabditis elegans molecular genetics has allowed analysis of many of its heterotrimeric GTP-binding proteins (G proteins; α, β, γ) and small GTP-binding proteins. The roles of its 20 alpha, two beta and two gamma heterotrimeric G protein subunits have been studied by gene knockouts, overexpression analysis, gene interaction experiments and by analysis of expression patterns. The G alpha subunits that are highly conserved with mammalian G protein alpha subunits are broadly expressed and are involved in neuromuscular regulation. The less conserved G alpha subunits are expressed in small subsets of sensory neurons and play a role in sensory behaviour. The study of small G proteins such as ras in *C. elegans* has helped to elucidate their cellular functions. Maintenance of signalling specificity and interactions between G protein signalling pathways can be studied when the G proteins function in the same processes. In principle, one can hope to understand the role of each G protein in every cell in *C. elegans*, and thus unravel cell type-specific variation in signalling pathways and the full spectrum of the cellular roles for these crucial regulatory proteins.

1. Heterotrimeric G Proteins in *C. elegans*

Caenorhabditis elegans is a free living rhabditid nematode introduced as a promising model organism for genetic, developmental and behavioural investigation by Sydney Brenner in 1965. It can be propagated either clonally by self-fertilizing hermaphrodites, or by outcrossing hermaphrodites with males, and thus mutants with severe locomotion or mating defects can be maintained. It has a short three-day life cycle under favourable conditions. It is small (1 mm as adult, with 959 cells in the hermaphrodite and 302 neurons), transparent and easily cultivated in the laboratory. *C. elegans* has an essentially invariant cell lineage (Sulston and Horvitz, 1977; Kimble and Hirsh, 1979; Sulston *et al.*, 1983) and its complete anatomy has been determined by serial section electron micrograph reconstructions (Albertson and Thomson, 1976; White *et al.*, 1976, 1986; White, Southgate and Durbin, 1988). Its compact genome, at 100 Mb (only twenty times that of *E. coli*), led to the efficient completion of its genomic sequence (The *C. elegans* Sequencing Consortium, 1998). All of these features, coupled with the ability to use reverse genetics by making transgenic worms (Mello and Fire, 1995), RNAi technology (Fire *et al.*, 1998; Tabara, Grishok and Mello, 1998) and gene deletion selection technology (Plasterk, 1995; Jansen

I. Fairweather (ed.), Cell Signalling in Prokaryotes and Lower Metazoa, 195–242.
© 2004 *Kluwer Academic Publishers. Printed in Great Britain.*

et al., 1997), have facilitated significant insights into cellular signal transduction pathways, such as the Ras pathway, and pathways that regulate apoptosis. Ongoing research relating to a variety of different behavioural pathways, including response to heavy and light touch (see Driscoll and Kaplan, 1997 for review), temperature, volatile odourants, osmotic and ionic strength, food and other nematodes (see Bargmann and Mori, 1997 for review), suggest that *C. elegans* will provide insight into the molecular and genetic basis for behaviour. This review will focus on the knowledge accumulated on the role of heterotrimeric and small GTP-binding proteins (G proteins) in development, differentiation and behavioural response pathways in *C. elegans.*

Heterotrimeric GTP-binding proteins mediate intracellular responses to a wide variety of extracellular stimuli, including neurotransmitters, hormones and sensory stimuli via their interaction with seven-pass transmembrane receptors (7-TMRs) (see Gilman, 1987; Bourne, Sanders and McCormick, 1990, 1991; Dohlman *et al.*, 1991 for reviews). Heterotrimeric GTP-binding proteins are present in all eukaryotic organisms and are composed of α, β and γ subunits. In the inactive state, Gα subunits interact with GDP and are tightly associated with βγ subunits and receptors (see Fig. 1). Interaction of a receptor with an appropriate agonist induces a conformational change in the Gα subunit, which promotes the exchange of GDP for GTP, liberating and separating the trimeric αβγ complex from the receptor. Free GTP-bound alpha subunits and βγ dimers propagate signals by interacting with effector molecules such as phospholipase C, adenylate cyclase or, in the case of βγ, directly with ion channels. The signalling cascade is terminated upon hydrolysis of GTP. Gα subunits have an intrinsic GTPase activity which is further stimulated by "regulators of G protein signalling" molecules (RGS proteins) (Berman, Kozasa and Gilman, 1996; Hunt *et al.*, 1996; Faurobert and Hurley, 1997; and see Watson *et al.*, 1996 for review).

Studies conducted in mammalian systems suggest that at least eight distinct

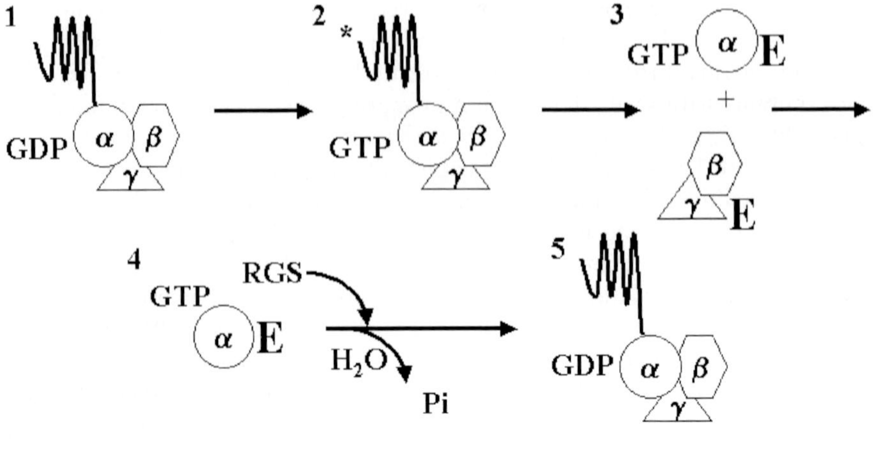

Figure 1. The G protein cycle. Activated receptor is indicated with an asterisk. Effector is indicated with an "E".

potassium currents, two calcium currents and a single sodium current are regulated by 7-TMRs (see Brown and Birnbaumer, 1990; Hille, 1994 for reviews). However, the molecular identity of the regulated channel has been determined in only a few cases (Curtis and Catterall, 1985; Kleuss *et al.*, 1991; Reuveny *et al.*, 1994). Mechanisms proposed for modulation of ion channels by 7-TMRs include direct binding of ion channels to G protein subunits, to cyclic nucleotides, to intracellular calcium ions or phospholipids, or phosphorylation or dephosphorylation of ion channels.

C. elegans expresses 20 Gα, 2 Gβ and 2 Gγ subunits. To date, mammals are known to express 16 Gα, 5 Gβ and 12 Gγ subunits. The 16 mammalian Gα subunits have been grouped into four families, based on amino acid identity: G_q, $G_{i/o}$, G_s and $G_{12/13}$. *C. elegans* has one clear homologue of each of the four G protein alpha families (see Table 1). *egl-30* encodes a G_q family homologue, *goa-1* encodes a G_o family homologue, *gsa-1* encodes a G_s family homologue and *gpa-12* encodes a $G_{12/13}$ homologue. The 16 other Gα subunits do not contain extensive sequence identity with mammalian Gα subunits, nor with Gα subunits from other organisms, and are not clearly related to each other (Fino Silva and Plasterk, 1990; Lochrie *et al.*, 1991; Zwaal *et al.*, 1997; Roayaie *et al.*, 1998; Jansen *et al.*, 1999).

Both reverse genetics and forward genetic screens have obtained insight into

Table 1. *Caenorhabditis elegans* G alpha subunits homologous to mammalian G alpha subunits

C. elegans protein	Mammalian protein	Identity	Reduction-of-function	Gain-of-function	Expression
GDA-1	$G\alpha_o$	82%	hyperactive, early eggs, defective male turning response, slow pumping	lethargic, late-stage eggs	all neurons, vulvul muscles, pharyngeal muscle, male diagonal muscles
EGL-30	$G\alpha_q$	82%	lethargic, late-stage eggs, weak pumping	hyperactive, early eggs	axons, some cell bodies in nervous system, pharyngeal muscle
GSA-1	$G\alpha_s$	66%	lethargic, late-stage eggs (mosaic animals) loss-of-function is lethal	hyperactive, early egg-laying, neural degeneration	most neurons and muscle cells
GPA-12	$G\alpha_{12}$	51%			pharynx ventral and lateral hypodermis
GPA-7	Gαo	57% to human GαoB	late-stage eggs, defects in water-source chemo-aversion	early egg-laying, multiple defects in chemo-sensation	many neurons and muscle cells

G protein subunit functions and signal transduction pathways. The *goa-1*, *gsa-1*, *gpa-12* and most of the *C. elegans*-specific G protein alpha subunit genes were all identified based on their homology with mammalian G protein alpha subunits. Their functions have been assessed by generating knockout alleles of these genes, by overexpression analysis and by analysis of phenotypes conferred by overexpression of GTPase-deficient mutant proteins. Worms that overexpress GTPase-deficient Gα subunits have been used successfully for suppressor screens to identify genes which function in the same pathway and have also identified candidate genes that might integrate different G protein-coupled pathways (Barger, Hart and Kaplan, 1998; Korswagen, Van der Linden and Plasterk, 1998; Hadju-Cronin *et al.*, 1999). Mutations in *egl-30* and in some of the other genes that encode components of G protein signalling pathways were first identified in forward behavioural genetic screens (Trent, Tsung and Horvitz, 1983; Thomas, 1990; Avery, 1993). Reporter gene fusions with green fluorescent protein (GFP) have aided in the analysis of expression patterns of heterotrimeric G protein subunits (Chalfie *et al.*, 1994) and so have facilitated an understanding of pathway interaction.

From these studies, *C. elegans* G protein alpha functions can be subdivided into three functional classes: those involved in neuromuscular regulation, those involved in cell division and cellular morphogenesis and those involved in chemosensation. In general, the mammalian homologues are primarily incorporated into the first two functional classes, while the *C. elegans*-specific Gα subunits are primarily involved in chemosensation (Zwaal *et al.*, 1997; Roayaie *et al.*, 1998; Jansen *et al.*, 1999). The broad expression patterns of *goa-1*, *egl-30* and *gsa-1* are consistent with their pleiotropic functions (see Table 1). Possible integration between these signalling pathways might be investigated by the study of behaviours regulated by more than one G alpha subunit. Egg-laying, pharyngeal pumping and movement (see Table 2 for a summary of neurons and muscles controlling these behaviours) are examples of three behaviours that all appear to be regulated by EGL-30, GOA-1 and GSA-1. *gpa-12*, on the other hand, does not seem to have a broad expression pattern (see Jansen *et al.*, 1999). Gene fusions with *GFP* indicate that *gpa-12* is primarily expressed in the pharyngeal muscle and in hypodermal cells.

The expression patterns for 15 of the 16 *C. elegans*-specific Gα subunits are consistent with a primarily chemosensory role for these proteins. *GFP* reporter gene fusions indicate that the *C. elegans*-specific Gα subunits are expressed almost exclusively in a small subset of the chemosensory neurons (Zwaal *et al.*, 1997; Roayaie *et al.*, 1998; Jansen *et al.*, 1999). The *C. elegans*-specific G protein system will be useful for exploring uncharacterized G protein-mediated signal transduction pathways, maintenance of signalling specificity between G protein pathways and possibly conserved mechanisms for the integration of signals transduced by different G proteins within the same cell.

C. elegans also expresses homologues of small GTP-binding proteins, such as Ras, Rho, Rac. The Ras pathway has been genetically defined from the receptor to nuclear transcription factors with respect to the cell fate and pattern

Table 2. An overview of behaviours of *Caenorhabditis elegans* and the senosory neurons that are associated with them (see Bargmann and Horvitz, 1991a; Kaplan, 1996; Bargmann and Mori, 1997)

Behaviour	Sensory neuron
Egg laying:	ASK
Dauer development:	
Dauer formation	ADF, ASG, ASI
Dauer recovery	ASJ
Olfaction:	
Chemotaxis	AWA, AWC, ASE (major neurons)
	ADF, ASG, ASI (minor neurons)
Chemical avoidance	AWB, ASH, ADL
Gentle body touch:	ALM, AVM (anterior body touch)
	PLM (posterior body touch)
Thermotaxis:	AFD

formation of the vulva; it also mediates cell fate and cell migration of many other cell types in *C. elegans* (see Sternberg and Han, 1998 for review). Mutations in *mig-2*, which contains similarity with Rho, were originally isolated in genetic screens for mutants with altered cell migrations (Hedgecock *et al.*, 1987; Forrester and Garriga, 1997; Zipkin, Kindt and Kenyon, 1997). *unc-73*, a guanine nucleotide exchange factor for Rac, is required in migrating and dividing cells for the spatial regulation of actin and cytoskeletal polarity (Hedgecock *et al.*, 1987; Desai *et al.*, 1988; Manser and Wood, 1990; Siddiqui, 1990; Siddiqui and Culotti, 1991; McIntire *et al.*, 1992; Chen, Branda and Stern, 1997; Forrester and Garriga, 1997; Wightman, Baron and Garriga, 1997; Steven *et al.*, 1998). Thus, the proteins encoded by these genes appear functionally equivalent to their mammalian counterparts. A genetic characterization of the signalling pathways mediated by these proteins will facilitate further understanding of such pathways in mammals.

2. Genetic Analysis of Heterotrimeric G Protein Function

2.1. GOA-1

The GOA-1 protein, which is 82% identical to human $G\alpha_o$, was originally identified by virtue of its sequence homology with mammalian $G\alpha_o$ subunits

(Lochrie *et al.*, 1991). GOA-1 was the first heterotrimeric Gα subunit functionally and genetically characterized in *C. elegans* (Mendel *et al.*, 1995; Segalat, Elkes and Kaplan, 1995). Mutations that either enhance or diminish GOA-1 signalling have opposite effects on egg laying, movement and pharyngeal pumping (Mendel *et al.*, 1995; Segalat, Elkes and Kaplan, 1995). The phenotypes exhibited by *goa-1* mutants indicate that GOA-1 plays a role in negatively regulating the activity of neurons and muscles. For example, while wild-type worms normally lay eggs approximately two to three hours after fertilization during the gastrulation phase of development (Sulston *et al.*, 1983), *goa-1* null and reduction-of-function mutants typically lay eggs during early embryogenesis. *goa-1* null and reduction-of-function mutants initiate sinusoidal movements with increased frequency and move with exaggerated body bends compared to wild-type (Mendel *et al.*, 1995; Segalat, Elkes and Kaplan, 1995). *goa-1* null mutations also cause worms to pump their pharynges more slowly (Segalat, Elkes and Kaplan, 1995). In males, *goa-1* reduction-of-function mutants exhibit defects in a stereotypical male turning response that is initiated in wild-type worms upon contact with hermaphrodites. Spicule insertion behaviour is also defective (Mendel *et al.*, 1995).

In direct contrast to phenotypes exhibited by mutants that have null or reduction-of-function mutations in *goa-1*, overexpression of *goa-1* causes worms to become lethargic and retain embryos well after the onset of gastrulation. These effects can be induced in adult animals by placing *goa-1* under the control of a heat-shock inducible promoter element (Mendel *et al.*, 1995). Since the effects of *goa-1* expression do not depend on early expression of *goa-1*, GOA-1 is involved in the inhibition of mature neuron and muscle activity, rather than being required for the development of these cell types.

Some phenotypes conferred by *goa-1* null mutations are similar to phenotypes exhibited in worms that have mutations in genes required for the biosynthesis of serotonin (Segalat, Elkes and Kaplan, 1995). In some cases, *goa-1* null mutants cannot respond normally to exogenous serotonin (Mendel *et al.*, 1995; Segalat, Elkes and Kaplan, 1995). In *C. elegans*, exogenous serotonin (5-HT) stimulates egg-laying and pharyngeal pumping and inhibits locomotion and defaecation (Horvitz *et al.*, 1982; Segalat, Elkes and Kaplan, 1995). *goa-1* null mutants are partially resistant to the inhibitory effects of 5-HT on movement and the inhibitory effect of 5-HT on enteric muscle contractions (Segalat, Elkes and Kaplan, 1995). However, *goa-1* null mutants remain sensitive to the stimulatory effects of 5-HT on pharyngeal pumping (Segalat, Elkes and Kaplan, 1995). Mutants that overexpress a GTPase-deficient form of GOA-1 are also partially resistant to the stimulatory effects of exogenous 5-HT on egg-laying behaviour (Mendel *et al.*, 1995). Defective response to serotonin of *goa-1* mutants indicates that *goa-1* acts downstream of, or in parallel to, serotonin with respect to movement, defaecation and egg-laying. *goa-1* null mutants are fully sensitive to transmitters such as octopamine and dopamine (Segalat, Elkes and Kaplan, 1995), so it is possible that *goa-1* might act downstream of serotonin to regulate movement and defaecation. However, the effect of *goa-1* on egg-laying is opposite to the effect of serotonin, so *goa-1* likely acts in parallel with serotonin to affect egg-laying behaviour.

goa-1 is expressed in most neurons, in the muscles involved in egg-laying and in male mating, in pharyngeal muscles, in the intestinal muscles and in the anal depressor and anal sphincter muscles (Mendel *et al.*, 1995; Segalat, Elkes and Kaplan, 1995). Genetic ablation of the HSNs in wild-type worms, the primary serotonergic neurons that control egg-laying (see Fig. 2), causes animals to lay late-stage eggs. However, HSN ablation does not suppress the early egg-laying phenotype of *goa-1* reduction-of-function mutant animals, indicating that these animals are insensitive to reduced levels of serotonin input to the vulval and uterine muscles (Segalat, Elkes and Kaplan, 1995). Therefore, GOA-1 probably acts either in the vulval muscles or in the VC motor neurons which regulate egg-laying (see Fig. 2).

In insects and mammals, $G\alpha_o$ is expressed in neurons and in developing neuronal growth cones (Huff, Axton and Neer, 1985). Given that *goa-1* is broadly expressed in the *C. elegans* nervous system, it was initially a surprise to find that *goa-1* null mutants had no gross defects with respect to development of the nervous system. However, in cultured vertebrate neurons, $G\alpha_o$ has also been shown to inhibit neural activity by either of two mechanisms: downstream activation of voltage-dependent potassium channels or downstream inhibition of voltage-dependent calcium channels, so *C. elegans* GOA-1 and mammalian $G\alpha_o$ probably share some equivalent functions.

2.2. EGL-10, A REGULATOR OF *GOA-1* FUNCTION

The *egl-10* gene was originally identified in a screen for mutations that have a defect in egg-laying (Egl-d) (Trent, Tsung and Horvitz, 1983). *egl-10* mutants

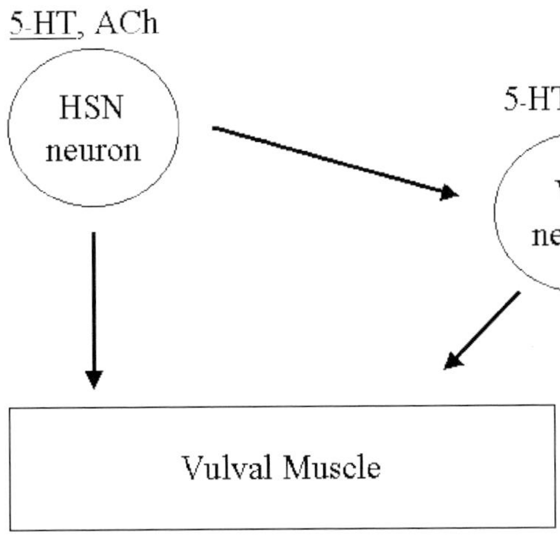

Figure 2. The egg-laying system in *Caenorhabditis elegans* (adapted from White, Southgate and Durbin, 1988, p. 452).

respond to serotonin agonists, but some *egl-10* alleles do not respond to imipramine, an inhibitor of serotonin reuptake at synaptic junctions (Trent, Tsung and Horvitz, 1983; Desai *et al.*, 1988; Desai and Horvitz, 1989). After cloning *egl-10*, it was discovered that EGL-10 has a region of similarity with the *Saccharomyces cerevisiae* SST2 protein (Koelle and Horvitz, 1996). SST2 encodes an RGS protein (regulator of G protein signalling), a negative regulator of the Gα protein subunit that mediates the yeast pheromone response (Blinder and Jenness, 1989; Dohlman *et al.*, 1995, 1996). Some RGS proteins have been shown to bind to activated Gα protein subunits and act as accelerators of GTP hydrolysis, a function referred to as GAP activity (Berman, Kozasa and Gilman, 1996; Berman, Wilkie and Gilman, 1996; Chen, Wieland and Simon, 1996; Hunt *et al.*, 1996; Watson *et al.*, 1996).

Given the fact that loss-of-function mutations in *egl-10* confer phenotypes that are opposite from those conferred by *goa-1* loss-of-function mutations with respect to egg-laying, *egl-10* mutant phenotypes were examined further for their relationship to *goa-1* mutant phenotypes. *egl-10* loss-of-function mutants initiate body bends less frequently, similar to *goa-1* gain-of-function mutants. Foraging movements are also less frequent in *egl-10* reduction-of-function mutants. In contrast, overexpression of *egl-10* causes phenotypes similar to *goa-1* loss-of-function phenotypes. Worms that overexpress *egl-10* lay early eggs and also initiate body-bends more frequently than wild type (Koelle and Horvitz, 1996).

Genetic analysis indicates that *egl-10* might negatively regulate *goa-1*, since *goa-1* null mutations are epistatic to *egl-10* null mutations. Mutations in both *goa-1* and *egl-10* have egg-laying and movement behaviours that are practically indistinguishable from *goa-1* single mutants (Koelle and Horvitz, 1996). This is consistent with *goa-1* functioning downstream of, or in parallel to, *egl-10*. Immunolocalization of the EGL-10 protein indicates that expression patterns of *egl-10* and *goa-1* do overlap. *egl-10* is expressed in the nerve ring, ventral nerve cord and dorsal nerve cord and is primarily localized at chemical synaptic junctions in the animal. *egl-10* expression is observed throughout development. Like *goa-1*, most or all neurons express *egl-10*. However, expression in the vulval muscles was not observed, even in animals that overexpress *egl-10* (Koelle and Horvitz, 1996).

Epistasis analysis and expression patterns of *egl-10* and *goa-1* are consistent with the possibility that these two proteins interact to regulate movement. However, *egl-10* and *goa-1* mutants vary with respect to serotonin response. *goa-1* mutants are resistant to serotonin. The ability of *egl-10* mutants to respond to serotonin, but not to imipramine, suggests that HSN function is compromised in *egl-10* mutants. Since ablation of the HSN neurons in *goa-1* mutants has no effect, *goa-1* functions redundantly with the HSN neurons to regulate egg-laying behaviour. One must be cautious when interpreting epistasis of broadly expressed genes, because the inferred pathways extend through many cells (for example, interneuron to motor neuron to muscle). It remains unclear whether the epistasis observed between *egl-10* and *goa-1* reflects the fact that *goa-1* acts in a cell downstream of *egl-10's* site of action (that is, vulval muscles) to regulate egg-laying, rather than in the same molecular pathway.

The *C. elegans* genome encodes 12 RGS proteins (Bargmann, 1998) and 19 RGS proteins have been identified in mammals (Berman and Gilman, 1998). *In vitro*, RGS proteins are rather promiscuous with respect to their ability to interact with a variety of Gα subunits (see Arshavsky and Pugh, 1998 for review). It remains to be determined whether the same holds true *in vivo*. Also, in addition to their experimentally assessed role in the negative regulation of Gα subunits, it has been proposed that RGS proteins might act as scaffolding proteins with a role in maintaining the specificity of G protein-coupled signalling pathways (Zeng *et al.*, 1998; Xu *et al.*, 1999). In addition to GAP (GTPase-activating) domains, some RGS proteins contain DEP (Dsh, EGL-10 and pleckstrin) and GGL (G-gamma-like, discussed later) and membrane association (Snow *et al.*, 1998; Chen *et al.*, 1999; De Vries and Farquhar, 1999; Levay *et al.*, 1999; Siderovski, Stockbine and Behe, 1999) domains. The functions and the regulatory roles for these domains are questions that will be most convincingly addressed by using genetic approaches, coupled with biochemical analyses.

2.3. EGL-30, THE Gα_q HOMOLOGUE

A *C. elegans* Gα_q subunit was cloned because of its sequence homology with mammalian Gα_q and was shown to correspond with the previously identified genetic locus *egl-30* (Brundage *et al.*, 1996). In mammals, Gα_q activates all isoforms of PLCb (see Fig. 3 for pathway). EGL-30 subunits can activate mammalian PLCβ in transfected cells, which indicates that EGL-30 is most likely functionally equivalent to mammalian G_q. Strong loss-of-function alleles of *egl-30* are semilethal, with a paralyzed phenotype similar to mutants that are incapable of contracting body-wall muscles. Weaker alleles of *egl-30* display weak pharyngeal pumping, retain late stage eggs and are lethargic (Trent, Tsung and Horvitz, 1983; Brundage *et al.*, 1996). Phenotypes of overexpression and loss-of-function mutants are opposite to phenotypes conferred by *goa-1* mutations (Brundage *et al.*, 1996). All of the *egl-30* mutant phenotypes are consistent with EGL-30 positively regulating neuronal and/or muscle function. Expression of a *GFP* gene fusion with *egl-30* indicates that *egl-30* functions in the nervous system, and is not expressed in vulval muscles (Bastiani, in prep.). However, since severe *egl-30* mutants are not responsive to the stimulatory effects of serotonin in egg-laying assays, it is possible that *egl-30* functions in both the nervous system and in the vulval muscle to regulate egg-laying. *egl-30* is strongly expressed in the pharyngeal muscle (Lackner, Nurrish and Kaplan, 1999; Bastiani, in prep.) and this is consistent with a previously defined role for *egl-30* in this muscle type (Brundage *et al.*, 1996).

Failure of *egl-30* mutants to respond to serotonin in egg-laying assays suggests that *egl-30* acts downstream of serotonin. Additionally, a role for *egl-30* in acetylcholine signalling has been suggested by genetic interactions with

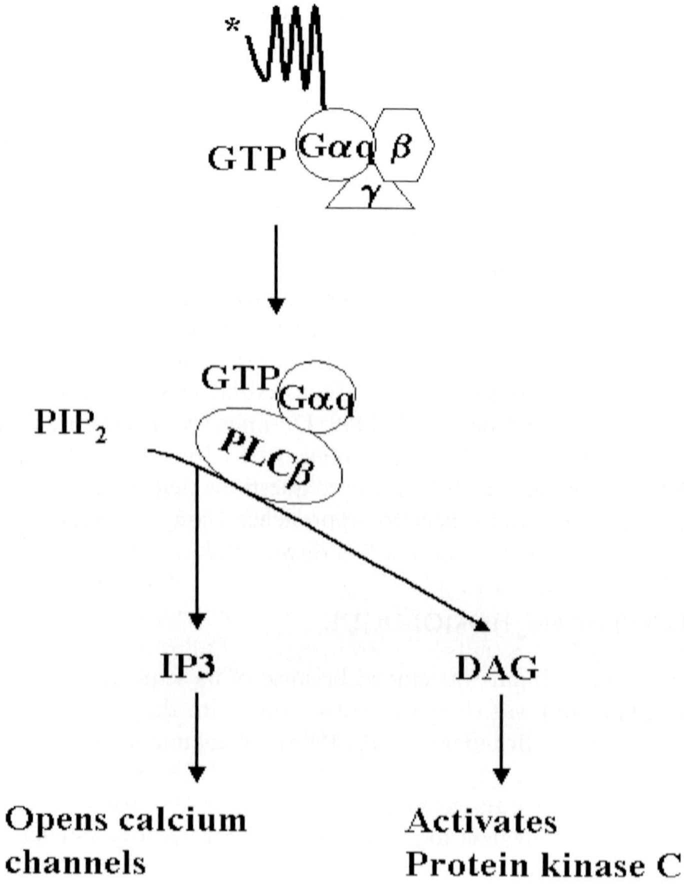

Figure 3. The mammalian $G\alpha_q$ signalling pathway.

eat-11 (Brundage *et al*., 1996), which encodes a putative negative regulator of *egl-30*, and by resistance of *egl-30* mutants to growth arrest by aldicarb (Ngyuyen *et al*., 1995), an acetylcholinesterase inhibitor.

A candidate RGS protein with specificity for EGL-30 has been identified. Interestingly, this protein might represent a link between the GOA-1- and EGL-30-mediated signalling pathways and will be discussed later in this context. In addition, mutations in other genes that are predicted to be downstream of *egl-30* have been isolated (see Fig. 3). The *C. elegans* genome sequence predicts one isoform of PLCβ. In a screen for mutants that are aldicarb-resistant and share movement and egg-laying phenotypes in common with *egl-30* mutants, *egl-8* was identified (Miller, Emerson and Rand, 1999). Mutations in *egl-8* had previously been identified in forward genetic screens for mutants defective in egg-laying (Trent, Tsung and Horvitz, 1983) and defaecation (Thomas, 1990). *egl-8* was subsequently found to encode the

C. elegans homologue of PLCβ (Miller, Emerson and Rand, 1999). However, a comparison of the severity of *egl-30* mutant phenotypes with mutant phenotypes conferred by an allele of *egl-8*, which disrupts the reading frame before the catalytic Y domain, suggests that *egl-30* has one or more additional functions that are not shared with *egl-8* (Miller, Emerson and Rand, 1999). Identification of effector signalling pathways downstream of *egl-30*, and in parallel to *egl-8* pathways, might lead to the identification of additional candidate effectors for mammalian $G\alpha_q$.

A null allele of the only *C. elegans* IP_3 receptor homologue, encoded by *itr-1/dec-4/lfe-1*, which would also be expected to be downstream of *egl-30* and of *egl-8* (see Fig. 3), has also been isolated (see Dal Santo *et al.*, 1999). The phenotypic consequences that result from mutations in *egl-8* and mutations in *itr-1* indicate that EGL-8 and ITR-1 are both involved in defaecation (Thomas, 1990; Iwasaki, Liu and Thomas, 1995). *egl-8* and *egl-30* share movement and egg-laying phenotypes (Trent, Tsung and Horvitz, 1983; Brundage *et al.*, 1996; Miller, Emerson and Rand, 1999) and *egl-30* and *itr-1* are important for normal rates of growth of *C. elegans* (see Brundage *et al.*, 1996; Dal Santo *et al.*, 1999). Although there is evidence that *egl-8* is downstream of *egl-30* with respect to movement, it remains to be tested whether shared functions between these genes reflect similar pathways in all cases. However, the fact that some functions are not shared indicates that a genetic analysis of all of these genes will most likely uncover novel signalling pathways.

2.4. GSA-1, THE $G\alpha_s$ HOMOLOGUE

gsa-1 was also identifiedby virtue of its amino acid identity with mammalian $G\alpha_s$, but is only 66% identical to mammalian $G\alpha_s$ subunits. A direct effector of $G\alpha_s$ in mammalian cells is adenylyl cyclase which $G\alpha_s$ stimulates to generate cAMP (Sunahara, Dessauer and Gilman, 1996, and see Fig. 4). This second messenger subsequently mediates a wide range of responses through activation of protein kinase A (PKA) (Walsh and Van Patten, 1994) and by modulating cyclic nucleotide-gated ion channels. $G\alpha_s$ may also directly regulate some ion channels. L-type voltage gated Ca^{++} channels in skeletal muscle cells are activated by $G\alpha_s$ and $G\alpha_s$ inhibits cardiac Na^+ channels (see Wickman and Clapham, 1995a,b for reviews).

Animals with *gsa-1* loss-of-function mutations die as first-stage larvae with a shrunken, withered tail phenotype similar to the lethal phenotype in which the excretory canal-associated neurons (CAN) have been killed by laser ablation (Korswagen *et al.*, 1997). The excretory cell, believed to regulate osmotic content, is located beneath the posterior pharyngeal bulb. The cell sends processes (canals) left and right to the lateral ridge, where they branch and extend almost the entire length of the worm (Nelson and Riddle, 1984). The lethality and associated phenotype of *gsa-1* loss-of-function mutations indicates that $G\alpha_s$ plays an essential role in development or homeostasis.

To analyze the effects of both overexpression and loss of GSA-1 signalling

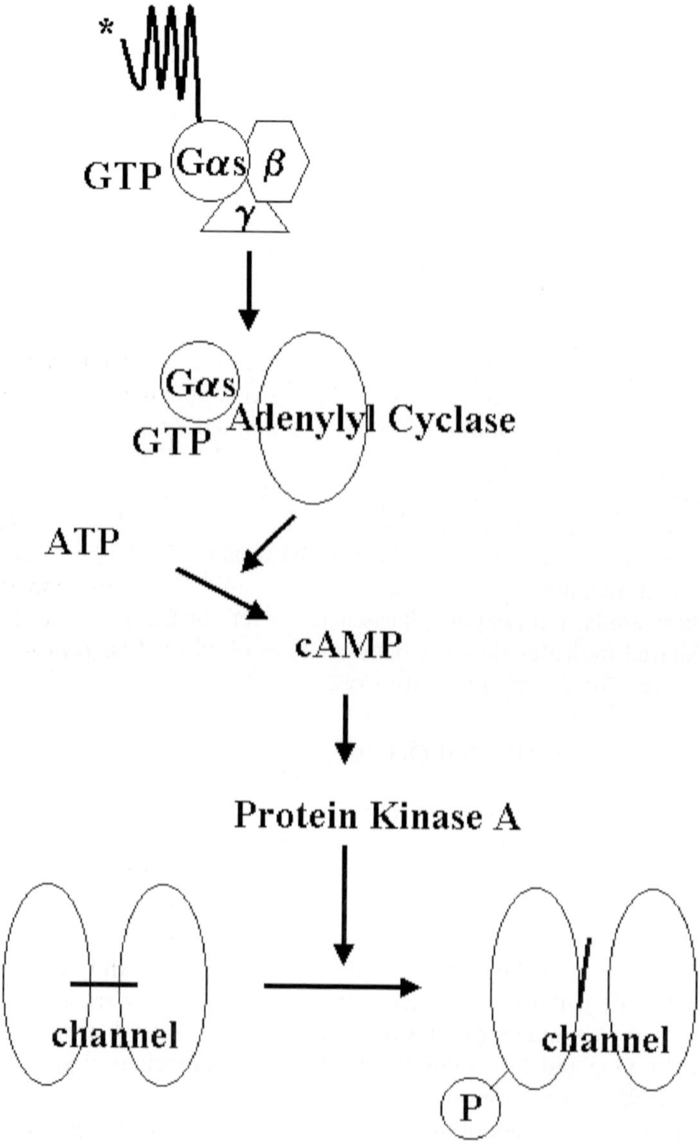

Figure 4. The mammalian $G\alpha_s$ signalling pathway.

in adult animals, transgenic animals were generated which overexpress *gsa-1* in a *gsa-1* mutant animal. When a transgene is not integrated into the genome, transgenic animals express the transgene in a mosaic fashion. In this case, behavioural effects in adult animals that have lost the transgene in some of the somatic cells can be assessed, since the transgene rescues the lethality caused by loss of *gsa-1* function.

Loss of *gsa-1* expression or overexpression of wild-type *gsa-1* have opposite effects on movement, egg-laying and defaecation. Overexpression of *gsa-1* results in hyperactive movement and early egg-laying (Korswagen *et al.*, 1997). The early egg-laying phenotype is not dependent upon the HSNs, indicating that *gsa-1* might function in the vulval muscles (Korswagen *et al.*, 1997) or in (or via) the VC motor neurons. Partial translational gene fusions of *gsa-1* with *GFP* and *lacZ* indicate that *gsa-1* is expressed throughout the nervous system, in muscles, and in intestinal cells (Korswagen *et al.*, 1997; Park *et al.*, 1997). *gsa-1*, like *egl-30*, positively regulates the activity of neurons and/or muscles.

Overexpression of constitutively active (GTPase-deficient) $G\alpha_s$ subunits induces novel phenotypes not evident when wild-type *gsa-1* gene is overexpressed. Ubiquitous overexpression induces body-wall muscle hypercontraction and induces a subset of expressing ventral nerve cord motor neurons and some neurons in the head and tail ganglia to undergo a necrotic pattern of cell death, characterized by cell swelling and lysis (Korswagen *et al.*, 1997). To characterize the mechanism of $G\alpha_s$-induced neurotoxicity, mutants that affect cell death pathways or mutations in genes that encode molecules that are possibly downstream of $G\alpha_s$, were tested for their ability to suppress GSA-1-induced neurodegeneration. In addition, suppressor screens were conducted. Mutations that reduce neural activity (for example, *unc-36*, a gene that encodes an α2-subunit of an L-type calcium channel) are partially protective against $G\alpha_s$-induced neurotoxicity (Barger, Hart and Kaplan, 1998). Analysis of double mutants with the *ced* apoptosis genes indicates that the normal programmed cell death pathway is not required for cytotoxicity (Korswagen *et al.*, 1997; Barger, Hart and Kaplan, 1998). In suppressor screens, mutations in the *sgs-1/acy-1* gene, which encodes a membrane-bound adenylyl cyclase, completely blocked both the swelling and cytotoxicity of activated forms of $G\alpha_s$ (Korswagen *et al.*, 1997; Barger, Hart and Kaplan, 1998), indicating that the cytotoxic effects of $G\alpha_s$ are mediated by increases in intracellular cAMP. Mutations in *sgs-1/acy-1* also partially and variably block $G\alpha_s$-induced body-wall hypercontraction (Korswagen *et al.*, 1997).

sgs-1/acy-1 is most similar to mammalian adenylyl cyclase type IX. *sgs-1/acy-1* knockout animals do not have obvious developmental or behavioural defects. Therefore, SGS-1/ACY-1 is not the only target for GSA-1. Null alleles were generated in another gene that encodes an adenylyl cyclase homologue, *acy-2*. *acy-2* is specifically expressed in the CAN cells, the same cells that are affected in *gsa-1* null mutants, and a null mutation in this gene results in a larval lethal phenotype similar to phenotypes conferred by *gsa-1* null mutations. These results provide strong evidence that the effector pathway for $G\alpha_s$ is conserved between worms and mammals. It will be interesting to find out whether the necrotic death of neuronal cells, caused by overexpression of an activated mutant of *gsa-1*, will provide an experimental model for neurodegenerative human diseases that are characterized by similarly specific neuronal cell deaths.

2.5. GPB-1 AND GPB-2, Gβ HOMOLOGUES

There are two genes that encode Gβ homologues in *C. elegans* versus five in mammals, and one each in the yeast *Saccharomyces cerevisiae* and in *Dictyostelium*. The first one to be identified, *gpb-1*, was identified based on its homology to mammalian G protein β subunits and reveals a very specific role for G proteins in early cell divisions of the embryo. The second, *gpb-2*, has also been identified based on its homology to mammalian G protein β subunits. Interestingly, GPB-2 shares the highest level of amino acid identity with mammalian $G\beta_5$, which is highly expressed in the mammalian brain (Betty *et al.*, 1998). $G\beta_5$ has recently been shown to interact specifically with certain RGS proteins to regulate their function (Snow *et al.*, 1998; Levay *et al.*, 1999).

GPB-1 shares 86% amino acid identity with mammalian Gβ subunits (Van der Voorn *et al.*, 1990). Mutations that inactivate *gpb-1* are lethal, with an early larval arrest phenotype similar to that of *gsa-1* mutants. Disruption of both zygotic and maternal *gpb-1* in genetic mosaic animals reveals that in the absence of maternal rescue, inactivation of *gpb-1* leads to abnormalities early in embryogenesis. In such animals, defects are observed in the stereotypic orientations of the mitotic spindle during early embryonic divisions. Mosaic rescue of *gpb-1* reveals sterile phenotypes in some adult animals and an abnormal germ line, suggesting that the *gpb-1* may also play a role in germ line development. Overexpression of *gpb-1* produces lethargic locomotion, and causes animals to retain late-stage eggs (Zwaal *et al.*, 1996). These effects are similar to the effects elicited by overexpression of *goa-1* (Mendel *et al.*, 1995; Segalat, Elkes and Kaplan, 1995), consistent with GOA-1 and GPB-1 acting together.

Immunolocalization of GPB-1 is consistent both with its roles in early cell divisions and for a role in adult animals (Zwaal *et al.*, 1996). GPB-1 is present at all developmental stages. GPB-1 is present in early embryos and seen as early as the one-cell stage. GPB-1 is detected most strongly at membranes, with staining concentrated at the contact between cells. Staining also colocalizes with the asters (arrays of microtubules emanating from the centrosomes) just before and during early cell divisions. Once tissue differentiation occurs, staining is brightest in neuronal cells. In larval and adult stages, GPB-1 expression is seen in most or all neurons, including the nerve ring, the dorsal and ventral cords, the preanal ganglion, the somatic gonad, vulva and hypodermal seam cells. The intestine, pharynx and body wall muscles appear to have a low level of staining. *gpb-1* is expressed throughout development in nearly all somatic tissues and in the germ line (Zwaal *et al.*, 1996).

The phenotypic effects of a *gpb-1* knockout and its expression pattern implicate GPB-1 as an integral component of signal transduction pathways mediated by possibly all of the Gα subunits. The role for GPB-2 might be quite different. In mammals, the role for $G\beta_5$ is not well understood. Earlier reports suggest that it can interact with $G\gamma_2$ in Sf9 insect cells and the $G\beta_5\ \gamma_2$ complex was shown to interact specifically with $G\alpha_q$ *in vitro* (Fletcher *et al.*, 1998). $G\beta_5\ \gamma_2$

dimers have also been shown to be less effective than $G\beta_5\gamma_1$ dimers at activating type II adenylyl cyclase when expressed in Sf9 cells, and $G\beta_5\gamma_1$ dimers modulated the ability of $G\alpha_q$ to couple to the mammalian ET_B receptor *in vitro* (Lindorfer *et al.*, 1998). More recent reports implicate $G\beta_5$ as a regulator of signal transduction pathways by interacting with RGS proteins that contain a GGL (G-gamma-like) domain (Snow *et al.*, 1998; Levay *et al.*, 1999). The weak Egl-d and lethargic phenotypes (personal communication, Sander van der Linden and Ronald Plasterk) conferred by a knockout allele of *gpb-2* might indicate that GPB-2 is involved in the regulation, rather than being an integral component, of G protein-mediated pathways involved in egg-laying and movement. There are two, previously discussed, RGS proteins expressed in *C. elegans* that contain regions with high similarity to GGL domains, namely, EGL-10 and EAT-16. Both of these genes are involved in the regulation of egg-laying behaviour and movement (Trent, Tsung and Horvitz, 1983; Koelle and Horvitz, 1996; Hadju-Cronin *et al.*, 1999) and so might conceivably interact with GPB-2 to regulate these behaviours.

2.6. OTHER *C. ELEGANS* Gα PROTEINS

The *C. elegans*-specific Gα subunit genes are not clearly related to any other Gα genes expressed in the yeast, *Dictyostelium* or in plants (Wilkie and Yokoyama, 1994). In general, G protein alpha subunits in the same class share >80% identity and G protein alpha subunits that are 40–50% identical have not been observed to interact with the same receptor or effector and are considered to have distinct functions (Lochrie *et al.*, 1991). Given their divergence from G protein alpha subunits in other organisms, the *C. elegans*-specific alpha subunits were probably acquired after the nematode lineage diverged from the vertebrate lineage. Genomic clustering and sequence comparisons also indicate that these Gα subunits were probably not acquired by gene duplication (see Wilkie, 1999).

The first *C. elegans*-specific G proteins to be identified, *gpa-1*, *gpa-2* and *gpa-3*, were cloned because of their homology with mammalian Gα genes (Fino-Silva and Plasterk, 1990; Lochrie *et al.*, 1991). Like GOA-1, GPA-1 and GPA-3 retain the carboxyl cysteine that, in mammalian $G\alpha_i$ subunits, is ADP-ribosylated by pertussis toxin (Gilman, 1987). In addition, the majority of the intron/exon boundaries are identical between *goa-1*, *gpa-1*, *gpa-2* and *gpa-3* and between the G_i class (Lochrie *et al.*, 1991). Using reverse genetics, *gpa-2* and *gpa-3* have been found to regulate entry and exit from an alternative larval stage, the dauer larvae (Zwaal *et al.*, 1997). ODR-3 is a G_i/G_o-like (47% amino acid identity with G_{i2}) Gα protein that was identified in forward genetic screens for animals that failed to be attracted to benzaldehyde (Roayaie *et al.*, 1998).

The identification of the remaining Gα subunits was facilitated by sequence information obtained from the *C. elegans* Genome Sequencing Project and their expression was characterized using *GFP* as a reporter gene (Jansen *et al.*, 1999). Loss-of-function and constitutively active mutants were generated to assess function (Jansen *et al.*, 1999). GPA-7, which shares 57% sequence identity with $G\alpha_o$, is the only *C. elegans*-specific $G\alpha$ subunit that displays a wide pattern of expression (see Table 1). *gpa-7* is expressed throughout the nervous system and in muscle cells and, like the *C. elegans* homologues of mammalian G proteins, appears to be involved in neuromuscular regulation (Jansen *et al.*, 1999, and see Table 1 for summary). Mutations in *gpa-7* confer phenotypic effects solely on egg-laying and these effects are opposite from mutations in *goa-1*. Null mutations in *gpa-7* cause worms to lay eggs at a developmental stage slightly later than wild-type animals. Overproduction of constitutively active GPA-7 confers reciprocal phenotypes (Jansen *et al.*, 1999). *gpa-7* null mutants are defective in aversion to 4% SDS. Worms that overproduce GTPase-deficient GPA-7 have multiple chemo-attractive and chemo-aversive defects (Jansen *et al.*, 1999).

Interestingly, there are two other G protein alpha subunits that share high sequence similarity with the $G\alpha_i/G\alpha_o$ family. Like GPA-7, they contain consensus myristoylation signals at the amino terminus and arginine residues that might be substrates for cholera toxin. However, unlike mammalian $G\alpha_i/G\alpha_o$ proteins, the predicted proteins do not contain C-terminal cysteines that might be substrates for pertussis toxin. GPA-16, which is 65% identical with $G\alpha_i$ and 61% identical to $G\alpha_o$, is expressed in a few sensory neurons, and is weakly expressed in the pharynx, the body wall muscle and in the vulval muscle (Jansen *et al.*, 1999). GPA-4, which shares 62% sequence identity with $G\alpha_i$ and 51% identity with $G\alpha_o$, is reported to be expressed specifically in the ASI amphid neurons (Jansen *et al.*, 1999). No phenotypic effects were reported for knockout alleles of *gpa-4* and *gpa-16*. However, it remains to be seen whether these genes share redundant functions with each other or with *goa-1*.

GFP reporter gene fusions indicate that the remaining *C. elegans*-specific $G\alpha$ subunits are expressed almost exclusively in a small subset of the chemosensory neurons in *C. elegans* (Jansen *et al.*, 1999). Therefore, they are most likely to be involved in sensory perception (of volatile attractants and repellents, water-soluble attractants and repellents, dauer pheromone, nose touch, temperature and mating partners). Some genes also showed expression in phasmid neurons, which are presumed to be sensory neurons. In addition to the expression seen in hermaphrodites, male-specific expression was seen for four of the *C. elegans*-specific G alpha subunits. A knockout in each of the *C. elegans*-specific $G\alpha$ subunits does not produce a lethal phenotype, as would be predicted from expression pattern analysis (Jansen *et al.*, 1999).

Chemosensory neurons make up about 10% of the *C. elegans* nervous system. Three types of sensory organs in the hermaphrodite appear to be chemosensory, based on their associations with openings in the cuticle. They are the bilaterally symmetrical amphids in the animal's head, the bilaterally symmetrical phasmids in the animal's tail and the six inner labial sensory organs in the head (see Bargmann and Mori, 1997). The major chemosensory organs in *C. elegans* are the bilateral amphid sensilla located at the anterior tip of the worm. Each amphid contains 12 neurons that have been subdivided according to their morphology. Neurons that have cilia directly exposed to the environment detect mostly water-soluble chemicals (Bargmann, Thomas and Horvitz, 1990; Bargmann and Horvitz, 1991a,b; Kaplan and Horvitz, 1993; Troemel *et al.*, 1995). Neurons that have cilia near the amphid pore, but are enclosed by the amphid sheath cells, detect volatile chemicals (Bargmann, Hartweig and Horvitz, 1993) and one type of neuron has a sensory ending embedded in the sheath cell that detects thermal cues (Mori and Ohshima, 1995). Chemosensory neurons can respond independently to different stimuli perceived within the same cell (Bargmann, 1993; Bargmann, Hartweig and Horvitz, 1993; Colbert and Bargmann, 1995), indicating that multiple signalling pathways occur within the same cell.

As single mutations, phenotypic effects from loss-of-function alleles were observed for only five other *C. elegans*-specific subunit genes. Defects were observed in chemotaxis to water-soluble compounds, avoidance behaviour and olfaction. In contrast, eight of ten of the gain-of-function mutants tested showed defects in the above assays and two of these mutants exhibited reduced male potency. Some loss-of-function alleles did show interactions with *odr-3* mutations. *odr-3* mutants are mildly defective in butanone chemotaxis, but animals that are doubly mutant for *odr-3* and *gpa-2* are severely defective in butanone chemotaxis. *odr-3* mutant animals are defective in 2,4,5-trimethyl thiazole (tmt) perception, while *odr-3*; *gpa-5* double mutants have a chemotaxis index similar to that in wild-type worms (Jansen *et al.*, 1999). Therefore, *gpa-5* might negatively regulate tmt perception of the AWA neurons.

Consequently, it is clear that there are multiple signal transduction pathways that occur within the same chemosensory neuron in *C. elegans*. Furthermore, the interaction between *odr-3* and *gpa-5* provides evidence for interaction between these signalling pathways (Jansen *et al.*, 1999). Since chemosensory neurons are not required for viability, it is possible to ablate neurons with redundant functions and examine simple chemosensory behaviours in the context of different genetic backgrounds. In this way, both maintenance of signal specificity, as well as cross-talk between signalling pathways, can be examined in exquisite detail.

2.6.1. *gpa-2 and gpa-3*

gpa-2 and *gpa-3* share 45–55% identity with all members of the mammalian subclasses. These two *C. elegans*-specific G protein alpha subunits appear to regulate entry into an alternative larval stage, called the dauer phase. Sensory signals regulate entry into the dauer phase, which can substitute for the third larval stage (Cassada and Russell, 1975). Dauer larvae are formed under adverse conditions of low food, high nematode density and high temperature (Golden and Riddle, 1984a; Riddle, 1988; Thomas, Birnby and Vowels, 1993). Dauer larvae are specialized for survival and dispersal. They are resistant to drying and high temperature, are arrested in development and aging, have an altered energy metabolism and do not feed. They have an extended life span compared to non-dauer animals. The presence of dauer-inducing pheromone is both necessary and sufficient for entry into the dauer phase (Golden and Riddle, 1982, 1984b). The decision to enter dauer phase results from integration of both chemosensory and thermosensory signals, which are recognized and transduced through the ciliated chemosensory neurons of the amphids (see Table 2). The dauer stage is completely reversible. Animals resume development when circumstances are favourable (Cassada and Russell, 1975; Klass and Hirsh, 1976). Recovery from the dauer phase involves the same environmental cues that also control dauer formation (Golden and Riddle, 1982, 1984a,b).

Loss of both *gpa-2* and *gpa-3* results in a diminished response to dauer pheromone (Zwaal *et al.*, 1997). Double mutants are more resistant to pheromone than each single mutant. However, double mutants are not completely resistant, so there is another functional pheromone response pathway. Constitutive activation of *gpa-2* and *gpa-3* induces dauer larvae formation in the absence of pheromone and renders worms insensitive to the exogenous addition of pheromone. However, appropriate responses to food and temperature are retained in these double mutants, because restricting food and elevating temperature led to the formation of more dauer larvae.

Exit from the dauer stage is repressed by dauer pheromone in wild type. However, dauer larvae lacking *gpa-2* were able to recover in the presence of dauer pheromone. This implicates *gpa-2* in both recovery from the dauer phase as well as entry into the dauer phase (Zwaal *et al.*, 1997). Both genes are expressed in chemosensory neurons, and both interact genetically with the many genes known to affect dauer formation (Riddle, Swanson and Albert, 1981; Swanson and Riddle, 1981; Georgi, Albert and Riddle, 1990; Vowels and Thomas, 1992; Estevez *et al.*, 1993; Thomas, Birnby and Vowels, 1993). Since expression of *gpa-2* and *gpa-3* do not overlap, it is probable that the dauer-defective (Daf-d) phenotypes of *gpa-2* and *gpa-3* mutants are exerted in parallel by different neurons (Zwaal *et al.*, 1997). The increased longevity of dauer larvae has established this alternative developmental programme as a model for understanding the aging process. How different signalling pathways are integrated to regulate entry into, and exit from, dauer formation is expected to have wide-ranging applications.

2.6.2. *ODR-3 and its Candidate Receptor, ODR-10*

ODR-3 is a G_i/G_o-like (47% amino acid identity with G_{i2}) Gα protein that was identified in forward genetic screens for animals that failed to be attracted to benzaldehyde (Bargmann, Hartweig and Horvitz, 1993; Roayaie *et al.*, 1998). *odr-3* null mutants are defective in chemotaxis to all volatile attractants sensed by the AWA and AWC neurons and are defective in the osmotic avoidance responses mediated by ASH (see Table 2). However, high concentrations of these odourants are still detected, indicating that there is functional redundancy with a distinct signal transduction pathway. The ODR-3 protein is localized to the sensory cilia. In addition to its role in chemosensory function, ODR-3 is also required for the formation and maintenance of the fan-like structure of the AWC olfactory cilia. There is no previous evidence for G protein function in cilium development, but G proteins in other systems can contribute to cell polarity and morphogenesis (Klotz and Jesaitis, 1994; Parent and Devroetes, 1996; Zwaal *et al.*, 1996). Since the olfactory function of the AWC olfactory neurons requires *odr-3*, a transmembrane guanylyl cyclase encoded by *daf-11* (Birnby *et al.*, 2000), and a cyclic nucleotide-gated channel encoded by *tax-2* and *tax-4* (Coburn and Bargmann, 1996; Komatsu *et al.*, 1996), it is possible that ODR-3 regulates cyclic nucleotide production or degradation (Roayaie *et al.*, 1998).

odr-10 mutants were identified in forward genetic screens based on an inability of the AWA neurons to sense diacetyl, but were not affected in their ability to sense other AWA-sensed odourants. *odr-10* encodes a predicted seven-transmembrane receptor that is expressed at high levels in the sensory cilia of the AWA neurons. Overexpression of *odr-10* in an animal that has lost its sensitivity to several odourants results in a specific restoration of the diacetyl response (Sengupta, Chou and Bargmann, 1996). Expression of *odr-10* in *odr-7* mutant animals, which are defective in the sensation of most AWA-specific odourants, specifically restores sensitivity to diacetyl. Expression of *odr-10* in human cells confers sensitivity to this odourant (Zhang *et al.*, 1997), indicating that ODR-10-mediated signalling pathways are most likely conserved in vertebrates. It is possible that ODR-3 is directly coupled to the ODR-10 receptor.

2.7. GENETIC ANALYSIS OF INTERACTIONS BETWEEN G PROTEIN SIGNALLING PATHWAYS

2.7.1. *goa-1, eat-16, dgk-1/sag-1, and egl-30*

Mutagenesis screens designed to identify and characterize genes that function in a pathway with heterotrimeric GTP-binding subunits have also identified candidates required for regulatory cross-talk between different G protein signalling pathways. In a screen for mutants that can suppress the egg-laying defective and lethargic phenotypes conferred by overexpression of activated GOA-1, loss-of-function alleles of *eat-16*, a gene previously identified as a regulator of pharyngeal pumping behaviour (Avery, 1993), were isolated (Hadju-Cronin *et al.*, 1999). *eat-16* loss-of-function mutants resemble *goa-1* loss-of-function mutants and *egl-30* gain-of-function mutants with respect to early egg-laying behaviour and hyperactive locomotion phenotypes.

Analysis of animals defective in both *eat-16* and *egl-30* indicates that *eat-16* is a negative regulator of *egl-30*. *eat-16* can partially suppress the effects of a reduction-of-function allele of *egl-30*, but cannot suppress the effects of a putative null allele of *egl-30*. These results are consistent with the hypothesis that EGL-30 is a direct target of negative regulation by EAT-16, an RGS protein. In support of this hypothesis, EAT-16 can reduce activation of PLCβ activity by endogenous Gα_q in transfected COS-7 cells (Hadju-Cronin *et al.*,

1999). Analysis of the *eat-16* expression pattern using partial *eat-16* translational fusions with GFP indicate that the expression patterns of *goa-1*, *eat-16* and *egl-30* overlap and thus could interact in the same cells. *eat-16* is expressed in most or all neurons and in the vulval and pharyngeal muscles (Hadju-Cronin *et al.*, 1999).

Epistasis of *egl-30* over *goa-1* indicates that *goa-1* and *egl-30* might function in a linear pathway, with *egl-30* functioning downstream of *goa-1*. The reciprocal phenotypes of *goa-1* and *egl-30* mutants then implicate GOA-1 as a negative regulator of EGL-30 signalling. Since *eat-16* was isolated as a suppressor of activated GOA-1, it is either downstream of *goa-1*, or it acts in parallel to *goa-1*. Finally, since biochemical and genetical analyses implicate EAT-16 as a negative regulator of EGL-30, it is conceivable that GOA-1 negatively regulates the EGL-30 pathway via activation of EAT-16.

Mutations in *dgk-1*, a gene that encodes diacylglycerol kinase (Hadju-Cronin *et al.*, 1999; Nurrish, Segalat and Kaplan, 1999), were also recovered as suppressors of activated GOA-1 (Hadju-Cronin *et al.*, 1999) and independently as suppressors of serotonin-induced inhibition of locomotion (Nurrish, Segalat and Kaplan, 1999). *dgk-1* mutants have similar phenotypes to *eat-16* mutants, but although they completely suppress locomotion defects conferred by overexpression of activated *goa-1*, these mutations only partially suppress the egg-laying defect (Hadju-Cronin *et al.*, 1999) and mutations do not suppress response to serotonin in egg-laying assays (Nurrish, Segalat and Kaplan, 1999). Therefore, pharmacological and double mutant analyses suggest that *dgk-1* functions downstream of *goa-1* and serotonin with respect to movement (Hadju-Cronin *et al.*, 1999; Nurrish, Segalat and Kaplan, 1999), but not with respect to egg-laying behaviour. *dgk-1* is also able to partially suppress the aldicarb resistance of *egl-30* mutants, suggesting that *dgk-1* acts antagonistically to *egl-30*, downstream of *egl-30*, or in parallel to *egl-30* (Miller, Emerson and Rand, 1999). Therefore, GOA-1 might negatively regulate the EGL-30 signalling pathway via two previously unpredicted molecular targets, EAT-16 or DGK-1 (see Fig 5).

2.7.2. *From Pharmacology to Genes to Molecules: Players in Neurotransmission*

When wild-type worms are treated with aldicarb, it causes hypercontracted paralysis, as well as cessation of growth, feeding, development and reproduction (Rand and Russell, 1995). In a screen for mutants that are resistant to growth and reproductive inhibition by aldicarb (an inhibitor of acetylcholinesterase that hydrolyzes acetylcholine after its release into the synaptic cleft), mutations in genes (Ric mutants, resistant to inhibitors of acetylcholinesterase) which are potentially involved in G protein signalling pathways were uncovered. They were *unc-13*, *egl-10*, *egl-30* and *egl-8* (Nguyen *et al.*, 1995; Miller *et al.*, 1996). Mutations in *egl-8* were further characterized because of the fact that they share phenotypes in common with *egl-30* reduction-of-function mutants, such

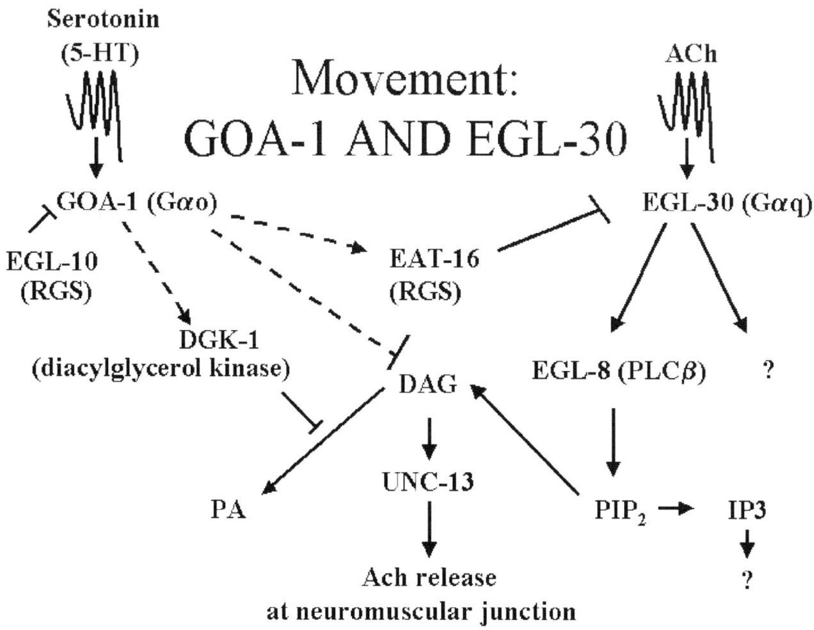

Figure 5. Proposed interactions between EGL-30 and GOA-1 signalling pathways in *Caenorhabditis elegans* with respect to movement. Dashed lines indicate interactions that have been proposed based on genetic analysis.

as lethargic movement and egg-laying defective phenotypes (Miller, Emerson and Rand, 1999). Cloning of *egl-8* revealed that *egl-8* encodes the only PLCβ homologue in the worm and consequently is a likely direct effector of EGL-30 (Miller, Emerson and Rand, 1999). As discussed previously, it is unlikely that EGL-8 is the only direct effector for *egl-30*. *egl-30* loss-of-function mutants have much more severe locomotion and egg-laying defective phenotypes than *egl-8* loss-of-function mutants (Miller, Emerson and Rand, 1999) and *egl-8* loss-of-function mutants suppress only very weakly the egg-laying constitutive phenotypes of *egl-30* gain-of-function mutations (Bastiani, in prep.). Expression patterns of *egl-30* and *egl-8* do overlap in the nervous system. *egl-8* is expressed in most neurons. *egl-8* is also expressed near intestinal cell junctions. The expression of *egl-8* in the intestine could be related to its role in the posterior body wall muscle contraction step of the defaecation cycle (Thomas, 1990), a phenotype not shared with *egl-30* mutants.

A role for serotonin as a neuromodulator was investigated because serotonin and acetylcholine have opposite effects on movement in worms (Nurrish, Segalat and Kaplan, 1999) and because serotonin has been shown to act as a neuromodulator in other systems. It was speculated that serotonin might act as a neuromodulator and inhibit movement by inhibiting release of the stimulatory neurotransmitter, acetylcholine, by the ventral cord motor neurons. In support of this model, when wild-type animals are treated with serotonin, they become resistant to aldicarb (Nurrish, Segalat and Kaplan, 1999). However, serotonin did not affect the ability of muscles to respond to levamisole, a nicotinic acetylcholine agonist.

goa-1 and *dgk-1* mutants, both isolated in screens for suppressors of inhibition of locomotion by serotonin (Segalat, Elkes and Kaplan, 1995; Nurrish, Segalat and Kaplan, 1999), were then examined with respect to aldicarb response (Nurrish, Segalat and Kaplan, 1999). All of these mutants are hypersensitive to aldicarb, which is consistent with an increase in acetylcholine release in these mutants. A possible link between GOA-1 and acetylcholine release was identified when it was discovered that mutants that lack GOA-1 accumulate abnormally high levels of UNC-13 at motor neuron nerve terminals. This suggests that serotonin couples to GOA-1 and inhibits synaptic transmission by decreasing the abundance of UNC-13 at release sites (Nurrish, Segalat and Kaplan, 1999).

C. elegans unc-13 is important for normal neurotransmitter release, as is the mammalian homologue, Munc-*13*. Mutations in *unc-13* result in severely uncoordinated movement, increased accumulation of the neurotransmitter acetylcholine and resistance to acetylcholine inhibitors (Brenner, 1974; Rand and Russell, 1985; Hosono, Sassa and Kuno, 1989; Nguyen *et al.*, 1995; Miller *et al.*, 1996). *unc-13* mutants have more synaptic vesicles associated with both the GABAergic and cholinergic presynaptic specializations (Richmond, Davis and Jorgensen, 1999). Mutants do not respond post-synaptically to depolarization of the ventral cord motor neurons and exhibit a severe decrease in the frequency of endogenous spontaneous currents in the muscle that are induced from both GABAergic and cholinergic transmission. Since *unc-13* mutants retain the same percentage of docked vesicles relative to the total number of vesicles present at the synapse, it was inferred that in *unc-13* mutants docked vesicles are defective in the fusion step with the plasma membrane (Richmond, Davis and Jorgensen, 1999).

UNC-13 contains one C1 and two C2 homology domains which enable the protein to bind DAG and phorbol esters and to bind phospholipids in a calcium-dependent manner (Maruyama and Brenner, 1991; Ahmed *et al.*, 1992; Kazanietz *et al.*, 1995). The presence of phorbol esters induces overexpressed Munc-*13* to become associated with the plasma membrane in *Xenopus* motor neurons (Betz *et al.*, 1998). It is presumed that *goa-1* decreases the abundance of DAG at nerve terminals, because increasing levels of DAG with exogenous phorbol esters restores aldicarb sensitivity to *goa-1* mutant worms. Also, by mutating a conserved histidine residue within the C1 domain of UNC-13, which is required for binding phorbol esters, it was shown that localization of *unc-13* at nerve terminals in *goa-1* knock-out animals is dependent upon interaction with DAG.

Conversely, as discussed, *egl-30* and *egl-8* mutants are also resistant to the effects of aldicarb (Miller *et al.*, 1996; Miller, Emerson and Rand, 1999). Worms that express a GTPase-deficient form of *egl-30* under the control of a promoter that confers expression in ventral cord motor neurons are hypersensitive to aldicarb (Lackner, Nurrish and Kaplan, 1999). Mutations in *egl-8* can partially suppress this effect. Consistent with this observation, mutations in *egl-8* can partially suppress the hyperactive movement conferred by *egl-30* gain-of-function mutations (Bastiani, in prep.). *egl-30*, like *goa-1*, affects localization of UNC-13 at synaptic terminals. Treatment of worms with arecoline, an acetylcholine agonist, results in a 40% increase in *unc-13* associated with punctate structures in the nerve terminals of the ventral cord (Lackner, Nurrish and Kaplan, 1999). However, this effect is not apparent when *egl-30* reduction-of-function mutants are treated with arecoline.

These studies indicate a pathway and a common mechanism by which EGL-30 and GOA-1 antagonistically regulate movement. GOA-1 decreases the abundance of DAG at nerve terminals (Nurrish, Segalat and Kaplan, 1999), either by directly suppressing the EGL-30 pathway via EAT-16 (Hadju-Cronin *et al.*, 1999), or indirectly via activation of DGK-1. Conversely, activation of the EGL-30 pathway leads to accumulation of DAG at nerve terminals via activation of PLCβ. Accumulation of DAG localizes UNC-13 to nerve terminals, whereas depletion of DAG prevents localization of UNC-13 at nerve terminals and prevents fusion of synaptic vesicles with the plasma membrane.

However, it is clear that this is not the only pathway or mechanism to explain antagonism between *goa-1* and *egl-30*. Mutations in *egl-8* only partially suppress hypersensitivity to aldicarb conferred by a GTPase-deficient allele of *egl-30* (Lackner, Nurrish and Kaplan, 1999) and only partially suppress the hyperactive locomotion of *egl-30* gain-of-function mutants. Furthermore, the pathway for egg-laying behaviour appears to be distinct, because the effects of serotonin on egg-laying are opposite from the effects of *goa-1* on egg-laying. Mutations in *egl-8* very weakly suppress the early egg-laying phenotype of *egl-30* gain-of-function mutants (Bastiani, in prep.). Characterization of *dgk-1* mutants again reveals that there are distinct pathways for egg-laying behaviour and movement, because *dgk-1* mutants can respond to serotonin in egg-laying assays, but are insensitive to serotonin with respect to movement (Nurrish, Segalat and Kaplan, 1999).

It is significant that analyses of single behaviours in *C. elegans* has allowed for the study of signal integration and will allow for the dissection of parallel pathways downstream of a single G protein. It is clear that a single G protein, EGL-30, regulates behaviours via multiple signalling pathways. Additional suppressor screens and epistasis analysis with other mutants that affect these simple behaviours in *C. elegans* will undoubtedly lend further insight into the complexity of these signalling pathways as related to behaviour.

2.7.3. *Serotonin, Protein Kinase C and Egg Laying*

By careful analysis of egg-laying patterns, it has been possible to assign specific roles to neurons and to neurotransmitters that affect egg-laying behaviour (Waggoner *et al.*, 1998). Specifically, animals appear to fluctuate between discrete behavioural states: an active state, during which eggs are laid in clusters, and an inactive state, during which eggs are retained. Serotonin appears to increase the rate of egg-laying by switching the animal from a quiescent to an active egg-laying phase. This switch appears to be regulated by the HSN neurons and may be mediated through a protein kinase C-dependent (PKC-dependent) signalling pathway (*egl-30*, *egl-4*, *egl-11*, *egl-21*, *egl-24* also are implicated in this response; see Waggoner *et al.*, 1998). On the other hand, acetylcholine was identified as a candidate neurotransmitter that might induce individual egg-laying events within the active phase.

egl-19, a gene that encodes an $\alpha 1$ subunit of an L-type voltage-gated calcium channel, was identified as a gene which might be required both for initiation of the active phase and for stimulation of individual egg-laying events. In vertebrate cells, G protein-coupled receptors have been shown to facilitate PKC-dependent phosphorylation of L-type calcium channels, which enhances the coupling between excitation and channel opening (Boland, Allen and Dingledine, 1991). It is possible that PKC phosphorylates EGL-19 in response to serotonin, making channels more likely to open in response to voltage changes. PKC has been shown to be downstream of $G\alpha_q$ in mammals. The overlap between phenotypes conferred by reduction in EGL-30 function and loss of PKC suggests that they might function in a common pathway for egg-laying (see Fig. 6 for a tentative model).

2.7.4. *lfe-1/itr-1/dec-4*

Mutations in the *C. elegans* IP_3 receptor homologue have been independently identified in three forward genetic screens. The *C. elegans* protein is 38% identical to the mouse type 1 IP_3 receptor and it is the only IP_3 receptor sequence present in the *C. elegans* genome. It contains the three functional

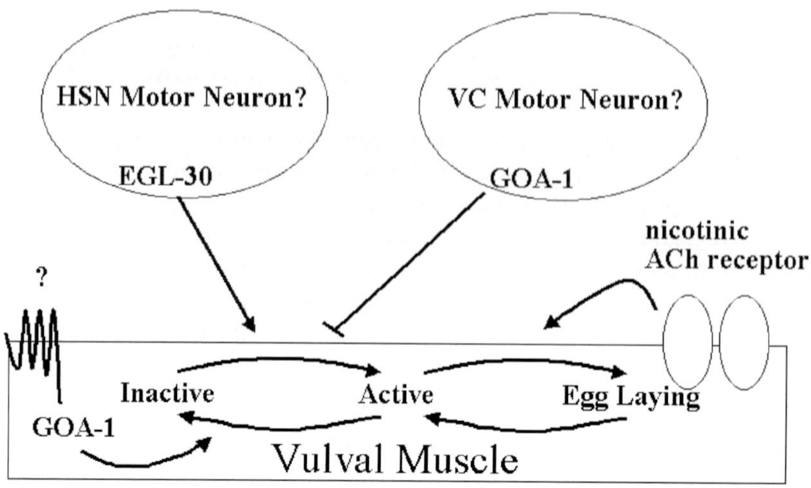

Figure 6. Proposed roles for EGL-30 and GOA-1 on egg-laying behaviour in *Caenorhabditis elegans.*

domains present in mammalian IP_3 receptors: an amino-terminal ligand binding domain, a central modulatory domain and six carboxyl-terminal membrane-spanning regions (Mikoshiba, Furuichi and Miyawaki, 1994).

Mutations in the *C. elegans* IP_3 receptor were originally identified in a screen for mutations that could suppress the sterility caused by a *lin-3* reduction-of-function mutation (Clandinin, DeModena and Sternberg, 1998). *lin-3* encodes a *C. elegans* epidermal growth factor (EGF)-like protein and *let-23* encodes its cognate receptor tyrosine kinase (see Sternberg and Han, 1998 for review). As will be reviewed later, *lin-3/let-23* have multiple functions in *C. elegans* which are Ras-dependent. These Ras-dependent functions can be rescued in a *let-23* null background by a gain-of-function mutation in the Ras homologue, *let-60*. However, since Ras cannot rescue sterility caused by *lin-3* and *let-23* mutations, it was presumed that there is a tissue-specific Ras-independent pathway downstream of *lin-3/let-23* for fertility.

A screen for extragenic suppressors of *lin-3* sterility identified gain-of-function mutations in *lfe-1* (Clandinin, DeModena and Sternberg, 1998). The mutations are strong semi-dominant suppressors of *lin-3* sterility and can bypass the requirement for *let-23* in the gonad. The mutations suppress defective ovulation caused by *let-23* mutations. Specifically, proper dilation of the spermatheca, where fertilization in hermaphrodites takes place, is rescued by *lfe-1* gain-of-function mutations. *lfe-1* mutants are wild-type for viability and vulval induction. However, they display slightly reduced fertility. By analogy with other systems, LET-23-mediated signalling probably promotes spermathecal dilation by increasing intracellular calcium levels via a PLCγ-like homologue (reviewed in Berridge, 1993).

In a screen for mutants with abnormal defaecation cycles, a reduction-of-function allele in *lfe-1* was isolated (Iwasaki, Liu and Thomas, 1995). Calcium oscillations occur in the intestine with peak calcium levels just prior to the first muscle contraction of the defaecation cycle. IP_3 receptor-mediated calcium release in a single cell type, the intestine, coordinates the multicellular responses of defaecation (Dal Santo *et al.*, 1999). The IP_3 receptor is an essential component of the timekeeper for the defaecation cycle and demonstrates a novel mechanism for the control of behavioural rhythms. Mosaic analysis indicates that IP_3 receptor expression in the intestinal cells is necessary and sufficient to provide normal behavioural rhythms (Dal Santo *et al.*, 1999).

In another screen for lethal mutations, a null allele of *lfe-1* was isolated (see Dal Santo *et al.*, 1999). This mutation is not lethal, but these mutants develop at a slow rate, are sterile, have no observable defaecation cycle and so are constipated. These animals actively forage, the pharynx pumps in the presence of food and locomotion is comparable to wild-type animals of the same age. A transcriptional *GFP* fusion indicates expression in the intestine, the rectal epithelial cells, the pharynx, the somatic gonad and vulval hypodermal cells (Dal Santo *et al.*, 1999).

2.7.5. *Flp-1*

C. elegans also expresses neuropeptide signalling molecules (Rosoff, Burglin and Li, 1992; Rosoff *et al.*, 1993; Nelson *et al.*, 1998). FMRFamide-related neuropeptides (FaRPs) represent a large family of peptides implicated as neurotransmitters or neuromodulators in many invertebrate and vertebrate behaviours, including muscular control, cardioregulation, pain modulation and learning (Wong, Greenberg and Tse, 1985; Mackey *et al.*, 1987; Small, Kandel and Hawkins, 1989; Kavaliers, 1990; Kavaliers and Colwell, 1993; and for reviews see Raffa, 1988; Maule *et al.*, 1996). The FMRFamide (Phe-Met-Arg-Phe-amide)-related neuropeptide gene family in *C. elegans* is composed of at least 18 genes that may encode 53 distinct FMRFamide-related peptides. Disruption of one of these genes, *flp-1*, causes numerous behavioural defects in mutant worms, including movement body bends, hyperactivity, osmotic avoidance defects, slight reduction in sensitivity to nose touch and insensitivity to the exogenous effects of serotonin on movement (Nelson, Rosoff and Li, 1998). The movement phenotypes conferred by *flp-1* mutants are reminiscent of *goa-1* loss-of-function mutants. Overexpression of *flp-1* results in phenotypes that are reciprocal with loss-of-function phenotypes. Double mutant analysis suggests that G protein pathways are epistatic to *flp-1* mutants (Nelson, Rosoff and Li, 1998), so it is possible that G proteins are directly coupled to the FLP-1 receptor. Many FaRPs are likely to signal through a heterotrimeric GTP-binding protein coupled receptor (Voltera and Siegelbaum, 1988; Man Son Hing *et al.*, 1992; Payza and Yang, 1993; Wang, Lange and Orchard, 1995).

2.8. OTHER

2.8.1. *General*

The genome sequence predicts the expression of a large number of genes homologous to mammalian genes known to be involved in heterotrimeric G protein signalling (Robertson, 2000). 5% of all *C. elegans* genes encode G protein-coupled receptors. There are predicted to be 20 neuropeptide genes, more than thirty PDZ domain proteins, 12 RGS regulators, 3 adenylyl cyclases, 8 PLCs, and 4 cGMP PDEs. There are a plethora of potential G protein-coupled receptor-encoding genes: there are 18 class A amine receptors, 50 class B peptide receptors, 4 metabotropic glutamate receptors, 3 GABA-B receptors. Interestingly, there are an amazing number of predicted G protein coupled orphan receptors, 970 in all. There are also predicted to be a total of 31 potential receptor or soluble guanylyl cyclase genes. 6 cyclic nucleotide regulated channels, 11 TRP-related channels, one CREB and many CREB regulatory pathway genes (see Bargmann, 1998).

2.8.2. *7-TM Receptors*

Most worm-specific G protein-coupled receptors are thought to encode chemoreceptors, while the others are thought to encode neurotransmitter receptors (Bargmann, 1998). In the case of olfactory receptors, although the *C. elegans* and vertebrate olfactory receptors are relatively unrelated in sequence, they have similar properties. In addition to the *odr-10* family (*str*), three further families of candidate chemosensory receptor genes have been identified (*sra*, *sro*, and *srg* gene families), bringing the total number of functional chemoreceptor genes to about 800 (Troemel *et al.*, 1995; Sonnhammer and Durbin, 1997; Robertson, 1998, 2000). The chemosensory receptors are often found in clusters that contain as many as 10 or 15 closely related genes. A single neuron can express at least four *sr* genes, consistent with the observation from cell-killing experiments that single neurons detect many chemicals (Bargmann and Horvitz, 1991a; Bargmann, Hartweig and Horvitz, 1993; Troemel *et al.*, 1995). The *sr* genes are expressed in chemosensory neurons that detect water-soluble attractants, pheromones, volatile repellents and water-soluble repellents. A few *sr* genes are expressed in the male-specific chemosensory neurons in the male mating structures, which suggests that the *sr*'s might detect signals involved in mating. Candidate receptor genes are in clusters of two to nine similar genes (Troemel *et al.*, 1995). Some of them might arise from the same transcript. Linked genes are the most closely related (Troemel *et al.*, 1995), indicating expansion of the genes during the evolution of *C. elegans*.

2.8.3. *NPY Receptor*

Natural variation in one predicted G protein-coupled receptor in *C. elegans* appears to be the basis for natural variation observed in nematode social and feeding behaviour. Wild-type worms of the laboratory strain N2 move slowly and disperse in response to food, but aggregate when food is limiting, and assemble into small clumps as dauer larvae. Mutations in some of the genes involved in dauer formation can cause N2 to aggregate, even in the presence of abundant food, indicating that integration of sensory inputs might regulate the decision to remain solitary or to aggregate (Thomas, Birnby and Vowels, 1993). A regulated swarming response has also been observed in plant-parasitic nematodes, marine nematodes and various rhabditid species, indicating that control of this behaviour is common in nematodes (McBride and Hollis, 1966; Croll, 1970). However, some wild strains of *C. elegans* have been observed to aggregate even in plentiful food and do not slow down in response to food. Failure to respond to food does not appear to stem from an inability to sense food, because these worms appear to have essentially comparable chemoattraction responses to olfactory cues. These strains are considered to be social strains. Two natural isoforms of the receptor NPY-1 have been identified that differ at a single amino acid residue. One isoform is found solely in social strains and the other is found exclusively in solitary strains. Expression of the "solitary isoform" in wild social strains can induce solitary feeding behaviour (de Bono and Bargmann, 1998).

Mutants which contain loss-of-function mutations in the *npy-1* gene induce a clumping response stronger than that of the dauer-constitutive clumping mutants (see de Bono and Bargmann, 1998). The average speed of these worms is indistinguishable from N2 worms in the absence of food, but the speed of their movement is barely reduced in the presence of food. These behaviours resemble behaviours observed in wild social strains, providing further evidence that polymorphisms at this single genetic locus can account for differences in feeding and social behaviours. The difference in their behaviours appears to stem from a difference in response to other animals and does not require a clumping-induced signal from other animals. The two different isoforms differ at NPY-1 residue 215, which in other seven-transmembrane receptors might contribute to the specificity of G protein coupling (Bluml, Mutschler and Wess, 1994a,b; Gudermann, Schoneberg and Schultz, 1997). The "social" NPY-1 isoform probably represents a less active version of the receptor, since high copy expression of this isoform can partially rescue *npy-1* mutants and inhibit social behaviour and/or activate solitary behaviour. GFP fusions are expressed in the nervous system.

NPY-1 is most closely related to the mammalian NPY receptors Y1, Y4 and Y6. Mammalian NPY receptors have been implicated in the control of feeding behaviour, blood pressure, anxiety, memory retention, hormone release and food intake and, in mammals, are most likely coupled to pertussis toxin-sensitive G proteins of the G_i/G_o class (for recent reviews see Colmers and Bleakman, 1994; Blomqvist and Herzog, 1997). The NPY receptor family is defined by six subtypes (Wan and Lau, 1995; Blomqvist and Herzog, 1997) in mammals and the *C. elegans* genome includes at least six potential members of the NPY receptor family besides *npy-1* (de Bono and Bargmann, 1998).

3. Small GTP-Binding Proteins are Involved in Development, Axon Guidance, Neuronal Signalling, Tissue Contraction and Cell Shape

3.1. Ras, MULTIPLE ROLES IN DEVELOPMENT

Ras was identified in *C. elegans* by several genetic screens. Loss-of-function mutations in the gene that encodes Ras, *let-60*, were identified as recessive lethal mutations. Subsequently, more than 30 mutations in *let-60* have been identified in forward genetic screens primarily focused on delineating the genetic pathway for establishing vulval cell fate and pattern formation in *C. elegans*. This pathway has been defined from the LET-23 EGF receptor homologue to nuclear transcription factors and key elements of this pathway are conserved in mammals. *let-60* is expressed broadly during *C. elegans* development and plays a role in many other processes. Analysis of *let-60* mutants, coupled with mosaic analysis, has revealed a number of other roles for *let-60* and in some cases distinct signalling pathways appear to be involved. Ras is important for fate specification of the excretory duct cell, the male spicules and the preanal ganglion. In addition, Ras plays roles in sex-myoblast migration and germ-cell development (see Sternberg and Han, 1998 for review).

3.2. Rho, Rac, Cdc42

Ras-related GTPases of the Rho family (Rho, Rac, Cdc42) function in the receptor-mediated signalling cascades that regulate actin dynamics in a way to provide the cell with polarity and the force for motility (reviewed in Nobes and Hall, 1994; Ridley, 1995; and Hall, 1998). In mammalian cultured cells, Rac induces the formation of lamellipodia (Ridley *et al.*, 1992). Activated Rac affects both axon outgrowth and dendritic spine morphogenesis in mouse Purkinje cells (Luo *et al.*, 1996). Rho is responsible for the formation of stress fibres (Ridley and Hall, 1992; Nobes and Hall, 1995). Cdc42 promotes the formation of filopodia (Kozma *et al.*, 1995; Nobes and Hall, 1995). In non-vertebrate multicellular organisms such as *Drosophila melanogaster*, the Rho subfamily has been shown to be involved in various aspects of actin-dependent morphogenesis, including dorsal (Harden *et al.*, 1995) and epithelial (Eaton *et al.*, 1995) cell shape changes. Mutations in Rho family members can also affect cell migrations and axon outgrowth *in vivo*. Activated or dominant negative forms of Dracl in *Drosophila* truncate axon growth, and activated Dcdc42 inhibits both axon and dendrite outgrowth (Luo *et al.*, 1994). Dominant negative Rac inhibits border cell migration during *Drosophila* oogenesis (Murphy and Montell, 1996).

3.3. Rho/Rac (MIG-2), AXON GUIDANCE AND DEVELOPMENT

Mutations in *mig-2* were originally isolated in genetic screens for mutants with altered cell migrations (Hedgecock *et al.*, 1987; Forrester and Garriga, 1997; Zipkin, Kindt and Kenyon, 1997). MIG-2 is most closely related to Rac and CDC42, but shares the most identity with Rho family members in regions important for GTP hydrolysis and in the putative effector binding domains. Gain-of-function and loss-of-function mutations in *mig-2* inhibit cell migrations of several neuronal cells, as well as some migratory mesodermal cells. Only a small subset of the migrations inhibited by activating mutations (Q lineage and migratory coelomocytes) are inhibited by null mutations, suggesting that there is considerable functional redundancy in this system. Activated alleles also cause misguided axon growth of the HSN neurons. This effect is cell-autonomous, suggesting that Rho family GTPases may couple guidance cues to process outgrowth.

Functional *mig-2::GFP* fusions are first visible in early embryos and at hatching in many cell types that express *mig-2*. It is highly expressed in cells that undergo long-range migrations such as the neurons HSN, CAN and ALM. Throughout development, the fusion protein is clearly localized at the cell periphery and no asymmetric subcellular localization is observed. Isolated extragenic suppressors of activated *mig-2* alleles act in a cell-specific fashion and might lead to the identification of additional Rho family GTPases, as well as components of the biochemical pathways in which they function (see Zipkin, Kindt and Kenyon, 1997).

3.4. UNC-73, A GUANINE NUCLEOTIDE EXCHANGE FACTOR FOR Rac

Cell growth cone migrations in animals are guided by extracellular molecules that provide polarity information to the cell in the form of substrate-bound or diffusable guidance cues (Tessier-Lavigne and Goodman, 1996). *unc-73* is required in migrating and dividing cells for spatial regulation of actin and cytoskeletal polarity that occurs in response to extracellular polarity cues. Mutations in *unc-73* cause many reported defects in cell and axon growth cone migrations and these effects include both pioneer and non-pioneer axon guidance (Hedgecock *et al.*, 1987; Desai *et al.*, 1988; Manser and Wood, 1990; Siddiqui, 1990; Siddiqui and Culotti, 1991; McIntire *et al.*, 1992; Chen, Branda and Stern, 1997; Forrester and Garriga, 1997; Wightman, Baran and Garriga, 1997). A range of phenotypes are exhibited by *unc-73* mutations, from near paralysis to weak uncoordination. Paralyzed worms often have withered tails, ectopic vulvae and egg-laying defects (see Hedgecock, Culotti and Hall, 1990). The most severe animals exhibit maternally rescued embryonic and larval lethality.

The *unc-73* locus encodes overlapping isoforms, UNC-73A and UNC-73B, related to vertebrate Trio and Kalirin proteins, respectively (Debant *et al.*, 1996; Alam *et al.*, 1997). The predicted UNC-73A protein has two Dbl/Pleckstrin homology (DH) domains that have the potential to activate Rho family GTPases (Hart *et al.*, 1994; Cerione and Zheng, 1996). One of these DH domains has GEF activity for mammalian Rac, but not Rho or Cdc42, *in vitro* and can stimulate actin polymerization at the plasma membrane when expressed in Rat2 cells (Steven *et al.*, 1998). Such a membrane ruffling defect is characteristic of injecting Rac but not Rho GTPases into fibroblasts (Debant *et al.*, 1996).

UNC-73A and UNC-73B are expressed ubiquitously in the early embryo, starting at about the 100-cell stage during gastrulation (Steven *et al.*, 1998). Later, expression is more limited. The strongest expression is observed in migrating cells and in growing neurons as these cells undergo movements on the epidermis. Later in development, *unc-73* is expressed in most neuron types which are known to have axon guidance defects in *unc-73* mutants (Steven *et al.*, 1998). Expression in the HSNs is absent in early larval stages, but begins late in the second larval stage, precisely when axon outgrowth is initiated from the cell bodies (Garriga, Guenther and Horvitz, 1993). *unc-73* appears to act cell autonomously to guide the PLM axons (Steven *et al.*, 1998).

To address whether UNC-73 might interact with a *C. elegans* Rho homologue (MIG-2), *unc-73*; *mig-2* double mutants were constructed (Zipkin, Kindt and Kenyon, 1997). These double mutants were observed to have a synthetic phenotype. Neither single mutant affected the migration of the P nuclei. However, in 70% of double mutants, one or more of the P cell nuclei failed to migrate into the ventral cord. Cell divisions and morphogenesis were observed to appear normal. In addition, many of the cell migration defects seen in *unc-73* mutants were exacerbated, or else other migrations were affected (Zipkin, Kindt and Kenyon, 1997). So, if UNC-73 interacts with MIG-2, it is clear that these two proteins also interact with distinct proteins that affect similar processes and these studies could lend insight into possible molecular interactions that occur in vertebrates. The *mig-2* expression pattern overlaps with *unc-73*. Also, gain-of-function mutations in *mig-2* produce many of the same effects as *unc-73* mutations.

3.5. LET-502 (Rho-BINDING KINASE) AND MEL-11 (MYOSIN PHOSPHATASE REGULATORY SUBUNIT), OPPOSING FUNCTIONS IN DEVELPOMENT

Embryogenesis in *C. elegans* can be divided into two phases. The first phase of seven hours is marked by rapid cell proliferation and two-thirds of the total adult somatic cells are generated. During the second seven hours of embryogenesis, morphogenesis is the most prominent event. No cell divisions are detected, but the embryo elongates by four-fold. During this elongation process, the embryo remains at a constant volume. Such a process involves dramatic cytoskeletal reorganization. Elongation occurs by squeezing the circumferentially-oriented microfilament bundles attached to adherens junctions at their apical membranes (Priess and Hirsh, 1986; Podbilewicz and White, 1994; Costa *et al.*, 1998). Within epithelial sheets, cell shape changes are often the driving force that produces higher order structure to tissues and organs (see Odell *et al.*, 1981; Fristrom, 1988 for reviews). The cell shape changes are driven by an actin-myosin-based contractile system. In *C. elegans*, this process requires LET-502 (Rho-binding kinase) and is opposed by the activity of a myosin phosphatase regulatory subunit (MEL-11).

MEL-11 belongs to a family of myosin phosphatase regulatory subunits that contain seven ankyrin repeats and a potential regulatory phosphorylation site near the C-terminus (Wissmann *et al.*, 1997). The myosin phosphatase holoenzyme consists of three subunits: a catalytic, a regulatory and a targeting subunit. As an effector of cell shape change, the myosin phosphatase holoenzyme dephosphorylates a serine residue of myosin regulatory light chain and counteracts phosphorylation of this residue by myosin light chain kinase. In this fashion, it regulates the level of actin-activated myosin Mg-ATPase activity (see Hartshorne, Ito and Erdodi, 1988 for review), which leads to muscle contraction. The sole non-myosin regulatory light chain gene in *C. elegans* is encoded by *mlc-4* which, like *let-502*, is required for embryonic elongation (Shelton *et al.*, 1999).

Intracellular calcium levels are the main regulator of myosin light chain kinase activity. Myosin phosphatase activity is regulated by phosphorylation of the regulatory subunit, which decreases activity of the holoenzyme (Trinkle-Mulcahy *et al.*, 1995; Ichikawa, Ito and Hartshorne, 1996). Inhibition is linked to GTP levels (Kitazawa, Masuo and Somlyo, 1991; Kubota *et al.*, 1992). The recently identified Rho-binding kinase is activated by Rho-GTP. Upon activation, it inactivates the myosin phosphatase holoenzyme by phosphorylation of the regulatory subunit (Kimura *et al.*, 1996).

Alleles of *let-502* were originally identified in saturation screens for lethal mutations on linkage group I (Howell and Rose, 1990) and another allele was obtained in a screen for zygotic lethal mutations that affect *C. elegans* morphogenesis (Wissmann *et al.*, 1997). The strongest alleles arrest during embryogenesis at the 1.5-fold stage with a fully penetrant terminal phenotype (Wissmann *et al.*, 1997). However, developmental processes such as pharyngeal pumping, muscle twitching, gut differentiation and cuticle formation appear to proceed normally and the embryos eventually hatch but do not grow. Weaker alleles lead to sterile adults, an apparently disorganized germ line and occasional dead embryos are detected. Adult animals are shorter than wild type, often show a Roller phenotype and abnormal alae (the longitudinal ridges of cuticle that cover the entire apical surface of the lateral seam cells in adult animals). All of the *let-502* mutations are gain-of-function alleles and show both maternal and zygotic effects (Wissmann *et al.*, 1997). Strong alleles contain missense mutations in highly conserved residues of the kinase domain and so have been proposed to inactivate the kinase domain (Wissmann *et al.*, 1997). Gain-of-function effects are attributed to preserved functional activity in another domain of the protein because they can be suppressed by a carboxyl-terminal deletion that removes the predicted Rho-binding site.

Genetic interactions between *mel-11* and components of the Rac signalling pathway suggest that Rho and Rac have antagonistic effects with respect to tissue contraction and relaxation during morphogenesis. Strong alleles of *let-502* can be suppressed by mutations in the *mel-11* gene. Mel-11 was originally identified in a screen for maternal-effect-lethal mutations on linkage group II (Kemphues, Kusch and Wolf, 1988). Three *mel-11* alleles have been characterized with distinct properties (Wissmann, Ingles and Mains, 1999). Near null recessive alleles confer a sterile phenotype and a fully penetrant maternal-effect-lethal phenotype at 25 °C. Another allele has a maternal-effect-lethal phenotype, but brood sizes are apparently normal. Another allele is recessive leaky for the maternal-effect-lethal phenotype and shows a modest decrease in brood size. *mel-11* maternal-effect mutants arrest at the 1.5- and 2-fold stages of embryonic elongation. Using temperature sensitive alleles, it was determined that *mel-11* is required for elongation at the same time that *let-502* becomes active (Wissmann *et al.*, 1997).

let-502 and *mel-11* have reciprocal expression patterns and reciprocal mutant phenotypes in both the epidermis during embryonic elongation and in the adult hermaphrodite spermatheca (Wissmann *et al.*, 1997; Wissmann, Ingles and Mains, 1999). In both tissues, *let-502* and *mel-11* mediate tissue contraction and relaxation, respectively. *mel-11* is expressed in most or all hypodermal cells at the 1.5-fold stage. Expression in lateral cells appears less relative to others. On the other hand, *let-502* is expressed strongly in the lateral cells, but is not detectable in the dorsal and ventral epidermis. At this time, expression of *mlc-4* is also apparent in the lateral hypodermal seam cells (Shelton *et al.*, 1999). A reciprocal expression pattern is similarly observed in the spermatheca. *mel-11* is expressed in the proximal spermatheca and weakly in many somatic gonad cells. *let-502* is expressed in the valve-like structures of the spermatheca.

Based on analogy with mammalian systems, a proposed pathway is: Rho activates LET-502, which in turn inhibits the activity of MEL-11 (Kimura *et al.*, 1996; Wissmann *et al.*, 1997). A mutant allele of *unc-73* is a strong enhancer of the *mel-11* phenotypes (Wissmann, Ingles and Mains, 1999). Worms exhibit more than an 80% decrease in hatching. No genetic interactions between *unc-73* and *let-502* were observed, indicating that *mel-11* has functions not shared with *let-502*. *mig-2* null alleles (similarity to Rho/Rac) result in impenetrant migration defects of some embryonic neuroblasts (Zipkin, Kindt and Kenyon, 1997) and *mig-2* mutations strongly enhance *unc-73* mutations (Zipkin, Kindt and Kenyon, 1997). *mig-2* also enhances the maternal-effect lethality conferred by *mel-11*, but not as robustly as the enhancement between *mel-11* and *unc-73* (Kemphues, Kusch and Wolf, 1988). Therefore, *unc-73* likely genetically interacts with *mel-11* at least partly through *mig-2*-independent mechanisms (other Rac homologues). No genetic interactions were detected between *mig-2* and *let-502* (see Wissmann, Ingles and Mains, 1999), again indicating that *mel-11* and *let-502* have some distinct functions. A possible model for the interactions between these proteins is illustrated in Fig. 7. However, it remains possible that the interactions between *unc-73* and *mel-11* do not reflect a common pathway and *unc-73* might enhance *mel-11* through parallel pathways.

3.6. Rac AND p21-ACTIVATED KINASE (CePAK)

A promising downstream target of the Rac/CDC-42 family of GTPases is the Rac/CDC42-activated serine/threonine kinase PAK (p21-activated kinase)

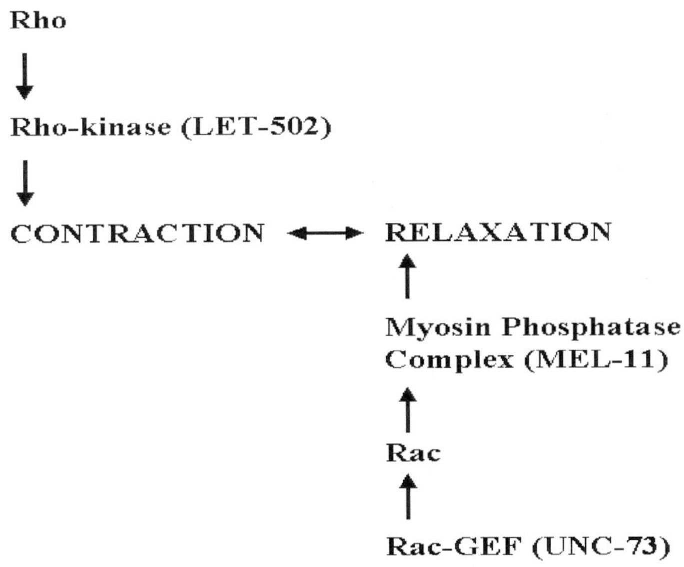

Figure 7. Possible model for interactions between LET-502 and MEL-11 in *Caenorhabditis elegans* (Wissmann, Ingles and Mains, 1999; Wissmann *et al.*, 1997).

(Manser *et al.*, 1994) and a homologue in *C. elegans* has been identified (Chen *et al.*, 1996). CePAK, the p21-activated kinase homologue in *C. elegans*, can bind to the active GTP-bound forms of either Rac or CDC42 and it exhibits kinase activity in the presence of *C. elegans* GTP-associated CDC42 and *C. elegans* GTP-associated Rac1 (Chen *et al.*, 1996).

Insight into a possible role for CePAK was obtained by analysis of its expression pattern (Chen *et al.*, 1996). Message levels of CePAK are high during embryogenesis and antibodies to CePAK reveal that CePAK is specifically localized at all hypodermal cell boundaries during the second phase of embryogenesis. *C. elegans* Rac and CDC42 are ubiquitously expressed throughout embryogenesis (see Chen *et al.*, 1996) and were found to colocalize with CePAK at cell boundaries when hypodermal cell fusion occurs. Therefore, these proteins interact with CePAK to affect cytoskeletal reorganization at this time.

3.7. RAB-3

The essential components of exocytosis are highly conserved between mammals and nematodes. One class of small GTP-binding proteins involved in secretory processes is the rab3 family (Simons and Zerial, 1993; Nuoffer and Blach, 1994). Members of the rab3 family are associated with synaptic vesicles in neurons and secretory vesicles in neuroendocrine cells (von Mollard *et al.*, 1994). rab3 mutations in *Saccharomyces cerevisiae* suggest that certain rab molecules are essential for specific vesicular transport steps, such as the formation of post-docking intermediates in the vesicle fusion process (see Ferro-Novick and Novick, 1993; Lian *et al.*, 1994; Sogaard *et al.*, 1994; and Sollner and Rothman, 1994 for reviews). In contrast, rab3A knock-out mice exhibit only a mild defect in excitatory cell secretion and both a stimulatory role as well as a inhibitory role for rab3 have been proposed for rab3 in mediating vesicle release (Oberhauser *et al.*, 1992; Padfield, Balch and Jamieson, 1992; Lledo *et al.*, 1993; Holz *et al.*, 1994; Johannes *et al.*, 1994; Piiper, Stryjek-Kaminska and Zeuzem, 1994; Geppert *et al.*, 1997), and so the role of rab3 in the secretory process remains unclear.

C. elegans expresses only one clear homologue of the *rab3* gene (Bargmann, 1998) and this gene is expressed in most or all of the synaptic-rich regions of the nervous system and colocalizes with the synaptic vesicle protein synaptotagmin, in accordance with its proposed role in synaptic function (Nonet *et al.*, 1997). Knockout of this gene in *C. elegans* reveals that RAB-3 function is not essential for the release of neurotransmitter (Nonet *et al.*, 1997). *rab-3* mutants have only mild behavioural defects. They move more slowly with sinusoidal tracks of increased amplitude and are only slightly impaired with respect to chemotaxis to the volatile odourant, isoamyl alcohol. Animals also have a slightly decreased pharyngeal pumping rate, a slightly longer defaecation cycle period and a slightly lower male mating efficiency. *rab-3* knockout animals are also more resistant to the acetylcholinesterase inhibitor, aldicarb, but not as resistant as animals carrying mutations in the gene encoding synaptotagmin

(Nguyen *et al.*, 1995; Nonet *et al.*, 1997). This suggests that there is a slight defect in cholinergic transmission. Electropharyngeograms, a method for recording currents produced by changes in pharyngeal muscle membrane potential (Raizen and Avery, 1994), of *rab-3* knock-out animals reveal defects in the generation of both polarizing and depolarizing currents in the pharyngeal muscle (Raizen and Avery, 1994). Analysis of mutant neuromuscular junctions by electron microscopy reveals that although the total number of vesicles is similar to wild-type, the terminals of *rab-3* mutant animals contain two- to three-fold fewer vesicles. Instead, vesicles accumulate lateral to the active zones (Nonet *et al.*, 1997). These observed morphological defects are consistent with a synaptic site of action for RAB-3, similar to its vertebrate counterpart. Given the conservation of the components of exocytosis between *C. elegans* and vertebrates, further genetic analysis of *rab-3* mutants in *C. elegans* is likely to prove fruitful in order to understand the discrepancies between the proposed roles for rab3A in vertebrates.

4. Conclusions

It is clear that during the course of study of signalling pathways in *C. elegans*, both similarities and differences with respect to vertebrate signalling pathways have and will continue to emerge. One initial surprise that came from the completion of the *C. elegans* genomic sequence was the fact that *C. elegans* expresses more heterotrimeric GTP-binding Gα subunits than vertebrates. The unexpectedly high number of Gα subunits expressed in *C. elegans* might have evolved in part because of the difference between *C. elegans* chemosensory neurons and mammalian neurons. Unlike *C. elegans*, each vertebrate olfactory epithelial neuron expresses a single olfactory receptor (see Buck, 1996; Mombaerts, 1996; and Mombaerts *et al.*, 1996 for reviews). Each mammalian olfactory neuron is thought to activate the same Gα subunit, Gα_{olf}, a Gα_s-like protein. Gα_{olf} activates adenylate cyclase III, which raises cAMP levels and causes a cyclic nucleotide-gated channel to open. The influx of cations through this channel leads to the formation of an action potential, which signals to discrete loci at the olfactory bulb glomeruli. Discrimination between odourants is possible since individual odourants interact with specific receptors and each cell has only one receptor. Therefore, individual odourants are associated with specific spatial patterns at the olfactory bulb (see Ebrahimi and Chess, 1998 for review).

C. elegans chemosensory neurons express more than one receptor per neuron (Troemel *et al.*, 1995) and specific responses are generated and possibly integrated via multiple G proteins at the level of the neurons, rather than in a complex structure such as the olfactory bulb. Such a strategy may have evolved due to the relatively small number of neurons in nematodes. Animals can discriminate between two odourants recognized by one sensory neuron and can adapt independently to two different types of chemicals that are detected

by the same chemosensory neuron (Colbert and Bargmann, 1995). Regulatory events within the olfactory neuron probably contribute to sensory odourant discrimination (Colbert and Bargmann, 1995). On the other hand, the specificity of the behavioural output is determined by the synaptic contacts that the chemosensory neurons make with post-synaptic partners. ASH and ADL sensory neurons synapse onto the AVA and AVD interneurons and activate the motor neurons required for backward locomotion (White *et al.*, 1986). In spite of these differences, it is likely that chemosensation in *C. elegans*, a behaviour that has been extensively characterized genetically and at a cellular level through laser ablation of these neurons (Bargmann, Hartweig and Horvitz, 1993; Sengupta *et al.*, 1993), will provide a superb model for understanding how specificity and integration between G protein pathways and other signalling pathways are maintained. Undoubtedly, many of these strategies will be conserved across species.

Characterization of the predicted mammalian homologues of $G\alpha_q$ and $G\alpha_s$ subunits has so far confirmed that these G protein alpha subunits retain biochemical functions identical to their mammalian counterparts. In the case of the $G\alpha_q$ homologue, EGL-30, it is very likely that it can activate either other PLC homologues that are not clear homologues of PLCβ, or else it activates other signalling molecules. These possible additional functions might or might not be shared by mammalian $G\alpha_q$. However, the extremely high level of sequence identity between EGL-30 and mammalian $G\alpha_q$ suggests that they will be shared with mammalian $G\alpha_q$.

As in other organisms, the proteins involved in signalling processes in *C. elegans* often contain multiple domains and many of them are also found amongst the large repertoire of vertebrate signalling domains. It is thought that the acquisition of new functions during the evolution of multicellular eukaryotic signalling has relied more upon the expansion of existing domain families' novel combinations of associations, rather than on the development of novel domain types (Copley *et al.*, 1999). One example for this is demonstrated by *C. elegans* PLC210 (Shibatohge *et al.*, 1998), a novel phosphoinositide-specific phospholipase C (PI-PLC) molecule that contains conserved X and Y domains found in vertebrate PLCs. In addition, PLC210 contains an N-terminal domain with similarity to a family of guanine nucleotide exchange proteins for Ras (represented by *S. cerevisiae* CDC25) and a C-terminal Ras association domain (Ponting and Benjamin, 1996) that bears similarity to a domain conserved among Ras-binding regions of mammalian Ral-GDS and AF-6 (Shibatohge *et al.*, 1998). This domain can associate *in vitro* with human Ha-Ras in a GTP-dependent manner and competes with yeast adenylyl cylase for binding Ha-Ras. This molecule might represent a novel prototype for a class of PI-PLCs (Shibatohge *et al.*, 1998) and might indicate a possible direct mode of regulation by Ras of PI-PLC, which has been implied in mammalian cells (Fleischman, Chahwala and Cantley, 1986; Noh, Shin and Rhee, 1995).

In the case of mammalian GTP-binding proteins, it is likely that subtype

variability and differences with respect to the subcellular localization of components have provided for considerable signalling specificity. However, in order to achieve an accurate model of any signalling pathway, it is necessary to address questions such as how possibly competing interactions are regulated *in vivo* and how these interactions affect the velocity, rate and amplitude of specific responses. For example, seven transmembrane receptors can couple to multiple GTP-binding proteins. Many regulators of both heterotrimeric and small GTP-binding proteins (GTPase-activating, GDP dissociation inhibitors and guanine nucleotide exchange factors) contain other domains that can act as scaffolding proteins, regulate localization of signalling components to the membrane, or regulate interactions with other proteins (Huang *et al.*, 1997, 1999; Hart *et al.*, 1998; Kozasa *et al.*, 1998). These issues are difficult to address in more complex organisms and are often masked by overexpression of genes encoding signalling molecules *in situ*, since regulatory molecules are expected to be limiting factors. The powerful genetic techniques in *C. elegans*, coupled with the ability to dissect behavioural and developmental processes in detail in the intact organism, provides a feasible approach in order to address questions of this nature.

Acknowledgements

The authors thank Dr. Paul Mains for critical reading of section 3.5. of this Chapter.

References

Ahmed, S., Maruyama, I.N., Kozma, R., Lee, J., Brenner, S., and Lim, L. (1992) The *Caenorhabditis elegans unc-13* gene product is a phospholipid-dependent high-affinity phorbol ester receptor. *Biochemical Journal* **287**, 995–999.

Alam, M.R., Johnson, R.C., Darlington, D.N., Hand, T.A., Mains, R.E., and Eipper, B.A. (1997) Kalirin, a cytosolic protein with spectrin-like and GDP/GTP exchange factor-like domains that interacts with peptidylglycine alpha-amidating monooxygenase, an integral membrane peptide-processing enzyme. *Journal of Biological Chemistry* **272**, 12667–12675.

Albertson, D.G., and Thomson, J.N. (1976) The pharynx of *Caenorhabditis elegans. Philosophical Transactions of the Royal Society of London B* **275**, 299–325.

Arshavsky, V.Y., and Pugh, E.N., Jnr. (1998) Lifetime regulation of G protein-effector complex: emerging importance of RGS proteins. *Neuron* **20**, 11–14.

Avery, L. (1993) The genetics of feeding in *Caenorhabditis elegans. Genetics* **133**, 897–917.

Barger, A.J., Hart, A.C., and Kaplan, J.M. (1998) Gα_s-induced neurodegeneration in *Caenorhabditis elegans. Journal of Neuroscience* **18**, 2871–2880.

Bargmann, C.I. (1993) Genetic and cellular analysis of behavior in *C. elegans. Annual Review of Neuroscience* **16**, 47–71.

Bargmann, C.I. (1998) Neurobiology of the *Caenorhabditis elegans* genome. *Science* **282**, 2028–2033.

Bargmann, C.I., and Horvitz, H.R. (1991a) Chemosensory neurons with overlapping functions direct chemotaxis to multiple chemicals in *C. elegans. Neuron* **7**, 729–742.

Bargmann, C.I., and Horvitz, H.R. (1991b) Control of larval development by chemosensory neurons in *Caenorhabditis elegans. Science* **251**, 1243–1246.

Bargmann, C.I., and Mori, I. (1997) Chemotaxis and Thermotaxis. In: D.L. Riddle, T. Blumenthal, B.J. Meyer and J.R. Priess (eds.) *C. ELEGANS II*. Cold Spring Harbor Laboratory Press, Cold Spring Harbor, pp. 717–737.

Bargmann, C.I., Thomas, J.H., and Horvitz, H.R. (1990) Chemosensory cell function in the behavior and development of *Caenorhabditis elegans*. *Cold Spring Harbor Symposia on Quantitative Biology* **55**, 529–538.

Bargmann, C.I., Hartweig, E., and Horvitz, H.R. (1993) Odorant-selective genes and neurons mediate olfaction in *C. elegans*. *Cell* **74**, 515–527.

Berman, D.M., Kozasa, T., and Gilman, A.G. (1996) The GTPase-activating protein RGS4 stabilizes the transition state for nucleotide hydrolysis. *Journal of Biological Chemistry* **271**, 27209–27212.

Berman, D.M., Wilkie, T.M., and Gilman, A.G. (1996) GAIP and RGS4 are GTPase-activating proteins for the G_i subfamily of G protein α subunits. *Cell* **86**, 445–452.

Berman, D.M., and Gilman, A.G. (1998) Mammalian RGS proteins: barbarians at the gate. *Journal of Biological Chemistry* **273**, 1269–1272.

Berridge, M.J. (1993) Inositol trisphosphate and calcium signaling. *Nature* **361**, 315–325.

Betty, M., Harnish, S.W., Rhodes, K.J., and Cockett, M.I. (1998) Distribution of heterotrimeric G-protein beta and gamma subunits in the rat brain. *Neuroscience* **85**, 475–486.

Betz, A., Ashery, U., Rickman, M., Neher, E., Sudlof, T.C., Rettig, J., and Brose, N. (1998) Munc13-1 is a presynaptic phorbol ester receptor that enhances neurotransmitter release. *Neuron* **21**, 123–136.

Birnby, D.A., Link, E.M., Vowels, J.J., Tian, H., Colacurcio, P.L., and Thomas, J.H. (2000) A transmembrane guanylyl cyclase (DAF-11) and Hsp90 (DAF-21) regulate a common set of chemosensory behaviors in *Caenorhabditis elegans*. *Genetics* **155**, 85–104.

Blinder, D., and Jenness, D.D. (1989) Regulation of postreceptor signaling in the pheromone response pathway of *Saccharomyces cerevisiae*. *Molecular and Cellular Biology* **9**, 3720–3726.

Blomqvist, A.G., and Herzog, H. (1997) Y-receptor subtypes – how many more? *Trends in Neurosciences* **20**, 294–298.

Bluml, K., Mutschler, E., and Wess, J. (1994a) Identification of an intracellular tyrosine residue critical for muscarinic receptor-mediated stimulation of phosphatidylinositol hydolysis. *Journal of Biological Chemistry* **269**, 402–405.

Bluml, K., Mutschler, E., and Wess, J. (1994b) Functional role in ligand finding and receptor activation of an asparagine residue present in the 6th transmembrane domain of all muscarinic acetylcholine receptors. *Journal of Biological Chemistry* **269**, 18870–18876.

Boland, L.M., Allen, A.C., and Dingledine, R. (1991) Inhibition by bradykinin of voltage-activated barium current in a rat dorsal-root ganglion-cell line – role of protein kinase C. *Journal of Neuroscience* **11**, 1140–1149.

Bourne, H.R., Sanders, D.A., and McCormick, F. (1990) The GTPase superfamily – a conserved switch for diverse cell functions. *Nature* **348**, 125–132.

Bourne, H.R., Sanders, D.A., and McCormick, F. (1991) The GTPase superfamily – conserved structure and molecular mechanism. *Nature* **349**, 117–127.

Brenner, S. (1974) The genetics of *Caenorhabditis elegans*. *Genetics* **77**, 71–94.

Brown, A.M., and Bimbaumer, L. (1990) Ionic channels and their regulation by G protein subunits. *Annual Review of Physiology* **52**, 197–213.

Brundage, L., Avery, L., Katz, A., Kim, U.-J., Mendel, J.E., Sternberg, P.W., and Simon, M.I. (1996) Mutations in a *C. elegans* $G_q\alpha$ gene disrupt movement, egg-laying and viability. *Neuron* **16**, 999–1009.

Buck, L.B. (1996) Information coding in the vertebrate olfactory system. *Annual Review of Neuroscience* **19**, 517–544.

Cassada, R.C., and Russell, R.L. (1975) The dauer larva, a post-embryonic developmental variant of the nematode *Caenorhabditis elegans*. *Developmental Biology* **46**, 326–342.

Cerione, R.A., and Zheng, Y. (1996) The Db1 family of oncogenes. *Current Opinion in Cell Biology* **8**, 216–222.

Chalfie, M., Tu, Y., Euskirchen, G., Ward, W.W., and Prasher, D.C. (1994) Green fluorescent protein as a marker for gene expression. *Science* **263**, 802–805.

Chen, C., Seow, K.T., Guo, K., Yaw, L.P., and Lin, S.-C. (1999) The membrane association domain of RGS16 contains unique amphipathic features that are conserved in RGS4 and RGS5. *Journal of Biological Chemistry* **274**, 19799–19806.

Chen, C.K., Wieland, T., and Simon, M.I. (1996) RGS-r, a retinal specific RGS protein, binds an intermediate conformation of transducin and enhances recycling. *Proceedings of the National Academy of Sciences, USA* **93**, 12885–12889.

Chen, E.B., Branda, C.S., and Stern, M.J. (1997) Genetic enhancers of *sem-5* define components of the gonad-independent guidance mechanism controlling sex myoblast migration in *Caenorhabditis elegans* hermaphrodites. *Developmental Biology* **182**, 88–100.

Chen, W.N., Chen, S., Yap, S.F., and Lim, L. (1996) The *Caenorhabditis elegans* p21-activated kinase (CePAK) colocalizes with CeRac1 and CDC42Ce at hypodermal cell boundaries during embryo elongation. *Journal of Biological Chemistry* **271**, 26362–26368.

Clandinin, T.R., DeModena, J.A., and Sternberg, P.W. (1998) Inositol trisphosphate mediates a RAS-independent response to LET-23 receptor tyrosine kinase activation in *C. elegans*. *Cell* **92**, 523–533.

Coburn, C.M., and Bargmann, C.I. (1996) A putative cyclic nucleotide-gated channel is required for sensory development and function in *C. elegans*. *Neuron* **17**, 695–706.

Colbert, H.A., and Bargmann, C.I. (1995) Odorant-specific adaptation pathways generate olfactory plasticity in *C. elegans*. *Neuron* **14**, 803–812.

Colmers, W.F., and Bleakman, D. (1994) Effects of neuropeptide-Y on the electrical properties of neurons. *Trends in Neurosciences* **17**, 373–379.

Copley, R.R., Schultz, J., Ponting, C.P., and Bork, P. (1999) Protein families in multicellular organisms. *Current Opinion in Structural Biology* **9**, 408–415.

Costa, M., Raich, W., Agbunag, C., Leung, B., Hardin, J., and Priess, J.R. (1998) A putative catenin-cadherin system mediates morphogenesis of the *Caenorhabditis elegans* embryo. *Journal of Cell Biology* **141**, 297–308.

Croll, N.A. (1970) *The Behaviour of Nematodes*. Edward Arnold Publishers, Ltd., London, pp. 18–27.

Curtis, B.M., and Catterall, W.A. (1985) Phosphorylation of the calcium-antagonist receptor of the voltage-sensitive calcium-channel by cAMP-dependent protein-kinase. *Proceedings of the National Academy of Sciences, USA* **82**, 2528–2532.

Dal Santo, P., Logan, M.A., Chisholm, A.D., and Jorgensen, E.M. (1999) The inositol trisphosphate receptor regulates a 50-second behavioral rhythm in *C. elegans*. *Cell* **98**, 757–767.

Debant, A., Serra-Pages, C., Seipel, K., O'Brien, S., Tang, M., Park, S.H., and Streuli, M. (1996) The multidomain protein Trio binds the LAR transmembrane tyrosine phosphatase, contains a protein kinase domain, and has separate rac-specific and rho-specific guanine nucleotide exchange factor domains. *Proceedings of the National Academy of Sciences, USA* **93**, 5466–5471.

de Bono, M., and Bargmann, C.I. (1998) Natural variation in a neuropeptide Y receptor homolog modifies social behavior and food response in *C. elegans*. *Cell* **94**, 679–689.

Desai, C., and Horvitz, H.R. (1989) *Caenorhabditis elegans* mutants defective in the functioning of the motor neurons responsible for egg laying. *Genetics* **121**, 703–721.

Desai, C., Garriga, G., McIntire, S.L., and Horvitz, H.R. (1988) A genetic pathway for the development of the *Caenorhabditis elegans* HSN motor neurons. *Nature* **336**, 638–646.

De Vries, L., and Farquhar, M.G. (1999) RGS proteins: more than just GAPs for heterotrimeric G proteins. *Trends in Cell Biology* **9**, 138–144.

Dohlman, H.G., Thorner, J., Caron, M.G., and Lefkowitz, R.J. (1991) Model systems for the study of 7-transmembrane-segment receptors. *Annual Review of Biochemistry* **60**, 653–688.

Dohlman, H.G., Apaniesk, D., Chen, Y., Song, J., and Nusskern, D. (1995) Inhibition of G-protein signaling by dominant gain-of-function mutations in SST2P, a pheromone desensitization factor in *Saccharomyces cerevisiae*. *Molecular and Cellular Biology* **15**, 3635–3643.

Dohlman, H.G., Song, J., Ma, D.R., Courchesne, W.E., and Thorner, J. (1996) Sst2, a negative regulator of pheromone signaling in the yeast *Saccharomyces* cerevisiae: expression, localization, and genetic interaction and physical association with $G_{pa}1$ (the G-protein alpha subunit). *Molecular and Cellular Biology* **16**, 5194–5209.

Driscoll, M., and Kaplan, J. (1997) Mechanotransduction. In: D.L. Riddle, T. Blumenthal, B.J. Meyer, and J.R. Priess (eds.) *C. ELEGANS II*. Cold Spring Harbor Laboratory Press, Cold Spring Harbor, pp. 645–677.

Eaton, S., Auvinen, P., Luo, L.Q., Jan, Y.N., and Simons, K. (1995) CDC42 and Rac1 control different actin-dependent processes in the *Drosophila* wing disc epithelium. *Journal of Cell Biology* **131**, 151–164.

Ebrahimi, F.A.W., and Chess, A. (1998) Olfactory G proteins: simple and complex signal transduction. *Current Biology* **8**, R431–R433.

Estevez, M., Attisano, L., Wrana, J.L., Albert, P.S., Massague, J., and Riddle, D.L. (1993) The *daf-4* gene encodes a bone morphogenetic protein receptor controlling *C. elegans* dauer larva development. *Nature* **365**, 644–649.

Faurobert, E., and Hurley, J.B. (1997) The core domain of a new retina specific RGS protein stimulates the GTPase activity of transducin *in vitro*. *Proceedings of the National Academy of Sciences, USA* **94**, 2945–2950.

Ferro-Novick, S., and Novick, P. (1993) The role of GTP-binding proteins in transport along the exocytotic pathway. *Annual Review of Cell Biology* **9**, 575–599.

Fino-Silva, I., and Plasterk, R.H.A. (1990) Characterization of a G-protein α-subunit gene from the nematode *Caenorhabditis elegans*. *Journal of Molecular Biology* **215**, 483–487.

Fire, A., Xu, S.Q., Montgomery, M.K., Kostas, S.A., Driver, S.E., and Mello, C.C. (1998) Potent and specific genetic interference by double-stranded RNA in *Caenorhabditis elegans*. *Nature* **391**, 806–811.

Fleischman, L.F., Chahwala, S.B., and Cantley, L. (1986) Ras-transformed cells – altered levels of phosphatidylinositol-4,5-bisphosphate and catabolites. *Science* **231**, 407–410.

Fletcher, J.E., Lindorfer, M.A., DeFilippo, J.M., Yasuda, H., Guilmard, M., and Garrison, J.C. (1998) The G protein β_5 subunit interacts selectively with the $G_q\alpha$ subunit. *Journal of Biological Chemistry* **273**, 636–644.

Forrester, W.C., and Garriga, G. (1997) Genes necessary for *C. elegans* cell and growth cone migrations. *Development* **124**, 1831–1843.

Fristrom, D. (1988) The cellular basis of epithelial morphogenesis – a review. *Tissue and Cell* **20**, 645–690.

Garriga, G., Guenther, C., and Horvitz, H.R. (1993) Migrations of the *Caenorhabditis elegans* HSNs are regulated by *egl-43*, a gene encoding two zinc-finger proteins. *Genes and Development* **7**, 2097–2109.

Georgi, L.L., Albert, P.S., and Riddle, D.L. (1990) *daf-1*, a *C. elegans* gene controlling dauer larva development, encodes a novel receptor protein kinase. *Cell* **61**, 635–645.

Geppert, M., Goda, Y., Stevens, C.F., and Sundlof, T.C. (1997) The small GTP-binding protein Rab3A regulates a late step in synaptic vesicle fusion. *Nature* **387**, 810–814.

Gilman, A.G. (1987) G proteins: transducers of receptor-generated signals. *Annual Review of Biochemistry* **56**, 615–649.

Golden, J.W., and Riddle, D.L. (1982) A pheromone influences larval development in the nematode *Caenorhabditis elegans*. *Science* **218**, 578–580.

Golden, J.W., and Riddle, D.L. (1984a) The *Caenorhabditis elegans* dauer larva: developmental effects of pheromone, food, and temperature. *Developmental Biology* **102**, 368–378.

Golden, J.W., and Riddle, D.L. (1984b) A pheromone-induced developmental switch in *Caenorhabditis elegans*: Temperature-sensitive mutants reveal a wild-type temperature-dependent process. *Proceedings of the National Academy of Sciences, USA* **81**, 819–823.

Gudermann, T., Schonberg, T., and Schulz, G. (1997) Functional and structural complexity of signal transduction via G-protein-coupled receptors. *Annual Review of Neurosciences* **20**, 399–427.

Hadju-Cronin, Y.M., Chen, W.J., Patikoglou, G., Koelle, M.R., and Sternberg, P.W. (1999) Antagonism between $G_o\alpha$ and $G_q\alpha$ in *Caenorhabditis elegans*: the RGS protein EAT-16 is necessary for $G_o\alpha$ signaling and regulates $G_q\alpha$ activity. *Genes and Development* **13**, 1780–1793.

Hall, A. (1998) Rho GTPases and the actin cytoskeleton. *Science* **279**, 509–514.

Harden, N., Loh, H.Y., Chia, W., and Lim, L. (1995) Dominant inhibitory version of the small GTP-binding protein Rac disrupts cytoskeletal structures and inhibits developmental cell-shape changes in *Drosophila*. *Development* **121**, 903–914.

Hart, M.J., Eva, A., Zangrilli, D., Aaronson, S.A., Evans, T., Cerione, R.A., and Zheng, Y. (1994) Cellular-transformation and guanine-nucleotide exchange activity are catalyzed by a common domain on the *dbl* oncogene product. *Journal of Biological Chemistry* **269**, 62–65.

Hart, M.J., Jiang, X.J., Kozasa, T., Roscoe, W., Singer, W.D., Gilman, A.G., Sternweiss, P.C., and Bollag, G. (1998) Direct stimulation of the guanine nucleotide exchange activity of p115 RhoGEF by $G\alpha_{13}$. *Science* **280**, 2112–2114.

Hartshorne, D.J., Ito, M., and Erdodi, F. (1998) Myosin light chain phosphatase: subunit composition, interactions and regulation. *Journal of Muscle Research and Cell Motility* **19**, 325–341.

Hedgecock, E.M., Culotti, J.G., Hall, D.H., and Stern, B.D. (1987) Genetics of cell and axon migrations in *Caenorhabditis elegans. Development* **100**, 365–382.

Hedgecock, E.M., Culotti, J.G., and Hall, D.H. (1990) The *unc-5*, *unc-6*, and *unc-40* genes guide circumferential migrations of pioneer axons and mesodermal cells on the epidermis in *C. elegans. Neuron* **4**, 61–85.

Hille, B. (1994) Modulation of ion-channel function by G-protein-coupled receptors. *Trends in Neurosciences* **17**, 531–536.

Holz, R.W., Brondyk, W.H., Senter, R.A., Kuizon, L., and Macara, I.G. (1994) Evidence for the involvement of Rab3A in Ca^{2+}-dependent exocytosis from adrenal chromaffin cells. *Journal of Biological Chemistry* **269**, 10229–10234.

Horvitz, H.R., Chalfie, M., Trent, C., Sulston, J.E., and Evans, P.D. (1982) Serotonin and octopamine in the nematode *Caenorhabditis elegans. Science* **216**, 1012–1014.

Hosono, R., Sassa, T., and Kuno, S. (1989) Spontaneous mutations of trichlorfon resistance in the nematode *Caenorhabditis elegans. Zoological Sciences* **6**, 697–708.

Howell, A.M., and Rose, A.M. (1990) Essential genes in the *hDf6* region of chromosome-1 in *Caenorhabditis elegans. Genetics* **126**, 583–592.

Huang, L.J.S., Durick, K., Weiner, J.A., Chun, J., and Taylor, S.S. (1997) D-AKAP2, a novel protein kinase A anchoring protein with a putative RGS domain. *Proceedings of the National Academy of Sciences, USA* **94**, 11184–11189.

Huang, L.J.S., Wang, L., Ma, Y.L., Durick, K., Perkins, G., Deerinck, T.J., Ellisman, N.H., and Taylor, S.S. (1999) NH2-terminal targeting motifs direct dual specificity A-kinase-anchoring protein 1 (D-AKAP1) to either mitochondria or endoplasmic reticulum. *Journal of Cell Biology* **145**, 951–959.

Huff, R.M., Axton, J.M., and Neer, E.J. (1985) Physical and immunological characterization of a guanine nucleotide-binding protein purified from bovine cerebral-cortex. *Journal of Biological Chemistry* **260**, 864–871.

Hunt, T.W., Fields, T.A., Casey, P.J., and Peralta, E.G. (1996) RGS10 is a selective activator of $G\alpha_i$ GTPase activity. *Nature* **383**, 175–177.

Ichikawa, K., Ito, M., and Hartshorne, D.J. (1996) Phosphorylation of the large subunit of myosin phosphatase and inhibition of phosphatase activity. *Journal of Biological Chemistry* **271**, 4733–4740.

Iwasaki, K., Liu, D.W.C., and Thomas, J.H. (1995) Genes that control a temperature-compensated clock in *Caenorhabditis elegans. Proceedings of the National Academy of Sciences, USA* **92**, 10317–10321.

Jansen, G., Hazendonk, E., Thijssen, K.L., and Plasterk, R.H.A. (1997) Reverse genetics by chemical mutagenesis in *Caenorhabditis elegans. Nature Genetics* **17**, 119–121.

Jansen, G., Thijssen, K.L., Werner, P., van der Horst, M., Hazendonk, E., and Plasterk, R.H.A. (1999) The complete family of genes encoding G proteins of *Caenorhabditis elegans. Nature Genetics* **21**, 414–419.

Johannes, L., Lledo, P.M., Roa, M., Vincent, J.D., Henry, J.P., and Darchen, F. (1994) The GTPase Rab3A negatively controls calcium-dependent exocytosis in neuroendocrine cells. *EMBO Journal* **13**, 2029–2037.

Kaplan, J.M. (1996) Sensory signaling in *Caenorhabditis elegans. Current Opinion in Neurobiology* **6**, 494–499.

Kaplan, J.M., and Horvitz, H.R. (1993) A dual mechanosensory and chemosensory neuron in *Caenorhabditis elegans. Proceedings of the National Academy of Sciences, USA* **90**, 2227–2231.

Kavaliers, M. (1990) Inhibitory influences of mammalian FMRFamide (Phe-Met-Arg-Phe-amide)-related peptides on nociception and morphine-induced and stress-induced analgesia in mice. *Neuroscience Letters* **115**, 307–312.

Kavaliers, M., and Colwell, D.D. (1993) Neuropeptide-FF (FLQPQRFamide) and IgG from neuropeptide-FF antiserum affect spatial learning in mice. *Neuroscience Letters* **157**, 75–78.

Kazanietz, M.G., Lewin, N.E., Bruns, J.D., and Blumberg, P.M. (1995) Characterization of the cysteine-rich region of the *Caenorhabditis elegans* protein Unc-13 as a high-affinity phorbol ester receptor – analysis of ligand-binding interactions, lipid cofactor requirements, and inhibitor sensitivity. *Journal of Biological Chemistry* **270**, 10777–10783.

Kemphues, K.J., Kusch, M., and Wolf, N. (1988) Maternal-effect lethal mutations on linkage group-II of *Caenorhabditis elegans*. *Genetics* **120**, 977–986.

Kimble, J., and Hirsh, D. (1979) The postembryonic cell lineages of the hermaphrodite and male gonads in *Caenorhabditis elegans*. *Developmental Biology* **70**, 396–417.

Kimura, K., Ito, M., Amano, M., Chihara, K., Fukata, Y., Nakafuku, M., Yamamori, B., Feng, J., Nakano, T., Okawa, K., Iwamatsu, A., and Kaibuchi, K. (1996) Regulation of myosin phosphatase by Rho and Rho-associated kinase (Rho-kinase). *Science* **273**, 245–248.

Kitazawa, T., Masuo, M., and Somlyo, A.P. (1991) G-protein-mediated inhibition of myosin light-chain phosphatase in vascular smooth muscle. *Proceedings of the National Academy of Sciences, USA* **88**, 9307–9310.

Klass, M.R., and Hirsh, D. (1976) Nonaging developmental variant of *Caenorhabditis elegans*. *Nature* **260**, 523–525.

Kleuss, C., Hescheler, J., Ewel, C., Rosenthal, W., Schultz, G., and Wittig, B. (1991) Assignment of G-protein subtypes to specific receptors inducing inhibition of calcium currents. *Nature* **353**, 43–48.

Klotz, K.N., and Jesaitis, A.J. (1994) Neutrophil chemattractant receptors and the membrane skeleton. *Bioessays* **16**, 193–198.

Koelle, M.R., and Horvitz, H.R. (1996) EGL-10 regulates G protein signalling in the *C. elegans* nervous system and shares a conserved domain with many mammalian proteins. *Cell* **84**, 115–125.

Komatsu, H., Mori, I., Rhee, J.-S., Akaike, N., and Ohshima, Y. (1996) Mutations in a cyclic nucleotide-gated channel lead to abnormal thermosensation and chemosensation in *C. elegans*. *Neuron* **17**, 707–718.

Korswagen, H.C., Van der Linden, A.M., and Plasterk, R.H. (1998) G protein hyperactivation of the *Caenorhabditis elegans* adenylyl cyclase SGS-1 induces neuronal degeneration. *EMBO Journal* **17**, 5059–5065.

Korswagen, H.C., Park, J.H., Ohshima, Y., and Plasterk, R.H.A (1997) An activating mutation in a *Caenorhabditis elegans* G_s protein induces neural degeneration. *Genes and Development* **11**, 1493–1503.

Kozasa, T., Jiang, X.J., Hart, M.J., Sternweis, P.M., Singer, W.D., Gilman, A.G., Bollag, G., and Sternweis, P.C. (1998) p115 RhoGEF, a GTPase activating protein for $G\alpha_{12}$ and $G\alpha_{13}$. *Science* **280**, 2109–2111.

Kozma, R., Ahmed, S., Best, A., and Lim, L. (1995) The Ras-related protein Cdc42Hs and bradykinin promote formation of peripheral actin microspikes and filopodia in Swiss 3T3 fibroblasts. *Molecular Cell Biology* **15**, 1942–1952.

Kubota, Y., Nomura, M., Kamm, K.E., Mumby, M.C., and Stull, J.T. (1992) GTPγ S-dependent regulation of smooth muscle contractile elements. *American Journal of Physiology* **262**, C405–C410.

Lackner, M.R., Nurrish, S.J., and Kaplan, J.M. (1999) Facilitation of synaptic transmission by EGL-30 $G_q\alpha$ and EGL-8 PLC β: DAG binding to UNC-13 is required to stimulate acetylcholine release. *Neuron* **24**, 335–346.

Levay, K., Cabrera, J.L., Satpaev, D.K., and Slepak, V.Z. (1999) Gβ5 prevents the RGS7-Gαo interaction through binding to a distinct G gamma-like domain found in RGS7 and other RGS proteins. *Proceedings of the National Academy of Sciences, USA* **96**, 2503–2507.

Lian, J.P., Stone, S., Jiang, Y., Lyons, P., and Ferro-Novick, S. (1994) Ypt1p implicated in v-SNARE activation. *Nature* **372**, 698–701.

Lindorfer, M.A., Myung, C.S., Savino, Y., Yasuda, H., Khazan, R., and Garrison, J.C. (1998) Differential activity of the G protein $\beta_5\gamma_2$ subunit at receptors and effectors. *Journal of Biological Chemistry* **273**, 34429–34436.
Lledo, P.-M., Vernier, P., Vincent, J.-D., Mason, W.T., and Zorac, R. (1993) Inhibition of Rab3B expression attenuates Ca^{++}-dependent exocytosis in rat anterior pituitary cells. *Nature* **364**, 540–544.
Lochrie, M.A., Mendel, J.E., Sternberg, P.W., and Simon, M.I. (1991) Homologous and unique G protein alpha subunits in the nematode *Caenorhabditis elegans*. *Cell Regulation* **2**, 135–154.
Luo, L., Liao, Y.J., Jan, L.Y., and Jan, Y.N. (1994) Distinct morphogenetic functions of similar small GTPases: *Drosophila* Dracl is involved in axonal outgrowth and myoblast fusion. *Genes and Development* **8**, 1787–1802.
Luo, L., Hensch, T.K., Ackerman, L., Barbel, S., Jan, L.Y., and Jan, Y.N. (1996) Differential effects of the Rac GTPase on Purkinje cell axons and dendritic trunks and spines. *Nature* **379**, 837–840.
Mackey, S.L., (1987) Tail shock produces inhibition as well as sensitization of the siphon-withdrawal reflex of *Aplysia* – possible behavioral role for presynaptic inhibition mediated by the peptide Phe-Met-Arg-Phe-NH_2. *Proceedings of the National Academy of Sciences, USA* **84**, 8730–8734.
Manser, E., Leung, T., Salihuddin, H., Zhao, Z.S., and Lim, L. (1994) A brain serine/threonine protein kinase activated by Cdc42 and Rac1. *Nature* **367**, 40–46.
Manser, J., and Wood, W.B. (1990) Mutations affecting embryonic cell migrations in *Caenorhabditis elegans*. *Developmental Genetics* **11**, 49–64.
Man Song Hing, H.J., Codina, J., Abramowitz, J., and Haydon, P.G. (1992) Microinjection of the alpha-subunit of the G-protein G_{o2}, but not G_{o1}, reduces a voltage-sensitive calcium current. *Cell Signalling* **4**, 429–441.
Maruyama, I.N., and Brenner, S. (1991) A phorbol ester/diacyglycerol-binding protein encoded by the *unc-13* gene of *Caenorhabditis elegans*. *Proceedings of the National Academy of Sciences, USA* **88**, 5729–5733.
Maule, A.G., Bowman, J.W., Thompson, D.P., Marks, N.J., Friedman, A.R., and Geary, T.G. (1996) FMRFamide-related peptides (FaRPs) in nematodes: occurrence and neuromuscular physiology. *Parasitology* **113**, S119–S135.
McBride, J.M., and Hollis, J.P. (1966) Phenomenon of swarming in nematodes. *Nature* **211**, 545–546.
McIntire, S.L., Garriga, G., White, J., Jacobson, D., and Horvitz, H.R. (1992) Genes necessary for directed axonal elongation or fasciculation in *C. elegans*. *Neuron* **8**, 307–322.
Mello, C., and Fire, A. (1995) DNA transformation. *Methods in Cell Biology* **48**, 451–482.
Mendel, J.E., Korswagen, K.S., Liu, K.S., Hadju-Cronin, Y.M., Simon, M.I., Plasterk, R.H.A., and Sternberg, P.W. (1995) Participation of the protein G_o in multiple aspects of behavior in *C. elegans*. *Science* **267**, 1652–1655.
Mikoshiba, K., Furuichi, T., and Miyawaki, A. (1994) Structure and function of IP3 receptors. *Seminars in Cell Biology* **5**, 273–281.
Miller, K.G., Emerson, M.D., and Rand, J.B. (1999) G_o alpha and diacylglycerol kinase negatively regulate the G_q alpha pathway in *C. elegans*. *Neuron* **24**, 323–333.
Miller, K.G., Alfonso, A., Nguyen, M., Crowell, J.A., Johnson, C.D., and Rand, J.B. (1996) A genetic selection for *Caenorhabditis elegans* synaptic transmission mutants. *Proceedings of the National Academy of Sciences, USA* **93**, 12593–12598.
Mombaerts, P. (1996) Targeting olfaction. *Current Opinion in Neurobiology* **6**, 481–486.
Mombaerts, P., Wang, F., Dulac, C., Vassar, R., Chao, S.K., Nemes, A., Mendelsohn, M., Edmondson, J., and Axel, R. (1996) The molecular biology of olfactory perception. *Cold Spring Harbor Symposia on Quantitative Biology* **61**, 135–145.
Mori, I., and Ohshima, Y. (1995) Neural regulation of thermotaxis in *Caenorhabditis elegans*. *Nature* **376**, 344–348.
Murphy, A.M., and Montell, D.J. (1996) Cell type-specific roles for Cdc42, Rac, and RhoL in *Drosophila* oogenesis. *Journal of Cell Biology* **133**, 617–630.
Nelson, F.K., and Riddle, D.L. (1984) Functional study of the *Caenorhabditis elegans* secretory-excretory system using laser microsurgery. *Journal of Experimental Zoology* **231**, 45–56.

Nelson, L.S., Rosoff, M.L., and Li, C. (1998) Disruption of a neuropeptide gene, *flp-1*, causes multiple behavioral defects in *Caenorhabditis elegans*. *Science* **281**, 1686–1690.

Nelson, L.S., Kim, K., Memmott, R.E., and Li, C. (1998) FMRFamide-related gene family in the nematode, *Caenorhabditis elegans*. *Molecular Brain Research* **58**, 103–111.

Nguyen, M., Alfonso, A., Johnson, C.D., and Rand, J.B. (1995) *Caenorhabditis elegans* mutants resistant to inhibitors of acetylcholinesterase. *Genetics* **140**, 527–535.

Nobes, C., and Hall, A. (1994) Regulation and function of the Rho subfamily of small GTPases. *Current Opinion in Genetics and Development* **4**, 77–81.

Nobes, C.D., and Hall, A. (1995) Rho, Rac, and Cdc42 GTPases regulate the assembly of multimolecular focal complexes associated with actin stress fibers, lamellipodia, and filopodia. *Cell* **81**, 53–62.

Noh, D.Y., Shin, S.H., and Rhee, S.G. (1995) Phosphoinositide-specific phospholipase-C and mitogenic signaling. *Biochimica et Biophysica Acta* **1242**, 99–113.

Nonet, M.L., Staunton, J.E., Kilgard, M.P., Fergestad, T., Hartweig, E., Horvitz, H.R., Jorgensen, E.M., and Meyer, B.J. (1997) *Caenorhabditis elegans* rab-3 mutant synapses exhibit impaired function and are partially depleted of vesicles. *Journal of Neuroscience* **17**, 8061–8073.

Nuoffer, C., and Blach, W.E. (1994) GTPases – multifunctional molecular switches regulating vesicular traffic. *Annual Review of Biochemistry* **63**, 949–990.

Nurrish, S., Segalat, L., and Kaplan, J.M. (1999) Serotonin inhibition of synaptic transmission: $G\alpha_o$ decreases the abundance of UNC-13 at releases sites. *Neuron* **24**, 231–242.

Oberhauser, A.F., Monck, J.R., Balch, W.E, and Fernandez, J.M. (1992) Exocytotic fusion is activated by Rab3a peptides. *Nature* **360**, 270–273.

Odell, G.M., Oster, G., Alberch, P., and Burnside, B. (1981) The mechanical basis of morphogenesis. 1. Epithelial folding and invagination. *Developmental Biology* **85**, 446–462.

Padfield, P.J., Balch, W.E., and Jamieson, J.D. (1992) a synthetic peptide of the Rab3A effector domain stimulates amylase release from permeabilized pancreatic acini. *Proceedings of the National Academy of Sciences, USA* **89**, 1656–1660.

Parent, C.A., and Devreotes, P.N. (1996) Molecular genetics of signal transduction in *Dictyostelium*. *Annual Review of Biochemistry* **65**, 411–440.

Park, J.H., Ohshima, S., Tani, T., and Ohshima, Y. (1997) Structure and expression of the *gsa-1* gene encoding a G protein α_s subunit in *Caenorhabditis elegans*. *Gene* **194**, 183–190.

Payza, K., and Yang, H.Y.T. (1993) Modulation of neuropeptide-FF receptors by guanine nucleotides and cations in membranes of rat brain and spinal cord. *Journal of Neurochemistry* **60**, 1894–1899.

Piiper, A., Stryjek-Kaminska, D., and Zeuzem, S. (1994) Synthetic *Rab*3A effector domain peptide STIR stimulates inositol 1,4,5-trisphosphate production in various permeabilized cells. *Biochemical and Biophysical Research Communications* **203**, 756–762.

Plasterk, R.H.A. (1995) Reverse genetics: from gene sequence to mutant worm. *Methods in Cell Biology* **48**, 59–80.

Podbilewicz, B., and White, J.G. (1994) Cell fusions in the developing epithelia of *C. elegans*. *Developmental Biology* **161**, 408–424.

Ponting, C.P., and Benjamin, D.R. (1996) A novel family of Ras-binding domains. *Trends in Biochemical Sciences* **21**, 422–425.

Priess, J.R., and Hirsh, D.I. (1986) *Caenorhabditis elegans* morphogenesis: the role of the cytoskeleton in elongation of the embryo. *Developmental Biology* **117**, 156–173.

Raffa, R.B. (1988) The action of FMRFamide (Phe-Met-Arg-Phe-NH_2) and related peptides on mammals. *Peptides* **9**, 915–922.

Raizen, D.M., and Avery, L. (1994) Electrical activity and behavior in the pharynx of *Caenorhabditis elegans*. *Neuron* **12**, 483–495.

Rand, J.B., and Russell, R.L. (1985) Molecular basis of drug-resistance mutations in *C. elegans*. *Psychopharmacology Bulletin* **21**, 623–630.

Reuveny, E., Slesinger, P.A., Inlese, J., Morales, J.M., Iniguezlluhi, J.A., Lefkowitz, R.J., Bourne, H.R., Jan, Y.N., and Jan, L.Y. (1994) Activation of the cloned muscarinic potassium channel by G protein $\beta\gamma$ subunits. *Nature* **370**, 143–146.

Richmond, J.E., Davis, W.S., and Jorgensen, E.M. (1999) UNC-13 is required for synaptic vesicle fusion in *C. elegans*. *Nature Neuroscience* **2**, 959–964.

Riddle, D.L. (1988) The dauer larva. In: W.B. Wood and the Community of *C. elegans* Researchers (eds.) *The nematode Caenorhabditis elegans*. Cold Spring Harbor Laboratory, Cold Spring Harbor, New York, pp. 393–412.

Riddle, D.L., Swanson, M.M., and Albert, P.S. (1981) Interacting genes in nematode dauer larva formation. *Nature* **290**, 668–671.

Ridley, A.J. (1995) Rho-related proteins: actin cytoskeleton and cell cycle. *Current Opinion in Genetics and Development* **5**, 24–30.

Ridley, A.J., and Hall, A. (1992) The small GTP-binding protein rho regulates the assembly of focal adhesions and actin stress fibers in response to growth factors. *Cell* **70**, 389–399.

Ridley, A.J., Paterson H.F., Johnston, C.L., Dieckmann, D., and Hall, A. (1992) The small GTP-binding protein rac regulates growth factor-induced membrane ruffling. *Cell* **70**, 401–410.

Roayiae, K., Crump, J.G., Sagasti, A., and Bargmann, C.I. (1998) The Gα protein ODR-3 mediates olfactory and nociceptive function and controls cilium morphogenesis in *C. elegans* olfactory neurons. *Neuron* **20**, 55–67.

Robertson, H.M. (1998) Two large families of chemoreceptor genes in the nematodes *Caenorhabditis elegans* and *Caenorhabditis briggsae* reveal extensive gene duplication, diversification, movement, and intron loss. *Genome Research* **8**, 449–463.

Robertson, H.M. (2000) The large *srh* family of chemoreceptor genes in *Caenorhabditis* nematodes reveals processes of genome evolution involving large duplications and deletions and intron gains and losses. *Genome Research* **10**, 192–203.

Rosoff, M.L., Burglin, T.R., and Li, C. (1992) Alternatively spliced transcripts of the *flp-1* gene encode distinct FMRFamide-like peptides in *Caenorhabditis elegans*. *Journal of Neuroscience* **12**, 2356–2361.

Rosoff, M.L., Doble, K.E., Price, D.A., and Li, C. (1993) The *flp-1* propeptide is processed into multiple, highly similar FMRFamide-like peptides in *Caenorhabditis elegans*. *Peptides* **14**, 331–338.

Schmidt, C.J., Garen-Fazio, S., Chow, Y.K., and Neer, E.J. (1989) Neuronal expression of a newly identified *Drosophila melanogaster* G-protein α_o subunit. *Cell Regulation* **1**, 125–134.

Segalat, L., Elkes, D.A., and Kaplan, J.M. (1995) Modulation of serotonin-controlled behaviors by G_o in *Caenorhabditis elegans*. *Science* **267**, 1648–1651.

Sengupta, P., Chou, J.C., and Bargmann, C.I. (1996) *odr-10* encodes a seven transmembrane domain olfactory receptor required for responses to the odorant diacetyl. *Cell* **84**, 899–909.

Sengupta, P., Colbert, H.A., Kimmel, B.E., Dwyer, N., and Bargmann, C.I. (1993) The cellular and genetic basis of olfactory responses in *Caenorhabditis elegans*. *CIBA Foundation Symposia* **179**, 235–244.

Shelton, C.A., Carter, J.C., Ellis, G.C., and Bowerman, B. (1999) The nonmuscle myosin regulatory light chain gene *mlc-4* is required for cytokinesis, anterior-posterior polarity, and body morphology during *Caenorhabditis elegans* embryogenesis. *Journal of Cell Biology* **146**, 439–451.

Shibatohge, M., Kariya, K., Liao, Y.H., Hu, C.D., Watari, Y., Goshima, M., Shima, F., and Kataoka, T. (1998) Identification of PLC210, a *Caenorhabditis elegans* phospholipase C, as a putative effector of Ras. *Journal of Biological Chemistry* **273**, 6218–6222.

Siddiqui, S.S. (1990) Mutations affecting axonal growth and guidance of motor neurons and mechanosensory neurons in the nematode *Caenorhabditis elegans*. *Neuroscience Research* **13** (supplt.), S171–S190.

Siddiqui, S.S., and Culotti, J.G. (1991) Examination of neurons in wild type and mutants of *Caenorhabditis elegans* using antibodies to horseradish peroxidase. *Journal of Neurogenetics* **7**, 193–211.

Siderovski, D.P., Strockbine, B., and Behe, C.I. (1999) Whither goest the RGS proteins? *Critical Reviews in Biochemistry and Molecular Biology* **34**, 215–251.

Simons, K., and Zerial, M. (1993) Rab proteins and the road maps for intracellular transport. *Neuron* **11**, 789–799.

Small, S.A., Kandel, E.R., and Hawkins, R.D. (1989) Activity-dependent enhancement of presynaptic inhibition in *Aplysia* sensory neurons. *Science* **243**, 1603–1606.

Snow, B.E., Krumins, A.M., Brothers, G.M., Lee, S.F., Wall,. M.A., Chung, S., Mangion, J., Arya, S., Gilman, A.G., and Siderovski, D.P. (1998) A G protein γ subunit-like domain shared between RGS11 and other RGS proteins specifies binding to $G_{\beta 5}$ subunits. *Proceedings of the National Academy of Sciences, USA* **95**, 13307–13312.

Sogaard, M., Tani, K., Geromanos, S., Tempst, P., Kirchhausen, T., Rothman, J.E., and Sollner, T. (1994) A Rab protein is required for the assembly of SNARE complexes in the docking of transport vesicles. *Cell* **78**, 937–948.

Sollner, T., and Rothman, J.E. (1994) Neurotransmission harnessing fusion machinery at the synapse. *Trends in Neurosciences* **17**, 344–348.

Sonnhammer, E.L.L., and Durbin, R. (1997) Analysis of protein domain families in *Caenorhabditis elegans*. *Genomics* **46**, 200–216.

Sternberg, P.W., and Han, M. (1998) Genetics of RAS signaling in *C. elegans*. *Trends in Genetics* **14**, 466–472.

Steven, R., Kubiseski, T.J., Zheng, H., Kulkarni, S., Mancillas, J., Morales, A.R., Hogue, C.W.V., Pawson, T., and Culotti, J. (1998) UNC-73 activates the Rac GTPase and is required for cell and growth cone migrations in *C. elegans*. *Cell* **92**, 785–795.

Strittmatter, S.M., Valenzuela, D., Kennedy, T.E., Neer, E.J., and Fishman, M.C. (1990) G_o is a major growth cone protein subject to regulation by GAP-43. *Nature* **344**, 836–841.

Sulston, J.E., and Horvitz, H.R. (1977) Post-embryonic cell lineages of the nematode, *Caenorhabditis elegans*. *Developmental Biology* **56**, 110–156.

Sulston, J.E., Schierenberg, E., White, J.G., and Thomson, J.N. (1983) The embryonic cell lieage of the nematode *Caenorhabditis elegans*. *Developmental Biology* **100**, 64–119.

Sunahara, R.K., Dessauer, C.W., and Gilman, A.G. (1996) Complexity and diversity of mammalian adenylyl cyclases. *Annual Review of Pharmacology and Toxicology* **36**, 461–480.

Swanson, M.M., and Riddle, D.L (1981) Critical periods in the development of the *C. elegans* dauer larva *Developmental Biology* **84**, 27–40.

Tabara, H., Grishok, A., and Mello, C.C. (1998) RNAi in *C. elegans*: soaking in the genome sequence. *Science* **282**, 430–431.

Tessier-Lavigne, M., and Goodman, C.S. (1996) The molecular biology of axon guidance. *Science* **274**, 1123–1133.

The *C. elegans* Sequencing Consortium (1998) Genome sequence of the nematode *Caenorhabditis elegans*: a platform for investigating biology. *Science* **282**, 2012–2018.

Thomas, J.H. (1990) Genetic analysis of defecation in *Caenorhabditis elegans*. *Genetics* **124**, 855–872.

Thomas, J.H., Birnby, D.A., and Vowels, J.J. (1993) Evidence for parallel processing of sensory information controlling dauer formation in *Caenorhabditis elegans*. *Genetics* **134**, 1105–1117.

Trent, C., Tsung, N., and Horvitz, H.R. (1983) Egg-laying defective mutants of the nematode *Caenorhabditis elegans*. *Genetics* **104**, 619–647.

Trinkle-Mulcahy, L., Ichikawa, K., Hartshorne, D.J., Siegman, M.J., and Butler, T.M. (1995) Thiophosphorylation of the 130 kDa subunits is associated with a decreased activity of myosin light-chain phosphatase in alpha-toxin-permeabilized smooth muscle. *Journal of Biological Chemistry* **270**, 18191–18194.

Troemel, E.R., Chou, J.H., Dwyer, N.D., Colbert, H.A., and Bargmann, C.I. (1995) Divergent seven transmembrane receptors are candidate chemosensory receptors in *C. elegans*. *Cell* **83**, 207–218.

van der Voorn, L., Gebbink, M., Plasterk, R.H.A., and Ploegh, H.L. (1990) Characterization of a G-protein β-subunit gene from the nematode *Caenorhabditis elegans*. *Journal of Molecular Biology* **213**, 17–26.

Van Dongen, A.M.J., Codina, J., Olate, J., Mattera, R., Johr, R., Birnbaumer, L., and Brown, A.M. (1988) Newly identified brain potassium channels gated by the guanine nucleotide binding protein G_o. *Science* **242**, 1433–1437.

Volterra, A., and Siegelbaum, S.A. (1988) Role of 2 different guanine nucleotide-binding proteins in the antagonistic modulation of the S-type K^+ channel by cAMP and arachidonic acid metabolites in *Aplysia* sensory neurons. *Proceedings of the National Academy of Sciences, USA* **85**, 7810–7814.

von Mollard, G.F., Stahl, B., Li, C., Sudhof, T.C., and Jahn, R. (1994) Rab proteins in regulated exocytosis. *Trends in Biochemical Sciences* **19**, 164–168.

Vowels, J.J., and Thomas, J.H. (1992) Genetic analysis of chemosensory control of dauer formation in *Caenorhabditis elegans. Genetics* **130**, 105–123.

Waggoner, L.E., Zhou, G.T., Schafer, R.W., and Schafer, W.R. (1998) Control of alternative behavioral states by serotonin in *Caenorhabditis elegans. Neuron* **21**, 203–214.

Walsh, D.A., and Van Platten, S.M. (1994) Multiple pathway signal-transduction by the cAMP-dependent protein-kinase. *FASEB Journal* **8**, 1227–1236.

Wan, C.P., and Lau, B.H.S. (1995) Neuropeptide-Y receptor subtypes. *Life Sciences* **56**, 1055–1064.

Wang, Z.X., Lange, A.B., and Orchard, I. (1995) Coupling of a single receptor to 2 different G-proteins in the signal-transduction of FMRFamide-related peptides. *Biochemical and Biophysical Research Communications* **212**, 531–538.

Watson, N., Linder, M.E., Druey, K.M., Kehrl, J.H., and Blumer, K.J. (1996) RGS family members: GTPase-activating proteins for heterotrimeric G-protein alpha-subunits. *Nature* **383**, 172–175.

White, J., Southgate, E., and Durbin, R. (1988) Appendix 2. Neuroanatomy. In: W.B. Wood and the Community of *C. elegans* Researchers (eds.) *The nematode Caenorhabditis elegans.* Cold Spring Harbor Laboratory, Cold Spring Harbor, New York, pp. 433–455.

White, J.G., Southgate, E., Thomson, J.N., and Brenner, S. (1976) The structure of the ventral nerve cord of *Caenorhabditis elegans. Philosophical Transactions of the Royal Society of London B* **275**, 327–348.

White, J.G., Southgate, E., Thomson, J.N., and Brenner, S. (1986) The structure of the nervous system of the nematode *C. elegans. Philosophical Transactions of the Royal Society of London B* **314**, 1–340.

Wickman, K., and Clapham, D.E. (1995a) Ion-channel regulation by G-proteins. *Physiological Reviews* **75**, 865–885.

Wickman, K., and Clapham, D.E. (1995b) G-protein regulation of ion channels. *Current Opinion in Neurobiology* **5**, 278–285.

Wightman, B., Baran, R., and Garriga, G. (1997) Genes that guide growth cones along the *C. elegans* ventral nerve cord. *Development* **124**, 2571–2580.

Wilkie, T.M., and Yokoyama, S. (1994) Evolution of the G protein alpha subunit multigene family. *Society of General Physiologists Series* **49**, 249–270.

Wilkie, T.M. (1999) G proteins, chemosensory perception, and the *C. elegans* genome project. An attractive story. *Bioessays* **21**, 713–717.

Wissman, A., Ingles, J., and Mains, P.E. (1999) The *Caenorhabditis elegans mel-11* myosin phosphatase regulatory subunit affects tissue contraction in the somatic gonad and the embryonic epidermis and genetically interacts with the Rac signaling pathway. *Developmental Biology* **209**, 111–127.

Wissman, A., Ingles, J., McGhee, J.D., and Mains, P.E. (1997) *Caenorhabditis elegans* LET-502 is related to Rho-binding kinases and human myotonic dystrophy kinase and interacts genetically with a homolog of the regulatory subunit of smooth muscle myosin phosphatase to affect cell shape. *Genes and Development* **11**, 409–422.

Wong, T.M., Greenberg, M.J., and Tse, S.Y.H. (1985) Cardiovascular effects of intraventricular injection of FMRFamide, Met-enkephalin and their common analogs in the rat. *Comparative Biochemistry and Physiology* **81C**, 175–179.

Xu, X., Zeng, W., Popov, S., Berman, D.M., Davignon, I., Yu, K., Yowe, D., Offermanns, S., Muallem, S., and Wilkie, T.M. (1999) RGS proteins determine signaling specificity of G_q-coupled receptors. *Journal of Biological Chemistry* **274**, 3549–3556.

Zeng, W., Xu, X., Popov, S., Mukhopadhyay, S., Chidiac, P., Swistok, J., Danho, W., Yagaloff, K.A., Fisher, S.L., Ross, E.M., Muallem, S., and Wilkie, T.M. (1998) The N-terminal domain of RGS4 confers receptor-selective inhibition of G protein signaling. *Journal of Biological Chemistry* **273**, 34687–34690.

Zhang, Y., Chou, J.H., Bradley, J., Bargmann, C.I., and Zinn, K. (1997) The *Caenorhabditis elegans* seven-transmembrane protein ODR-10 functions as an odorant receptor in mammalian cells. *Proceedings of the National Academy of Sciences, USA* **94**, 12162–12167.

Zipkin, I.D., Kindt, R.M., and Kenyon, C.J. (1997) Role of a new Rho family member in cell migration and axon guidance in *C. elegans. Cell* **90**, 883–894.

Zwaal, R.R., Ahringer, J., van Luenen, H.G.A.M., Rushforth, A., Anderson, P., and Plasterk, R.H.A. (1996) G proteins are required for spatial orientation of early cell cleavages in *C. elegans* embryos. *Cell* **86**, 619–629.

Zwaal, R.R., Mendel, J.E., Sternberg, P.W., and Plasterk, R.H.A. (1997) Two neuronal G proteins are involved in chemosensation of the *Caenorhabditis elegans* Dauer-inducing pheromone. *Genetics* **145**, 715–727.

CHAPTER 8

ELECTROPHYSIOLOGICAL AND PHARMACOLOGICAL STUDIES ON EXCITABLE TISSUES IN NEMATODES

Nematode Electrophysiology and Pharmacology

ROBERT J. WALKER, CANDIDA M. ROGERS, CHRISTOPHER J. FRANKS and LINDY HOLDEN-DYE

School of Biological Sciences, University of Southampton, Biomedical Sciences Building, Bassett Crescent East, Southampton SO16 7PX

Summary

Nematodes include both major parasites of humans, livestock and plants and free-living species such as *Caenorhabditis elegans*. The nematode nervous system (especially in *C. elegans*) is exceptionally well defined in terms of the number, location and projections of the small number of neurons in the nervous system and their integration into circuits involved in regulatory behaviours vital to their survival. This chapter will summarize what is known about the biological activity of neurotransmitters in nematodes: the biosynthetic pathways and genes involved, their receptors, inactivation mechanisms and second messenger signalling systems. It will cover the "classical" transmitters, such as acetylcholine (ACh), GABA, glutamate, serotonin, dopamine, octopamine, noradrenaline and nitric oxide. The localization of peptides throughout the nematode nervous system is summarized, in addition to the isolation of nematode neuropeptides by both traditional biochemical techniques and more modern genetic means. The major contribution of the completion of the *C. elegans* genome-sequencing programme is highlighted throughout. Efforts to unravel neurotransmitter action in various physiological actions such as locomotion, feeding and reproduction are detailed, as well as the various inactivation mechanisms for the current complement of the nematode transmitters.

1. Introduction

Nematodes are a highly successful phylum and have adapted to occupy a wide range of habitats. Many of them are parasitic and are of key importance in animal husbandry, food crops and can be found in many areas of the human body, inducing a number of serious clinical conditions. The nematode body is covered in a thick, colourless, acellular cuticle which contains diagonally arranged collagen fibres. Nematodes are dioecious and the sexes are often morphologically distinct. The male usually has copulatory spicules at the posterior end. Around 14 species of nematode are mainly parasitic of humans, but up to an additional 50 species can parasitise humans though humans are not their primary host. Nematode infections can be widespread and approximately one quarter of the world human population is infected by *Ascaris* (Muller and Baker, 1990). Since the human population now exceeds six billion, this represents a huge medical, economic and social problem. Very broadly,

I. Fairweather (ed.), Cell Signalling in Prokaryotes and Lower Metazoa, 243–301.
© 2004 *Kluwer Academic Publishers. Printed in Great Britain.*

parasitic nematodes can be divided into intestinal nematodes, such as *Ascaris lumbricoides* and *Ascaris suum* and tissue nematodes, such as *Brugia malayi* and *Wuchereria bancrofti.* A key member of the nematodes is the free living *Caenorhabditis elegans*, for which the entire genome has now been sequenced (Hodgkin *et al.*, 1998). A considerable volume of research has been undertaken using the two species of *Ascaris.* It has been proposed that they are sibling species that probably do not cross breed or commonly infect the alternative host (Anderson, 2000).

There have been a number of reviews on various aspects of the physiology and pharmacology of nematodes over the past five years and the reader is referred to these for further information on the subject of this chapter (Brownlee *et al.*, 1996; Davis and Stretton, 1996; Martin *et al.*, 1996; Maule *et al.*, 1996; Brownlee and Fairweather, 1999; Day and Maule, 1999; Sattelle and Culetto, 1999; Brownlee, Holden-Dye and Walker, 2000). This chapter will consider the basic structure of the nematode nervous and muscle systems, the electrophysiological properties of these systems, the classical transmitters involved in nerve-nerve and nerve-muscle transmission and the possible roles of nitric oxide and neuroactive peptides in nematodes.

2. Basic Anatomy of Nerve and Muscle System

The nervous system of nematodes is constructed around a relatively small number of neurones. They are arranged into ganglia in the head and tail and along the nerve cord. The structure of the head ganglia was first described by Goldschmidt (1908) (Fig. 1). He reported 23 ganglia containing 162 neurones making up the head ganglia complex. The retrovesicular ganglion has been described by Angstadt , Donmoyer and Stretton (1989) and by Angstadt and Stretton (1989) and contains 13 neurones. Four of these neurones are motoneurones, two project to the dorsal cord and two to the ventral cord, the rest are interneurones. In *Ascaris suum* there are approximately 300 neurones in total, of which around 90 make up the somatic motoneurone system (Stretton *et al.*, 1978). In the case of the free living nematode, *C. elegans*, the structure of the entire nervous system has been determined and in the hermaphrodite contains 302 neurones (White *et al.*, 1976,1986). In *A. suum* there is no neuropile in the nerve cords and synapses are made by axo-axonal contact or onto short spines, 1–2 μm in length (Stretton *et al.*, 1978). The motoneurone system in *A. suum* is composed of five repeating sets of motoneurones, each set making up a functional segment. Each of these segments contains eleven motoneurones and six interneurones. Stretton and his group have shown that the motoneurones can be divided into seven different types, each having a specific morphology and axon projection (Fig. 2). The nematode nervous system has both a ventral and dorsal nerve cord (Fig. 3), but interestingly all the motoneurone cell bodies are located in the ventral cord (Fig. 2). Their geometry is simple. They all have a process in the ventral cord but, in addition, five of the seven

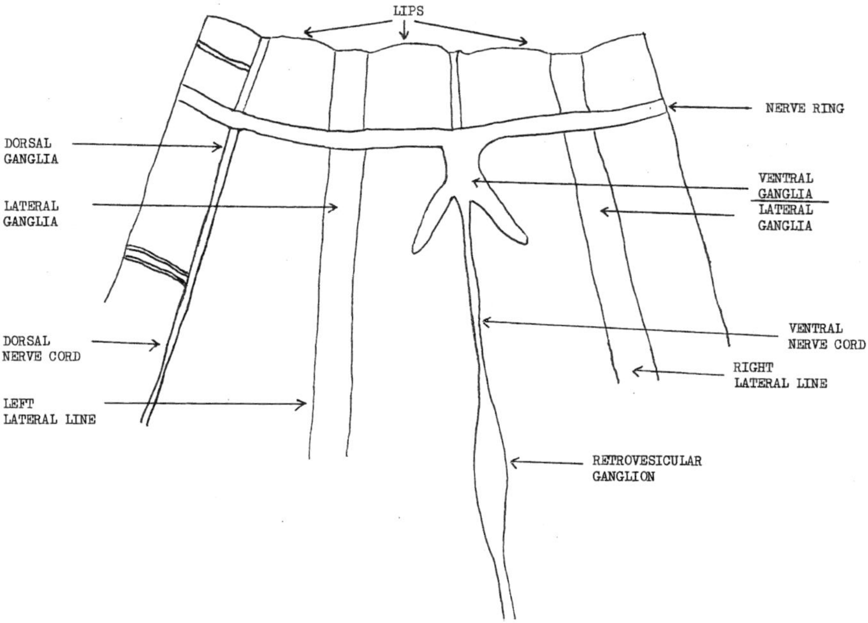

Figure 1. The anterior end of *Ascaris suum* opened out to show the anterior ganglia and nerve ring. (after Goldschmidt, 1908 and Angstad, Donmoyer and Stretton, 1989)

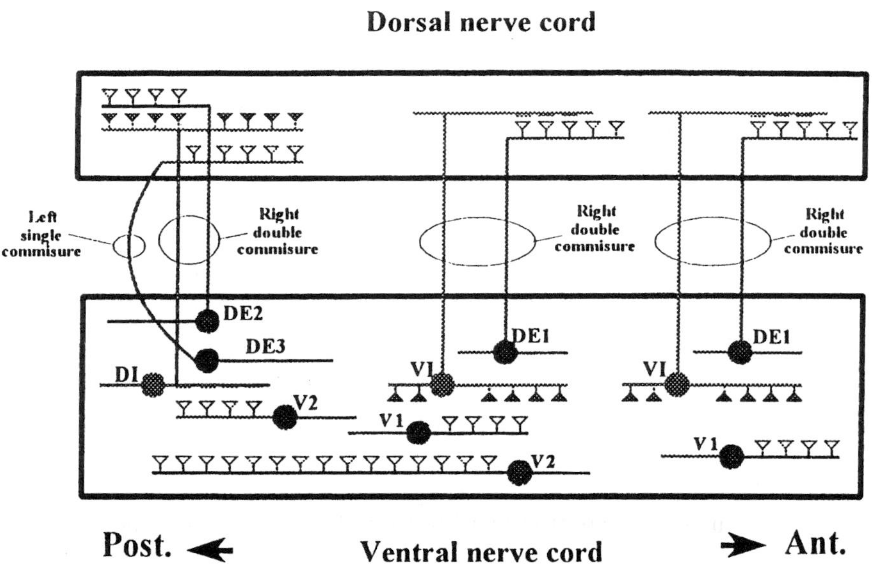

Figure 2. Schematic diagram showing the organisation of inhibitory (I) and excitatory (E) motoneurones in one body segment of *Ascaris suum.* There are seven anatomically distinct motoneurone types and a total of eleven motoneurones per segment. DE-1, DE-2, DE-3, dorsal excitors; D-1, dorsal inhibitor; V-1 and V-2, ventral excitors. (from Franks, 1996, after Martin *et al.*, 1991)

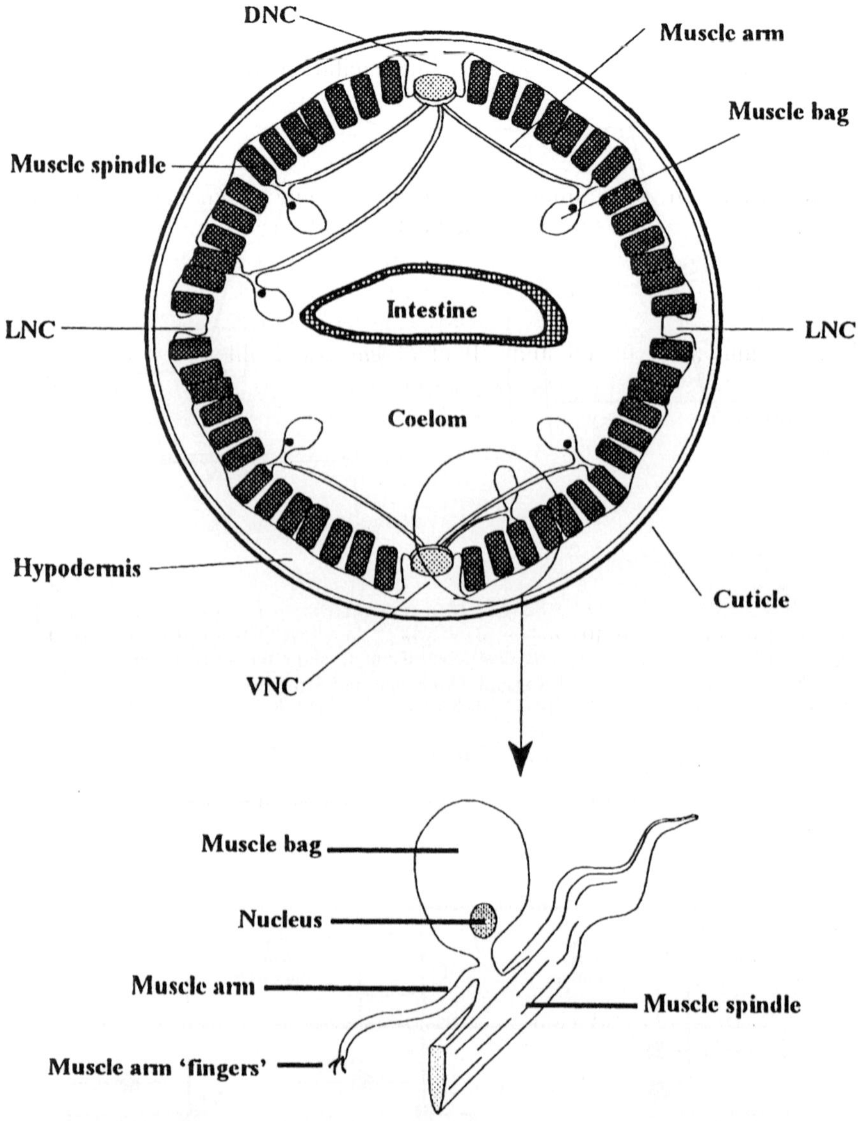

Figure 3. Diagrammatic representation of the body wall in *Ascari suum*, which shows the general structure of a somatic muscle cell. The muscle spindle, muscle bag and muscle arm are all part of a single muscle cell. Muscle arms project to either the dorsal (DNC) or ventral nerve cord (VNC), but do not cross the lateral line/nerve cord (LNC). (from Franks, 1996, after Croll and Matthews, 1977).

types also have processes in the dorsal cord (Fig. 2). These processes are linked through commissures which connect the two cords. The seven types of motoneurones have been divided functionally into excitatory and inhibitory motoneurones and designated DE-1, DE-2 and DE-3, VI and DI and V-1 and V-2. The first three are excitatory motoneurones which innervate the dorsal somatic musculature. V-1 and V-2 are excitatory onto the ventral somatic muscles, while VI and DI are inhibitory motoneurones which innervate, respectively, the ventral and dorsal somatic muscles. The inhibitory motoneurones contain GABA while the excitatory motoneurones contain ACh (Johnson and Stretton, 1985, 1987). DE-1, DE-2 and DE-3 are equivalent to DAS, DB and DA of *C. elegans*, D-I and V-I are equivalent to DD and VD of *C. elegans* and V-1 and V-2 are equivalent to VA and VB of *C. elegans*. Unlike most other nerve-muscle systems, the motoneurones do not send axons to the muscle cells. In nematodes, the muscle fibres have 1–5 arms which extend and make contact with the motoneurone fibre in the cord (Figs. 3 and 4).

3. Electrophysiological Studies

The first intracellular recordings from nematode somatic muscle were performed by Jarman (1959) using *A. lumbricoides*. In this study he recorded a resting membrane potential of around −30 mV upon which were imposed membrane oscillations. Some examples of somatic muscle activity are shown in Fig. 5. Jarman's pioneering research led to a detailed analysis of the ions involved in the generation of these po tentials. Unlike the situation with many excitable cells, the extracellular potassium concentration can be varied with relatively little effect on the membrane potential (Del Castillo, De Mello and Morales, 1964a; Brading and Caldwell, 1971). Their results will now be summarised. Increasing external potassium from 3 mM to 35 mM, whilst lowering external sodium, resulted in a non-linear hyperpolarization of the membrane potential of about 5 mV.

Above 45 mM potassium, a linear depolarization occurred with a gradient of 23 mV for a ten-fold change in potassium (Del Castillo, De Mello and Morales, 1964a). This is very different from the theoretical value of a 61 mV change in membrane potential for a ten-fold change in external potassium at 37°C. Brading and Caldwell (1964, 1971) found only a 1.5 mV change in membrane potential for a ten-fold change in external potassium through the range of 1–40 mM. When external chloride was increased ten-fold, there was a hyperpolarization in the membrane potential of 13 mV. Interestingly, if external potassium was increased while external chloride was decreased, then there was a much greater change in membrane potential: 59 mV change for a ten-fold change in chloride. Changing external sodium appeared to have little effect on the membrane potential, although zero sodium caused a hyperpolarization of 17 mV which suggests a reasonable resting permeability to this ion. A number of mechanisms were suggested to account for the membrane potential, including

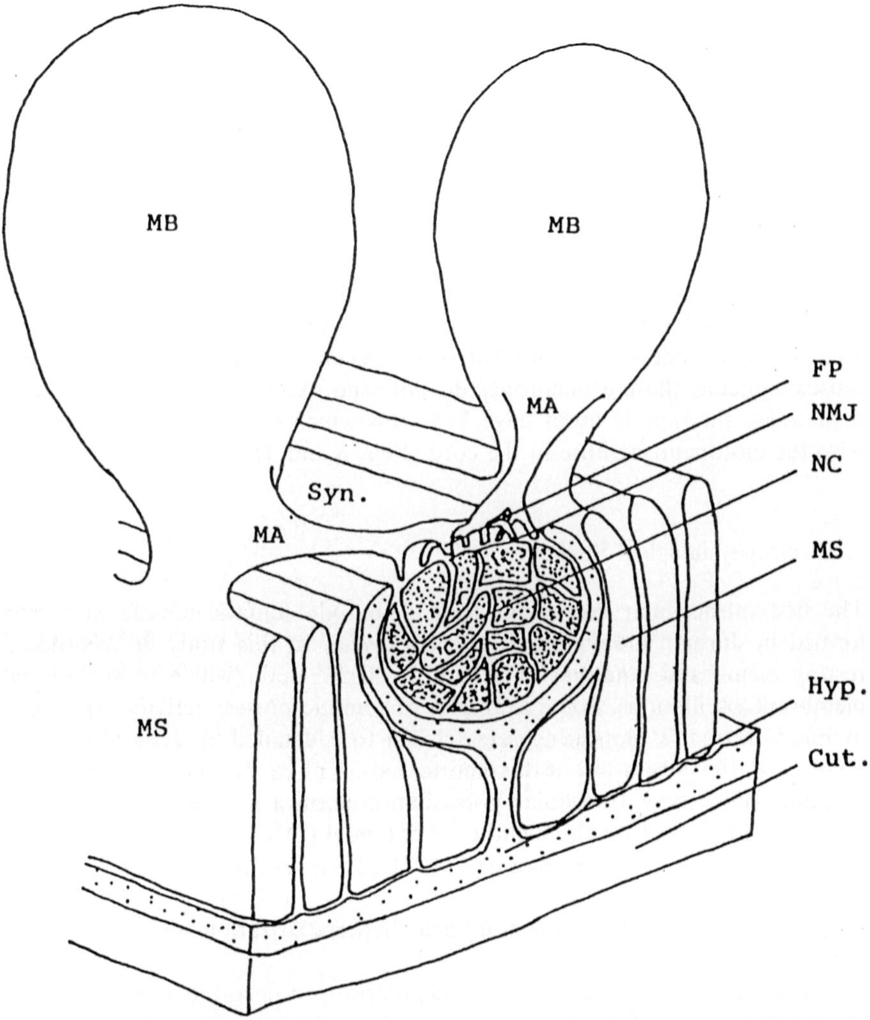

Figure 4. Diagram which illustrates neuromuscular innervation in *Ascaris suum*. The muscle spindles (MS) run longitudinally along the animal. The muscle bags (MB) send out muscle arms (MA) which terminate in fingerlike projections (FP). These intertwine with projections from other muscle cells and electrically couple to form a syncytium (Syn). The fingers synapse at the nerve cord (NC) to form the neuromuscular junction (NMJ). The nerve cord lies in a channel formed by the hypodermis (Hyp) which lies adjacent to the cuticle (Cut). (from Parri, 1991)

a role for organic anions which are present in *Ascaris* extracellular fluid or an electrogenic pump. Using labelled material, Caldwell and Ellory (1968) determined the relative permeabilities of the muscle membrane as 1:!:7 in terms of potassium, sodium and chloride, respectively. Weisblat and Russell (1976) observed that 2 mM ouabain depolarized the membrane potential by up to 12 mV, suggesting an electrogenic component.

The earlier work described above also reported the presence of oscillations and spike activity superimposed on the somatic muscle membrane potential (Fig. 5). This activity was further analysed by Weisblat and Russell (1976) who classified it into three forms of activity. For example, 30 mV action potentials with a duration of 10–50 msec, 15 mV slow waves of up to one second in duration and 5 mV modulatory waves lasting 3–20 seconds. The action potentials were mainly dependent on external calcium (Fig. 6) (Weisblat, Byerly and Russell, 1976) and were blocked by 5 mM cobalt. Changing external sodium did not appear to have any effect. However, removal of external calcium did not block the slow wave activity (Weisblat and Russell, 1976). In order to block this slow wave activity the removal of sodium was also required. It is likely that these muscle cells are myogenic. Muscle action potentials are temperature-dependent, where above 34°C their amplitude increases with temperature, while below 33°C their activity ceases (Wann, 1987).

Byerly and Masuda (1979) recorded from *A. lumbricoides* pharyngeal muscle under voltage clamp to analyse the potassium current that produced a negative-going regenerative action potential in the muscle. Constant current clamp experiments showed that the pharyngeal muscle could produce both positive-going and negative-going action potentials. These authors recorded a resting membrane potential of between −35 and −45 mV with positive-going action potentials, with an overshoot of 30–40 mV above zero potential. The membrane potential remained positive until returned to the negative resting level by the negative-going action potential. The level the membrane potential returned to depended on external potassium: for example, at 24 mM, the negative spike reached −45 mV, but in 2 mM external potassium it reached −110 mV. The

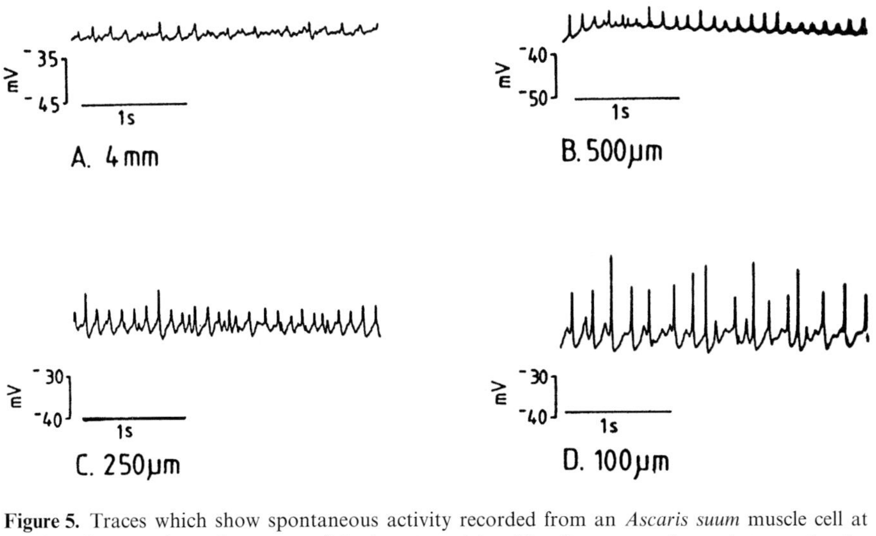

Figure 5. Traces which show spontaneous activity recorded from an *Ascaris suum* muscle cell at varying distances from the source of the largest activity. The distances under each trace give the distance from the site which produced the largest depolarizing spikes. (from Hewitt, 1987)

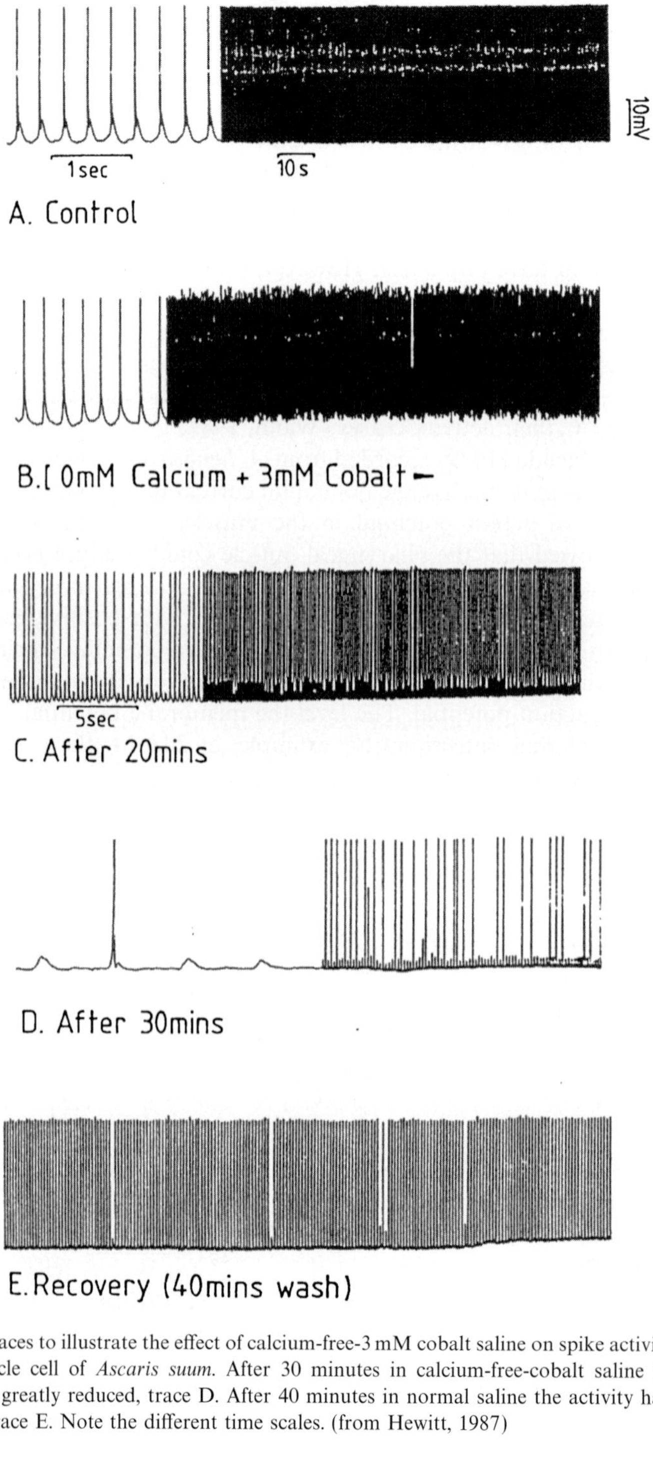

Figure 6. Traces to illustrate the effect of calcium-free-3 mM cobalt saline on spike activity recorded from a muscle cell of *Ascaris suum*. After 30 minutes in calcium-free-cobalt saline the spiking activity was greatly reduced, trace D. After 40 minutes in normal saline the activity had partially recovered, trace E. Note the different time scales. (from Hewitt, 1987)

negative-going spike is not coupled to the positive-going action potential, but is independently triggered. The negative spike current was not particularly reduced by 10 mM TEA, 10 mM 4-AP or 1 mM barium. However, 10 mM rubidium and 10 mM caesium showed voltage-dependent block. Thus, in this muscle the voltage-dependent inward current drives the transition from resting membrane potential to a positive plateau and the negative spike current drives the membrane potential from the positive plateau back to the negative resting level.

Raizen and Avery (1994) recorded electrical activity in the form of an electropharyngeogram (EPG) which could be correlated with pharyngeal pumping in *C. elegans*. The EPG could be divided into four phases: E, the initial excitatory phase; R, which correlates with relaxation of the pharynx; between E and R is the P (plateau) phase which contains negative transients; and finally the I (interpump) phase which does not correlate with pharyngeal activity. The pharynx is innervated by 20 neurones but will continue to pump when all 20 neurones are destroyed (Avery and Horvitz, 1989) and so is capable of myogenic activity. EPG recordings from mutant *C. elegans* can demonstrate the effects of mutations on pharyngeal muscle electrical activity (David *et al.*, 1995). These authors have also recorded intracellularly from pharyngeal muscle cells and found resting membrane potentials of around −45 mV with muscle action potentials up to 40 mV in amplitude. *eat-6* mutants, with reduced sodium pump function, have less-negative membrane potentials, around −35 mV, and reduced action potential amplitude, up to +15 mV. More recently, Franks, Walker and Holden-Dye (1997) have also made intracellular recordings from *C. elegans* pharyngeal muscle and found membrane potentials which were more negative than those reported by David *et al.*(1995). In these *al.* (1995). In these studies the membrane potential was around −68 mV with action potentials of around 85 mV total amplitude and duration of around 250 msec.

Intracellular recordings have been made from motoneurone commissures from *A. lumbricoides* (Stretton *et al.*, 1985). Their membrane resistance values were high, Em being 60–300 kΩ cm^2 with a membrane capacitance, Cm of 0.5 μF/cm^2. Stretton *et al.* found no evidence for voltage-sensitive channels equivalent to those producing delayed rectification, i.e., potassium channels. Strong depolarizing currents produced small onset transients which were reversibly blocked by cobalt, suggesting that they are due to calcium. No evidence was found for all-or-none action potentials in these neurones. It is likely that signal propagation in these cells is determined by the passive membrane properties of fibres, i.e. spike signals are graded and can be conducted over long distances. However, voltage-sensitive channels are present in commissural membranes. The membrane potentials of these cells are in the range of −25 to −44 mV, with little mean difference between the different motoneurones and are similar to the membrane potentials recorded from somatic muscle cells (Davis and Stretton, 1989a). These motoneurones possess spontaneous excitatory postsynaptic potentials (epsps) and inhibitory postsynaptic potentials (ipsps) and spontaneous oscillatory potentials. Davis and Stretton (1989b) found evidence for the tonic release of transmitter from motoneurones. Slow active potentials were recorded in V-I motoneurone which were blocked by cobalt or cadmium but not by TTX or by reducing external sodium (Angstadt and Stretton, 1989). These potentials were up to 20 mV in amplitude and of several hundred msec duration. Both amplitude and duration were increased in the presence of barium, strontium or TEA, 5 mM. It was concluded that these slow potentials were calcium mediated.

Recently, Goodman *et al.* (1998) have developed a tight-seal, whole cell patch-clamp recording method from identified *C. elegans* neurones *in situ*. These authors found no evidence for classical sodium action potentials in an identified chemosensory cell, ASER, or in a number of unidentified neurones. Thus, it would appear that these neurones can signal effectively without sodium action potentials. All neurones exhibited an outward potassium current, I_K, activated by depolarization. In ASER neurones, I_K was balanced by an inward calcium current which was increased when barium replaced all external divalent cations.

4. Classical Transmitters in Nematodes

It is likely that most, if not all, of the classical transmitters found in other invertebrate phyla, such as arthropods, annelids and molluscs, also occur in nematodes. However research on these compounds has not been uniform, with major studies involving acetylcholine (ACh) and gamma-aminobutyric acid (GABA). Possible physiological roles for 5-hydroxytryptamine (5-HT, serotonin) and glutamate have been investigated to a lesser extent. In addition it is likely that both dopamine and octopamine occur in nematodes and have physiological functions. The evidence for noradrenaline, adrenaline, histamine and glycine is less convincing. The role for each of these ligands will be reviewed.

4.1. ACETYLCHOLINE (ACh)

It has been known since the pioneering studies of Baldwin and Moyle (1949) that ACh can excite *A. lumbricoides* body wall muscle strips and it was then found that ACh is present in nematodes (Mellanby, 1955). It was later shown that the excitatory effect of ACh could be blocked by tubocurarine (Norton and DeBeer, 1957). Later electrophysiological recordings from somatic muscle confirmed that ACh depolarized the muscle membrane potential (Del Castillo, De Mello and Morales, 1963a). This depolarization is associated with an increase in conductance to both sodium and potassium ions, i.e., through a non-selective cation channel (Harrow and Gration, 1985), with a reversal potential around zero mV. The presence of both acetylcholinesterase (AChE) (Lee, 1962) and choline acetyltransferase (Johnson and Stretton, 1985) has also been confirmed. To complete the evidence, excitatory junction potentials (ejps) can also be recorded from somatic muscle following stimulation of the nerve cord and these ejps can be blocked by cholinergic antagonists, such as mecamylamine and benzoquinonium (Holden-Dye *et al.*, 1995). An example of benzoquinonium blocking an ejp is shown in Fig. 7.

All these data provide excellent evidence for ACh to be the major fast-acting excitatory transmitter onto somatic body wall muscle in nematodes, it being released from excitatory motoneurones DE-1–3 and V1–2.

Detailed analysis of the pharmacological profile of the ACh receptor has been performed by Colquhoun, Holden-Dye and Walker (1991). The most potent agonists were m-hydroxyphenylpropyltrimethylammonium (HPPT) and dimethylphenylpiperazinium (DMPP), both nicotinic agonists at mammalian autonomic ganglia. Both HPPT and DMPP were more potent than ACh. Nicotine was an agonist but around five times less potent compared to ACh. Muscarinic agonists were very weak or even induced a slight hyperpolarization, e.g. muscarine, pilocarpine and McN-A-343. An example of the effect of pilocarpine is shown in Fig. 8, where 10 mM slows spontaneous activity of the muscle cell while 1 mM induces a clear hyperpolarization. In this respect the nematode somatic muscle resembles that of the leech where ACh is the excitatory transmitter linked to a nicotinic receptor but where muscarinic agonists are weakly inhibitory (Woodruff, Walker and Newton, 1971). The most potent antagonists were benzoquinonium and mecamylamine, respectively striated muscle ACh receptor and ganglion ACh receptor blockers. Figure 9 shows the effect of

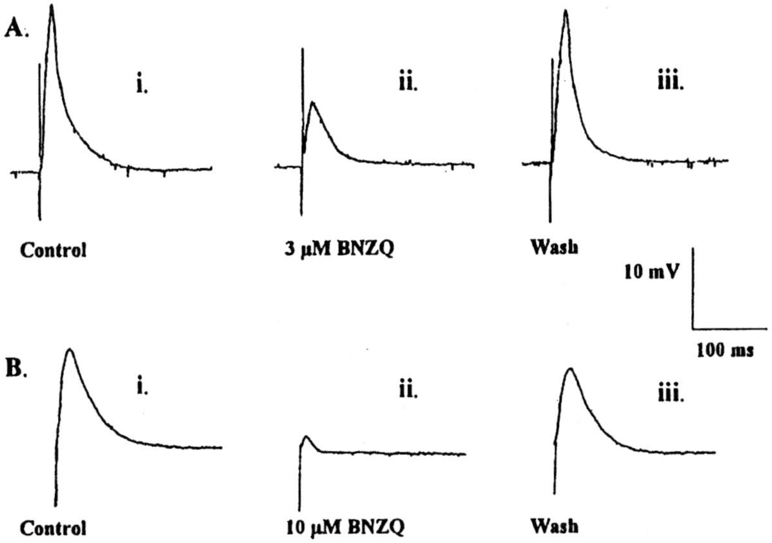

Figure 7. Traces to illustrate the effect of benzoquinonium (BNZQ) on the amplitude of excitatory junction potentials (ejps) recorded from a muscle cell from *Ascaris suum*. Each trace represents the average of eight consecutive ejps recorded (i) prior to addition of 3 μM BNZQ, (ii) during the maximum amplitude reduction in the presence of either 3 or 10 μM BNZQ and (iii) following a wash period of 10–15 minutes. (from Franks, 1996).

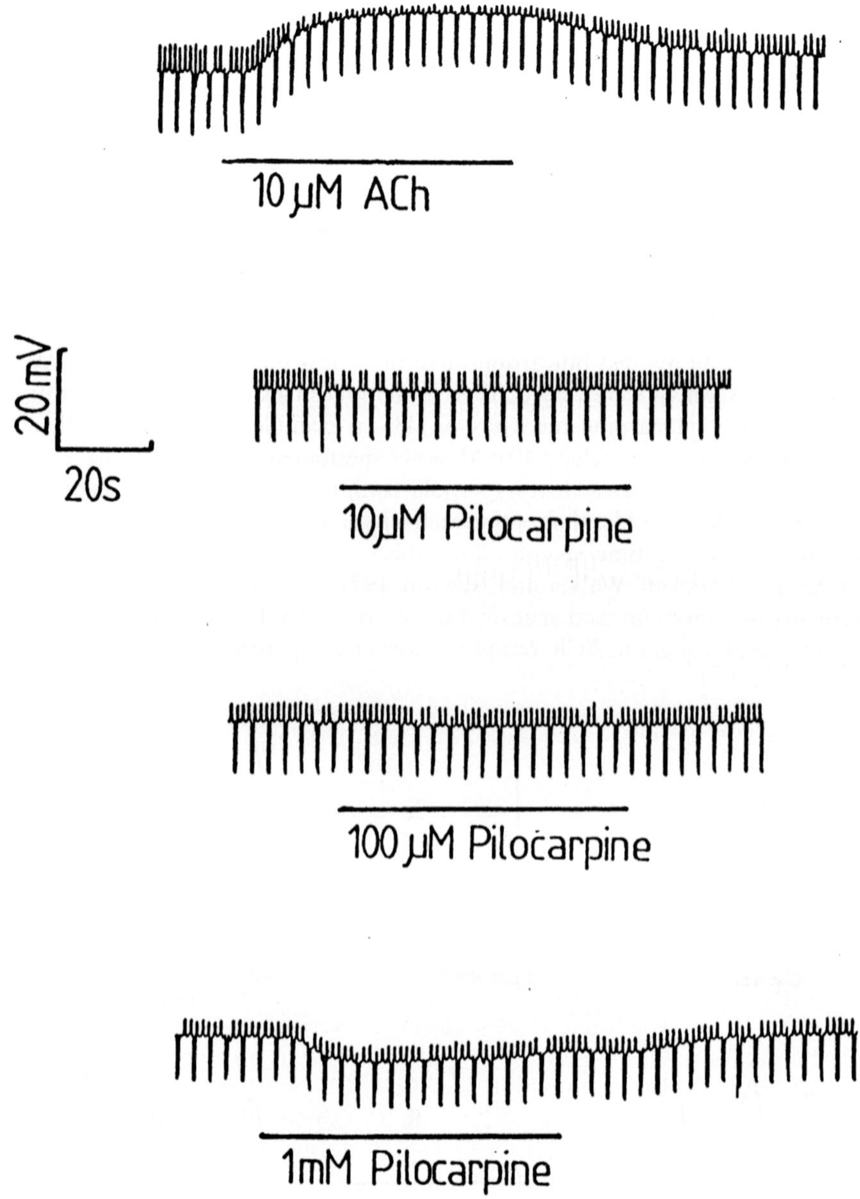

Figure 8. Traces to illustrate the action of 10 µM acetylcholine and increasing concentrations of pilocarpine on the membrane potential of an *Ascaris suum* muscle cell. In contrast to the depolarization produced by a low concentration of acetylcholine, a high concentration of pilocarpine was required to produce a hyperpolarization. (from Colquhoun, 1991).

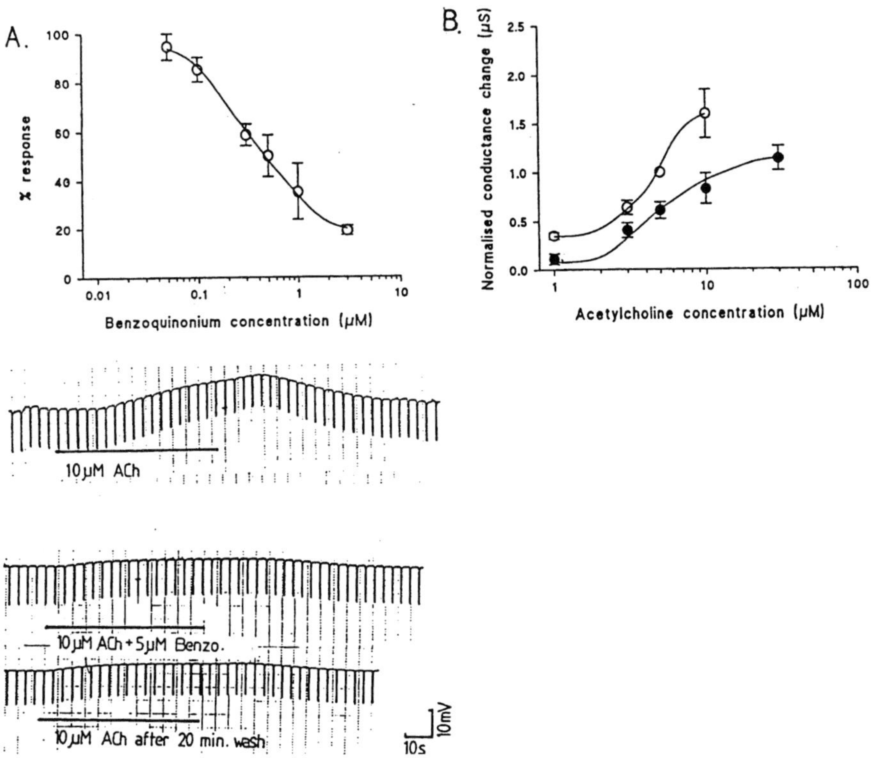

Figure 9. Traces and graphs to show the interaction beTween benzoquinonium chloride and acetylcholine on an *Ascaris suum* muscle cell. In graph (A) increasing concentrations of benzoquinonium gradually reduce the response to a standard concentration of acetylcholine. Graph (B) shows the effect of 10 μM benzoquinonium on a concentration curve to acetylcholine; control O, presence of antagonist ●. The traces show the blocking action of 5 μM benzoquinonium (Benzo) on the response to 10 μM acetylcholine (ACh). (from Colquhoun, 1991).

benzoquinonium on responses to ACh. Graph A shows the effect of increasing concentrations of benzoquinonium on a standard ACh response, while B shows concentration-response curves to ACh, control and in presence of antagonist. Interestingly, another ganglion blocker, dihydro-β-erytroidine, was inactive while α-bungarotoxin was relatively weak. In summary, this study suggested that the Ach receptor on somatic muscle was nicotinic and ganglion-like. In a further study (Colquhoun, Holden-Dye and Walker, 1992), a series of toxins was tested on the ACh receptor. Their order of decreasing potency as blockers was neuronal bungarotoxin > neosurugatoxin > N-methyllycaconitine = histrionicotoxin ≫ alpha-bungarotoxin ≫ conotoxin. By comparison with vertebrate ACh receptor studies, this profile would suggest that the *A. suum* ACh receptor most resembled the $\alpha_3\beta_2$ configuration of the nicotinic receptor. A further toxin, lophotoxin, which blocks vertebrate nicotinic ACh receptors, failed to block *A. suum* ACh receptors (Tornoe *et al.*, 1995). In the *C. elegans*

UNC-38 a subunit, tyrosine 190 is replaced by a proline and it is possible the same mayoccur in the *A. suum* ACh receptor α subunit. This could explain the lack of antagonism of lophotoxin at the *A. suum* ACh receptor. Two analogues of the wasp venom philanthotoxin were tested on *A. suum* nicotinic ACh receptors. Both were reversible antagonists with philanthotoxin 433, IC-50 $0.25 \pm 0.13\,\mu M$, being more potent than philanthotoxin 343, IC-50 $1.55 \pm 0.4\,\mu M$. Figure 10 shows the effect of philanthotoxin 433 on the ACh

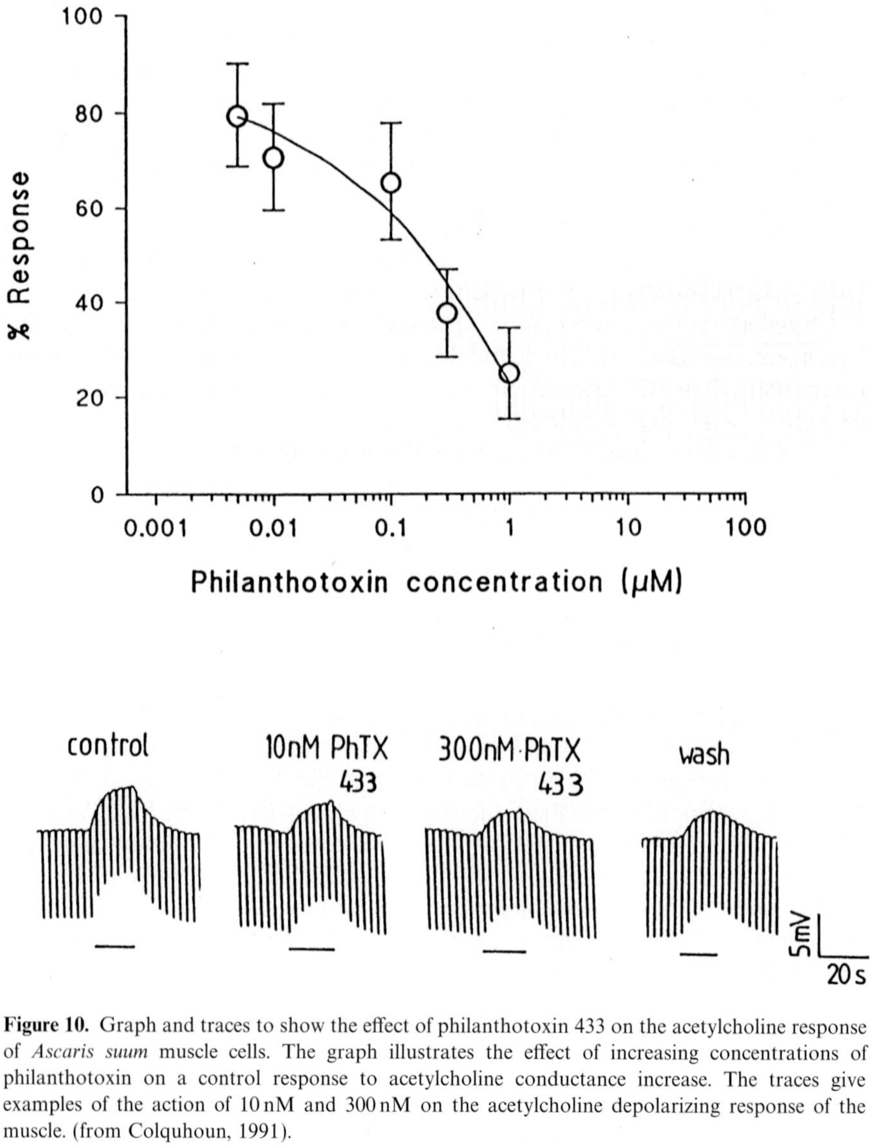

Figure 10. Graph and traces to show the effect of philanthotoxin 433 on the acetylcholine response of *Ascaris suum* muscle cells. The graph illustrates the effect of increasing concentrations of philanthotoxin on a control response to acetylcholine conductance increase. The traces give examples of the action of 10 nM and 300 nM on the acetylcholine depolarizing response of the muscle. (from Colquhoun, 1991).

response. In this figure, the graph shows the effect of increasing concentrations of toxin on a standard 5 μM ACh response. The traces show examples of the change in the ACh depolarization in the presence of 10 nM and 300 nM philanthotoxin 433.

The pharmacology of the ACh receptors on somatic muscle and neurones of *A. suum* has also been investigated by Segerberg and Stretton (1993). These authors found that both tubocurarine, 100 μM, and N-methyl-scopolamine, 100 μM, decreased the excitatory response to ACh of somatic muscle, though tubocurarine was more potent as an antagonist. N-methyl-scopolamine had a direct depolarizing action on the muscle. It was concluded that ACh receptors were present on DE-1, DE-2, D-I and V-I motoneurones since ACh reduced their input resistance. These neuronal receptors were blocked by tubocurarine. However, N-methyl-scopolamine directly lowered the input resistance of DE-1 but not of the other motoneurones tested. This direct effect of N-methyl-scopolamine was resistant to tubocurarine and may not be a specific effect on ACh receptors.

The action of ACh has been investigated on the vagina vera which is the distal part of the ovijector of *A. suum* (Fellowes *et al.*, 2000a). These authors found that ACh excited this muscle with a threshold of around 1 μM.

A number of current anthelmintics act at the nicotinic ACh receptor including levamisole, pyrantel, morantel and bephenium (Martin *et al.*, 1996). These compounds depolarize the muscle membrane potential in a similar manner to ACh and act through the non-selective cation channel (Harrow and Gration, 1985). While these ligands are agonists at low concentrations, at high concentrations their depolarization action is reduced. These compounds are large organic cations and probably try to enter the ion channel pore when it is open. They are too large to pass through and so block the open nicotinic ACh receptor channel (Martin *et al.*, 1996). This type of block is enhanced when the membrane potential is enhanced when the membrane potential is hyperpolarized. There is evidence for two blocking sites within the ion channel (Martin *et al.*, 1996).

Studies using *C. elegans* have clearly confirmed the importance of the cholinergic system in nematodes. These studies have been very well reviewed by Sattelle and Culetto (1999) where a number of genes have been identified which are involved in the synthesis and transport of ACh and also genes involved in nicotinic ACh receptor and AChE proteins. Thus, *C. elegans* is an ideal model to perform a functional analysis of genes encoding cholinergic synaptic proteins. Over 30 genes have been identified which play a role in the *C. elegans* ACh system. Around 20 nicotinic ACh receptor subunits have been identified in *C. elegans*, of which 13 are α subunits.

4.2. 5-HYDROXYTRYPTAMINE (5-HT, SEROTONIN)

The indolealkylamine 5-HT occurs widely in thc animal kingdom and is probably a transmitter in most, if not all, phyla (Walker, Brooks and Holden-Dye, 1996). It certainly has a transmitter role in the major phyla that have been studied, such as arthropods, molluscs and annelids. Evidence for a transmitter role in nematodes is based on chemical identification, synthesis, immunocytochemical localization in identified neurones, physiological actions and molecular biology identification of receptors (Chaudhuri, Martin and Donahue, 1988a; Chaudhuri and Donahue, 1989; Barreteau *et al.*, 1991; Brownlee *et al.*, 1994a; Brownlee, Holden-Dye and Walker, 1995; Johnson *et al.*, 1996; Olde and McCrombie, 1997; Hamdan *et al.*,1999; Huang *et al.*, 1999). 5-HT plays a transmitter or neurohormonal role in key physiological processes in nematodes, viz., feeding, locomotion, egg-laying and metabolic regulation.

It is 20 years since Donahue *et al.* (1981) demonstrated that 5-HT activates glycogenolysis in *A. suum*, acting as a neurohormone. 5-HT can both be synthesised from tryptophan by nematodes and absorbed from exogenous sources (Chaudhuri, Martin and Donahue, 1988a, b). Specific binding of 5-HT occurs in *A. suum* tissues: for example, specific binding was three times greater in somatic muscle compared with intestinal muscle. Ketanserin, a 5-HT-2 antagonist, reacted differently in the two tissues, possibly suggesting two distinct receptor proteins. Evidence from studies using mammalian 5-HT agonists and antagonists suggested a greater affinity for 5-HT-2-like receptors on somatic muscle than for 5-HT-1-like receptors. There is a single 5-HT receptor on somatic muscle which is linked to cAMP. However, Williams *et al.* (1992) were unable to associate this receptor with any of the mammalian 5-HT subtypes and preferred to designate it as a 5-HT-N (N for nematode) receptor.

Using immunocytochemical techniques, 5-HT has been localized in the pharynx of *A. suum* (Stretton, 1992) and is localized in specific neurones in the enteric nervous system (Brownlee *et al.*, 1994a). Reinitz and Stretton (1996) demonstrated that injection of 5-HT into *A. suum* perienteric fluid caused paralysis with an increase in body length. 5-HT can also inhibit ejps recorded from somatic muscle in this species. 5-HT can likewise reduce ACh-induced contractions of *A. suum* somatic muscle (Trim *et al.*, 2001) and this effect may be associated with stimulation of adenylate cyclase since 5-HT increases resting levels of cAMP in this muscle. The action of 5-HT to reduce ACh contractions is mimicked by 8-OH-DPAT and 5-carboxyamidotryptamine, both 5HT-1 agonists, tryptamine and α-methyl-5-HT. Figure 11 is an example where α-methyl-5-HT, a 5HT-2 agonist, reduced ACh-induced contractions of *A. suum* somatic muscle. Ketanserin partially blocked this effect of α-methyl-5-HT on ACh contractions. Interestingly, fluoxetine, the 5-HT uptake blocker, potentiated the inhibitory effect of 5-HT on ACh-induced contractions. These pharmacological studies suggest a 5-HT receptor with a 5-HT-2-like profile, but which stimulates adenylate cyclase.

There is evidence from studies using *C. elegans*, that 5-HT interacts with

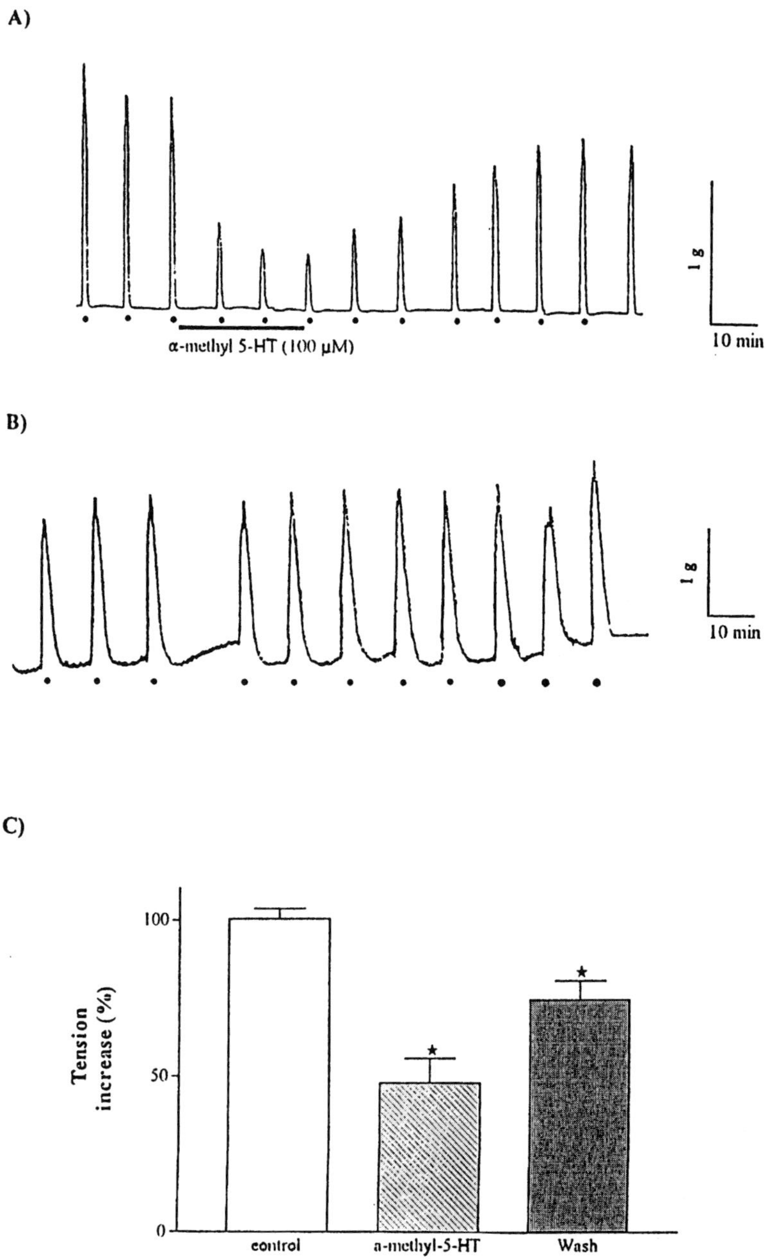

Figure 11. Traces and a histogram to show the effect of α-methyl 5-HT on contractions of a muscle strip of *Ascaris suum* induced by 30 μM acetylcholine. Trace (A) shows that the response to acetylcholine is reversibly reduced in the presence of α-methyl 5-HT. Trace (B) is a muscle strip from the same animal responding to acetylcholine in the absence of α-methyl 5-HT. In (C) the data were normalised and expressed as a percentage of the muscle tension increase in the presence of acetylcholine. (from Brooman, 1998).

ACh in this species. For example, 5-HT inhibits the release of ACh from motoneurones onto body wall muscle (Nurrish, Segalat and Kaplan, 1999). This 5-HT effect is inhibited by two 5-HT antagonists, methiothepin and ketanserin. Interestingly, these authors found no evidence that 5-HT inhibited the direct action of ACh on somatic muscle. In another study using *C. elegans*, Schafer, Sanchez and Kenyon (1996) found evidence for ACh inhibiting the excitatory effect of 5-HT on vulva muscle and so affecting egg laying.

5-HT failed to stimulate adenylate cyclase in pharyngeal muscle but did increase inositol phosphate levels in this muscle (Trim *et al.*, 2001). 5-HT stimulates pumping of the *A. suum* pharynx, EC-50 of 44 μM, an effect which is antagonised by ketanserin. This provides some evidence for a 5-HT-2-like receptor on pharyngeal muscle. However, it must be remembered that ketanserin is not specific for 5-HT-2-like receptors in mammals but will interact with other receptors, including adrenoceptors. It can also be used to distinguish between 5-HT-1-B and 5-HT-1-D receptors. In contrast to 5-HT, tryptamine failed to stimulate pumping, indicating the importance of the hydroxyl group on position 5 of the indole nucleus. When low mM tryptamine was perfused with 5-HT, pumping became irregular, suggesting an interaction between tryptamine and the 5-HT receptor. Sumatriptan, 8-OHDPAT and 5-carboxy amido -tryptamine all failed to stimulate pumping.

5-HT occurs in identified neurones in *C. elegans* within the enteric nervous system, in the male tail and in the female reproductive system (Schafer, Sanchez and Kenyon, 1996; Nurrish, Segalat and Kaplan, 1999; Sze *et al.*, 2000). The rate limiting enzyme for 5-HT synthesis in vertebrates and presumably nematodes is tryptophan hydroxylase. There is one probable orthologue of mammalian tryptophan hydroxylase in the *C. elegans* genome, ZK1290.2, which is called *tph-1* (Sze *et al.*, 2000). *tph-1* shares 71% homology with mammalian tryptophan hydroxylase in the 330 amino acid catalytic domain. *tph-1* is expressed in the NSM neurosecretory cells in the pharynx; ADF cells; hermaphrodite-specific HSN neurones, which innervate the vulva muscles; male- specific CP cells which have a role in male mating behaviour; and rarely in AIM and RIH cells. Sze *et al.* (2000) generated a *tph-1* deletion mutant which was viable but exhibited a range of behavioural and metabolic defects. These animals did not appear to accumulate any detectable 5-HT but possessed a normal dopamine distribution. The pharynx of these mutants pumped slowly and they retained more eggs than wild type *C. elegans*. In the absence of food the pharynx pumping rate of tph-1 was similar to the wild type, but the rate did not increase in the presence of food. This provides further evidence that 5-HT is required for enhanced pharyngeal pumping but not for basal activity. These *tph-1* mutants also exhibited an altered metabolic state and it is likely that 5-HT plays a role in both the insulin-like and DAF-7 transforming growth factor (TGF) b signalling pathways. An interesting parallel is that 5-HT enhances insulin synthesis, release and target tissue sensitivity in mammals (Breum *et al.*, 1995). The link between 5-HT and food appears to have evolved at an early stage in animals.

There have been three recent molecular biology studies involving 5-HT receptors in nematodes. Huang *et al.* (1999) prepared cDNA from poly A^+ RNA isolated from *A. suum* pharyngeal muscle. From this material they identified a putative 5-HT receptor cDNA, designated AS-1. This receptor showed greatest homology with mammalian 5-HT-2 receptors. AS-1 has seven transmembrane domains and so acts through a G protein. When expressed in COS-7 cells, the receptor showed high affinity, saturable binding of (^{125}I) LDS. 5-HT was the only amine which inhibited this binding, with an apparent K_i of around 0.4 μM. Neither dopamine nor octopamine, 100 μM, had any effect on this binding. Recently, this group has found evidence for a number of 5-HT receptor isoforms which are probably generated by alternative-splicing (Huang *et al.*, 2000). Sequence analysis suggests that these AS isoforms are related to the *C. elegans* 5-HT-2Ce receptor. AS-1, AS-2 and AS-3 isoforms have different C terminals, while AS-4Δ4 and AS-2Δ4 lack 42 amino acids at the C terminus of the third intracellular loop. AS-1 and AS-2 are expressed in somatic muscle and pharynx, while AS-1Δ4 and AS-3 only occur in the pharynx and AS-2Δ4 only occurs in somatic muscle. The pharmacological profiles of AS-1, AS-1Δ4 and AS-2 are similar when expressed in HEK 293 cells, but differ from the 5-HT-2Ce pharmacological profile. In HEK 293 cells expressing AS-1 and AS-2 receptors, 5-HT stimulates inositol triphosphate (Ins $(1,4,5)P_3$) production. AS-1?4 does not appear to couple to phosphoinositol turnover, indicating the importance of the C terminal of the third intracellular loop for G-protein coupling.

In a second study, Hamdan *et al.* (1999) cloned a novel receptor cDNA from *C. elegans.* This receptor had high sequence homology with mammalian 5-HT-2 receptors and so they called it 5-HT-2Ce. Unfortunately, when expressed in COS-7 cells, 5-HT-2Ce exhibited a pharmacological profile which resembled both mammalian 5-HT-1 and 5-HT-2 receptors. Activation of 5-HT-2Ce resulted in a rise in intracellular calcium in COS-7 cells.

So far, these studies suggest the presence of a 5-HT-2-like receptor in nematodes. Not surprisingly, considering the time scale since chordates and nematodes separated from a common helminth stock, ligands designed for mammalian receptors fail to be selective for nematode receptors. However, it should be possible in the light of molecular biology studies to attempt to design ligands selective for nematode 5-HT receptors and to distinguish between different nematode receptor subtypes.

Recently, an ionotropic 5-HT receptor has been identified in *C. elegans* (Ranganathan, Cannon and Horvitz, 2000). The gene for this receptor, designated as *mod-1* (for modulation of locomotion defect), is expressed in neurones in the head and tail ganglia and in the ventral cord. No expression was observed in muscle cells. The receptor protein, MOD-1, is similar to members of the GABA and glycine family which gate chloride. When the gene is expressed in *Xenopus* oocytes, the receptor is chloride-dependent, where a ten-fold change in external chloride results in a change in membrane potential of 53 mV. Interestingly, MOD-1 is not blocked by vertebrate 5-HT-3 receptor antagonists, but by vertebrate metabotropic 5-HT receptor antagonists, such as mianserin and methiothepin. *mod-1* mutants are defective in a 5-HT-mediated experience-dependent behaviour and 5-HT fails to affect their swimming behaviour.

There are a number of mechanisms whereby 5-HT and other monoamines may be inactivated. These include re-uptake systems, monoamine oxidase (MAO) and N-acetyltransferase (NAT) (Isaac, MacGregor and Coates, 1996). N-acetylation appears to be an important method for 5-HT inactivation in nematodes (Smart, 1988; Isaac, Muimo and MacGregor, 1990; Isaac *et al.*, 1991). There is some evidence for an uptake system for 5-HT in nematodes. For example, the 5-HT uptake blocker imipramine stimulates pharyngeal pumping in *C. elegans* (Avery and Horvitz, 1990). Weinshenker, Garriga and Thomas (1995) concluded that *C. elegans* has a fluoxetine-sensitive 5-HT re-uptake system where the effect of fluoxetine on egg laying depends on the presence of 5-HT. Fluoxetine appears to have a direct inhibitory effect on ACh-induced contractions of *A. suum* somatic muscle (Trim *et al.*, 2001). This effect may be independent of any interaction with 5-HT uptake mechanisms.

4.3. DOPAMINE, NORADRENALINE AND OCTOPAMINE

The enzymes required for the synthesis of dopamine, noradrenaline, octopamine and 5-HT, i.e., tyrosine hydroxylase, dopamine-β-hydroxylase, aromatic amino acid decarboxylase and tryptophan hydroxylase, together with a monoamine transporter are all present in the *C. elegans* genome and are likely to occur in all nematodes (Bargmann, 1998). This would suggest that dopamine, noradrenaline and octopamine are present in nematodes though the levels appear to be variable between species, but this may in part be due to the detection methods employed. For example, in *Nippostrongylus brasiliensis*, dopamine was present in larvae and adults, 138.0 and 19.8 ng/g, respectively (Barreteau *et al.*, 1991), but octopamine and adrenaline could not be detected. Noradrenaline ran with the solvent front and so could not be assayed. In the filarial nematode, *Setaria cervia*, dopamine was the major amine but noradrenaline was also detected (Agarwal *et al.*, 1990). Fluorescence studies suggest that dopamine is the main catecholamine in *Prionchulus punctatus*, *Panagrellus redivivus* and *Aphelenchus avenae* (Wright and Awan, 1978). In the case of *A. avenae*, a dopamine content of 41ng/g wet weight was obtained. In fluorescence and labelled uptake studies evidence was obtained for a high affinity uptake of labelled noradrenaline into papillary nerves of *Phocanema decipiens* (Goh and Davey, 1976a,b). There was also evidence for dopamine being localized in the lateral nerves. In the second paper, Goh and Davey present further evidence for the occurrence of catecholamines in the central nervous system and a possible role in the control of ecdysis in *P. decipiens*; thus, application of noradrenaline to third instar larvae brings about ecdysis.

In an excellent study, Sharpe and Atkinson (1980) identified dopamine-containing neurones in the anterior nervous system of *P. redivivus* and

Trichostrongylus colubriformis. In female *P. redivivus*, a pair of dopamine-containing neurone cell bodies were identified in the region of the vulva in the tail region. These authors showed that the location of dopaminergic cells in *P. redivivus* and *T. colubriformis* resembled that in *C. elegans* described by Sulston, Dew and Brenner (1975). In *C. elegans* there are eight neurones in the hermaphrodite which contain dopamine, including the cephalic and deirid mechanosensory neurones. The male has six additional dopamine-containing neurones in the tail. There are five genes which affect the structure or localization of dopamine in the dopamine-containing neurones, viz., *cat*-1 to 5 (catecholamine deficient) (Sulston, Dew and Brenner, 1975). In *cat*-2 and *cat*-4 mutants, the level of dopamine is greatly reduced or absent (Rand and Nonet, 1997). Sulston, Dew and Brenner (1975) also studied the nervous system of *A. lumbricoides* and found a very similar fluorescence pattern of dopamine-containing cells to that of *C. elegans*.

The level of biogenic amines has been examined in *A. lumbricoides* and *A. galli* (Mishra, Sen and Ghatak, 1984). For dopamine and noradrenaline in *A. lumbricoides*, they found values in the range of 0.15–0.43 μg/g tissue and 0.015–0.09 μg/g tissue, respectively. For comparison, the levels of 5-HT were in the range of 0.73–2.75 μg/g tissue. As with *A. lumbricoides*, the levels of noradrenaline were in general lower than those for dopamine in *A. galli*. However, the position was reversed in the intestine. Evidence was found for noradrenaline, adrenaline and DOPA in the gonads of *A. galli* and for dopamine in both the intestine and gonads (Smart, 1988). *A. galli* was able to synthesise noradrenaline and adrenaline, but not dopamine, from DOPA. Tyrosine hydroxylase activity was demonstrated, indicating that *A. galli* can synthesise catecholamines from tyrosine.

N-Acetylation of biogenic amines has been investigated by Isaac, Muimo and MacGregor (1990) and Isaac *et al.* (1991) and amine metabolism in general reviewed in nematodes by Isaac, MacGregor and Coates (1996). *Brugia pahangi* and *A. galli* were both shown to be capable of forming N-acetyldopamine, N-acetyloctopamine and N-acetyl-5-HT and N-acetyltransferase activity was detected. Monoamine oxidase activity is present in some species of nematode, but may not be a major method for amine inactivation, although there is evidence for monoamine oxidase from the *C. elegans* genome. There is also some evidence for the presence of catechol-O-methyltransferase (COMT), but there have been few studies involving this enzyme in nematodes (Isaac, MacGregor and Coates, 1996).

Using HPLC, Frandsen and Bone (1987) have analysed homogenates from *T. colubriformis* for biogenic amines. These authors found evidence for a number of compounds, including N-acetyldopamine, 1.28 μg/g wet weight, dopamine, 0.94 μg/g wet weight, adrenaline, 5.81 μg/g wet weight, noradrenaline, 6.38 μg/g wet weight, octopamine, 2.46 μg/g wet weight and 5-HT, 0.046 μg/g wet weight. The levels for noradrenaline and adrenaline are particularly high in this study compared to other studies. In agreement with the work of Isaac, Muimo and MacGregor (1990) and Isaac *et al.* (1991), Frandsen and Bone found evidence for N-acetylation of biogenic amines in this species. In a second paper, Frandsen and Bone (1988) incubated labelled ligands in homogenates of *T. colubriformis* and found that vanillylmandelic acid (VMA), 4-hydroxy-3-methoxyphenyl glycol (HMPG) and metanephrine were formed from adrenaline. VMA, HMPG and normetanephrine were formed from noradrenaline. Octopamine was converted to noradrenaline, synephrine and adrenaline.

Dopamine can alter *C. elegans* behaviour: for example, it can inhibit both movement and egg laying, while 5-HT stimulates egg laying (Schafer and Kenyon 1995). Pretreatment with either dopamine or 5-HT prevents the effect on egg laying of the respective amine, i.e., induces adaptation or desensitization. Schafer and Kenyon found that the gene *unc-2* is required for adaptation of the two ligands. Interestingly, *unc-2* encodes a homologue of a voltage-sensitive calcium channel a1 subunit. Weinshenker, Garriga and Thomas (1995) have analysed the role of transmitters in regulating egg laying in *C. elegans*. They studied a mutant *egl-2* in which imipramine can stimulate egg laying but not 5-HT. It was found that dopamine antagonists of the mammalian D-2 type, such as chlorpromazine, haloperidol, butaclamol, droperidol and pimozide all stimulated egg laying and defaecation. Tricyclic antidepressants also stimulated egg laying in *egl-2* mutants. These authors suggested that both dopamine antagonists and tricyclics act on a common site.

Octopamine occurs in *C. elegans* (Horvitz *et al.*, 1982) at a concentration of 0.13 µg/g wet weight, a value similar to that determined for dopamine, 0.145 µg/g wet weight (Sulston, Dew and Brenner, 1975). Since noradrenaline appears to be absent in *C. elegans*, it is likely that dopamine and octopamine are synthesised in different neurones,i.e., one with a tyrosine hydroxylase and the other with tyramine-b- hydroxylase. Three *C. elegans* mutants, *def*-10, *che-3* and *osm-3* lack detectable levels of octopamine.

There have been relatively few physiological studies to investigate the role of catecholamines and octopamine in nematode function. Both dopamine and octopamine inhibit ejps recorded from *A. suum* somatic muscle cells (Chen, Holden-Dye and Walker, 1997). The IC-50 value for dopamine was 0.3 µM, while that for octopamine was greater than 10 µM. Similarly, noradrenaline was less potent in inhibiting ejps and the IC-50 was greater than 10 µM. At 37°C dopamine, 10 µM, caused a small hyperpolarization to the membrane potential without any clear effect on muscle cell input conductance. In spontaneously active muscle cells, dopamine, 50 µM, reduced muscle action potential frequency and amplitude. These studies provide evidence for a dopamine receptor within the *A. suum* motor nervous system which, when activated, should inhibit body movement and support the studies on *C. elegans* by Schafer and Kenyon (1995). Dopamine also inhibits pharyngeal pumping in *A. suum* but is less potent than GABA (Brownlee, unpublished observation). In contrast, dopamine excites pharyngeal pumping in *C. elegans* (Rogers, unpublished). Using intracellular recording techniques, octopamine has been shown to have an inhibitory effect on pharyngeal pumping in *C. elegans*, threshold 1 µM (Rogers *et al.*, 2000). This octopamine effect does not appear to involve a change in pharyngeal muscle membrane potential and was absent in the mutant *goa-1*.

4.4. GLUTAMIC ACID

Some of the best evidence for glutamate as a transmitter in nematodes comes from the molecular biology studies of Cully and her colleagues using *C. elegans* (Cully *et al.*, 1996). This work was prompted by studies to determine the site of action of avermectins. *C. elegans* is very sensitive to avermectins, with an EC-50 value of around 0.1 μg/ml (~100 pM) for a ten minute incubation (Kass *et al.*, 1980). A specific binding site for avermectins has also been investigated using *C. elegans* membranes (Schaeffer and Haines, 1989). The affinity of avermectins for this binding site was 100-fold greater than for a site identified in rat brain. GABA, taurine, glutamate and dopamine had no effect on this avermectin binding site. In later studies, *Xenopus* oocytes injected with *C. elegans* poly(A)$^+$RNA expressed receptors which were activated by avermectin (Arena *et al.*, 1991). These responses were linked with chloride channels and were picrotoxin-sensitive, but failed to respond to GABA, muscimol, glycine or bicuculline. The avermectin-sensitive current was blocked by flufenamic acid. In related studies, these authors also found evidence for a glutamate-sensitive chloride current in *Xenopus* oocytes injected with *C. elegans* poly(A)$^+$RNA (Arena *et al.*, 1992). This current desensitised in the presence of glutamate and was directly activated by avermectin, 1 μM. Interestingly, low concentrations of avermectin, for example, 2 nM, also potentiated the response to glutamate. This current was activated by ibotenate and weakly by aspartate and AMPA. However, NMDA, quisqualate and kainate were inactive. The observation that this avermectin-sensitive current could also be activated by glutamate led Arena *et al.* (1995) to establish an excellent correlation between biological activity of avermectins and milbemycin and their ability to bind to a high affinity ivermectin binding site and so directly activate and potentiate this glutamate-gated chloride channel expressed in *Xenopus* oocytes. Arena *et al.* (1995) concluded that this is the site of nematocidal activity of avermectins.

Glutamate-gated chloride channel cDNA has been isolated from *C. elegans* by expression cloning in *Xenopus* oocytes (Cully *et al.*, 1994). Two subunits were identified and designated GluClα and GluClβ. These two proteins displayed approximately 21% homology with GABA-A and glycine receptors, the conservation being particularly good in the transmembrane domains and the N-terminal domain. In addition, a GluClα 2 cDNA has been isolated which shows 75% and 49% amino acid identity with GluClα and GluClβ subunits, respectively (Vassilatis *et al.*, 1997). The GluClα and GluClβ heteromeric channel responded to both glutamate and ivermectin. The glutamate response was reversible and desensitised in the presence of high glutamate concentrations, while that to ivermectin was irreversible. The individual subunits GluClα and GluClβ expressed functional homomeric channels. GluClα responded selectively to ivermectin, while GluClβ responded to glutamate. In addition to a direct effect, low nM ivermectin potentiated the action of glutamate on both GluClβ homomeric and on GluClα/GluClβ heteromeric channels. The EC-50 for ivermectin on either the GluClα homeric or the heteromeric channels was in the range 140–190 nM, while the EC-50 for glutamate on GluClβ channels was 380 μM, which increased to 1360 μM in the case of heteromeric channels. The only full agonist for GluClβ channels was ibotenate. There is some evidence that the N terminal of GluClα contains a glutamate binding site. When GluClα threonine at 308 in the M2 domain is replaced with a proline, glycine or alanine, glutamate is able to gate the channel (Etter *et al.*, 1996). Interestingly, an ivermectin resistant *C. elegans* mutant, *avr-15*, displays abnormal eating behaviour and the pharyngeal muscle of this mutant does not respond to glutamate. This suggests that the mutants lack neurotransmission by the glutamate-containing M3 motoneurone to the pharynx muscle cell pm4 (Cully *et al.*, 1996). The *avr-15* gene encodes GluClα2 protein and expression of this protein in pharyngeal muscle restores M3 transmission and avermectin sensitivity (Dent, Davis and Avery, 1997). These authors conclude that M3 releases glutamate which acts as a fast inhibitory transmitter onto pharyngeal muscle. Both M3 and glutamate shorten the duration of muscle contractions during pumping. The *avr-15* mutant lacks M3-generated inhibitory junctional potentials (ijps) and is insensitive to glutamate. Using reporter gene constructs, Laughton, Lunt and Wolstenholme (1997) have presented evidence that GluClβ is expressed solely on the pharyngeal muscle cells.

Intracellular recordings have been made from the pharynx of *C. elegans* to investigate the properties of the native glutamate receptors using wild and mutant strains (Pemberton *et al.*, 2001). In this study, the muscle resting potential was -78 ± 1 mV with action potentials at 1–2Hz. These action potentials were coupled 1:1 with contractions of the terminal bulb of the pharynx. Glutamate depolarized this muscle through an increase in permeability to chloride. This depolarization was associated with inhibition of the action potentials. In the wild type, glutamate and ibotenate have EC-50 values of 166 μM and 17.8 μM, respectively. In contrast, the EG-50 for ivermectin was 2.7 nM, five orders of magnitude more potent than glutamate. The threshold for ivermectin was 10 pM compared with 10–100 nM at the GluClα1 and GluClα2B subunits expressed in *Xenopus* oocytes (Cully *et al.*, 1994; Vassilatis *et al.*, 1997). While *avr-14* was responsive to ivermectin, *avr-15* failed to respond to ivermectin up to 1 μM. In contrast to earlier findings, *avr-15* did respond to glutamate though the EC-50 was increased to 1.15 mM. While the wild type glutamate response desensitised, that of the *avr-15* mutant did not desensitise. *avr-14* was less sensitive to glutamate than the wild type, but more sensitive to glutamate than *avr-15*. The EC-50 for glutamate for avr-14 was 355 μM. *avr-15* is expressed in the muscle while *avr-14* is expressed in pharyngeal motoneurones and so makes an indirect contribution to the pharyngeal glutamate response. *avr-14* is possibly expressed in M-4.

Recently, Dent *et al.* (2000) found that simultaneous mutations of three genes in *C. elegans*, i.e., *avr-14* which encodes GluClα3 (GBR-2) in around 40 extrapharyngeal neurones, *avr-15* which encodes GluClα2 in pharyngeal muscle and *glc-1* which encodes GluClα1 in unidentified neurones, confer high level (4,000-fold) resistance to ivermectin. The neurones which express GluClα3 are mainly in the ring ganglia of the head but also in some motoneurones in the ventral cord and in some mechanosensory neurones. cDNAs encoding a putative GluCl subunit (HG4) have been cloned from *H. contortus* (Delaney, Laughton and Wolstenholme, 1998). HG4 protein has an 82% amino acid identity with *C. elegans* GluClβ subunit. These authors failed to find evidence for expression of HG4 on pharyngeal muscle cells. Recently, Jagannathan *et al.* (1999) have isolated cDNAs encoding the product of the *gbr-2/avr-14* orthologues from *H. contortus* and *A. suum*.

The actions of two non-specific chloride channel blockers, 5-nitro-2-(3-phenypropylamino) benzoic acid (NPPB) and 4,4-dinitrostilbene-2,2-disulphonic acid (DNDS) have been tested on the GluClα/GluClβ heteromer channel expressed in *Xenopus* oocytes (Bush *et al.*, 1997). Both compounds decreased the size of glutamate-induced responses, with IC-50 values of 588 nM for DNDS and 42.1 μM for NPPB. In this system picrotoxin had an IC-50 value of 23.8 μM. Interestingly, two barbiturates, amobaritone and pentobarbitone also reduced the glutamate response, with IC-50 values of 27.5 μM and 36.2 μM, respectively. On GABA-A gated chloride channels in mammals barbiturates enhance the GABA response.

The characteristics of a putative glutamate binding site have been described in *Haemonchus contortus* (Paiement, Prichard and Ribeiro, 1999). This binding site was sensitive to ivermectin. In a subsequent study, both *H. contortus* and *Telodorsagia circumcincta* membrane preparations were found to contain high amounts of low affinity specific L-glutamate binding sites. The number of these sites was increased in two isolates of ivermectin-resistant *H. contortus* and one isolate of ivermectin-resistant *T. circumcincta* (Hejmadi *et al.*, 2000). There were around 100 times more binding sites for glutamate compared to ivermectin binding sites. L-Glutamate binding was blocked by ibotenic acid, kynurenic acid (an NMDA antagonist) and β-hydroxy aspartate (a glutamate transporter inhibitor), but not blocked by ivermectin, argiopine, kainic acid, quisqualic acid or NMDA.

Glutamate inhibits pumping of the pharynx of *A. suum* with an IC-50 value of 491 μM (Brownlee, Holden-Dye and Walker, 1997). This value is very comparable to that mentioned above for *C. elegans* GluClβ subunit homomer. Ivermectin in the low pM range potentiated this inhibitory action of glutamate on *A. suum* pharyngeal pumping. This glutamate response was only weakly antagonised by picrotoxin, for example, 500 μM picrotoxin partially reduced the response to 100 μM glutamate. Using a two electrode current-clamp recording technique, Martin and Robertson (2000) have confirmed that glutamate hyperpolarizes the pharyngeal muscle of *A. suum* through an increase in chloride conductance and that ivermectin produced a slower hyperpolarization.

The action of glutamic acid has been investigated on the vagina vera, which is the distal part of the ovijector of *A. suum* (Fellowes *et al* 2000a). These authors found that glutamate, threshold 5 μM, inhibited the activity of this muscle. There was also evidence for a glutamate innervation of the vagina vera, suggesting that glutamate plays a physiological role in the regulation of its activity.

A number of naturally occurring compounds have been employed as anthelmintics, including kainate, domoate and quisqualate (Takemoto, 1978). When either kainate, final concentration in animal 25 μM, or glutamate, final concentration in animal 500 μM, were injected into *A. suum* there was a distinct change in body posture compared to control (Davis, 1998a). Glutamate, 100–5000 μM, depolarized and excited the DE-2 motoneurone of *A. suum*. The membrane potential of these neurones is around −30 mV and glutamate depolarization reached the reversal potential of spontaneous epsps recorded in DE-2, i.e., −5 to −15 mV. Kainate, domoate, bromowillardine and AMPA all depolarized and excited DE-2 motorneurone. The relative potencies were domoate > kainate ≫ glutamate ∼ bromowillardine ≥ aspartate. The EC-50 values for domoate, kainate, glutamate and aspartate were approximately 0.3 μM, 10 μM, 2000 μM and 5000 μM, respectively. NMDA failed to excite this cell. Quisqualate and AMPA appeared to be only partial agonists. However, the graphs presented did not show a maximum response to either compound, though high concentrations up to 30 mM were applied. Sodium-free saline blocked both epsps and glutamate depolarization recorded from DE-2. Experiments suggested that glutamate receptors were localized to the region where DE-2 received synaptic input from VNC neurones. Davis concluded that the epsps recorded from DE-2 are glutamatergic, since in the presence of glutamate epsps were lost even if the membrane potential was hyperpolarized back to the resting level. This suggested that the epsps and glutamate were acting at a common site through a common mechanism. Application of non-NMDA antagonists, such as DNQX or CNQX, reduced both the EPSPS and glutamate or domoate responses. In a second paper, Davis (1998b) presented evidence for a glutamate transporter both in the DE-2 motoneurone and in the hypodermis. Glutamate induced non-conductance mediated depolarization of the hypodermis which was sodium-dependent but calcium- independent. Interestingly, the hypodermis glutamate receptor only responded to glutamate and aspartate, both ligands being equiactive, with apparent EC-50 values in the low mM range. DNQX failed to antagonise the glutamate-induced depolarization of the hypodermis.

Analysis of the *C. elegans* genome indicates there are up to ten excitatory glutamate receptors (Bargmann, 1998). There are six of the AMPA/kainate class for which mutations have been described for one, *glr-1*. These mutants have defects in their locomotory response to mechanosensory stimuli. *glr-1* is expressed in motoneurones and interneurones, including neurones involved in locomotion (Maricq *et al.*, 1995; Hart, Sims and Kaplan, 1995). ASH sensory neurones transduce mechanical stimuli and transmit them through the release of glutamate onto certain of these *glr-1*-expressing interneurones. ASH cells also transduce chemical repellent stimuli and transmit them onto follower neurones, but through the release of a different chemical transmitter. There are two potential NMDA type receptors and two unusual outlier genes related to glutamate receptors. It is very likely that the excitatory receptors described by Davis (1998a,b) in *A. suum* are related to two of these *C. elegans* glutamate receptors.

4.5. GAMMA-AMINOBUTYRIC ACID (GABA)

GABA is an inhibitory transmitter in many phyla, including nematodes (Walker, Brooks and Holden-Dye, 1996). GABA was first shown to hyperpolarize *A. lumbricoides* somatic muscle membrane potentials by Del Castillo, De Mello and Morales (1964b). This group also demonstrated that the anthelmintic piperazine, which relaxes the body wall muscles, mimicked the effect of GABA (Del Castillo, Morales and Sanchez, 1963). The action of both GABA and piperazine were further analysed by Martin (1980, 1982). Both events were associated with an increase in permeability to chloride ions, though piperazine was 200–500 times less potent than GABA. Figure 12 shows the effect of low external chloride on the GABA response. The value for the reversal potential

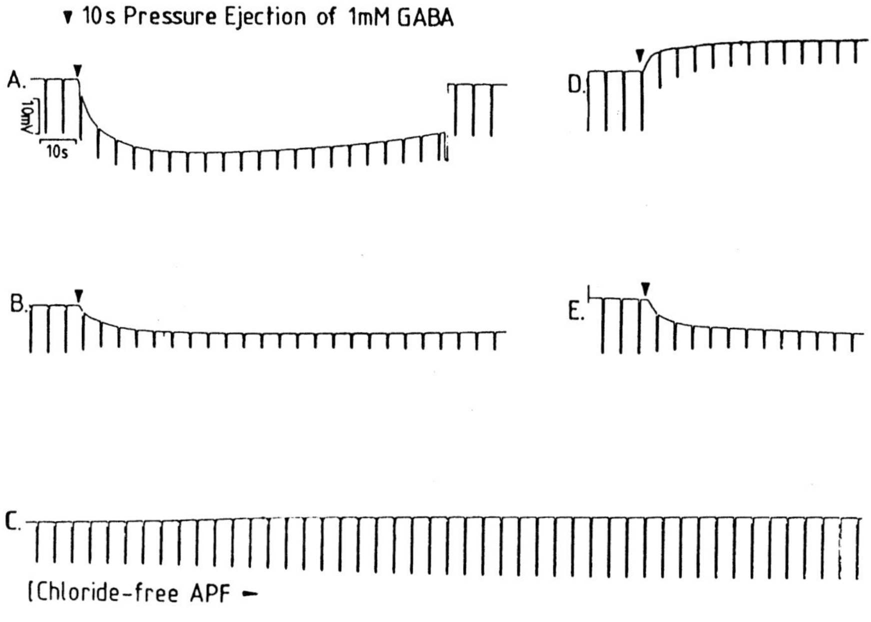

Figure 12. The effect of chloride-free saline on the response of an *Ascaris suum* muscle cell to GABA. Trace (A) is the first application of GABA and trace (B) is the fifth application of GABA (note fading of the response). In (C) the cell is perfused with chloride-free saline which produces an increase in cell resistance and depolarizes the membrane potential. Application of GABA in chloride-free saline causes a depolarization, trace (D). On returning to normal saline, the GABA response reverts to a hyperpolarization, trace (E). GABA was applied by pressure ejection of 1 mM GABA. (from Hewitt, 1987).

for GABA depends on the concentration of chloride in the external saline. If all the sodium is sodium chloride, 135 mM, then the reversal potential for GABA is around −65 mV. However, if sodium is 50% chloride and 50% acetate, then the equilibrium potential for chloride and hence reversal potential for GABA is less negative, i.e. around −55 mV. A number of ions will substitute for chloride in the external saline to induce a GABA hyperpolarization (Parri, Holden-Dye and Walker, 1991). These ions include thiocyanate, bromide, iodide, chlorate and nitrate.

GABA is present in the inhibitory motoneurones, D-I and V-I, of *A. suum* (Stretton *et al.*, 1985; Johnson and Stretton, 1987). In addition to these inhibitory motoneurones, Guastella, Johnson and Stretton (1991) have shown that GABA is present in ten cephalic and one dorsal rectal ganglion neurones. The distribution of the GABA uptake site in *A.* suum has also been investigated by Guastella and Stretton (1991). These authors found evidence for three types of neurone: neurones which contain GABA and possess a GABA uptake system, Type 1; neurones which contain GABA but are devoid of a GABA uptake system, Type 2; and neurones which lack GABA but possess a GABA uptake system, Type 3.

The agonist profile for the *A. suum* somatic muscle GABA receptor has been studied in detail by Holden-Dye *et al.* (1989). These authors concluded that this nematode GABA receptor had certain mammalian GABA-A-like properties based on a similar chloride-mediated ionic mechanism and a similar agonist profile. For example, the mammalian GABA-A agonist muscimol was only slightly less potent than GABA. Using pairs of enantiomers of structural analogues of GABA, a very good correlation was found between the stereoselectivity of these compounds on the mammalian GABA-A receptor and the GABA receptor on *A. suum* somatic muscle. The most potent of these ligands was S-(+)-dihydromuscimol, which was more potent than GABA and around ten times more potent than its R-(−)-isomer. Apart from the sulphonic derivatives of GABA, there was a very good correlation between the potencies of the GABA agonists tested on *A. suum* GABA receptors and the ability of the same ligands to displace GABA-A receptor binding in the mammalian brain. In contrast, the GABA antagonist profile was very different between mammalian GABA-A receptors and the nematode GABA receptor. For example, the potent mammalian GABA-A receptor antagonist, SR 95531, did not significantly block the nematode GABA receptors at concentrations below 100 μM. The *A. suum* GABA receptor did not recognize t-butylbicyclophosphoro-thionate (TBPS) or the insecticide, dieldrin. This nematode GABA receptor was not potentiated by pentobarbital or flurazapam. Bicuculline was also inactive at this receptor.

The interactions between a series of azole derivatives and GABA receptors of *A. suum* somatic muscle have been investigated (Bascal *et al.*, 1996a). The lead compound, SN606078 (2-(2,6-dichloro-4-trifluoremethylphenyl)-4-(4,5-dicyano-1H-imidazol-2-yl)-2H-1,2,3-triazole) and four related azoles reversibly blocked the GABA response, with IC-50s of less than 10 μM.

Evidence suggests that these compounds may act as non-competitive antagonists, possibly acting as open-channel blockers. Another group of compounds, the arylaminopyridazines, also block *A. suum* GABA receptors (Duittoz and Martin, 1991). Interestingly, compounds in this series which were active on vertebrate GABA-A receptors had low activity against *A. suum* GABA receptors. The reverse also occurred, where an analogue with low potency at vertebrate GABA-A receptors was highly active against *A. suum* GABA receptors. In contrast to the azole derivatives, the arylaminopyridazines were competitive antagonists.

There are 26 GABA-containing neurones in *C. elegans*, 6 DD motoneurones, 13 VD motoneurones, 4 RME head foraging motoneurones, one AVL enteric muscle motoneurone, one DVB enteric muscle motoneurone and one RIS (McIntire, Jorgensen and Horvitz, 1993; McIntire *et al.*, 1993).

Molecular biology studies to determine the genes involved in GABA function in *C. elegans* have been undertaken by McIntire and Jorgensen (McIntire, Jorgensen and Horvitz, 1993; McIntire *et al.*, 1993; Bamber *et al.*, 1999). A key receptor gene appears to be *unc-49* which encodes multiple subunits of a heteromultimeric GABA receptor (Bamber *et al.*, 1999). *unc-49* contains a single copy of a GABA receptor N terminus, followed by three tandem copies of a GABA receptor C terminus. *unc-49* can generate three distinct GABA-A receptor-like subunits, UNC-49A, UNC-49B and UNC-49C. However, only UNC-49B and UNC-49C are expressed at high levels and it is likely that normally these two subunits co-assemble to form a functional heteromeric receptor. Functional receptors could be expressed in *Xenopus* oocytes where the UNC-49B homomer was slightly more potent than UNC-49B/C heteromer. UNC-49C did not form a homeric GABA receptor. Mutation in gene *unc-25* eliminates GABA expression in all 26 neurones and is likely to be responsible for glutamic acid decarboxylase enzyme. Two genes, *unc-46* and *47*, are responsible for GABA release (McIntire, Jorgensen and Horvitz, 1993).

Recently, the responses of a GABA receptor on *C. elegans* somatic muscle has been investigated electrophysiologically (Richmond and Jorgensen, 1999). The GABA response persisted in zero calcium/0.5 mM EGTA, indicating that it was post-junctional on the muscle fibres. This GABA hypolarization is a chloride-mediated event. There was no response to GABA in *unc-49* mutants; *unc-49* encodes at least two GABA subunits located on somatic muscle. Muscimol also failed to hyperpolarize *unc-49* mutant somatic muscle cells.

Ivermectin (22,23 dihydroavermectin B1a) has been shown to be a potent anthelmintic (Campbell, 1989) and its mode of action associated with chloride conductance increase either linked to GABA or glutamate receptors, or through a direct effect on muscle to enhance chloride conductance (Rohrer and Arena, 1995). Electrophysiological studies have shown that ivermectin and related analogues will block the inhibitory action of GABA on *A. suum* somatic muscle but μM values are required (Holden-Dye and Walker, 1990) and this action is non-competitive. Ivermectin has potent effects on nematode pharyngeal muscle when nM ivermectin will induce complete inhibition of pumping. For example, an EC-50 of 5 nM will block feeding in *C. elegans* (Avery and Horvitz, 1990), while in *Haemonchus contortus* ivermectin induces paralysis of the pharynx with an EC-50 of 1–10 pM (Geary *et al.*, 1993). This is also true in *A. suum* where 1–500 nM ivermectin has a direct inhibitory effect on pharyngeal pumping (Brownlee, Holden-Dye and Walker, 1997). Concentrations of ivermectin in this range were irreversible which *in vivo* would lead to permanent cessation of feeling. Both GABA and glutamate inhibit *A. suum* pharyngeal pumping, with EC-50 values of 23 μM and 491 μM, respectively (Brownlee, Holden-Dye and Walker, 1997). Ivermectin, 1 pM, will enhance the sensitivity of the pharynx to both GABA and glutamate. For example, in the presence of 1 pM ivermectin, GABA threshold for inhibition of pumping increases from control of around 1 μM to 0.1 μM. Similarly, with glutamate, the threshold can be enhanced from a control of 10 μM to 1 μM. Picrotoxin in the 500–1000 μM range reduced or completely blocked the GABA response, 10 μM, on the pharynx. Picrotoxin is therefore very insensitive as a GABA antagonist in the pharynx.

Dihydroavermectin depressed the opening of GABA channels of outside out patches from *A. suum* somatic muscle (Martin and Pennington, 1989). In outside out patches, 1–100 pM opened 9–15 pS channels which were permeable to chloride. These authors found no evidence for potentiation of GABA or for this compound acting as a GABA agonist. On the contrary, dihydroavermectin opened non-GABA-activated chloride channels and non-selective cation channels.

The action of GABA has been examined on the vagina vera of *A. suum* (Fellowes *et al.*, 2000a). GABA, threshold 10 μM, inhibited the activity of this muscle and induced long term paralysis. Histochemical and immunochemical analysis showed the presence of GABA innervation of the vagina vera.

5. Nitric Oxide (NO)

There is now overwhelming evidence that NO acts as a transmitter or modulator in a wide range of animal groups (Vincent, 1995; Muller, 1997; Colasanti and Venturini, 1998; Beck *et al.*, 1999; Moroz, Gillette and Sweedler, 1999). A considerable amount of the evidence for NO function depends on the localization of NADPH diaphorase as an indicator for nitric oxide synthase (NOS). There is a family of NOS enzymes which includes neuronal NOS (type I, nNOS), inducible NOS (type II iNOS) and endothelial NOS (type III eNOS). NOS requires NADPH for activation and has NADPH diaphorase activity. Under certain conditions, NADPH diaphorase can be taken as an indicator of nNOS (Brehmer and Stack, 1997).

The first histochemical localization of NADPH diaphorase in nematodes was demonstrated by Bascal *et al.* (1995). Positive staining for NADPH diaphorase was obtained in various parts of the central nervous system of *A. suum*, including neurone cell bodies and fibres in the ventral ganglion, retrovesicular ganglion, ventral and dorsal nerve cords and sublateral lines. Not all the neurones stained with the same degree of intensity. For example, in the ventral ganglia, four nerve cells stained intensely for NADPH diaphorase with 2–5 others staining to a lesser extent. The other cells in the ventral ganglion did not stain. In the dorsal ganglion, of the 3 cell bodies, one stained intensely, one less so and the third cell showed no reaction. Out of the 13 cell bodies in the retrovesicular ganglion, 3–5 stained for diaphorase. Intensive staining also occurred in the motoneurone commissures, suggesting that NO might have a role in nerve-muscle transmission and so locomotion in nematodes. NADPH diaphorase staining was observed in close proximity to both dorsal and ventral cords in the area which includes the muscle arms. There was specific staining in particular areas of the somatic muscle cells which possibly suggests that NO might diffuse from the muscle cells to the motoneurones to influence their activity. Since the motoneurone commissures also stain for NADPH diaphorase it is possible that NO may also pass from motoneurones to the muscle cells. The amphidial commissures which are also part of the sensory system also stain for NADPH diaphorase, which would suggest a role in the transmission of sensory information. The enteric nervous system which regulates pharyngeal pumping also stains for NADPH diaphorase. This staining was localised to two neurones in the pharynx and their axons which innervate the pharyngeal muscle system. This again suggests that NO plays a role in pharyngeal pumping.

In a second paper, Bascal *et al.* (1996b) investigated possible histochemical evidence for co-localization between NADPH diaphorase and two antisera, one raised against KYSALMFamide and one raised against calcitonin-gene-related peptide (CGRP) in the nervous system of *A. suum*. Distinct groups of neurones in the ventral, dorsal and lateral ganglia displayed staining to NADPH diaphorase and immunoreactivity to KYSALMFamide antiserum. However, this only represented a small percentage of the total number of neurones which reacted to either NADPH diaphorase or KYSALMFamide antiserum. For example, in the ventral ganglion, four cell bodies were positive to both as was a single cell body in both left and right lateral lines. There was also some localization in the commissures. Very few neurones demonstrated co-localization between NADPH diaphorase and CGRP. For example, one neurone in the dorsal ganglia co-expressed CGRP and NADPH diaphorase and one cell body in the retrovesicular ganglion, one cell body in the lateral ganglia and two cell bodies at the junction of the nerve ring and lateral line.

The possibility that two species of filarial parasites, *Dirofilaria immitis* and *Brugia pahangi*, make NO has been investigated (Kaiser, Geary and Williams, 1998). Both species make NO, with production in the range of 50–150 pmol/ml/min. PF-1 enhanced release of NO from *D. immitis*. The NO inhibitor L-NMMA, 100 μM, reduced PF-1-induced release of NO from *D. immitis*. These results demonstrate that filarial nematodes produce and release NO which may influence tone of host blood and lymphatic vessels. In a second paper using filarial nematodes, Pfarr and Fuhrman (2000) detected the presence of NOS in *Brugia malayi* extracts. Using NADPH diaphorase staining and immunohistochemistry with a polyclonal antibody made to a conserved mouse

NOS isoform, NOS was localized in body wall muscle cells and the hypodermis. NADPH diaphorase was also detected in early and late stage embryos, in oocytes and spermatozoa.

The interaction between PF-1, PF-2 and NO has been investigated by Bowman *et al.* (1995) who present evidence to suggest these peptides exert their relaxing/inhibitory effect on *A. suum* somatic muscle by activating synthesis and release of NO. This evidence was based on the finding that PF-1 reversal of levamisole contractions was blocked by pretreatment with NOS inhibitors, such as NNLA, NMLA and L-NAME, all 100 μM. Methylene blue, 1 mM, an NO scavenger, had a similar effect. In contrast, an NO donor, such as sodium nitroprusside, 1 mM, mimicked the action of PF-1 although the maxium relaxation was less. NOS activity was measured in *A. suum* tissues where the activity in hypodermis/cuticle, muscle and ovary was 0.19, 0.11 and 0.002 pmol/min/mg protein, respectively. For comparison, NOS activity in whole *P. redivivus* supernate was 0.47 pmol/min/mg protein. NNLA and NMLA, 100 μM, both reduced NOS activity to *P. redivivus* by about 50%. While these results are indicative, considerable more work is required before one can be certain that PF-1 and PF-2 are primarily acting on somatic muscle by stimulating NOS activity in nematodes.

A partial biochemical characterization of *A. suum* NOS has been performed by Bascal *et al.* (2001). Labelled citrulline was detected using HPLC following incubation of a cytosol extract of *A. suum* nerve cords with labelled arginine and NADPH. The following cofactors were also included: calmodulin, BH_4 (tetrahydro-L-bipterin), FAD (flavin adenine dinucleotide) and FMN (flavin mononucleotide). There was a correlation between the occurrence of positive NADPH diaphorase staining and production of citrulline in the different tissues examined. For example, in the intestine where there is no NADPH diaphorase localization, there was no detectable citrulline produced. In contrast, nerve cords were able to synthesise citrulline. Bascal *et al.* (2001) found that *A. suum* nerve cord extract only produced citrulline at about 100 times less than rat cerebellum, that is, 3.3 pmol/mg protein/min compared to 333 pmol/mg protein/min. Interestingly, when NADPH, calcium, FAD, FMN and BH_4 were omitted from the incubation medium and W-13 (N-(4-aminobutyl)-5-chloro-2-naphthalene-sulphonamide), 1 mM, a calmodulin antagonist was added, citrulline production was only reduced by around 30%. The same procedure with rat cerebellum resulted in a fall in citrulline production of 99.6%. This difference could be due to a difference in the co-factor requirements of *A. suum* NOS compared to rat cerebellar NOS. Alternatively, since the *A. suum* extract is relatively crude it could contain endogenous co-factors which are sufficient to maintain NOS activity. Clearly, the *A. suum* extract needs to be purified and these experiments repeated. Surprisingly, in the presence of high levels of the calcium chelators EDTA and EGTA, 25 mM and 12.5 mM, respectively, citrulline production almost doubled. EDTA and EGTA virtually abolish citrulline production in rat cerebellar NOS homogenate. Another difference between the profile of rat cerebellum NOS and *A. suum* NOS was the effect of

NOS inhibitors. In the case of rat cerebellar NOS, addition of 1 mM S-methylisothiourea, N^G-nitro-L-arginine (NNA), N^G-dimethyl-L-arginine (ADMA) or L-N^G-(1-iminoethyl) lysine (NIO) reduced citrulline production by over 90%, while with *A. suum* extract NOS activity was only reduced by up to 15%. Incubation of *A. suum* nerve cord extract with oxyhaemoglobin resulted in the formation of methaemoglobin which indicates NO formation, 11 pmol NO/mg protein/min. Calcium enhanced methaemoglobin formation, while the NOS inhibitor L-NMMA reduced it. This work provides clear evidence for NOS in *A. suum* but suggests that its properties differ from vertebrate NOS, certainly from that in the rat cerebellum. A further complication is that a NOS gene has not yet been identified from *C. elegans*. If NOS occurs in *A. suum* and other nematodes it is to be expected that it would also occur in *C. elegans*.

The effect of NO donors on basal tone and ACh-induced contractions of somatic muscle of *A. suum* has been investigated (Bascal *et al.* 2001). Addition of the NO donors sodium nitrite and hydroxylamine, 0.1–10 mM, resulted in an increase in basal tone of muscle strips. These NO donors also reduced the amplitude of ACh-induced control contractions of somatic muscle. These effects were reversible following washing and suggest a possible physiological role in the regulation of body wall tone by NO.

Apart from a role in the regulation of excitable tissues, NO appears to have a completely unrelated role in at least some parasitic nematodes. *A. lumbricoides* perienteric fluid contains an octameric haemoglobin which can bind oxygen around 25,000 times more tightly than human haemoglobin (Davenport, 1949). Minning *et al.* (1999) found that NO drives a reaction whereby haemoglobin consumes oxygen to keep the perienteric fluid hypoxic. Minning *et al.* suggest that *A. lumbricoides* haemoglobin is an NO-dependent deoxygenase which uses endogenously produced NO substrate to detoxify oxygen since adults metabolise anaerobically and so it is likely that free oxygen is toxic to them.

6. Neuropeptides

Our knowledge on the localization, physiology and molecular biology of neuroactive peptides in nematodes comes from work on relatively few species. However, in common with the general uniform structure in nematodes, it is likely that these species will generate information that can be applied across the whole phylum. The genome project on *C. elegans* provides excellent molecular biology data regarding the peptide sequences present and using both wild type and mutants, physiological data can be collected. A surprising finding is that the peptides so far identified belong to one family, the FaRP family, first described in lamellibranch molluscs by Price and Greenberg (1977). However, there is evidence, both direct and indirect, that other families of peptides are present in nematodes. For example, the *C. elegans* genome project has shown that there are at least 15 other genes which encode 58 non-FaRP neuropeptides

(Bargmann, 1998) and more recent evidence has raised this figure to around 100. Most of these are nematode-specific but some are related to families already described in other phyla. Interestingly, *C. elegans* does not have highly conserved homologues to a number of mammalian neuropeptides, including endorphins, enkephalins, dynorphins, Substance P and Neuropeptide Y. However, the *C. elegans* genome does have receptors which are related to mammalian receptors for these peptides and so there must be relevant ligands in *C. elegans* and other nematodes to activate the receptors (Bargmann, 1998).

The first clear evidence for neuroactive peptides in nematodes came from immunocytochemical studies which are summarised by Stretton *et al.* (1991), in which antisera raised against a wide range of mammalian peptides showed reactivity with nematode nervous tissue, particularly using *A. suum* tissue (Brownlee, Fairweather and Johnston, 1993; Brownlee *et al.*, 1993, 1994a,b, 1996a,b). During the 1980s and 1990s, immunocytochemical studies demonstrated positive FaRP staining in a number of species, including *C. elegans, B. pahangi, H. contortus, P. redivivus* and *T. colubriformis* (Atkinson *et al.*, 1988; Warbrick, Rees and Howells, 1992; Geary *et al.*, 1995; Keating *et al.*, 1995). For example, in *C. elegans* it is suggested that around 30 neurones, i.e. 10% of the total neuronal population, are FMRFamide-like immunoreactive (Schinkmann and Li, 1992). However, the number of neurones showing positive immunoreactivity depends on the antisera used and the number may be as high as 50% (Li *et al.*, 1999). The first peptide in this family to be extracted and sequenced was AF-1 from *A. suum* (Cowden, Stretton and Davis, 1989). Since then at least 22 FaRPs have been identified in *A. suum, P. redivivus* or *H. contortus*, with the majority from *A. suum* (Table 1). Molecular biology studies using *C. elegans* have shown the presence of an even greater number of FaRPs, currently around 60 predicted from 22 flp (FMRFamide-like peptide) genes (Nelson *et al.*, 1998; Day and Maule, 1999; Li, Kim and Nelson, 1999; Li *et al.*, 1999) (Table 2). This includes one peptide, RNKEFIRFamide, which is not encoded by any of the 22 flp genes. The first *C. elegans* FaRPgene, *flp-1*, was described by Li and colleagues (Rosoff, Burglin and Li, 1992; Rosoff *et al.*, 1993) and codes for seven PNFLRFamides.

Functional roles for FaRPs in *C. elegans* have been investigated starting with attempts to delete *flp-1* (Li, Kim and Nelson, 1999; Li *et al.*, 1999). These experiments demonstrated defects in *C. elegans* behaviour due to malfunction of its sensory system. Several motor defects were also observed. Some of the interneurones which express *flp-1* synapse onto neurones of the egg laying system and *flp-1* deletion animals were found to lay fewer viable eggs. Overexpression of the *flp-1* gene resulted in the opposite effect to that observed in *flp-1* deletions. To determine whether FaRPs in *C. elegans* function through a G protein pathway, Li, Kim and Nelson (1999) and Li *et al.* (1999) constructed double mutants carrying a *flp-1* mutation (either a deletion or overexpression allele) and a G-protein subunit mutation, either a loss- or gain-of-function allele. They found that the double mutant showed the movement defect of

Table 1. Peptides present in *Ascaris suum* (AF), *Panagrellus redivivus* (PF) and *Haemonchus contortus* (HcF) also occur in *Caenorhabditis elegans* (CF)

Peptide	Sequence
AF-1/CF-12	KNEFIRFa
AF-2/PF-/HcF-/CF-8	KHEYLRFa
AF-8/PF-3/CF-10	KSAYMRFa
AF-3	AVPGVLRFa
AF-4	GDVPGVLRFa
AF-10	GFGDEMSMPGVLRFa
AF-13	SDMPGVLRFa
AF-14	SMPGVLRFa
AF-20	GMPGVLRFa
AF-11	SDIGISEPNFLRFa
AF-6	FIRFa
AF-7	AGPRFIRFa
AF-9	GLGPRPLRFa
AF-16	ILMRFa
AF-15	AQTFVRFa
AF-17	FDRDFMHFa
AF-19	AEGLSSPLIRFa
AF-5	SGKPTFIRFa
PF-1/CF-1	SDPNFLRFa
PF-2/CF-2	SADPNFLRFa
PF-4	KPNFIRFa

animals with a single G-protein mutation. This suggests that the peptides were acting upstream of the G-proteins.

So far, PF-1, PF-2, PF-3 (AF-8), AF-1, AF-2 and AF-15 have been identified in *C. elegans* but, as can be seen from Tables 1 and 2, a number of other peptides are similar in structure, including AF-5, AF-10, AF-19 and AF-20, in terms of their six C terminal sequences. The Tables also show that the C terminal F of FxRFamide can be replaced by another aromatic amino acid, such as Y (tyrosine) but not by W (tryptophan). In fact, tryptophan never appears in any of the FaRP sequences in nematodes (Tables 1 and 2). The C terminal F of FxRFamide can also be replaced by a number of amino acids, i.e., V, P, L, T, M, or D and there is even an *A. suum* sequence which is not an RFamide, i.e., HFamide (AF-17). By comparing Tables 1 and 2, certain additional points can be made, for example, there is only one tetrapeptide and one pentapeptide, there are 25 hepta/octapeptides, and the C terminal sequence is often VRFamide, IRFamide, LRFamide or MRFamide. An RFamide-coding gene has been described in *A. suum*, *afp-1*, which encodes for six PGVLRFamides, AF-3, AF-4, AF-10, AF-13, AF-14 and AF-20 (Edison, Messinger and Stretton, 1997) (Table 1). Three of these peptides have been isolated from *A. suum*, viz., AF-3, AF-4, AF-10 (Cowden and Stretton, 1995).

Table 2. FaRP sequences present in *Caenorhabditis elegans*. Predicted FaRPs identified from 22 flp genes in *C. elegans* (after Li *et al.*, 1999)

Gene	Peptide	
flp-1	CF-1 (PF-1)	SDPNFLRFa
	CF-2 (PF-2)	SADPNFLRFa
	CF-3	SQPNFLRFa
	CF-4	ASGDPNFLRFa
	CF-5	AAADPNFLRFa
	CF-6	PNFLRFa
	CF-7	AGSDPNFLRFa
flp-2		SPREPIRFa
		LRGEPIRFa
flp-3		SPLGTMRFa
		TPLGTMRFa
		EAEEPLGTMRFa
		NPLGTMRFa
		ASEDALFGTMRFa
		EDGNAPFGTMRFa
		SAEPFGTMRFa
		SADDSAPFGTMRFa
		NPENDTPFGTMRFa
pf-4		PTFIRFa
		ASPSFIRFa
pf-5		GAKFIRFa
		AGAKFIRFa
		APKPKFIRFa
flp-6	CF-10 (AF-8/PF-3)	KSAYMRFa
flp-7		SPMQRSSMVRFa
		TPMQRSSMVRFa
		SPMERSAMVRFa
		SPMDRSKMVRFa
flp-8	CF-12 (AF-1)	KNERFIRFa
flp-9	CF-11	KPSFVRFa
flp-10		QPKARSGYIRFa
flp-11		AMRNALVRFa
		ASGGMRNALVRFa
		NGAPQPFVRFa
flp-12	CF-13	RNKFEFIRFa
flp-13		SDRPTRAMDSPLIRFa
		AADGAPLIRFa
	CF-9	APEASPFIRFa
		ASPSAPLIRFa
		SPSAVPLIRFa
		ASSAPLIRFa
flp-14	CF-8 (AF-2)	KHEYLRFa

Table 2. (Continued.)

Gene		Peptide
flp-15		GGPQGPLRFa
		RGPSGPLRFa
flp-16	(AF-15)	AQTFVRFa
		GQTFVRFa
flp-17		KSAFVRFa
		KSQYIRFa
flp-18		DVPGVLRFa
		SVPGVLRFa
		SEVPGVLRFa
		EMPGVLRFa
		EIPGVLRFa
		DFDGAMPGVLRFa
flp-19		WANQVRFa
		ASWASSVRFa
flp-20		AMMVRFa
flp-21		GLGPRPLRFa
flp-22		SPSAKWMRFa

In addition, these authors identified AF-12 which Edison, Messinger and Stretton (1997) query as to whether it is a distinct peptide since it only differs from AF-10 by the absence of glycine (Table 1). In addition, Cowden and Stretton identified AF-5, AF-7, AF-9 and AF-11 (Table 1). The physiological actions of a number of these peptides will be reviewed.

6.1. AF-1 (KNEFIRFamide) AND AF-2 (KHEYLRFamide)

Both peptides have potent actions on *A. suum* somatic muscle and were the first nematode FaRPs to be sequenced (Cowden, Stretton and Davis, 1989; Cowden and Stretton, 1993). Cowden, Stretton and Davis (1989) calculated that each *A. suum* neurone which showed FMRFamide-like immunoreactivity contained around 5 fmol FMRFamide-like peptide. When 0.1 ml of 1 μM AF-1 was injected into the anterior end of *A. suum*, it blocked locomotory movements in the region of the injection. AF-1 also abolished excitatory potentials recorded in VI and DI motoneurones, with a threshold in some instances as low as 1 nM. AF-1, 100 nM, reversibly reduced the input resistance of the neurones. Interestingly, AF-1 had little effect on the input resistance of excitatory motoneurones, such as DE-1. Excitatory synaptic transmission did not appear to be affected by AF-1. The major action of AF-1 and AF-2 on somatic muscle is excitation, though in both cases an inhibitory or relaxing component can be observed, but this is generally quickly replaced by prolonged excitation (Maule *et al.*, 1995a; Pang *et al.*, 1995; Bowman *et al.*, 1996). The size of the initial relaxation is probably related to the muscle tension prior to peptide application. Under similar conditions, both AF-1 and AF-2 produce threshold responses in the range 10–100 nM, though in general AF-2 has a lower threshold than AF-1. A value as low as 10 pM has been observed for AF-2, but this is likely to be influenced by the level of peptidase activity (Sajid, Isaac and Harrow, 1997). These authors obtained evidence for an amino-peptidase which cleaved between the N terminal lysine and the second amino acid. AF2 was cleaved at a faster rate than AF-1. With AF-1, the excitation often involves a distinct contraction of the muscle, while with AF-2 the excitatory response is generally phasic without a clear increase in tone (Pang *et al.*, 1995). Figure 13 is an example showing the action of AF-1 on a body wall muscle strip of *A. suum*. The excitatory action of AF-2 on somatic muscle is blocked in the presence of mecamylamine, a nicotinic ACh antagonist, which blocked the action of ACh on *A. suum* somatic muscle (Pang *et al.*, 1995). These authors also found that low concentrations of AF-2, for example 100 nM, potentiated the excitatory action of ACh on somatic muscle, but failed to alter ACh depolarizations when muscle membrane potential recordings were made. Although AF-2 failed to enhance ACh depolarizations in the presence of 100 nM peptide, ACh depolarizations which previously failed to generate muscle action potentials, now generated action potentials at the same level of depolarization. This could indicate a key role in muscle contraction. Muscle membrane potential recordings also showed that AF-2, threshold 100 nM, dose-dependently depolarized the muscle resting potential. In addition, AF-2, 10–100 nM, increased the amplitude of ejps recorded from the muscle which clearly demonstrates that AF-2 has both pre- and postsynaptic actions on *A. suum* somatic muscle. The proportion of each will determine the overall response observed. Both caffeine and theophylline have been shown to inhibit AF-2-induced potentiation of ACh contractions of *A. suum* body wall muscle (Keating, Holden-Dye and Walker, 1996). Although in this study no evidence was found to link AF-2 potentiation of ACh contractions with a cAMP pathway, later experiments demonstrated that IBMX, 500 μM, forskolin, 10 μM, and dibutyryl cAMP, 100 μM, all reduced AF-2-induced potentiation of ACh contractions (Keating, 1996). This would suggest a role for cAMP in this effect and so further studies are required to resolve the issue.

Maule *et al.* (1995a) and Bowman *et al.* (1996) have examined the actions of a number of analogues of AF-1 on *A. suum* somatic muscle activity, including analogues where alanine was substituted for other amino acids. Removal of the amide at the C terminal or the N terminal lysine rendered the compound inactive. FIRFamide was also inactive. KNEFIAFamide had no direct activity, but at 10 μM it completely blocked the response to 1 μM AF-1. Interestingly, this concentration of KNEFIAFamide failed to block the action of AF-2, suggesting that the two peptides act on different receptors. KNEAIRFamide only excited the preparation, while KNAFIRFamide and KNEFARFamide were only inhibitory. Chimaera peptides between AF-1 and AF-2 were also synthesised and tested (Bowman *et al.*, 1996). Of these, KHEYIRFamide was

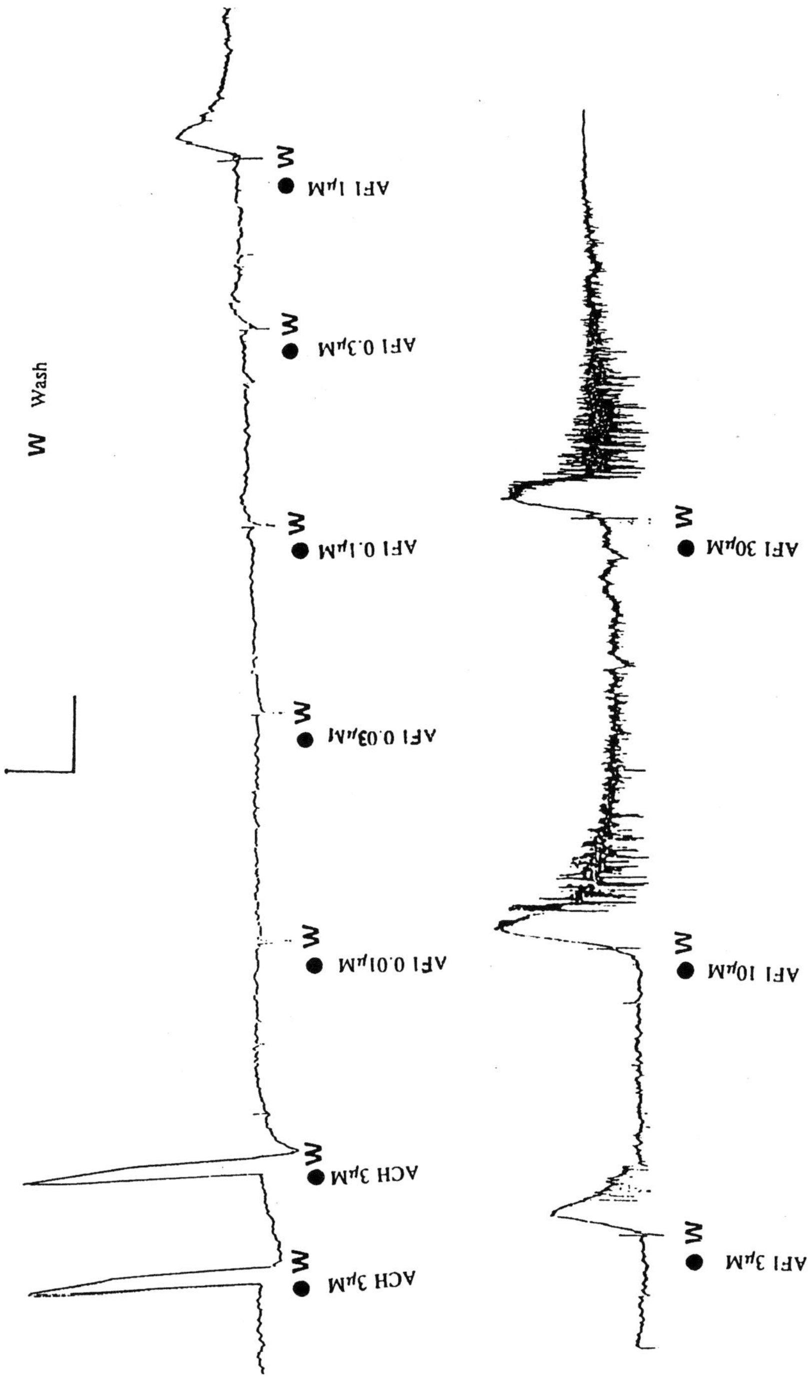

Figure 13. Traces taken from *Ascaris suum* muscle strip recording which compare the contraction to acetylcholine and AF-1. In this preparation the threshold for AF-1 was around 100 nM. Vertical scale: 5g; Horizontal scale: 30 min. (from Pang, 1992).

the most potent with a threshold of 10 nM. Both KNEYIRFamide and KNEFLRFamide were significantly less potent than the parent peptides, with thresholds of 3.2 μM.

AF-1, 1 μM, and AF-2, 10 μM, have also been tested on the vagina vera where they both inhibited spontaneous activity (Fellowes *et al.*, 1998). The vagina vera is a muscular tube connecting the uterus and vagina uteri to the gonopore and probably regulates egg output. Surprisingly, AF-2 at 1 μM increases contraction frequency, while 0.1 μM has an even greater excitatory action. This would reinforce the idea that AF-2 excites spontaneous activity at low concentrations, but as the concentration rises so the excitation is displaced by high threshold inhibition. This is not seen with AF-1, which suggests two separate receptors for AF-1 and AF-2 and possibly two receptors for AF-2. Both AF-1 and AF-2 also depress the amplitude of spontaneous contractions and increase the overall tone of the muscle.

The actions of AF-1 and AF-2 have also been tested on pharyngeal pumping in *A. suum* (Brownlee, Holden-Dye and Walker, 1995; Brownlee and Walker, 1999). AF-1 threshold 1 nM, inhibits concentration-dependently the 5-HT-induced pumping of this muscle. In contrast, AF-2 excited pumping. Intracellular recordings have been made from the pharyngeal muscle of *C. elegans* and AF-1 and AF-2 actions examined where both peptides excited the muscle, with EC-50 values in the nM range.

6.2. AF-3 (AVPGVLRFamide) AND AF-4 (GDVPGVLRFamide)

AF-3 and AF-4 form part of a subfamily of peptides whose C terminal sequence is PGVLRFamide and the N terminal is extended by up to eight additional amino acids and one encoded by *afp-1* (Edison, Messinger and Stretton, 1997) (Table 1). This subfamily is encoded by *flp-18* in *C. elegans* for three peptides (Table 2). There is one peptide in *C. elegans* which only differs from AF-4 by the absence of an N terminal glycine. The main study using this group has involved only AF-3 and AF-4, but it is possible the other PGVLRFamides act in a similar way. Initial bioactivity of this group was demonstrated by Cowden and Stretton (1995) who found that AF-4, 100 nM, strongly contracted *A. suum* body wall muscle strip and in some preparations also induced rhythmic contractions. The actions of both AF-3 and AF-4 have been studied in detail using *A. suum* somatic muscle and, to a lesser extent, *A. galli* somatic muscle (Trim *et al.*, 1997, 1998). The interaction with ACh was also investigated. Both peptides contracted body wall muscle strips with EC-50s of 24 ± 6 nM and 37 ± 2 nM, respectively, for AF-3 and AF-4. For comparison, the EC-50 value for ACh was 13 ± 1 μM, approximately 1000 times less potent. ACh contractions were more rapid with a higher maximum than those of the peptides. Intracellular recording from muscle cells showed that both peptides depolarized the cells with EC-50 values of around 700 nM and 900 nM, respectively, for AF-3 and AF-4. Again, ACh depolarizations were much faster, of a shorter duration and greater amplitude than those of the peptides (Fig. 14). In contrast

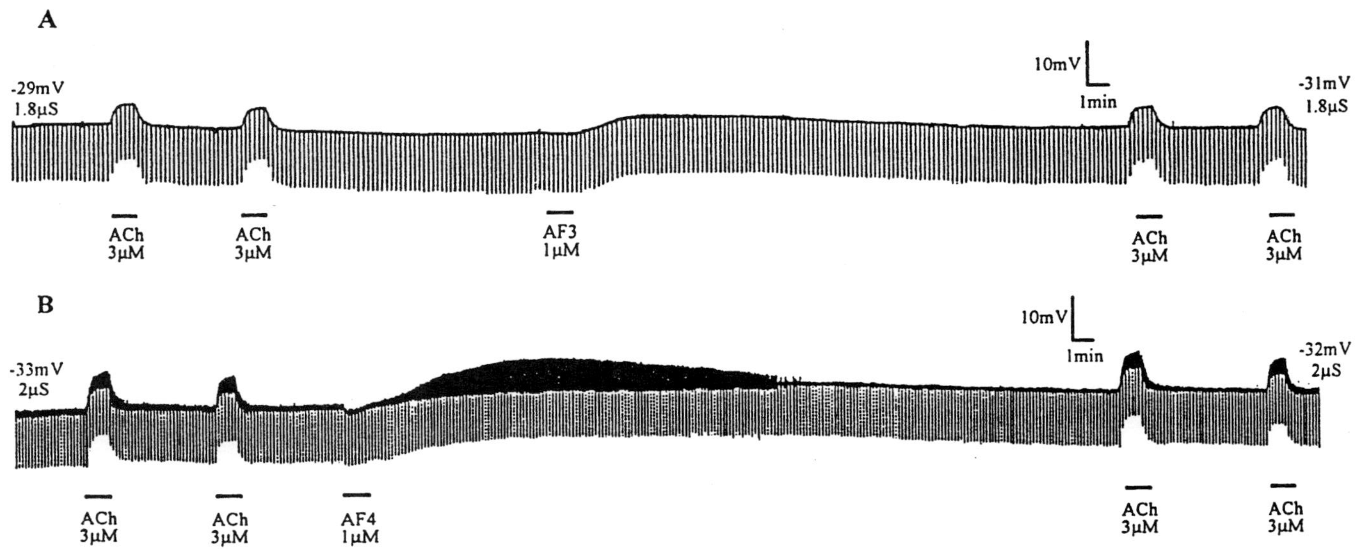

Figure 14. Traces to show the effects of acetylcholine, AF-3 and AF-4 on the membrane potential of an *Ascaris suum* muscle cell. Neither peptide had any obvious effect on the acetylcholine response. (from Trim, 1998)

to the situation with AF-1 and AF-2, the contractions due to AF-3 and AF-4 were not blocked by the nicotinic antagonist mecamylamine. This suggests that both peptides are probably acting directly on the muscle independently of ACh. There also appeared to be a difference between the ionic mechanisms for depolarizations for ACh and AF-3 and AF-4. For example, ACh depolarisation was more dependent on sodium than the peptide depolarization, while the latter was more calcium-dependent than ACh depolarization.

AF-3 and AF-4 had a similar effect on *A. galli* somatic muscle, though both peptides were less potent, particularly in terms of contraction of the body wall preparation.

AF-3 appears to sensitise the *A. suum* somatic muscle to ACh, since standard contractions to ACh were potentiated following application of peptide (Fig. 15). The interactions between cAMP and AF-3 contractions have been investigated (Trim *et al.*, 1998) and the results summarised in Table 3. AF-3 contractions were reduced in the presence of forskolin, 10 μM, and by IBMX (isobutylmethylxanthine), 500 μM. The effect of forskolin on the AF-3 response is shown in Fig. 15. DibutyrylcAMP, 100 μM, also reduced AF-3-induced contractions, while diBromo-cGMP, 100 μM, had no effect. AF-3, 1 μM, decreased cAMP levels in *A. suum* somatic muscle from a basal level of 1721–134 pmol/mg protein to 1148–133 pmol/mg protein. AF-3, 1 μM, also reduced forskolin-induced potentiation of cAMP levels in *A. suum* and in *A. galli*. The basal levels of cAMP in *A. galli* were around 14-fold lower than in *A. suum*. These results clearly suggest that AF-3 excitation is linked with a fall in adenylate cyclase activation.

AF-3 and AF-4 have also been examined on two other *A. suum* muscles, pharyngeal muscle where neither peptide has any effect and vagina vera muscle where the peptides have a complex action. At relatively high concentrations, 100 nM–100 μM, there is a rapid contraction accompanied by a variable period of spastic paralysis, followed by a return to normal (Fellowes *et al.*, 2000b). At 1–10 nM, there was a clear increase in both frequency and amplitude of the contractions. The effects of AF-4 were similar to those of AF-3.

On all the preparations where AF-3 and AF-4 have been examined, the actions of both peptides are very similar qualitatively, suggesting a common site of action. Assuming this is true for the other members of this group, one wonders why six analogues of these peptides are required physiologically.

6.3. PF-1 (SDPNFLRFamide) AND PF-2 (SADPNFLRFamide)

PF-1 and PF-2 were first isolated from *P. redivivus* by Geary *et al.* (1992). Both peptides are encoded by the flp-1 gene of *C. elegans* and so probably occur widely in nematodes. Interestingly, a related peptide, SDRNFLRFamide, occurs in arthropods (Trimmer, Kobierski and Kravitz, 1987). Both PF-1 and SDRNFLRFamide have similar actions on identified *Helix aspersa* neurones and are assumed to act at the same receptor on these cells (Pedder and Walker, 1999). Although neither PF-1 nor PF-2 have yet been isolated from *A. suum*,

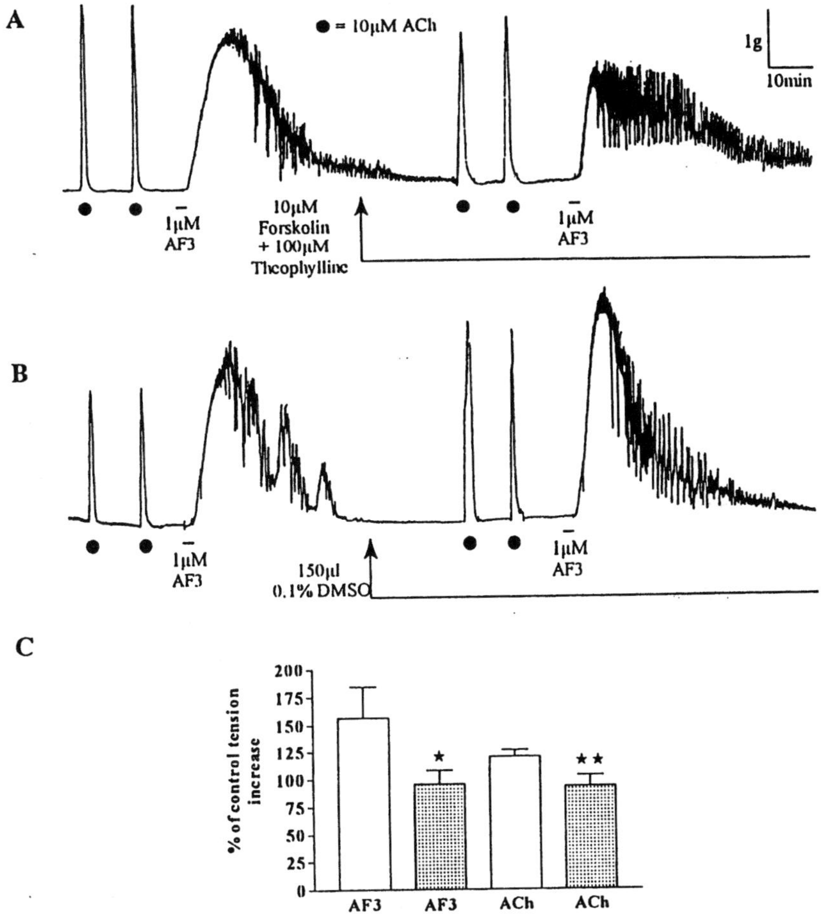

Figure 15. Traces and a histogram to show the effect of forskolin on the response of an *Ascaris suum* muscle strip to acetylcholine and AF-3. Trace (A) shows that in the presence of forskolin the response to AF-3 is reduced and the peptide fails to potentiate the acetylcholine-induced contraction. Trace (B) which is taken from another muscle strip prepared from the same animal shows that in the absence of forskolin, the AF-3 response is not reduced on a second applicaton and that AF-3 potentiates the action of acetylcholine. The histogram in (C) summarises the experiment. Open bars show the control responses to AF-3 and acetylcholine while the shaded bars show the responses in the presence of forskolin. Forskolin reduced both the maximum tension induced by AF-3 and AF-3 potentiation of acetylcholine contraction; $^*P<0.05$; $^{**}P<0.01$ (from Trim, 1998).

both peptides have potent relaxing actions on *A. suum* somatic muscle, threshold 1 nM (Franks *et al.*, 1994; Holden-Dye *et al.*, 1995; Bowman *et al.*, 1995). PF-1, threshold 1–10 nM, induces a slow hyperpolarization of *A. suum* somatic muscle cell membrane potential via a mechanism which is independent of GABA and is not chloride-mediated. The GABA antagonist, SN 612239 10 μM, reversibly blocks GABA 10 μM, while having no effect on PF-1, 200 nM,

Table 3. The effect of a number of drugs on the response to 1 μM AF3 and the potentiation of 10 μM ACh contractions by 1 μM AF3. Contractions in response to 1 μM AF3 (column 2) were reduced by a number of drugs. The contraction resulting from 10 μM ACh after exposure to 1 μM AF3 was always potentiated. This potentiation was reduced by some drugs (column 3). A reduction in response is indicated with its respective P value. nsd = no significant difference.

Drug	AF3 contraction	ACh potentiation
IBMX (500 μM)	reduced ($P < 0.05$)	reduced ($P < 0.001$)
forskolin (10 μM)		nsd
forskolin (10 μM) + theophylline (100 μM)	reduced ($P < 0.01$)	reduced ($P < 0.05$)
theophylline (100 μM)	nsd	nsd
dbcAMP (100 μM)	reduced ($P < 0.01$)	reduced ($P < 0.05$)
diBrcGMP (100 μM)	nsd	nsd
chelerythrine (10 μM)	nsd	nsd

hyperpolarizations. The EC-50 for PF-1-induced hyperpolarizations is 350 nM. Both PF-1 and PF-2 inhibit spontaneous somatic muscle cell activity (Fig. 16). Using body wall muscle preparations, PF-1 also reversibly reduced ACh-induced contractions. However, in contrast, PF-1 did not inhibit ACh-induced depolarizations of *A. suum* muscle cell membrane potentials. Both peptides reduce the amplitude of cholinergic ejps recorded from somatic muscle cells and have similar potencies, with an EC-50 of 311 nM. Since PF-1 does not reduce ACh-induced depolarizations of the muscle membrane potential, this effect of PF-1 on ejps is likely to be presynaptic. Interestingly, PF-1 has little apparent effect on the conductance of muscle cells. There is some preliminary evidence that PF-1 may act through an increase in permeability to potassium, since the response is enhanced in low external potassium saline and reversibly blocked in the presence of both 4-AP, 250 μM, and TEA, 5 mM (Fig. 17).

A possible role for NO in the mediations of PF-1 hyperpolarization was investigated by Bowman *et al.* (1995). These data have been reviewed in the NO section and provide some preliminary evidence for a possible role for NO in the action of PF-1 and PF-2 but further studies are required.

6.4. PF-3/AF-8 (KSAYMFRamide)

PF-3 was isolated from *P. redivivus* and tested for activity on *A. suum* somatic muscle (Maule *et al.*, 1994). PF-3, threshold 100 nM, produced an increase in muscle tension in innervated preparations. However, when the muscle was denervated, 10 μM, produced a small relaxation of the muscle. In a subsequent study, Maule *et al.* (1995a) found that the actions of PF-3 were more complex. For example, 10 μM, PF-3 contracted ventral body wall muscle strip but relaxed dorsal muscle strip. The threshold for contraction was 100 nM and that for relaxation was 1 μM. Removal of the ventral and dorsal nerve cords abolished consistent responses to PF-3. Pre-incubation with NOS inhibitors NNLA or

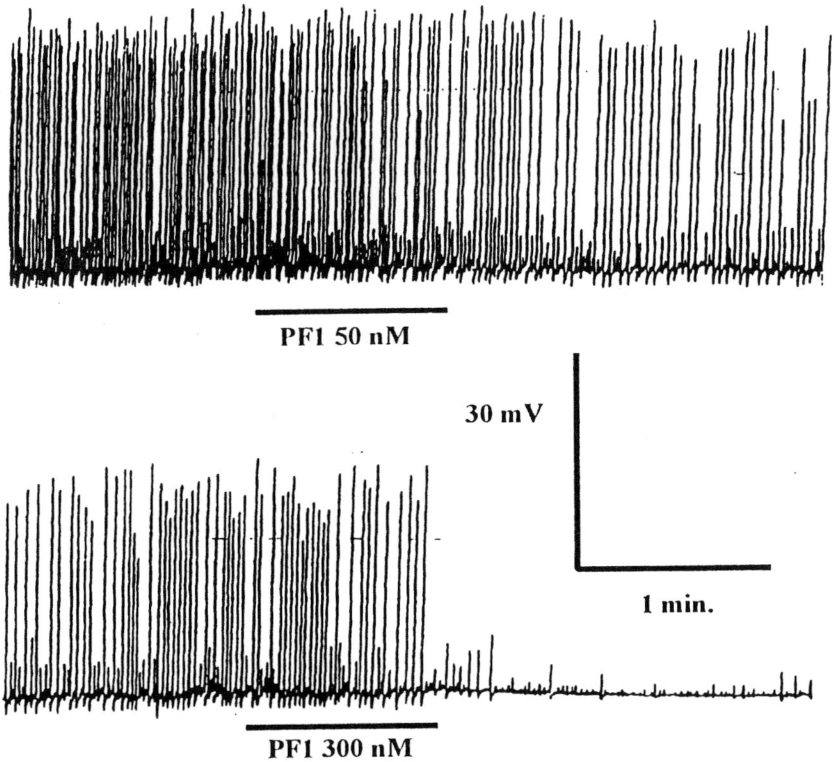

PF1(nM)	Hyperpolarisation (mV)	Spike min-1 (>20 mV)
0	0	60
30	.0	41
50	0	26
100	0	Slow wave
300	0	0

Figure 16. Traces to show the effect of PF-1 on the spontaneous activity recorded from an *Ascaris suum* muscle cell. The resting membrane potential was −36 mV with spike amplitude of around 30 mV. PF-1, 50–300 nM, reduced the frequency and amplitude of the spikes in a concentration-dependent manner. (from Franks, 1996).

NMLA, both 1 mM, had no effect on PF-3-induced relaxation. PF-3 has also been tested on *A. suum* vagina vera muscle and on pharyngeal pumping. Fellowes *et al.* (1998) found that PF-3 inhibited contractions of the muscle at 100 nM or higher, but at concentrations below 100 nM, the peptide increased the frequency of contractions. On *A. suum* pharyngeal muscle, PF-3 caused an initial increase in amplitude of the pumping strokes with an increase in frequency, threshold around 100 nM (Brownlee *et al.*, 1995; Brownlee and Walker,

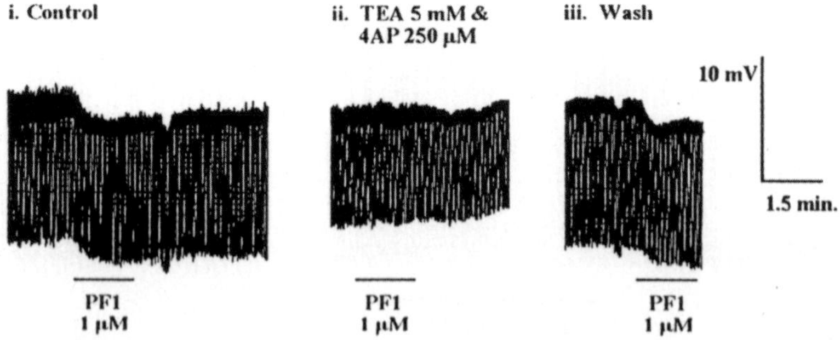

Figure 17. Traces to show the effects of the potassium channel blockers, tetraethylammonium and 4-aminopyridine (4-AP) (TEA) on the hyperpolarization induced by PF-1 in an *Ascaris suum* muscle cell. In the presence of 5 mM TEA and 250 μM 4-AP, the response to PF-1 was blocked, trace (ii). Following removal of the TEA and 4-AP, the peptide response partially returned, trace (iii). (from Franks, 1996).

1999). This excitation was followed by a reduction of the amplitude of contractions and a decrease in frequency. This biphasic effect occurred at threshold. As the concentration of peptide was increased, so the initial excitatory phase was reduced and the secondary inhibitory phase enhanced. So the major action of the peptide was to inhibit pumping. However, it achieved this by stopping the muscle in the contracted phase. The duration of this inhibition was concentration-dependent, with 1 μM PF-3 stopping the pumping for around 10 minutes. FMRFamide was inactive on the pumping pharynx, which would suggest that the N terminal amino acids are key for activity. The clear biphasic form of the PF-3 response would suggest that two receptors are involved in its activity in the pharynx. PF-3 also excites the pharynx of *C. elegans* but is less potent than AF-1.

6.5. PF-4 (KPNFIRFamide)

PF-4 was first isolated from *P. redivivus* (Maule *et al.*, 1995b). This peptide is a potent relaxing agent on *A. suum* body wall muscle, with a threshold of around 100 pM. This effect was observed using both innervated and denervated muscle preparations. Pre-incubation of the muscle with NOS inhibitors, such as NNLA or NMLA, both had no effect on PF-4 relaxation. A structure-activity study revealed the importance of proline for potent relaxing activity, since KANFIRFamide and KGNFIRFamide were less potent (Kubiak *et al.*, 1996). Interestingly, replacement of isoleucine by leucine to give KPNFLRFamide did not reduce activity. Since the effect of PF-4 is rapid, it resembles the action of GABA. Possible interactions between GABA and PF-4 and a possible chloride component in the PF-4 response have been investigated (Maule *et al.*, 1995b; Holden-Dye, Brownlee and Walker, 1997). PF-4, like GABA, hyperpolarizes the membrane potential of *A. suum* somatic muscle cells, with an EC-50 of 98 nM. This value is approximately 1000 times lower than the EC-50 for GABA, 59 μM. PF-4 increases input conductance of the muscle cells and this effect is additive with GABA, suggesting that PF-4 does not act through the release of GABA. The PF-4 hyperpolarization consists of two phases, a rapid early phase which is chloride – dependent and a later plateau phase which then polarizes to control membrane potential value. This later plateau component is unlikely to be due to chloride since in the presence of 1 mM GABA, which hyperpolarizes the membrane potential to around −62 mV, addition of 10 μM PF-4 results in a depolarization to −47.5 mV. This late plateau phase could be due to a decrease in potassium conductance or an increase in sodium conductance, since the equilibrium potential for sodium is +27 mV and for potassium is −90 mV (Brading and Caldwell, 1971). An interesting alternative is the involvement of an anion channel which is permeable to intracellular carboxylic acids which are produced by *A. suum* muscle cells (Saz and Weil, 1962; Tsang and Saz, 1973). Ivermectin failed to block the PF-4 hyperpolarization while blocking the GABA response, providing further evidence that the two compounds are acting at different receptors. The high potency of PF-4 on *A. suum* muscle would suggest that there is a specific receptor which is normally activated by an endogenous PF-4-like peptide in *A. suum*. PF-4 has not been tested on the pharynx of *A. suum* and it would be of interest to see if it is inhibitory. PF-4 has also been tested on the vagina vera muscle of *A. suum* where it inhibited spontaneous contractions (Marks *et al.*, 1999).

6.6. KPSFVRFamide

KPSFVRFamide is encoded by the *flp-9* gene of *C. elegans* (Nelson *et al.*, 1998). Since there is some similarity between this peptide and PF-4, KPNFIRFamide, Marks *et al.* (1999) compared the properties of both peptides and chimaeras of the two peptides. KPSFVRFamide, 10 μM, was inactive on *A. suum* somatic body wall muscle while, as previously shown, KPNFIRFamide produced a fast relaxing effect. The chimaera peptide, KPSFIRFamide, 10 μM, had no effect on somatic body wall muscle. In contrast, KPNFVRFamide, threshold 100 pM, had a rapid relaxing action on the somatic muscle preparation which was very similar to the PF-4 response. KPNFVRFamide was active on both innervated and denervated muscle strips. Interestingly, its relaxing action was reversed in chloride-free saline, suggesting that its effect involves chloride. KPSFVRFamide, 10 μM, inhibited spontaneous activity of *A. suum* vagina vera muscle but also increased baseline tone. Both KPSFIRFamide and KPNFVRFamide were inhibitory on the vagina vera muscle, but KPNFVRFamide was more potent, threshold 10 nM. KPSFVRFamide had no effect on resting membrane potential of *A. suum* somatic muscle cells. In contrast, both KPSFIRFamide and KPNFVRFamide hyperpolarized the muscle cells. Marks *et al.* (1999) obtained evidence that KPSFVRFamide was expressed during all larval stages but not by the adult *C. elegans*. They concluded that expression of *flp-9* was developmentally regulated and that any physiological role was restricted to the larval stages.

6.7. APEASPFIRFamide

APEASPFIRFamide has been isolated from *C. elegans* and some preliminary studies regarding its physiological actions investigated (Marks *et al.*, 1997). This peptide, 10 μM, caused a rapid relaxation when applied to innervated muscle strip preparation of *A. suum*. While the peptide still relaxed denervated muscle preparations, the degree of relaxation was less and in some cases appeared to have a slower time course. The peptide also inhibited spontaneous muscle activity. The threshold for relaxation was 10 nM and was blocked in high (125 mM) potassium saline. This peptide also inhibits action potential frequency of the pharynx of *C. elegans*.

6.8. SAEPFGTMRFamide

This is one of nine L or FGTMRFamides expressed by *flp-3* in *C. elegans* (Nelson *et al.*, 1998). *flp-3* is expressed in neurone ILID which has synaptic input onto the pharyngeal muscle via the circuit IL1D onto RIP onto I1 onto the pharynx. SAEPFGTMRFamide decreases pumping activity and action potential frequency concentration-dependently in *C. elegans* pharynx (Rogers *et al.*, 2000). The threshold for inhibition was 1 μM.

6.9. NON-FaRPs

The most recent estimation of the number of non-FaRPs present in *C. elegans* is around 100 which are encoded by *nlp* genes and of these 87 are unique to *C. elegans* (Li, Kim and Nelson, 1999; Li *et al.*, 1999). The 21 *nlp* genes are distributed on chromosomes I, II, III, IV, V and X of *C. elegans*. Expression patterns of 15 *nlp* genes have been investigated and all the reporter constructs, save for *nlp-4*, expressed in neurones. Expression also occurred in the gut, somatic gonad cells, pharyngeal muscle or vulva muscle. Expression of all the *nlp* genes, except *nlp-4* and *nlp-12*, occurred in the head. *nlp-3, nlp-8, nlp-14* and *nlp-15* were expressed in the pharynx. *nlp-1* encoded three putative neuropeptides which are most similar to *Aplysia californica* buccalin, MDSFAFAPGLamide. *nlp-2* encoded six neuropeptides which are most similar to *Lymnaea stagnalis* myomodulin, SLSMLRLamide. However, as can be seen from Table 4, the similarity is limited. *nlp-1* is expressed in five classes of neurones in the head and at least one in the tail. *nlp-2* is expressed in the gut

Table 4. Predicted NLP-1 and NLP-2 neuropeptides

	Peptide sequences from *C. elegans*	Peptide sequences from gastropod molluscs
NLP-1 peptides (3)	MDANAFRMSFa MDPNAFRMSFa VNLDPNSFRMSFa	*Aplysia californica* MDSFAFAPGLa (buccalin)
NLP-1 peptides (6)	SIALGRSGFRPa SMAMGRLGLRPa SMAYGRQGFRPa LYISRQGFRPa SMAIGRAGMRPa SIAIGRAGFRPa	*Lymnaea stagnalis* SLSMLRLa (myomodulin)

and two neurones in the head. Clearly, this is an area which is likely to develop rapidly. This is reinforced by the finding that there are now 32 identified *nlp* genes (Hart, personal communication). A gene has also been identified in nematodes which encodes GLWamides which are expressed in a pair of interneurons (Fujisawa *et al.*, 1999).

7. Conclusions

Cell signalling in nematodes uses many of the conventional molecules, including acetylcholine, 5-HT, dopamine, octopamine, glutamic acid and GABA. In addition, it is likely that NO and neuroactive peptides also play a role in cell signalling. Both acetylcholine and GABA act as neuromuscular excitatory and inhibitory transmitters, respectively, onto the somatic body wall muscle and regulate muscle tone. Thus, nematodes use the same somatic excitatory transmitter as vertebrates, but have evolved peripheral GABA-mediated inhibition as is also found in arthropods. Interestingly, acetylcholine is also used as the excitatory nerve-muscle transmitter in annelids and this phylum also has peripheral inhibition. Recent molecular biological evidence, however, suggests a link between nematodes and arthropods (Graham, 2000). In common with other invertebrate phyla, octopamine appears to replace noradrenaline as a key transmitter in nematodes, with dopamine as the only major catecholamine. Therefore, neurones do not contain both a beta hydroxylase and a tyrosine hydroxylase. In common with most phyla, 5-HT plays a key role in the normal physiology of nematodes, particularly in relation to locomotion, feeding, reproduction and metabolism, the major systems in nematodes.

Evidence collected over the past ten years clearly demonstrates that neuroactive peptides have important physiological actions in nematodes. Our current understanding regarding the roles for neuropeptides is greatly influenced by the research using FaRPs, a family which is also important in molluscs, annelids and arthropods. Clearly, there are non-FaRPs in nematodes and once they have been identified and their physiological actions characterised, the picture will become more balanced. Current evidence suggests that nematode neuropeptides have evolved separately from other phyla, including vertebrates, and so represent an excellent area for the development of novel anthelmintics. In common with peptides in general, it is likely that nematode peptides act through second messenger systems and there is some evidence in support of this. This is an important area for future research.

It would appear that after a long period of relative neglect, nematodes are beginning to provide an excellent model system, both in terms of genetic information and mutations using *C. elegans* and physiology using larger species, such as *A. suum*, although increasingly electrophysiological techniques are being applied to *C. elegans*.

References

Agarwal, A., Shukla, O.P., Ghatak, S., and Tekwani, B.L. (1990) Biogenic amines, metabolites and monoamine oxidase in filarial worm, *Setaria cervi*. *International Journal of Parasitology* **20**, 873–881.

Anderson, R.C. (2000) *Nematode Parasites of Vertebrates*. Second Edition. CABI Publishing, Oxford.

Angstadt, J.D., and Stretton, A.O.W. (1989) Slow active potentials in ventral inhibitory motor neurons of the nematode *Ascaris*. *Journal of Comparative Physiology A* **166**, 165–177.

Angstadt, J.D., Donmoyer, J.E., and Stretton, A.O.W. (1989) Retrovesicular ganglion of the nematode *Ascaris*. *Journal of Comparative Neurology* **284**, 374–388.

Arena, J.P., Liu, K.K., Paress, P.S., and Cully, D.F. (1991) Avermectin-sensitive chloride currents induced by *Caenorhabditis elegans* RNA in *Xenopus* oocytes. *Molecular Pharmacology* **40**, 368–374.

Arena, J.P., Liu, K.K., Paress, P.S., Schaeffer, J.M., and Cully, D.F. (1992) Expression of a glutamate-activated chloride current in *Xenopus* oocytes injected with *Caenorhabditis elegans* RNA: evidence for modulation by avermectin. *Molecular Brain Research* **15**, 339–348.

Arena, J.P., Liu, K.K., Paress, P.S., Frazier, E.G., Cully, D.F., Mrozik, H., and Schaeffer, J.M. (1995). The mechanism of action of avermectins in *Caenorhabditis elegans*: correlation between activation of glutamate-sensitive chloride current, membrane binding, and biological activity. *Journal of Parasitology* **81**, 286–294.

Atkinson, H.J., Isaac, R.E., Harris, P.D., and Sharpe, C.M. (1988) FMRFamide-immunoreactivity within the nervous system of the nematodes *Panagrellus redivivus*, *Caenorhabditis elegans* and *Heterodera glycines*. *Journal of Zoology* **216**, 663–671.

Avery, L. (1993) Motor neuron M3 controls pharyngeal muscle relaxation timing in *Caenorhabditis elegans*. *Journal of Experimental Biology* **175**, 283–297.

Avery, L., and Horvitz, H.R. (1989) Pharyngeal pumping continues after laser killing of the pharyngeal nervous system of *Caenorhabditis elegans*. *Neuron* **3**, 473–485.

Avery, L., and Horvitz, H.R. (1990) Effects of starvation and neuroactive drugs on feeding in *Caenorhabditis elegans*. *Journal of Experimental Zoology* **253**, 263–270.

Baldwin, E., and Moyle, V. (1949) A contribution to the physiology and pharmacology of *Ascaris lumbricoides* from the pig. *British Journal of Pharmacology* **4**, 145–152.

Bamber, B.A., Beg, A.A., Twyman, R.E., and Jorgensen, E.M. (1999) The *C. elegans unc-49* locus encodes multiple subunits of heteromultimeric GABA receptor. *Journal of Neuroscience* **19**, 5348–5359.

Bargmann, C.I. (1998) Neurobiology of the *Caenorhabditis elegans* genome. *Science* **282**, 2028–2033.

Barreteau, H., Trouvin, J.H., Goudey-Perriere, F., Jacquot, C., and Gayral, P. (1991) Biogenic amines and GABA in the larval and adult forms of the nematode *Nippostrongylus brasiliensis*. *Comparative Biochemistry and Physiology* **100C**, 445–449.

Bascal, Z.A., Montgomery, A., Holden-Dye, L., Williams, R.G., and Walker, R.J. (1995) Histochemical mapping of NADPH diaphorase in the nervous system of the parasitic nematode, *Ascaris suum. Parasitology* **110**, 625–637.

Bascal, Z.A., Holden-Dye, L., Willis, R.J., Smith, S.W.G., and Walker, R.J. (1996a) Novel azole derivatives are antagonists at the inhibitory GABA receptor on the somatic muscle cells of the parasitic nematode, *Ascaris suum. Parasitology* **112**, 253–259.

Bascal, Z.A., Montgomery, A., Holden-Dye, L., Williams, R.G., Thorndyke, M.C., and Walker, R.J. (1996b) NADPH diaphorase activity in peptidergic neurones of the parasitic nematode, *Ascaris suum. Parasitology* **112**, 125–134.

Bascal, Z.A., Cunningham, J.M., Holden-Dye, L., O'Shea, M., and Walker, R.J. (2001) Characterization of a putative nitric oxide synthase in the neuromuscular system of the parasitic nematode, *Ascaris suum. Parasitology* **122**, 219–231.

Beck, K.F., Eberhardt, W., Frank, S., Huwiler, A., Messmer, U.K., Muhl, H., and Pfeilschifter, J. (1999) Inducible NO synthase: Role in cellular signalling. *Journal of Experimental Biology* **202**, 645–653.

Bowman, J.W., Winterrowd, C.A., Friedman, A.R., Thompson, D.P., Klein, R.D., Davis, J.P., Maule, A.G., Blair, K.L., and Geary, T.G. (1995) Nitric oxide mediates the inhibitory effects of SDPNFLRFamide, a nematode FMRFamide-related neuropeptide, in *Ascaris suum. Journal of Neurophysiology* **74**, 1880–1888.

Bowman, J.W., Friedman, A.R., Thompson, D.P., Ichhpurani, A.K., Kellman, M.F., Marks, N., Maule, A.G., and Geary, T.G. (1996) Structure-activity relationships of KNEFIRFamide (AF1), a nematode FMRFamide-related peptide, on *Ascaris suum* muscle. *Peptides* **17**, 381–387.

Brading, A.F., and Caldwell, P.C. (1964) The effect of ions on the resting potential of muscle cells in *Ascaris lumbricoides. Journal of Physiology* **173**, 36P.

Brading, A.F., and Caldwell, P.C. (1971) The resting membrane potential of the somatic of *Ascaris lumbricoides. Journal of Physiology* **217**, 605–624.

Brehmer, A., and Stack, W. (1997) Morphological classification of NADPHd-positive and -negative myenteric neurons in the porcine small intestine. *Cell Tissue Research* **287**, 127–134.

Breum, L., Bierre, U., Bak, J.F., Jacobsen, S., and Astrup, A. (1995) Long-term effects of fluoxetine on glycemic control in obese patients with non-insulin-dependent diabetes mellitus or glucose intolerance: influence on muscle glycogen synthase and insulin receptor kinase activity. *Metabolism* **44**, 1570–1576.

Brooman, J.E. (1998) The characterisation of serotonin receptors in the parasitic nematode *Ascaris suum*. PhD Thesis, University of Southampton, U.K.

Brownlee, D.J.A., and Fairweather, I. (1999) Exploring the neurotransmittter labyrinth in nematodes. *Trends in Neurosciences* **22**, 16–24.

Brownlee, D.J.A., and Walker, R.J. (1999) Actions of nematode FMRFamide-related peptides on the pharyngeal muscle of the parasitic nematode, *Ascaris suum. Annals of the New York Academy of Sciences* **897**, 228–238.

Brownlee, D.J.A., Fairweather, I., and Johnston, C.F. (1993) Immunocytochemical demonstration of neuropeptides in the peripheral nervous system of the roundworm *Ascaris suum* (*Nematoda, Ascaroidea*). *Parasitology Research* **79**, 302–308.

Brownlee, D.J.A., Holden-Dye, L., and Walker, R.J. (1995) The action of serotonin and the nematode neuropeptide KSAYMRFamide on the pharyngeal muscle of the parasitic nematode *Ascaris suum. Parasitology* **111**, 379–384.

Brownlee, D.J.A., Holden-Dye, L., and Walker, R.J. (1997) Actions of the anthelmintic ivermectin on the pharyngeal muscle of the parasitic nematode, *Ascaris suum. Parasitology* **115**, 553–561.

Brownlee, D.J.A., Holden-Dye, L., and Walker, R.J. (2000) The Range and Biological Activity of FMRFamide-related Peptides and Classical Neurotransmitters in Nematodes. *Advances in Parasitology* **45**, 109–180.

Brownlee, D.J.A., Fairweather, I., Johnston, C.F., Smart, D., Shaw, C., and Halton, D.W. (1993) Immunocytochemical demonstration of neuropeptides in the central nervous system of the roundworm, *Ascaris suum* (Nematoda: Ascaroidea). *Parasitology* **106**, 305–316.

Brownlee, D.J.A., Fairweather, I., Johnston, C.F., and Shaw, C. (1994a) Immunocytochemical demonstration of peptidergic and serotonergic components in the enteric nervous system of the roundworm *Ascaris suum. Parasitology* **108**, 89–103.

Brownlee, D.J.A., Brennan, G.P., Halton, D.W., Fairweather, I., and Shaw, C. (1994b) Ultrastructural localization of pancreatic polypeptide- and FMRFamide-immunoreactivities within the central nervous system of the nematode, *Ascaris suum* (*Nematoda: Ascaroidea*). *Parasitology* **108**, 587–593.

Brownlee, D.J.A., Fairweather, I., Holden-Dye, L., and Walker, R.J. (1996a) Nematode Neuropeptides: Localization, Isolation and Functions. *Parasitology Today* **12**, 343–351.

Brownlee, D.J.A., Fairweather, I., Thorndyke, M.C., and Johnston, C.F. (1996b) Cellular and subcellular localization of SALMFamide (S1)-like immunoreactivity within the central nervous system of the nematode *Ascaris suum* (Nematoda: Ascaroidea). *Parasitology Research* **82**, 149–156.

Bush, E.R., Holden-Dye, L., Foreman, R.C., and Walker, R.J. (1997) Characterization of nematode glutamate receptors. *Journal of Physiology* **504P**, p25.

Byerly, L., and Masuda, M.O. (1979) Voltage-clamp analysis of the potassium current that produces a negative-going action potential in *Ascaris* muscle. *Journal of Physiology* **288**, 263–284.

Caldwell, P.C., and Ellory, J.C. (1968) Ion movements in the somatic muscle cells of *Ascaris lumbricoides. Journal of Physiology* **197**, 75–76P.

Campbell, W.C. (1989) *Ivermectin and Abamectin.* Springer-Verlag, New York.

Chaudhuri, J., and Donahue, M.J. (1989) Serotonin receptors in the tissues of adult *Ascaris suum. Molecular and Biochemical Parasitology* **35**, 191–198.

Chaudhuri, J., Martin, R.E., and Donahue, M.J. (1988a). Evidence for the absorption and synthesis of 5-hydroxytryptamine in perfused muscle and intestinal tissue and whole worms of adult *Ascaris suum. Parasitology* **96**, 157–170.

Chaudhuri, J., Martin, R.E., and Donahue, M.J. (1988b). Tryptophan hydroxylase and aromatic L-amino acid decarboxylase activities in the tissues of adult *Ascaris suum. International Journal of Parasitology* **18**, 318–324.

Chen, X.H., Holden-Dye, L., and Walker, R.J. (1997) Inhibitory actions of dopamine in the motor nervous system of the parasitic nematode, *Ascaris suum. Journal of Physiology* **504P**, 24–25P.

Colasanti. M., and Venturini, G. (1998) Nitric oxide in invertebrates. *Molecular Neurobiology* **17**, 157–174.

Colquhoun, L.M. (1991) Studies on the nicotinic receptor on the somatic muscle cells of the parasitic nematode *Ascaris suum.* PhD Thesis, University of Southampton, U.K.

Colquhoun, L.M., Holden-Dye, L., and Walker, R.J. (1991) The pharmacology of cholinoceptors on the somatic muscle cells of the parasitic nematode, *Ascaris suum. Journal of Experimental Biology* **158**, 509–530.

Colquhoun, L.M., Holden-Dye, L., and Walker, R.J. (1992) The actions of nicotinic receptor specific toxins on the somatic muscle cells of the parasitic nematode, *Ascaris suum. Molecular Neuropharmacology* **3**, 11–16.

Cowden, C., and Stretton, A.O.W. (1993) AF2, an *Ascaris* neuropeptide: Isolation, sequence, and bioactivity. *Peptides* **14**, 423–430.

Cowden, C., and Stretton, A.O.W. (1995) Eight novel FMRFamide-like neuropeptides isolated from the nematode *Ascaris suum. Peptides* **16**, 491–500.

Cowden, C., Stretton, A.O.W., and Davis, R.E. (1989) AF1, a sequenced bioactive neuropeptide isolated from the nematode *Ascaris suum. Neuron* **2**, 1465–1473.

Croll, N.A., and Matthews, B.E. (1977) *Biology of Nematodes.* Blackie & Sons, London.

Cully, D.F., Vassilatis, D.K., Liu, K.K., Paress, P.S., Van der Ploeg, L.H.T., Schaeffer, J.M., and Arena, J.P.

(1994) Cloning of an avermectin-sensitive glutamate-gated chloride channel from *Caenorhabditis elegans. Nature* **371**, 707–711

Cully, D.F., Wilkinson, H., Vassilatis, D.K., Etter, A., and Arena, J.P. (1996) Molecular biology and Electrophysiology of glutamate-gated chloride channels of invertebrates. *Parasitology* **113**, S191–S200.

Davenport, H.E. (1949) The haemoglobin of *Ascaris lumbricoides. Proceedings of the Royal Society* **B 136**, 255–270.

David, M.W., Somerville, D., Lee, R.Y.N., Lockery, S., Avery, L., and Fambrough, D.M. (1995) Mutations in the *Caenorhabditis elegans* Na, K-ATPase a subunit gene, *eat-6*, disrupts excitable cell function. *Journal of Neuroscience* **15**, 8408–8418.

Davis, R.E. (1998a) Neurophysiology of glutamatergic signalling and anthelmintic action in *Ascaris suum*: pharmacological evidence for a kainate receptor. *Parasitology* **116**, 471–486.

Davis, R.E. (1998b) Action of excitatory amino acids on hypodermis and the motornervous system of *Ascaris suum*: pharmacological evidence for a glutamate transporter. *Parasitology* **116**, 487–500.

Davis, R.E., and Stretton, A.O.W. (1989a) Passive membrane properties of motoneurones and their role in long-distance signalling in the nematode *Ascaris. Journal of Neuroscience* **9**, 403–414.

Davis, R.E., and Stretton, A.O.W. (1989b) Signalling properties of *Ascaris* motorneurones: Graded active responses, graded synaptic transmission and tonic transmitter release. *Journal of Neuroscience* **9**, 415–425.

Davis, R.E., and Stretton, A.O.W. (1996) The motonervous system of *Ascaris*: electrophysiology and anatomy of the neurons and their control by neuromodulators. *Parasitology* **113**, S97–S117.

Day, T.A., and Maule, A.G. (1999) Parasitic Peptides! The structure and function of neuropeptides in parasitic worms. *Peptides* **20**, 999–1019.

Delany, N.S., Laughton, D.L., and Wolstenholme, A.J. (1998) Cloning and localization of an avermectin receptor-related subunit from *Haemonchus contortus. Molecular and Biochemical Parasitology* **97**, 177–187.

Del Castillo, J., De Mello, W.C., and Morales, T. (1963) The physiological role of acetylcholine in the neuromuscular system of *Ascaris lumbricoides. Archives International Physiologie and Biochimie* **71**, 741–757.

Del Castillo, J., Morales, T., and Sanchez, V. (1963) Action of piperazine on the neuromuscular system of *Ascarislumbricoides. Nature* **200**, 706–707.

Del Castillo, J., De Mello, W.C., and Morales, T.A. (1964a) Influence of some ions on the membrane potential of *Ascaris* muscle. *Journal of General Physiology* **48**, 129–140.

Del Castillo, J., De Mello, W.C., and Morales, T.A. (1964b) Inhibitory action of g-aminobutyric acid (GABA) on *Ascaris* muscle. *Experientia* **20**, 141–145.

Dent, J.A., Davis, M.W., and Avery, L. (1997) *avr-15* encodes a chloride channel subunit that mediates inhibitory glutamatergic neurotransmission and ivermectin sensitivity in *Caenorhabditis elegans. European Molecular Biology Organization Journal* **16**, 5867–5879.

Dent, J.A., Smith, M.M., Vassilatis, D.K., and Avery, L. (2000) The genetics of ivermectin resistance in *Caenorhabditis elegans. Proceedings of the National Academy of Sciences U.S.A.* **97**, 2674–2679.

Donahue, M.J., Yacoub, N.J., Michnoff, C.A., Masaracchia, R.M., and Harris, B.G. (1981) Serotonin (5-hydroxytryptamine): A possible regulator of glycogenolysis in perfused muscle segments of *Ascaris suum. Biochemical Biophysical Research Communications* **101**, 112–117.

Duittoz, A.H., and Martin, R.J. (1991) Antagonist properties of arylaminopyridazine GABA derivatives at the *Ascaris* muscle GABA receptor. *Journal of Experimental Biology* **159**, 149–164.

Edison, A.S., Messinger, L.A., and Stretton, A.O.W. (1997) *afp-1*: A gene encoding multiple transcripts of a new class of FMRFamide-like neuropeptides in the nematode *Ascaris suum. Peptides* **18**, 929–935.

Etter, A., Cully, D.F., Schaeffer, J.M., Liu, K.K., and Arena, J.P. (1996) An amino acid substitution in the pore region of a glutamate-gated chloride channel enables the coupling of ligand binding to channel gating. *Journal of Biological Chemistry* **271**, 16035–16039.

Fellowes, R.A., Maule, A.G., Marks, N.J., Geary, T.G., Thompson, D.P., Shaw, C., and Halton, D.W. (1998) Modulation of the motility of the vagina vera of *Ascaris suum in vitro* by FMRFamide-related peptides. *Parasitology* **116**, 277–287.

Fellowes, R.A., Maule, A.G., Martin, R.J., Geary, T.G., Thompson, D.P., Kimber, M.J., Marks, N.J., and Halton, D.W. (2000a) Classical neurotransmitters in the ovijector of *Ascaris suum*: localization and modulation of muscle activity. *Parasitology* **121**, 325–336.

Fellowes, R.A., Maule, A.G., Marks, N.J., Geary, T.G., Thompson, D.P., and Halton, D.W. (2000b)

Nematode neuropeptide modulation of the vagina vera of *Ascaris suum*: *in vitro* effects of PF1, PF2, PF4, AF3 and AF4. *Parasitology* **120**, 79–89.

Frandsen, J.C., and Bone, L.W. (1987) Biogenic amines and their metabolites in *Trichostrongylus colubriformis*, a nematode parasitic of ruminants. *Comparative Biochemistry & Physiology* **87C**, 75–77.

Frandsen, J.C., and Bone, L.W. (1988) Derivatives of epinephrine, norepinephrine, octopamine and histamine formed by homogenates of *Trichostrongylus colubriformis*, a nematode parasite from ruminants. *Comparative Biochemistry and Physiology* **91C**, 385–387.

Franks, C.J. (1996) Studies on the actions of the nematode FMRFamide-like neuropeptide PF-1. PhD Thesis, University of Southampton, U.K.

Franks, C.J., Walker, R.J., and Holden-Dye, L. (1997) Pharyngeal muscle cell action potentials recorded from *Caenorhabditis elegans*. *Journal of Physiology* **504P**, 15P.

Franks, C.J., Holden-Dye, L., Williams, R.G., Pang, F.Y., and Walker, R.J. (1994) A nematode FMRFamide-like peptide SDPNFLRFamide (PF1) relaxes the dorsal muscle strip preparation of *Ascaris suum*. *Parasitology* **108**, 229–236.

Fujisawa, T., Takahashi, T., Yum, S., Halta, M., Shimizu, H., and Koizumi, O. (1999). Systematic identification of neuropeptides in Hydra. Brain Research **848**, P1–2, A21.

Geary, T.G., Price, D.A., Bowman, J.W., Winterrowd, C.A., MacKenzie, C.D., Garrison, R.D., Williams, J.F., and Friedman, A.R. (1992) Two FMRFamide-like peptides from the free-living nematode *Panagrellus redivivus*. *Peptides* **13**, 209–214.

Geary, T.G., Sims, S.M., Thomas, E.M., Vanover, L., Davis, J.P., Winterrowd, C.A., Klein, R.D., Ho, N.F.H., and Thompson, D.P. (1993) *Haemonchus contortus*: ivermectin-induced paralysis of the pharynx. *Experimental Parasitology* **77**, 88–96.

Geary, T.G., Bowman, J.W., Friedman, A.R., Maule, A.G., Davis, J.P., Winterrowd, C.A., Klein, R.D., and Thompson, D.P. (1995) The pharmacology of FMRFamide-related neuropeptides in nematodes: new opportunities for rational anthelmintic discovery? *International Journal of Parasitology* **25**, 1273–1280.

Goh, S.L., and Davey, K.G. (1976a) Localization and distribution of catecholaminergic structures in the nervous system of *Phocanema decipiens* (*Nematoda*). *International Journal of Parasitology* **6**, 403–411.

Goh, S.L., and Davey, K.G. (1976b) Selective uptake of noradrenaline, dopa and 5-hydroxytrypamine by the nervous system of *Phocanema decipiens* (*Nematoda*): a light autoradiographic and ultrastructural study. *Tissue and Cell* **8**, 421–435.

Goldschmidt, R. (1908) Das nervensystem von *Ascaris lumbricoides* und megalocephala. *Zeitchrift fur Wissenschaftliche Zoologie* **90**, 73–136.

Goodman, M.B., Hall, D.H., Avery, L., and Leckery, S.R. (1998) Active currents regulate sensitivity and dynamic range in *C. elegans* neurons. *Neuron* **20**, 763–772.

Graham, A. (2000) Animal phylogeny: Root and branch surgery. *Current Biology* **10**, R36–R38.

Guastella, J., and Stretton, A.O.W. (1991) Distribution of ^{3}H-GABA uptake sites in the nematode *Ascaris*. *Journal of Comparative Neurology* **307**, 598–608.

Guastella, J., Johnson, C.D., and Stretton, A.O.W. (1991) GABA-immunoreactive neurons in the nematode *Ascaris*. *Journal of Comparative Neurology* **307**, 584–597.

Hamdan, F.F., Ungrin, M.D., Abramovitz, M., and Ribeiro, P. (1999) Characterization of a novel serotonin

receptor from *Caenorhabditis elegans*: Cloning and expression of two splice variants. *Journal of Neurochemistry* **72**, 1372–1383.

Harrow, I.D., and Gration, K.A.F. (1985) Mode of action of the anthelmintics morantel, pyrantel and levamisole on muscle-cell membrane *Ascaris suum*. *Pesticide Science* **16**, 662–672.

Hart, A.C., Sims, S., and Kaplan, J.M. (1995) Synaptic code for sensory modalities revealed by *C. elegans* GLR-1 glutamate receptor. *Nature* **378**, 82–85.

Hejmadi, M.V., Jagannathan, S., Delany, N.S., Coles, G.C., and Wolstenholme, A.J. (2000) L-Glutamate binding sites of parasitic nematodes: an association with ivermectin resistance. *Parasitology* **120**, 535–545.

Hewitt, G.M. (1987) Electrophysiological studies on the *Ascaris* muscle GABA receptor. PhD Thesis, University of Southampton, U.K.

Hodgkin, J., Horvitz, H.R., Jasny, B.R., and Kimble, J. (1998) *C. elegans*: Sequence to Biology. *Science* **282**, 2011.

Holden-Dye, L., and Walker, R.J. (1990) Avermectin and avermectin derivatives are antagonists at the 4-aminobutyric acid (GABA) receptor on the somatic muscle cells of *Ascaris*: is this the site of anthelmintic action. *Parasitology* **101**, 265–271.

Holden-Dye, L., Brownlee, D.J.A., and Walker, R.J. (1997) The effects of the peptide KPNFIRFamide (PF4) on the muscle cells of the parasitic nematode *Ascaris suum*. *British Journal of Pharmacology* **120**, 379–386.

Holden-Dye, L., Krogsgaard-Larsen, P., Nielsen, L., and Walker, R.J. (1989) GABA receptors on the somatic muscle cells of the parasitic nematode *Ascaris suum*: Stereo selectivity indicates similarity to a GABA-A type agonist recognition site. *British Journal of Pharmacology* **98**, 841–850.

Holden-Dye, L., Franks, C.J., Williams, R.G., and Walker, R.J. (1995) The effect of the nematode peptides SDPNFLRFamide (PF1) and SADPNFLRFamide (PF2) on synaptic transmission in the parasitic nematode *Ascaris suum*. *Parasitology* 110, 449–455.

Horvitz, H.R., Chalfie, M., Trent, C., Sulston, J.E., and Evans, P.D. (1982) Serotonin and octopamine in the nematode *Caenorhabditis elegans*. *Science* **216**, 1012–1014.

Huang, X., Duran, E., Diaz, F., Xiao, H., Messer, W.S., and Komuniecki, R. (1999) Alternative splicing of serotonin receptor isoforms in the pharynx and muscle of the parasitic nematode, *Ascaris suum*. *Molecular and Biochemical Parasitology*, **101**, 95–106.

Huang, X., Xiao, H., Rex, E., Hinman, C., Messer, W.S., and Komuniecki, R.W. (2000) Alternative-splicing of 5HT-2 receptor isoforms in the pharynx and muscle of the parasitic nematode, *Ascaris suum*. *Society of Neuroscience Abstracts* **26**, 1425.

Isaac, R.E., Muimo, R., and MacGregor, A.N. (1990). N-acetylation of serotonin, octopamine and dopamine by adult *Brugia pahangi*. *Molecular and Biochemical Parasitology* **43**, 193–199.

Isaac, R.E., MacGregor, D., and Coates, D. (1996) Metabolism and inactivation of neurotransmitters in nematodes. *Parasitology* **113**, S157–S173.

Isaac, R.E., Eaves, L., Muimo, R., and Lamango, N. (1991) N-acetylation of biogenic amines in *Ascaridia galli*. *Parasitology* **102**, 445–450.

Jagannathan, S., Laughton, D.L., Cirtten, C.L., Skinner, T.M., Horoszok, L., and Wolstenholme, A.J. (1999) Ligand-gated chloride channel subunits encoded by the *Haemonchus contortus* and *Ascaris suum* orthologues of the *Caenorhabditis elegans gbr-2* (*avr-14*) gene. *Molecular and Biochemical Parasitology* **103**, 129–140.

Jarman, M. (1959) Electrical activity in the muscle cells of *Ascaris lumbricoides*. *Nature* **184**, 1244.

Johnson, C.D., and Stretton, A.O.W. (1985) Localization of choline acetyltransferase within identified motorneurones of the nematode *Ascaris*. *Journal of Neuroscience* **5**, 1984–1992.

Johnson, C.D., and Stretton, A.O.W. (1987) GABA immunoreactivity in inhibitory motorneurones of the nematode *Ascaris*. *Journal of Neuroscience* **7**, 223–235.

Johnson, C.D., Reinitz, C.A., Sithigorngul, P., and Stretton, A.O.W. (1996) Neuronal localization of serotonin in the nematode *Ascaris suum*. *Journal of Comparative Neurology* **367**, 352–360.

Kaiser, L., Geary, T.G., and Williams, J.F. (1998) *Dirofilaria immitis* and *Brugia pahangi*: Filarial parasites make nitric oxide. *Experimental Parasitology* **90**, 131–134.

Kass, I.S., Wang, C.C., Walrond, J.P., and Stretton, A.O.W. (1980) Avermectin B1a, a paralysing anthelmintic that affects interneurons and inhibitory motoneurons in *Ascaris*. *Proceedings of the National Academy of Sciences USA* **77**, 6211–6215.

Keating, C.D.I. (1996) The identification of, and studies on, a FMRF-NH2-related peptide from the nematode *Haemonchus contortus*. PhD Thesis, University of Southampton, U.K.

Keating, C.D., Holden-Dye, L.M., and Walker, R.J. (1996) Investigation of the mode of action of nematode neuropeptides. *Pesticide Science* **46**, 263–266.

Keating, C.D., Holden-Dye, L., Thorndyke, M.C., Williams, R.G., Mallett, A., and Walker, R.J. (1995) The FMRFamide-like neuropeptide AF2 is present in the parasitic nematode *Haemonchus contortus*. *Parasitology* **111**, 515–521.

Kubiak, T.M., Maule, A.G., Marks, N.J., Martin, R.A., and Wiest, J.R. (1996) Importance of the proline residue to the functional activity and metabolic stability of the nematode FMRFamide-related peptide, KPNFIRFamide (PF4). *Peptides* **17**, 1267–1277.

Laughton, D.L., Lunt, G.G., and Wolstenholme, A.J. (1997) Reporter gene constructs suggest that the *Caenorhabditis elegans* avermectin receptor β-subunit is expressed solely in the pharynx. *Journal of Experimental Biology* **200**, 1509–1514.

Lee, D.L. (1962) The distribution of esterase enzymes in *Ascaris lumbricoides*. *Parasitology* **52**, 241–260.

Li, C., Kim, K., and Nelson, L.S. (1999) FMRFamide-related neuropeptide gene family in *Caenorhabditis elegans*. Brain Research **848**, 26–34.

Li, C., Nelson, L.S., Kim, K., Nathoo, A., and Hart, A.C. (1999) Neuropeptide gene families in the nematode *Caenorhabditis elegans*. *Annals of the New York Academy of Sciences* **897**, 239–252.

Maricq, A.V., Peckol, E., Driscoll, M., and Bargmann, C.I. (1995) Mechanosensory signalling in *C. elegans* mediated by GLR-1 glutamate receptor. *Nature* **378**, 78–81.

Marks, N.J., Maule, A.G., Geary, T.G., Thompson, D.P., Davis, J.P., Halton, D.W., Verhaert, P., and Shaw, C. (1997) APEASPFIRFamide, a novel FMRFamide-related decapeptide from *Caenorhabditis elegans:* Structure and myoactivity. *Biochemical and Biophysical Research Communications* **231**, 591–595.

Marks, N.J., Maule, A.G., Li, C., Nelson, L.S., Thompson, D.P., Alexander-Bowman, S., Geary, T.G., Halton, D.W., Verhaert, P., and Shaw, C. (1999) Isolation, pharmacology and gene organization of KPSFVRFamide: A neuropeptide from *Caenorhabditis elegans*. *Biochemical and Biophysical Research Communications* **254**, 222–230.

Martin, R.J. (1980) The effect of g-aminobutyric acid on the input conductance and membrane potential of *Ascaris* muscle. *British Journal of Pharmacology* **71**, 99–106.

Martin, R.J. (1982) Electrophysiological effects of piperazine and diethylcarbamazine on *Ascaris suum* somatic muscle. *British Journal of Pharmacology* **77**, 255–26.

Martin, R.J., and Pennington, A.J. (1989) A patch-clamp study of effects of dihydroavermectin on *Ascaris* muscle. *British Journal of Pharmacology* **98**, 747–756.

Martin, R.J., and Robertson, A.P. (2000) Electrophysiological investigation of anthelmintic resistance. *Parasitology* **120**, S87–S94.

Martin, R.J., Pennington, A.J., Duittoz, A.H., Robertson, S., and Kusel, J.R. (1991) The physiology and pharmacology of neuromuscular transmission in the nematode parasite, *Ascaris suum*. *Parasitology* **102**, S41–S58.

Martin, R.J., Valkanov, M.A., Dale, V.M.E., Robertson, A.P., and Murray, I. (1996) Electrophysiology of *Ascaris* muscle and anti-nematodal drug action. *Parasitology* **113**, S137–S156.

Maule, A.G., Shaw, C., Bowman, J.W., Halton, D.W., Thompson, D.P., Geary, T.G., and Thim, L. (1994) KSAYMRFamide: A novel FMRFamide-related heptapeptide from the free-living nematode, *Panagrellus redivivus*, which is myoactive in the parasitic nematode, *Ascaris suum*. *Biochemical and Biophysical Research Communications* **200**, 973–980.

Maule, A.G., Geary, T.G., Bowman, J.W., Marks, N.J., Blair, K.L., Halton, D.W., Shaw, C., and Thompson, D.P. (1995a) Inhibitory effects of nematode FMRFamide-related peptides (FaRPs) on muscle strips from *Ascaris suum*. Invertebrate Neuroscience 1, 255–265.

Maule, A.G., Shaw, C., Bowman, J.W., Halton, D.W., Thompson, D.P., Thim, L., Kubiak, T.M., Martin, R.A., and Geary, T.G. (1995b) Isolation and preliminary biological characterization of KPNFIRFamide, a novel FMRFamide-related peptide from the free-living nematode, *Panagrellus redivivus*. *Peptides* **16**, 87–93.

Maule, A.G., Geary, T.G., Bowman, J.W., Shaw, C., Halton, D.W., and Thompson, D.P. (1996) The Pharmacology of Nematode FMRFamide-related Peptides. *Parasitology Today* **12**, 351–357.

McIntire, S.L., Jorgensen, E., and Horvitz, H.R. (1993a) Genes required for GABA function in *Caenorhabditis elegans*. *Nature* **364**, 334–337.

McIntire, S.L., Jorgensen, E., Kaplan, J., and Horvitz, H.R. (1993b) The GABAergic nervous system of *Caenorhabditis elegans*. *Nature* 364, 337–341.

Mellanby, H. (1955) Identification and estimation of acetylcholine in three parasitic nematodes

(*Ascaris lumbricoides, Litomosoides carnii* and the microfilariae of *Dirofilaria repens*). *Parasitology* **45**, 287–294.

Minning, D.M., Gow, A.J., Bonaventura, J., Braun, R., Dewhirst, M., Goldberg, M.E., and Stamler, J.S. (1999) *Ascaris* haemoglobin is a nitric oxide-activated 'deoxygenase'. *Nature* **401**, 497–502.

Mishra, S.K., Sen, R., and Ghatak, S. (1984) *Ascaris lumbricoides* and *Ascaridia galli*: biogenic amines in adults and developmental stages. *Experimental Parasitology* **57**, 34–39.

Moroz, L.L., Gillette, R., and Sweedler, J.V. (1999) Single-cell analyses of nitergic neurons in simple nervous systems. *Journal of Experimental Biology* **202**, 333–341.

Muller, R., and Baker, J.R. (1990) *Medical Parasitology*. Lippincott & Gower Medical Publishing, Philadelphia & London.

Muller, U. (1997) The nitric oxide system in insects. *Progress in Neurobiology* **51**, 363–381.

Nelson, L.S., Kim, K.Y., Memmott, R.E., and Li, C. (1998) FMRFamide-related gene family in the nematode, *Caenorhabditis elegans*. *Molecular Brain Research* **58**, 103–111.

Norton, S., and DeBeer, E.J. (1957) Investigation on the action of piperazine on *Ascaris lumbricoides*. *American Journal of Tropical Medicine* **6**, 898–905.

Nurrish, S., Segalat, L., and Kaplan, J.M. (1999) Serotonin inhibition of synaptic transmission: $G\alpha_o$ decreases the abundance of UNC-13 at release sites. *Neuron* **24**, 231–242.

Olde, B., and McCrombie, W.R. (1997) Molecular cloning and functional expression of a serotonin receptor from *Caenorhabditis elegans*. *Journal of Molecular Neuroscience* **8**, 53–62.

Paiement, J.P., Prichard, R.K., and Ribeiro, P. (1999) *Haemonchus contortus*: Characterization of a glutamate binding site in unselected and ivermectin-selected larvae and adults. *Experimental Parasitology* **92**, 32–39.

Pang, F.Y., Mason, J., Holden-Dye, L., Franks, C.J., Williams, R.G., and Walker, R.J. (1995) The effects of the nematode peptide, KHEYLRFamide (AF2) on the somatic musculature of the parasitic nematode *Ascaris suum*. *Parasitology* **110**, 353–362.

Parri, H.R. (1991) Studies on chloride in the muscle bag cells of *Ascaris suum*. PhD Thesis, University of Southampton, U.K.

Parri, H.R., Holden-Dye, L., and Walker, R.J. (1991) Studies on the ionic selectivity of the GABA operated chloride channel on the somatic muscle bag cells of the parasitic nematode, *Ascaris suum*. *Experimental Physiology* **76**, 597–606.

Pedder, S.M., and Walker, R.J. (1999) The actions of FXRFamide related neuropeptides on identified neurons from the snail, *Helix aspersa*. *Acta Biologica Hungarica* **50**, 185–198.

Pemberton, D.J., Franks, C.J., Walker, R.J., and Holden-Dye, L. (2001) Characterization of glutamate-gated chloride channels in the pharynx of wild-type and mutant *Caenorhabditis elegans* delineates the role of the subunit GluClα2 in the function of the native receptor. *Molecular Pharmacology* **59**, 1037–1043.

Pfarr, K.M., and Fuhrman, J.A. (2000) *Brugia malayi*: Localization of Nitric Oxide synthase in a lymphatic filariid. *Experimental Parasitology* **94**, 92–98.

Price, D.A., and Greenberg, M.J. (1977) Structure of a molluscan cardioexcitatory neuropeptide. *Science* **197**, 670–671.

Raizen, D.M., and Avery, L. (1994) Electrical activity and behaviour in the pharynx of *Caenorhabditis elegans*. *Neuron* **12**, 483–495.

Rand, J.B., and Nonet, M.L. (1997) Synaptic Transmission. In: D.L. Riddle, T. Blumenthal, B.J. Meyer, and J.R. Priess (eds.) *C. elegans II*. Cold Spring Harbor Laboratory Press, New York, pp. 611–644.

Ranganathan, R., Cannon, S.C., and Horvitz, H.R. (2000) MOD-1 is a serotonin-gated chloride channel that modulates locomotory behaviour in *C. elegans*. *Nature* **408**, 470–475.

Reinitz, C.A., and Stretton, A.O.W. (1996) Behavioural and cellular effects of serotonin on locomotion and male mating posture in *Ascaris suum*. *Journal of Comparative Physiology* **178**, 655–667.

Richmond, J.E., and Jorgensen, E.M. (1999) One GABA and two acetylcholine receptors function at the *C. elegans* neuromuscular junction. *Nature Neuroscience* **2**, 791–797.

Rogers, C.M., Franks, C.J., Walker, R.J., and Holden-Dye, L. (2000) A comparison of the actions of octopamine and the nematode neuropeptide SAEPFGTMRFamide on the pharyngeal muscle of the nematode *Caenorhabditis elegans*. *European Journal of Neuroscience* **12**, Supplement 11, 406.

Rohrer, S.P., and Arena, J.P. (1995) Ivermectin interactions with invertebrate ion channels. *American Chemical Society (ACS) Symposium Series* **591**, 264–283.

Rosoff, M.L., Burglin, T.R., and Li, C. (1992) Alternatively spliced transcripts of the *flp-1* gene encode distinct FMRFamide-like peptides in *Caenorhabditis elegans. Journal of Neuroscience* **12**, 2356–2361.

Rosoff, M.L., Doble, K.E., Price, D.A., and Li, C. (1993) The *flp-1* propeptide is processed into multiple, highly similar FMRFamide-like peptides in *Caenorhabditis elegans. Peptides* **14**, 331–338.

Sajid, M., Isaac, R.E., and Harrow, I.D. (1997) Purification and properties of a membrane aminopeptidase from *Ascaris suum* muscle that degrades neuropeptides AF1 and AF2. *Molecular and Biochemical Parasitology* **89**, 225–234.

Sattelle, D.B., and Culetto, E. (1999) Functional Genetics of Cholinergic Synaptic Transmission in *Caenorhabditis elegans.* In: D.J. Beadle (ed.) *Progress in Neuropharmacology and Neurotoxicology of Pesticides and Drugs.* S.C.I., London, pp 146–160.

Saz, N.J., and Weil, A. (1962) A pathway of formation of alpha-methylvalerate by *Ascaris lumbricoides. Journal of Biological Chemistry* **237**, 2053–2056.

Schaeffer, J.M., and Hains, H.W. (1989) Avermectin binding in *Caenorhabditis elegans.* A Two-state model for the avermectin binding site. *Biochemical Pharmacology* **38**, 2329–2338.

Schafer, W.R., and Kenyon, C.J. (1995) A calcium-channel homologue required for adaptation to dopamine and serotonin in *Caenorhabditis elegans. Nature* **375**, 73–78.

Schafer, W.R., Sanchez, B.M., and Kenyon, C.J. (1996) Genes affecting sensitivity to serotonin in *Caenorhabditis elegans. Genetics* **143**, 1219–1230.

Schinkmann, K., and Li, C. (1992) Localization of FMRFamide-like peptides in *Caenorhabditis elegans. Journal of Comparative Neurology* **316**, 251–260.

Segerberg, M.A., and Stretton, A.O.W. (1993) Actions of cholinergic drugs on the nematode *Ascaris suum*: Complex pharmacology of muscle and motorneurons. *Journal of General Physiology* **101**, 271–296.

Sharpe, M.J., and Atkinson, H.J. (1980) Improved visualization of dopaminergic neurons in nematodes using the glyoxylic acid fluorescence method. *Journal of Zoology* **190**, 273–284.

Smart, D. (1988) Investigations of the synthesis and metabolism of 5-hydroxytryptamine in *Ascaridia galli* (Nematoda). *International Journal of Parasitology* **18**, 747–752.

Stretton, A.O.W. (1992) Signalling systems in the nervous system of the nematode, *Ascaris.* In: I.R. Duce (ed.) *Neurotox 91*, Elsevier, pp. 124–138.

Stretton, A.O.W., Fishpool, R.M., Southgate, E., Donmoyer, J.E., Walrond, J.P., Moses, J.E.R., and Kass, I.S. (1978) Structure and physiological activity of the motoneurones of the nematode *Ascaris. Proceedings of the National Academy of Sciences U.S.A.* **75**, 3493–3497.

Stretton, A.O.W., Davis, R.E., Angstadt, J.D., Donmoyer, J.E., and Johnson, C.D. (1985) Neural control of behaviour in *Ascaris. Trends in Neuroscience* **8**, 294–300.

Stretton, A.O.W., Cowden, C., Sithigorngul, P., and Davis, R.E. (1991) Neuropeptides in the nematode *Ascaris suum. Parasitology* **102**, S107–S116.

Sulston, J.E., Dew, M., and Brenner, S. (1975) Dopaminergic neurons in the nematode *Caenorhabditis elegans. Journal of Comparative Neurology* **163**, 215–226.

Sze, J.Y., Victor, M., Loer, C., Shi, Y., and Ruvkun, G. (2000) Food and metabolic signalling defects in a *Caenorhabditis elegans* serotonin-synthesis mutant. *Nature* **403**, 560–564.

Takemoto, T. (1978) Isolation and structural identification of naturally occurring excitatory amino acids. In: McGeer, E.G., Olney, J.W., and McGeer, P.L. (eds.) *Kainic acid as a tool in Neurobiology.* Raven Press, New York, pp. 1–15.

Tornoe, C., Bai, D., Holden-Dye, L., Abramson, S.N., and Sattelle, D.B. (1995) Actions of neurotoxins (bungarotoxins, neosurugatoxin and lophotoxins) on insect and nematode nicotinic acetylcholine receptors. *Toxicon* **33**, 411–424.

Trim, N. (1998) Characterization of the effects of FMRFamide-like peptides on the somatic muscle of the parasitic nematodes *Ascaris suum* and *Ascaridia galli.* PhD Thesis, University of Southampton, U.K.

Trim, N., Holden-Dye, L., Ruddell, R., and Walker, R.J. (1997) The effects of the peptides AF3

(AVPGVLRFamide) and AF4 (GDVPGVLRFamide) on the somatic muscle of the parasitic nematodes *Ascaris suum* and *Ascaridia galli*. *Parasitology* **115**, 213–222.

Trim, N., Brooman, J.E., Holden-Dye, L., and Walker, R.J. (1998) The role of cAMP in the actions of the peptide AF3 in the parasitic nematodes *Ascaris suum* and *Ascaridia galli*. *Molecular and Biochemical Parasitology* **93**, 263–271.

Trim, J.E., Holden-Dye, L., Willson, J., Lockyer, M., and Walker, R.J. (2001) Characterization of 5-HT receptors in the parasitic nematode, *Ascaris suum*. *Parasitology* **122**, 207–217.

Trimmer, B.A., Kobierski, L.A., and Kravitz, E.A. (1987) Purification and characterization of FMRFamide-like immunoreactive substances from the lobster nervous system: Isolation and sequence of two closely related peptides. *Journal of Comparative Neurology* **266**, 16–26.

Tsang, V.C., and Saz, N.J. (1973) Demonstration and function of 2-methylbutyrate racemase in *Ascaris lumbricoides*. *Comparative Biochemistry and Physiology* **45B**, 617–623.

Vassilatis, D.K., Arena, J.P., Plasterk, R.H.A., Wilkinson, H.A., Schaeffer, J.M., Cully, D.F., and Van der Ploeg, L.H.T. (1997) Genetic and biochemical evidence for a novel avermectin-sensitive chloride channel in *Caenorhabditis elegans*. Isolation and characterization. *Journal of Biological Chemistry* **272**, 33167–33174.

Vincent, S.R. (1995) Localization of nitric oxide neurons in the central nervous system. In: S.R. Vincent (ed.), *Nitric Oxide in the Nervous System*. Academic Press, London, pp. 83–102.

Walker, R.J., Brooks, H.L., and Holden-Dye, L. (1996) Evolution and overview of classical transmitter molecules and their receptors. *Parasitology* **113**, S3–S33.

Wann, K.T. (1987) The electrophysiology of the somatic muscle cells of *Ascaris suum* and *Ascaridia galli*. *Parasitology* **94**, 555–566.

Warbrick, E.V., Rees, H.H., and Howells, R.E. (1992) Immunocytochemical localization of an FMRFamide-like peptide in the filarial nematodes *Dirofilaria immitis* and *Brugia pahangi*. *Parasitology Research* **78**, 252–256.

Weinshenker, D., Garriga, G., and Thomas. J.H. (1995) Genetic and pharmacological analysis of neurotransmitters controlling egg laying in *C. elegans*. *Journal of Neuroscience* **15**, 6975–6985.

Weisblat, D.A., and Russell, R.L. (1976) Propagation of electrical activity in the nerve cord and muscle syncytium of the nematode *Ascaris lumbricoides*. *Journal of Comparative Physiology* **107**, 293–307.

Weisblat, D.A., Byerly, L., and Russell, R.L. (1976) Ionic mechanisms of electrical activity in the somatic muscle of the nematode *Ascaris lumbricoides*. *Journal of Comparative Physiology* **111**, 93–113.

White, J.G., Southgate, E., Thompson, J.N., and Brenner, S. (1976) The structure of the ventral nerve cord of *Caenorhabditis elegans*. *Philosophical Transactions of the Royal Society London* **B 275**, 298–327.

White, J.G., Southgate, E., Thompson, J.N., and Brenner, S. (1986) The structure of the nervous system of *Caenorhabditis elegans*. *Philosophical Transactions of the Royal Society London* **B 314**, 1–340.

Williams, J.A., Shahkolaki, A.M., Abbassi, M., and Donahue, M.J. (1992) Identification of a novel 5-HT_N (Nematode) receptor from *Ascaris suum* muscle. *Comparative Biochemistry and Physiology* **101C**, 469–474.

Woodruff, G.N., Walker, R.J., and Newton, L.C. (1971) The actions of some muscarinic and nicotinic agonists on the Retzius cells of the leech. *Comparative and General Pharmacology* **2**, 106–117.

Wright, D.J., and Awan, F.A. (1978) Catecholaminergic structures in the nervous system of three nematode species, with observations on related enzymes. *Journal of Zoology* **185**, 477–489.

CHAPTER 9

EVIDENCE FOR AN ANNELID NEUROENDOCRINE SYSTEM

MICHEL SALZET, DIDIER VIEAU and CHRISTOPHE BRETON

Laboratoire d'Endocrinologie des Annélides, UPRESA CNRS 8017, SN3, Université des Sciences et Technologies de Lille, F-59655 Villeneuve d'Ascq Cédex, France

Summary

Signalling molecules such as angiotensin, oxytocin/vasopressin, RFamide peptides or opioids have been found in the central nervous system or immune system of annelids. Since these signalling molecules are found free in the haemolymph, they are considered as hormones. Hormonal mechanisms along with enzymatic processing similar to those found in vertebrates occur in annelids, for example, the renin-angiotensin system. Furthermore, amino acid sequence determination of annelid precursor gene products reveals the presence of the respective peptides that exhibit high sequence identity to their mammalian counterparts. These peptides are flanked by potential proteolytic signal sites for the various known enzymes, confirming that annelid neuropeptide precursors are processed in a similar manner to that described in mammals. Moreover, in leeches as in other parasites, we surmise that these signalling systems, because of their great similarity, are employed to escape host immunosurveillance. This further indicates that neuroendocrine processes are present within annelids. Taken together, the data indicate that, as in mammals, the signaling molecules can act as hormones that help modulate immune functions. Moreover, given their presence in neural tissue, they constitute a neuroendocrine system.

1. Introduction

Neuropeptides are widespread and play a major role in the animal kingdom, including in protozoa. The similar amino acid sequences of some of the neuropeptides, e.g., oxytocin/vasopressin and angiotensin in vertebrates and invertebrates, demonstrates that they are stable and old in evolutionary terms. These signalling peptides seem to have appeared before the development of a primitive nervous system, suggesting that they were first implicated in intercellular communication (Smith, 1994). Therefore, they may have evolved from a small number of signalling proteins encoded by ancestral genes that have been conserved and amplified over time (Stefano and Salzet, 1999).

2. What constitutes a neuroendocrine system?

In vertebrates, this system is based on chemical signalling between neural and endocrine structures. Final outcomes may be realized via the chemical messengers travelling from circulatory conduits to their specific target sites. This process may, in part, rely on neurosecretion of the signalling molecules (Stefano

I. Fairweather (ed.), Cell Signalling in Prokaryotes and Lower Metazoa, 303–324.
© 2004 *Kluwer Academic Publishers. Printed in Great Britain.*

et al., 1991). The complexity of this system can be visualized quickly when one considers its operating regulations that result from communication between classical neurotransmitters, cytokines, growth factors and hormones (including neuroendocrine, autocrine and paracrine messenger molecules). Apart from a neuroendocrine system, there also exists a neuroimmune communication system that consists of reciprocal signalling between neuroendocrine and immune cells which use the same molecules to coordinate their activity, e.g., adrenocorticotropin hormone (ACTH) (Stefano *et al.*, 1996). We have recently demonstrated that neurovascular regulation of neurosecretion occurs in rat median eminence fragments via a nitric oxide-mediated process (Stefano and Salzet, 1999). Thus, our concept of a neuroendocrine system is constantly growing, given its complexity, but may be simply summarized as the result of bi-directional communications between neural and endocrine structures over a distance greater than that occurring by synaptic communication.

3. Do annelids have a neuroendocrine system?

Annelid neural tissues do not contain anatomical correlates of the hypothalamus or pituitary. However, they possess localized ganglionic regions rich in mammalian-like neuroendocrine signalling molecules (Figs. 1 and 2), *e.g.* angiotensins (Salzet *et al.*, 1993a, 1995a), oxytocin/vasopressin peptides (Salzet *et al.*, 1993b; Oumi *et al.*, 1994; Satake *et al.*, 1999) and enkephalins (Salzet *et al.*, 1995b) (Tables 1 and 2). Furthermore, these molecules appear free in the haemolymph of annelids, demonstrating distant signalling (Salzet and Stefano, 1998; Satake *et al.*, 1999; Stefano and Salzet, 1999) via several target tissues, including immune cells (Stefano and Scharrer, 1991; Genedani *et al.*, 1994; Smith, 1994; Stefano and Salzet, 1999). A neuroendocrine system is also characterized by the presence of signalling molecule receptor(s) in target cells. To our knowledge, apart from pharmacological data, nothing is known about putative receptors in annelids. In the future, considerable efforts are needed to shed on light on the molecular structure of annelid receptors. Thus, the fact that a classical closed conduit system to carry signalling molecules does not exist in annelids should not detract from an endocrine presence because the baseline functioning of the system depends on distance between the origin of a signal molecule and its target tissue/receptor.

4. Are neuroendocrine signalling molecules in annelids and vertebrates similar?

This question represents an important aspect regarding the existence of an annelid neuroendocrine system. Among the 30 neuropeptides in annelids isolated so far (Tables 1 and 2), most of them already sequenced are related to the ones previously isolated in vertebrates. Signalling molecules can be divided into 4 groups, namely, the angiotensin, the oxytocin/vasopressin, the myotropic peptides and the opioid families. With the exception of the myotropic peptides,

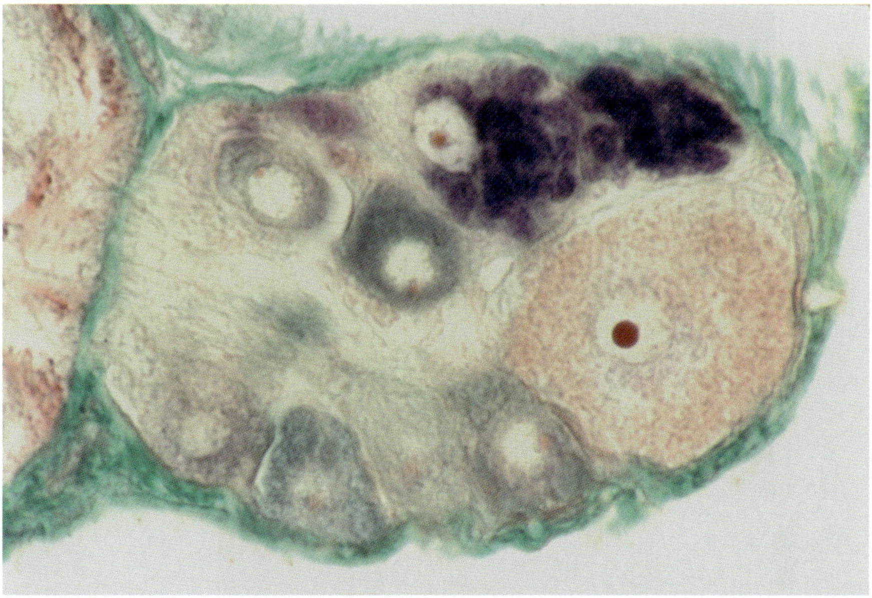

(a)

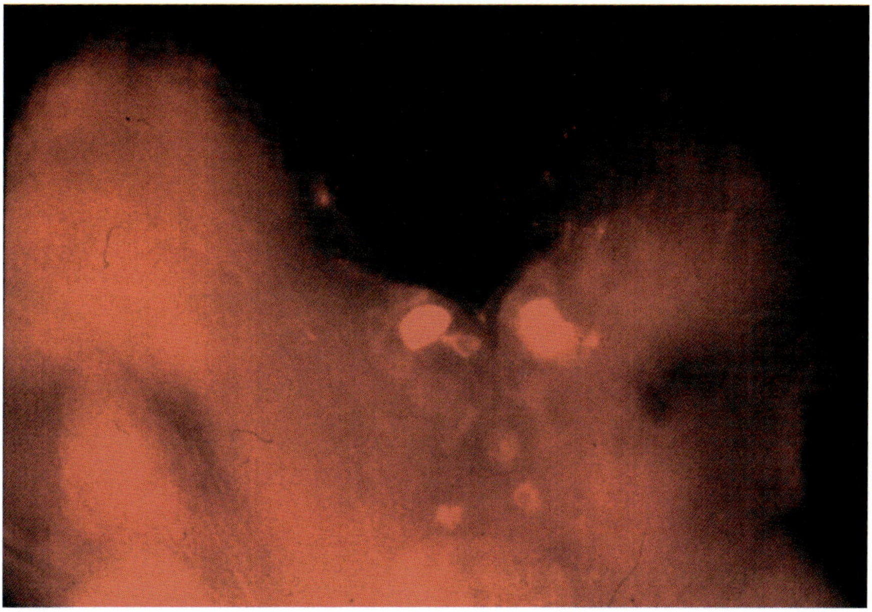

(b)

Figure 1. (a) Clark colouration of compartment 4 of the supraoesophageal ganglia of the leech, *Theromyzon tessulatum.* (b) Rhodamine fluorescent immunocytochemical labelling of angiotensin I-positive cells in the supra- oesophageal ganglia of the leech, *Theromyzon tessulatum.*

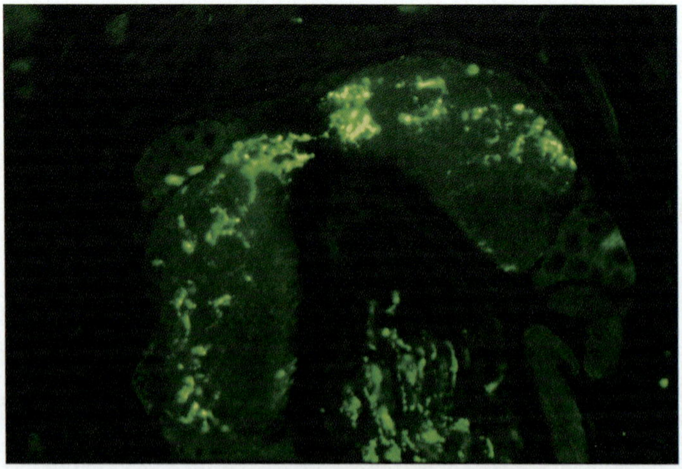

(c)

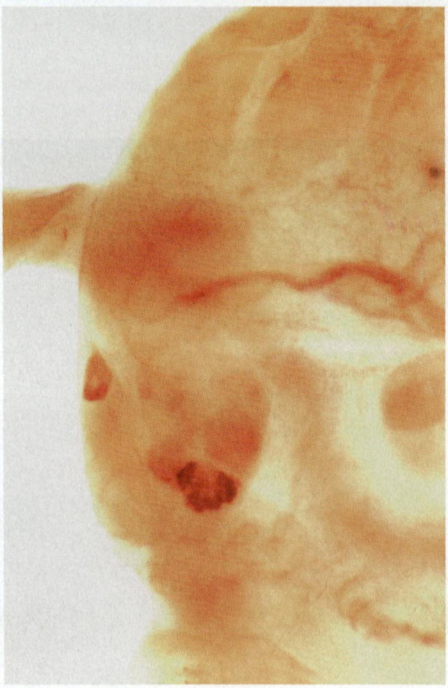

(d)

Figure 1 (*cont.*) (c) Angiotensin II-like immunoreactivity in the dorsal commissure and neurohaemal area of the brain of the leech, *Theromyzon tessulatum*. (d) Whole mount of compartment 4 and the dorsal commissure of the brain of the leech, *Theromyzon tessulatum* showing angiotensin II-immunoreactive cells.

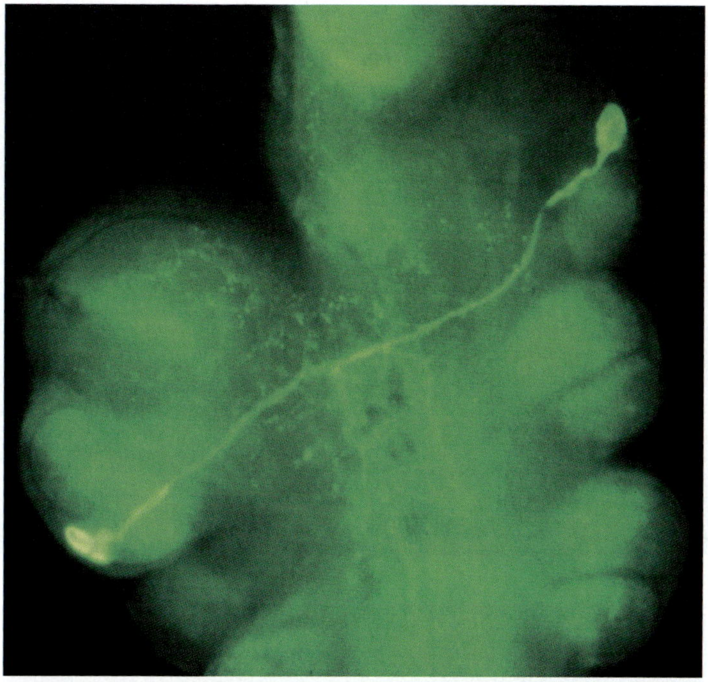

(a)

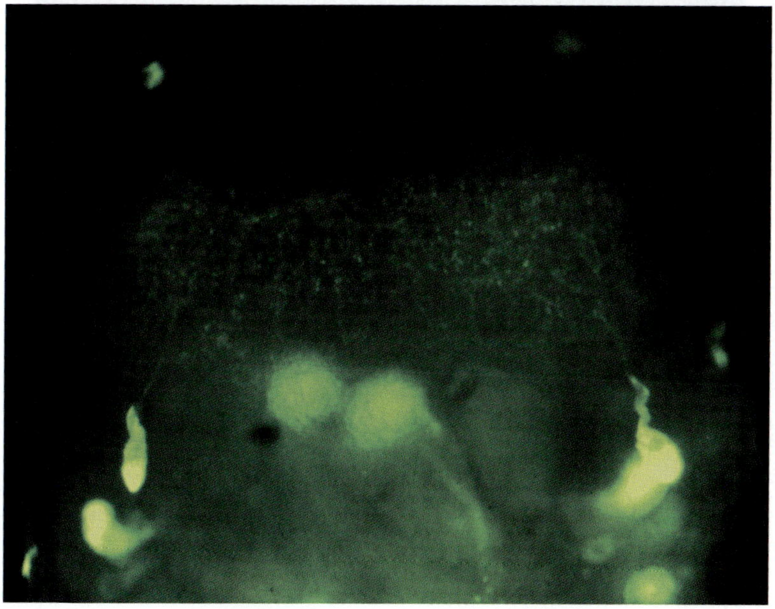

(b)

Figure 2. (a) Whole mount immunolabelling with anti-angiotensin II antibody in the brain of the leech, *Erpobdella octoculata. (b) Whole mount immunolabelling with anti-PLGamide II antibody in the brain of the leech, Theromyzon tessulatum.*

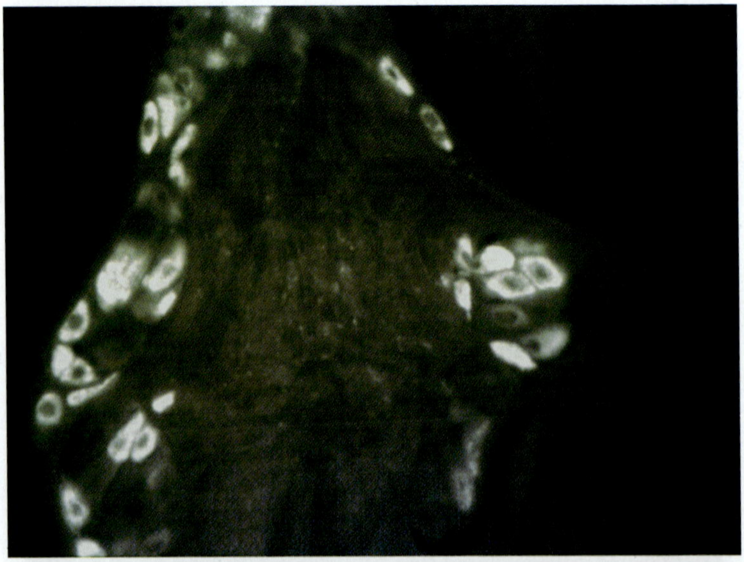

(c)

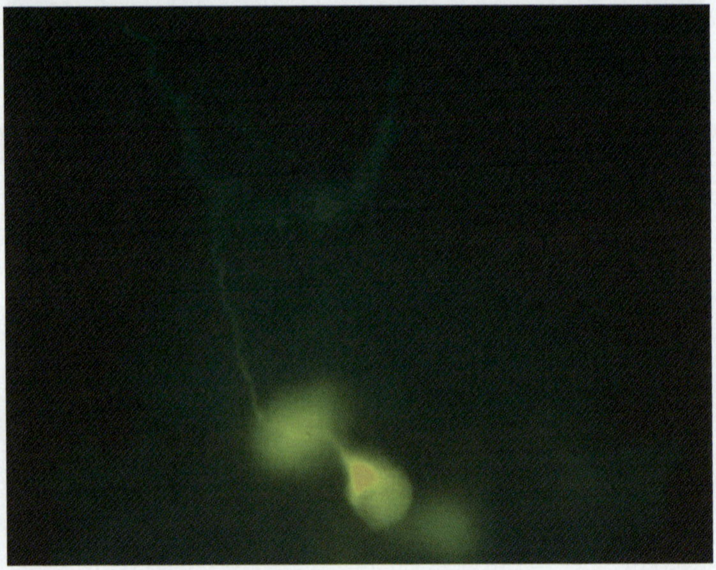

(d)

Figure 2 (*cont.*) (c) Fluorescence labelling for FMRFamide-immunoreactivity in the sex ganglia of *Erpobdella octoculata*. (d) Lucifer yellow injection into angiotensin-immunoreactive cells in compartment 4 of the supra- oesophageal ganglia of the leech, *Theromyzon tessulatum*.

Table 1. Characterized annelid neuropeptides from leeches

Species	Sequence	Name
Theromyzon tessulatum	SYVMEHFRWDKFGRKIKRRPIKVYPNGAEDESAEAFPLE	ACTH-like
	DRVYIHPFHLLXWG	Angiotensin I
	DRVYIHPF	Angiotensin II
	RVYIHPF	Angiotensin III
	IPEPYVWD	LORF
	FMRF-amide	FMRF-amide
	FLRF-amide	FLRF-amide
	GDPFLRFamide	GDPFLRF-amide
	PLG	MIF-1
	YGGFL	Leucine-enkephalin
	YGGFM	Met-enkephalin
	YGGFLRKYPK	α neoendorphin
	YVMGHFRWDKFamide	MSH-like peptide
	GSGVSNGGTEMIQLSHIRERQRYWAQDNLRRRFLEKamide	Leech egg laying hormone
Erpobdella octoculata	DRVYIHPF-amide	AngiotensinII-amide
	CFIRNCPKG-amide	Lysine-conopressin
	FMRF-amide; FM(O)RF-amide	FMRF-amide
	GDPFLRF-amide	GDPFLRF-amide
	FLRF-amide	FLRF-amide
	IPEPYVWD; IPEPYVWD-amide	LORF
Hirudo medicinalis	FMRF-amide, FM(O) RF-amide	FMRF-amide
	FLRF-amide	FLRF-amide
	AMGMLRM-amide	Myomoduline-like peptide
Hirudo nipponia *Whitmania pigra*	WRLRSDETVRGTRAKCEGEWAIHACLCLGGNamide	Leech excitatory peptide

which are more specific to the annelids, the other peptide families have also been identified in vertebrates in which they play crucial roles as neurohormones. This data further strengthens the existence of a neuroendocrine system in annelids.

4.1. ANGIOTENSIN-LIKE PEPTIDES

Biochemical identification of a "central" angiotensin II (AII)-like peptide (Fig. 1) in the leech *Erpobdella octoculata* was demonstrated and found to be amidated (Salzet *et al.* 1995a). This constituted the first characterization of an angiotensin-like peptide in an invertebrate, demonstrating its conservation during the course of evolution. Leech AII-amide exhibits, in its first eight amino acids, an almost identical sequence to the first eight amino acids of peripheral AI in human, horse, pig, mice, rat, rabbit, dog and guinea pig (Takei *et al.*,

Table 2. Characterized neuropeptides in annelids other than leeches

Species	Sequence	Name
Perinereis Vancaurica	AMGMLRMamide	Myomodulin-CARP
	WVGDVQ	Esophagus regulation peptides
	ATWLDT	
	WMVGDVQ	
	FYEGDVPY	
Nereis diversicolor	FMRF-amide, FM(O)RF-amide	FMRF-amide
	FTRF-amide	FTFR-amide
Nereis virens	FMRF-amide	FMRF-amide
	FLRF-amide	FLRF-amide
Eisenia Foetida	CFVRNCPTGamide	Annetocin
	APKCSGRWAIHSCGGNG	GGNG1
	GKCAGQWAIHACAGGNG	GGNG2
	RPKCAGRWAIHSCGGGNG	GGNG3

1993). Moreover, 6 out of these 8 amino acids are constant, with only amino acids in position 5 and 1 differing (Salzet *et al.*, 1995a).

Proteins immunoreactive to anti-AII have been identified both in brain extracts and in *in vitro* translated brain RNA products (Salzet *et al.*, 1993a). The pro-AII precursor detected in the brain extracts possesses a *ca* 19 kDa molecular mass and is also a "multiple hormone precursor" as it is also recognized by two other antisera: a polyclonal γ-MSH and a specific monoclonal antibody directed against leech neurons, Tt159 (Salzet *et al.*, 1993a). Furthermore, we found in leeches a *ca.* 11-kDa peptide with a sequence of DRVYIHPFHLLXWG, which exhibits a 78.5% sequence identity to the N-terminus of angiotensinogen and a 100% sequence identity to AI (Table 1). Interestingly, a 25.4 kDa protein immunoreactive to different antisera (e.g. anti-AII and anti-MSH antibodies) was found in leech haemocytes. This protein is related to proopiomelanocortin (POMC), with an homology of 25% (Fig. 1) (Salzet *et al.*, 1997). Thus, leeches might contain two putative angiotensin precursors, one in immune and the other in neural tissues. This finding highlights the importance of such molecules in the physiological regulations of these animals.

From experiments conducted on the biological activity of the AII-amide molecule isolated from *E. octoculata*, it appears that this peptide is involved in the control of hydric balance in leeches, exerting a diuretic effect (20% loss of mass on average for 1nmole of AII) after its injection into the animals (Salzet *et al.*, 1993a). Carboxyl-terminal amidation is very important because it has been demonstrated that the non-amidated peptide was 10 to 100 times less biologically active when used in the same experimental conditions (Salzet *et al.*, 1993a, 1995a). Because AII-amide injections at different doses in *Theromyzon tessulatum* suggest the existence of two different types of receptors, one at high and the other one at low affinities towards AIIamide (Salzet *et al.*, 1995a), we focussed our interest on the identification of the leech AIIamide receptors. Binding experiments on *T. tessulatum* brain membranes with mono [I^{125}]AIIamide reveals 70% specific binding and an IC_{50} of 10 nM (unpublished data). In addition, immunocytochemical experiments using a commercial anti-AT1 receptor antibody reveal neurons positively stained with the antiserum (unpublished data). These cells are also stained with an antibody to an AII and angiotensin converting enzyme. Taken together, the data suggest the presence of at least an AT1-like receptor in leech brain and the existence of a complete renin-angiotensin system in leeches.

Similarly, in the polychaete, *Nereis diversicolor*, injections of polyclonal antisera against AII provoked a partial inhibition of the increase in body weight in animals exposed to a hypo-osmotic medium. In a subsequent test, injections of synthetic AII-amide and, to a lesser extent AII, enhanced the increase in body weight and, therefore, strengthened the importance of these peptides in the neuroendocrine control of osmoregulation in *Nereis* (Dhainaut personnal communication).

4.2. OXYTOCIN/VASOPRESSIN PEPTIDES

The peptides of the vasopressin/oxytocin family have been discovered throughout the animal kingdom. They are alike, sharing at least five of nine residues and a disulphide-linked ring structure, which puts severe constraints on conformational flexibility (Acher *et al.*, 1985). In vertebrates, gene duplication gave rise to two distinct families, namely, the vasopressin and oxytocin families (Acher *et al.*, 1985). The differential binding of vasopressin and oxytocin to their respective receptors is largely due to the amino acid residue in position 8, that is, a basic one in vasopressin-related peptides and a neutral one in oxytocin-related peptides. In leeches, a vasopressin related peptide, lysine-conopressin has been isolated (Salzet *et al.*, 1993b). Oumi *et al.* (1994) have confirmed the existence of an oxytocin-related peptide in oligochaete annelids by the purification of annetocin (Table 2), a peptide having 7 residues in 9 common with the leech lysine-conopressin.

The annetocin cDNA has recently been cloned in a lumbricid earthworm, *Eisenia foetida*. This peptide, as in vertebrates, is associated with a neurophysin, having about 40% sequence identity to vertebrate neurophysin (Satake *et al.*, 1999). The data confirm the hypothesis of an ancestral gene common to both oxytocin and vasopressin peptide families dating back 500 million years. In annelids, both annetocin and lysine-conopressin may be involved in osmoregulation, via pulsatory contractions and bladder-shaking movement of the nephridia (Oumi *et al.*, 1994; Salzet *et al.*, 1993b). Lysine-conopressin inhibits the Na^+ amiloride-dependent transitory current before highly stimulating it in *Hirudo medicinalis* stomach or tegument preparations (Table 2). Moreover, lysine-conopressin and annetocin induce egg laying in earthworms, accompanied by secretion of mucus from the clitellum (Oumi *et al.*, 1994). This confirms that through the course of evolution the oxytocin/vasopressin peptide family has conserved its function on both osmoregulation and reproduction.

4.3. RFAMIDES

The family of RFamide peptides is a group of peptides sharing the carboxyl terminal sequence Arg-Phe-amide. These peptides are found throughout the metazoans (Walker, 1992) and notably in annelids (Baratte *et al.*, 1991; Evans *et al.*, 1991; Salzet *et al.*, 1995b). In the latter phylum, authentic FMRFamide is present in the polychaetes, *Nereis virens* and *Nereis diversicolor* (Walker, 1992). In the latter species, two other RFamide peptides (FM(O)RFamide and FTRFamide) have been isolated. Pharmacological data suggest that RFamide peptides are involved in the control of heartbeat and body wall tone in the polychaete, *Sabellastarte magnifica* Diaz-Miranda, Escalona de Motta and Garcia-Arraras, 1989) and in the oligochaete, *E. foetida* (Walker, 1992). In the earthworm, *E. foetida*, FMRFamide-like peptides are co-localized with serotonin, suggesting a role as neuromodulators influencing serotoninergic neurons (Lubics *et al.*, 1997). In *Lumbricus terrestris*, FMRFamide-like peptides seem to be involved in central integratory processes, neuromuscular regulation and sensory processes (Lubics *et al.*, 1997). In the Hirudinae, FMRFamide immunoreactivity has been demonstrated in cell bodies and neuronal processes of the central nervous system (Kuhlman, Li and Calabrese, 1985a,b). In the segmental ganglia of the ventral nerve cord, the immunoreactivity is localized in heart excitatory (HE) motor neurons, heart accessory (HA) modulatory neurons and several motor neurons innervating the longitudinal and medio-dorso ventral muscles (Kitazono *et al.*, 1995; Kim, Berliner and Nadler, 1996). Two RFamide peptides (FMRFamide and FLRFamide) Have been identified in *H. medicinalis* (Evans *et al.*, 1991) and four in E. *octoculata* (namely, FM(0)RFamide, FMRFamide, FLRFamide and GDPFLRFamide) from sex ganglia extracts (Salzet *et al.*, 1994). The presence of RFamide peptides has also been demonstrated in *T. tessulatum* in neurosecretory granules in nerve fibres within the neurohaemal area, suggesting that these RFamide are secreted into the dorsal vessel. So, the brain could exert a neuroendocrine control of certain functions via RFamide peptides. Electrophysiological experiments have demonstrated that GDPFLRFamide has a stimulating effect on Cl^- secretion across the caecal epithelium. While Na^+ absorption across the tissue remained unaffected by the peptide, Cl^- conductances were monophasically activated. When Cl^- ions are secreted, water follows passively, thereby causing a loss of water from the cells (Wenning and Calabrese, 1995; Zerbst-Boroffka, Bazin and Wenning, 1997). GDPFLRFamide left Na^+ and water absorption unchanged and stimulated water transport across the epithelium by stimulating Cl^- secretion. Consequently, the diuretic effect observed might be due, at least partially, to an action on the caecal epithelium (Vandenbulle *et al.*, 1997). By contrast, no electrophysiological effects on leech caecal epithelium were found for FLRFamide and FMRFamide, although it was shown that FMRFamide has an anti-diuretic action when injected into these animals Nandenbulle *et al.*, 1997). The anti-diuretic effect of FMRFamide is not due to a direct action on the caecal epithelium. Nevertheless, the control of the hydric balance might also be exerted directly on the nephridia. Indeed, Zerbst-Boroffka, Bazin and Wenning (1997) demonstrated that the nephridial nerve cells, which innervate the nephridia and contact the urine-forming cells, contain RFamide peptide(s) in *H. medicinalis.* Furthermore, Wenning and Calabrese (1995) showed that FMRFamide leads the hyperpolarization and decreases the rate of firing of the nephridial nerve cells, suggesting autoregulation of peptide release.

Besides RFamides, several related RXamides have been demonstrated in annelids. For example, a myomodulin-like peptide (GMGALRLamide) has been isolated from leech nerve cord (Table 2). In the polychaete, *Perinereis vancaurina* a myodulin-related peptide (AMGMLRMamide) has also been discovered (Minakata *et al.*, 1997). This peptide exerts the same function in both animals, namely, muscular contraction.

4.4. OPIOIDS

4.4.1. *Proenkephalin (pro-Enk)*

We recently demonstrated the presence of proenkephalin in the leech, *T. tessulatum* (Stefano and Salzet, 1999). Leech proenkephalin (Figs. 3 and 4) displays amino acid sequence similarity with amphibian proenkephalin (26.2%) and it contains Met- and Leu-enkephalins in a ratio of 1/2. It also possesses Met-enkephalin-Arg-Gly-Leu and Met-enkephalin-Arg-Phe that are flanked by dibasic amino acid residues, which are targets for proteolytic enzymes. Specific receptors for these peptides have been pharmacologically characterized in invertebrates, in both neural and immune systems, and they have been

```
G    65 CKDLLQVSKQELPQEGASSLRESGKQDESHLLSKKYGGFMKRYGGFMKKVDELYPVEPEE
L    36 -------------------------------IAKKYGGFMKRRRVKKRYGGFLKKI----

G 125 EANGGEILTKRYGGFMKKDAEDGDALANSSDLLKELLGTGDDRDRENHHQEGGDSDEGVS
L  60 ---------LRYGGFLK-----------------------------RLIQPGTDLMN---

G 185 KRYGGFMRGLKRSPQVEDEAKELQKRYGGFMRRVGRPEWWMDYQKRYGGFLKRFAEFLPS
L  80 KRYGGFMRGLRRS---------------------TRPEW------------KKFAEFLPS

G    245  DEEGESYSKEVPEMEKRYGGFMRF 268
L    107  EEEGESYSKEVPEMERRYGGFMRF 128
```

Figure 3. Sequence comparison of leech and guinea pig proenkephalins. G, guinea pig; L, leech. Putative cleavage sites are in bold.

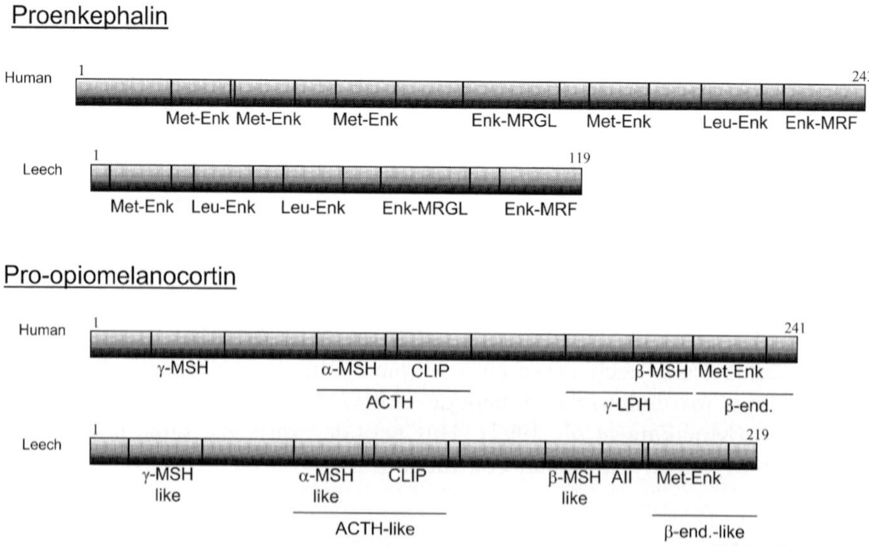

Figure 4. Opioid precursor structure of the leech, *Theromyzon tessulatum*. After its processing, proenkephalin generates opioid peptides such as Leucine-enkephalin (Leu-enk), Methionine-enkephalin (Met-enk) and other active molecules, including Methionine-enkephalin arginine glycine leucine (Enk-MRGL) and Methionine-enkephalin arginine phenylalanine (Enk-MRF). POMC maturation also releases opioids such as Met enk and β-endorphin (β-end). The structure comparison of the two species shows differences in the precursor structures. By contrast, the active peptides produced are similar. Methionine-enkephalin (Met-enk); Leucine-enkephalin (Leu-enk); Methionine-enkephalin arginine glycine leucine (Enk-MRGL); Methionine-enkephalin arginine phenylalanine (Enk-MRF); Melanocyte Stimulating Hormone (MSH); Adrenocorticotropic Hormone (ACTH); Corticotropin-Like Intermediate Lobe Peptide (CLIP); Lipotropin Hormone (LPH) and β-endorphine (β-end.). The size of the precursor is expressed in amino acid residues.

extensively reviewed elsewhere (Stefano *et al.*, 1999). These studies also suggest that the receptors have been particularly well conserved during evolution.

4.4.2. *Prodynorphin (pro-Dyn)*

Alpha-Neo-endorphin, exhibiting 100% sequence identity with the mammalian form, has been purified from *T. tessulatum* central nervous system and suckers (Salzet *et al.*, 1996), suggesting the presence of a larger precursor polypeptide similar to pro-Dyn of vertebrates. Subsequent characterization of the entire pro-Dyn opioid precursor (Fig. 2) revealed that it exhibits a 28.8% sequence identity to rat, and 22% to the human and pig precursors (Civelli *et al.*, 1985). Dynorphin A-like material and dynorphin B-like material shows a 50% and 76.6% sequence homology with the mammalian counterparts, respectively. Although the α-Neo-endorphin is identical to the one present in vertebrates, the dynorphins are slightly shorter. The number of copies of Leu-enkephalin is similar to that present in vertebrates, that is, 3 leech prodyn-derived peptides are found in positions flanked by potential proteolytic sites (basic amino acids).

4.4.3. *Proopiomelanocortin (POMC)*

In leeches, we also isolated and sequenced a POMC-like molecule (Salzet *et al.*, 1997). Of the six peptides, three showed high sequence similarity to their vertebrate counterparts, namely, met-enkephalin, α-MSH and ACTH (100, 84.6 and 70%, respectively), whereas γ-MSH, β-endorphin and γ-LPH exhibited only 45, 20 and 10% sequence identity, respectively. No dibasic amino acid residues were found at the C-terminus of the γ- and β-MSH peptides, suggesting that they are not produced in the leech or that they could be synthesized via an alternate biosynthetic pathway. In contrast, the leech α-MSH was flanked at its C-terminus by the Gly-Arg-Lys amidation signal.

Taken together, the results from leeches now demonstrate that those opioid precursors and many of the derived bioactive peptides, e.g., α-MSH and ACTH, that are important in mammalian neuroendocrine signalling are present in invertebrates as well. This adds to the growing body of evidence that a neuroendocrine system is also present in simple animals.

4.4.5. *Non- classical opioid peptides*

We have demonstrated the existence of PLGamide peptide, named MIF-1, in leeches (Salzet and Deloffre, 2000). Based on data obtained in vertebrates by Kastin and co-workers, we speculate that leech PLGamide can also be derived from the Tyr-MIF-1 peptide precursor, which has been characterized in human brain (Kastin *et al.*, 1995). MIF-1 (PLGamide), Tyr-MIF-1 (YPLGamide), Tyr-W-MIF-1 (YPWGamide) and Tyr-K-MIF-1 (YPKGamide) belong to the non-classical opioid family and are structurally related to the haemoglobin-derived opiate peptide, hemorphin-4 (YPWT) and the endomorphins (YPTFamide and YPFFamide) (Kastin *et al.*, 1995, 1998). The existence of MIF-1 in leeches allows us to speculate on the presence of such a family in annelids. All the molecules display similar properties in vertebrates, namely, long half-life of more than 4 hours, inhibition of ACE and a high affinity for the mu opiate receptor (Salzet and Joffre, 2000). In leeches, we have demonstrated that MIF-1 also acts as an opiate agonist/antagonist (Salzet and Joffre, 2000).

5. Are some neuroendocrine signalling molecules specific to annelids?

Regarding the Leech OsmoRegulator Factor (LORF) and myotropic peptides found in these animals, the answer would be yes. In early reports, we demonstrated that LORF is a peptidergic signalling molecule involved in osmoregulation (Salzet *et al.*, 1996; Salzet, Vandenbulcke and Verger-Bocquet, 1997). Furthermore, electrophysiological experiments conducted in the leech, *H. medicinalis* revealed an inhibition of the efficacy of Na^+ conductance in leech skin (Salzet *et al.*, 1996). Immunocytochemical studies with an antiserum against synthetic LORF revealed a large number of immunopositive neurons in all ganglia. The peptide was present in a single type of electron-dense secretory granule, 80–100 nm in diameter (Salzet, Vandenbulcke and Verger-Bocquet, 1997). However, we recently showed its presence in rat tissues, including discrete brain areas (Salzet *et al.*, 1998), which demonstrates that this peptide, originally isolated from the leech, is also present in mammals. Furthermore, in both the leech central nervous system and the rat brain (namely, the hypothalamus), LORF is coupled to nitric oxide (NO) release (Salzet *et al.*, 1998). It is also capable of stimulating NO release from human saphenous vein fragments (Salzet *et al.*, 1998). These results have shown that LORF is not specific to annelids, but they have also demonstrated that the leech represents a suitable model to isolate novel peptides also conserved in vertebrates.

The most specific peptides presently isolated in annelids are the ones related to the GGNG excitatory peptides (Fig. 5). In leeches, these peptides elicit muscular contractions of isolated preparations of penis and intestine, suggesting that they may play a role in reproductive behaviour in a similar manner to APGWamide in molluscs (Minakata *et al.*, 1997). In the earthworm, they elicit gut and esophageal muscular contractions. No peptides homologous to GGNG peptides have been isolated so far in any living organisms.

6. Are neuropeptide processing enzymes present in invertebrates?

Biologically active neuropeptides are produced via the proteolytic processing of precursor molecules. This post-translational step involves many types of enzymes that have been identified both in vertebrates and in invertebrates. Between them, enzymes such as neutral endopeptidase (NEP) (Shipp *et al.*, 1990, 1991; Laurent and Salzet, 1995a), ACE (Laurent *et al.*, 1996a,b; Laurent, Stefano and Salzet, 1998), renin (Laurent and Salzet, 1995b), aminopeptidases (Turner, Leung and Stefano, 1994), and serine proteases belonging to the subtilisin family (Seidah *et al.*, 1998) play an important role in neuropeptide precursors processing.

⇓

Leech mkllflvlcl sllhvtrgve **rr**y**r**lgsdeg **r**ie**rr***w***r***l***r***s* *detvrgt***r***akcegewaihac*
Eise m**rr**pvtniml lvicylwchi qstsg*gkcag* *qwaihacagg* *ng***kr***s*diddlvrptdlgtde

⇓ ⇑

Leech *lggng***kr**sqq **rr**nidsptee kdnlllnkqf rsvlptlsar lpfrveeksy y
Eise erpdiglqea rggergeggg g**r**egdgy**rgd** rdnddlrsel seg**r**qlled**k**

Eise **r**pptplesal grlrsllrtt dfdvd**rr**eee mkpaqwqeee tp**rr**kntdkn
srynikdslt qlldivn**rr**p mlrsilddie grldsqpvsi se**rk**t**r**gyad
dvendsyslq **kr**sgdq

Figure 5. Annelid excitatory peptide precursor sequences. Cleavage sites are marked by ⇓. Putative cleavage sites are in bold. Eise, *Eisenia*.

Neutral endopeptidase 24.11 (NEP, CD10, CALLA, enkephalinase) is found on the surface of granulocytes and in invertebrate immunocytes (Shipp *et al.*, 1990, 1991). In invertebrates, opioid precursor molecules are also found in the haemolymph (Stefano and Salzet, 1999). The role of NEP as a regulatory enzyme in neuropeptide processing has been reviewed elsewhere (Turner, Leung and Stefano, 1994; Stefano and Salzet, 1999). Briefly, NEP may not only be responsible for cleaving a neuropeptide precursor, e.g., POMC/ACTH, but the active processed peptides as well, *i.e.*, met-enkephalin and MSH (Duvaux-Miret *et al.*, 1990; Smith *et al.*, 1992; Stefano, 1999). This is a marvellously sophisticated processing mechanism because the same enzyme may break down the precursor to generate active peptides, then go on to inactivate the same active molecules, producing inactive products. This could represent a time-dependent multidimensional process, although the *in vivo* activity of NEP on invertebrate neuropeptide precursors has yet to be demonstrated.

Leech brain and immune tissues also contain RAS involving renin and ACE enzymes in AII biosynthesis and ACE, NEP and aminopeptidase in the catabolism of the angiotensins (Laurent and Salzet, 1996a). Leech renin has been detected in immune cells with other aspartyl proteases that would be involved in ACTH (1–39) cleavage to ACTH (1–24) (unpublished data). Interestingly, differential processing was revealed by opiate stimulation of ACTH processing by triggering ACE and NEP actions (Salzet *et al.*, 1997). This second mechanism involved the activation of the μ_3 morphine receptor-coupled NO release (Fig. 6) (Salzet *et al.*, 1997). Taken together, then, as is the case in mammals, differential processing of ACTH occurs in invertebrates. Additionally, invertebrate immunocytes are capable of displaying different responses to ACTH fragments (Genedani *et al.*, 1994), further supporting the differential processing pattern and its potential significance as a meaningful event.

As in invertebrates, it is important to note that RAS, i.e., the renin gene, has been detected in mammalian macrophage/monocyte cells, as was ACE and AII (Iwai *et al.*, 1996; Hernandez-Presa *et al.*, 1997). In man, AII may be implicated in the recruitment of monocytes into the blood vessel wall through the activation of NFkb and the induction of the monocyte chemoattractant protein-1 (Hernandez-Presa *et al.*, 1997). Angiotensin II also increases monocyte binding to endothelial cells (Iwai *et al.*, 1996). The ACE gene is expressed in macrophage-like U937 histiocytic lymphoma cells (Iwai *et al.*, 1996; Hernandez-Presa *et al.*, 1997).

In contrast, in leeches, AII has, in preliminary experiments, no immunocyte chemotactic effect. It also exerts an inhibitory effect via nitric oxide release during an inflammatory response (unpublished data). Immunocytochemical studies performed at the level of the immunocytes revealed a co-localization with AT1-like, AII, renin-like and ACE-like enzymes in leech immunocytes (unpublished data). These data were confirmed with binding experiments on immunocytes with ^{125}I-saralazine, confirming the presence of AT1-like receptors in the membrane of invertebrate immunocytes. Thus, these data suggest an inhibitory effect of the RAS on the invertebrate immune response (Fig. 4). This

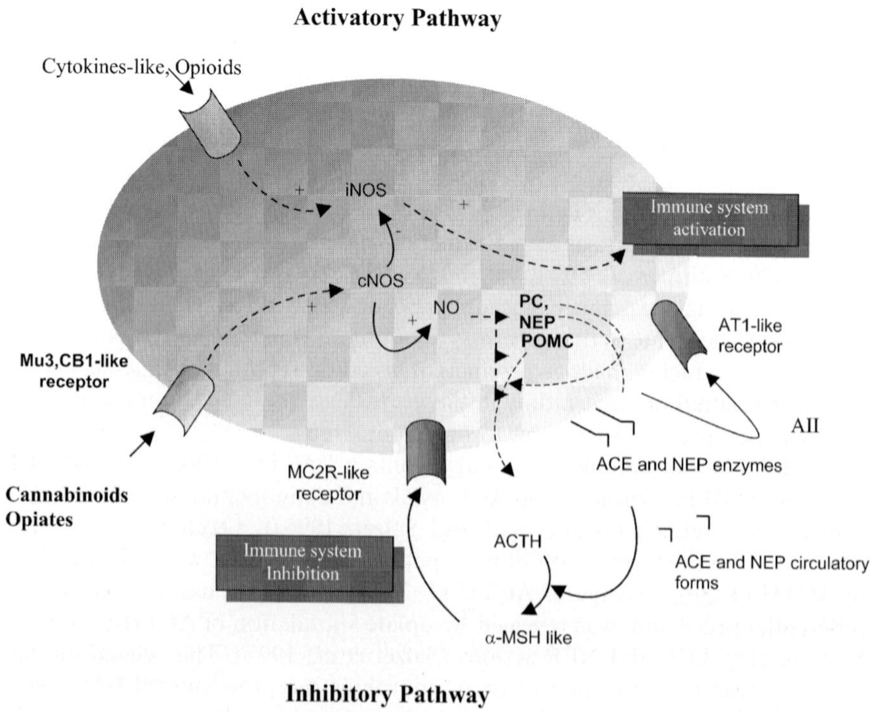

Figure 6. Immune response modulation in leeches: implication of the opiates-POMC derived peptides. Anandamide, after binding to its own receptor (CB1-like), rapidly stimulates NO production by the constitutive NO synthase (cNOS) enzyme. NO production stimulates Prohormorne Convertases (PC), POMC, ACE and NEP gene expression at both the transcriptional and translational level which, in turn, inhibits inducible NO synthase (iNOS) activity by inhibiting adenylyl cyclase. This leads to POMC maturation, ACTH release and α-MSH production. By either autocrine or paracrine action, or binding to its receptor (MC2-R), the α-MSH-like peptide down-regulates gene expression in invertebrate immunocytes. Furthermore, in this context, POMC-derived AII binds to its AT1-like receptor and acts in a similar manner by inhibiting previously activated immunocytes by cytokine-like molecules.

is opposite to what was recently found in man, nevertheless it shows that RAS is an ancient evolutionarily conserved hormonal system.

Among the enzymes involved in neuropeptide precursor processing, the serine proteases belonging to the subtilisin/kexin family and known as the proprotein convertases (PCs) are playing, at least in mammals, a crucial role. In mammals, this family includes seven members to date (PC1, PC2, furin, PC4, PACE4, PC5 and PC7) which are differentially involved in the processing, generally at single or dibasic residues, of a large number of protein precursors, including neuropeptides, growth factors, receptors, enzymes and viral envelope glycoproteins. Among the enzymes, only PC1 and PC2, which are abundantly expressed in neurons and endocrine cells, play an important role in neuropeptide precursor processing. Although the latter enzymes have been cloned in numerous species, including mammals, amphibians, molluscs, insects and nematodes, they have not yet been characterized in annelids (Seidah *et al.*, 1998). Preliminary results obtained in our laboratory (unpublished results) indicate that leech PC1 and PC2 are synthesized in leech brain. Future experiments will be necessary to evaluate the exact role(s) of these different enzymes (i.e., ACE, NEP and PCs) in the generation of biologically active neuropeptides.

7. Are naturally occurring enzyme inhibitors important in these processes?

Leeches produce a variety of serine protease inhibitors (Table 3). These enzymes are involved in either preventing coagulation during blood feeding or as inhibitors of leukocyte enzymes. In *T. tessulatum*, we have isolated both types of inhibitors (Table 3). Furthermore, these enzymes are also involved in immune response modulation, by inhibiting inflammation through serine protease inhibition. Moreover, some of the inhibitors can also act on some PCs such like furin, which has also been discovered in almost all invertebrate phyla.

Clearly, again, the significance of these actions becomes quite clear. Opioid neuropeptides are important for "awakening" the immune and neuroendocrine systems (Stefano and Salzet, 1999). However, they are, for the most part, "trapped" in their respective precursors. Immunocyte activation initiates precursor processing (Stefano and Salzet, 1999). Besides the arguments made in the Stefano and Salzet (1999) review, new evidence has come from the vast numbers of serine protease inhibitors found in the leech. We surmise that, in taking its lengthy blood meal, the leech injects these enzyme inhibitors into the host to prevent immune neuropeptide processing. Thus, eliminating the "early" alert system simultaneously also prevents a general neuroendocrine stress response (Stefano and Salzet, 1999). This hypothesis also takes into account the ability of the strategy to down-regulate the host endothelial surface, since it too responds to these common signaling molecules (Stefano *et al.*, 1996a,b; Stefano, 1999).

Recently, man has used this strategy with aprotinin, a serine protease inhibitor. Using this compound diminishes the diffuse inflammatory response associated with surgery (Soeparwata *et al.*, 1996; Chopin *et al.*, 1997, 1998; Chopin, Stefano and Salzet, 1998), indirectly demonstrating the significance of processing enzymes. In patients ready to undergo major heart surgery, we found that just before surgery, plasma ACTH levels dropped below the level of detection (Fricchione *et al.*, 1996), indicating the activation or inhibition of the processing enzymes (Stefano and Smith, 1996). In this regard, it is widely known that various immune and neural-type signalling molecules can up-regulate enzymes such as NEP (Shipp *et al.*, 1990, 1991; Stefano *et al.*, 1996a,b). Taken together, it is imperative that more studies emerge to examine the critical significance of peptide processing in various processes.

Table 3. Isolated anticoagulants from leeches

Species	Molecule	Biological action
Hirudo medicinalis	Calin	Platelet aggregating inhibitor factor
	Bdellins	Trypsin and plasmin inhibitors
	Bdellastasin	Tysin and plasmin inhibitor
	Destabilase	Fibrin de-polymerase factor
	Eglin	Elastase, cathepsin G, chymotrypsin, subtilisin inhibitor
	LDTI	Tryptase, trypsin, chymotrypsin inhibitor
	Hirudin	Thrombin inhibitor
	Hirustasin	Trypsin, chymotrypsin, cathepsin G, kallicrein inhibitor
Hirudo nipponia	Bdellin KL	Trypsin and plasmin inhibitors
	Granulin	Thrombin inhibitor
	Guamerin I	Elastase inhibitor
	Piguamerin	Trypsin and kallicrein inhibitor
Hirudinaria manillensis	Bufrudin	Thrombin inhibitor
	Gelin	Elstase, cathepsin G, chymoptrypsin, subtilisin inhibitor
Haemadipsa sylvestris	Haemadin	Thrombin inhibitor
Whitmania endulata	Guamerin II	Elastase, chymotrypsin inhibitor
Macrobdella decora	Decorsin	GPIIb/IIIa inhibitor
Haementeria officinalis	Antistasin	Factor Xa inhibitor
	LAPP	Platelet aggregating inhibitor factor
Haementeria ghilianii	Ghilanten	Factor Xa inhbitor
	Hementin	Fibrinolytic enzyme
Haementeria lutzi	Hementerin	Fibrinolytic enzyme
Haementeria depressa	Lefaxin	Factor Xa inhbitor
Placobdella ornata	Ornatin	GPIIb/IIIa inhibitor
Theromyzon tessulatum	Cytin	Chymotrypsin inhibitor
	Tessulin	Trypsin, chymotrypsin inhibitor
	Therin	Trypsin inhibitor
	Theromin	Thrombin inhibitor
	Therostasin	Factor Xa inhibitor

8. Conclusion

In conclusion, given the wealth of information now emerging on these mammalian-like neuroendocrine processes found in invertebrates, it would appear that this system, in all probability, originated in "simple" animals. Therefore, aside from its historical origin, it may be more appropriate to speak of the mammalian neuroendocrine system as annelid-like.

Acknowledgments

This work was in part supported by the Centre National de Recherche Scientifique and the NIH Fogarty INT 00045 grant.

References

Acher, R., Chauvet, J., Chauvet, M.T., and Hurpet, D. (1985) Evolution of neurohypophysial hormones and their precursors. In: B. Lofts and W.N. Holmes (eds.) *Current Trends in Comparative Endocrinology*. Hong Kong University Press, Hong Kong, pp. 1147–1152.

Baratte, B., Gras-Masse, H., Ricart, G., Bulet, P., and Dhainaut-Courtois, N. (1991) Isolation and characterization of authentic Phe-Met-Arg-Phe-NH2 and the novel Phe-Thr-ArgPhe-NH2 peptide from *Nereis diversicolor*. *European Journal of Biochem*istry **198**, 627–633.

Chopin, V., Stefano, G.B., and Salzet, M. (1998) Amino-acid-sequence determination and biological activity of tessulin, a naturally occurring specific trypsin-chymotrypsin inhibitor from the leech *Theromyzon tessulatum*. *European Journal of Biochemistry* **258**, 662–668.

Chopin, V., Bilfinger, T.V., Stefano, GB, Matias, I., and Salzet, M. (1997) Amino-acid-sequence determination of cytin, a naturally occurring specific chymotrypsin inhibitor from the leech *Theromyzon tessulatum*, *European Journal of Biochem*istry **249**, 733–738.

Chopin, V., Matias I, Stefano, G.B., and Salzet, M. (1998) Amino-acid-sequence determination and biological activity of therin, a naturally occurring specific trypsin inhibitor from the leech *Theromyzon tessulatum*. *European Journal of Biochemistry* **254**, 565–570.

Civelli, O., Douglass, J., Goldstein, A., and Herbert, E. (1985) Sequence and expression of the rat prodynorphin gene. *Proceedings of the National Academy of Sciences, USA*, **82**, 4291–4295.

Diaz-Miranda, L., Escalona de Motta, G., and Garcia-Arraras, J.E. (1989) Pharmacological effects of serotonin and FMRFamide on the longitudinal muscle of the polychaete *Sabellastarte magnifica*. *American Zoo*logist **29**, 51A.

Duvaux-Miret, O., Dissous, C., Guatron, J.P., Pattou, E., Kordon, C., and Capron, A. (1990). The helminth *Schistosoma* mansoni expresses a peptide similar to human beta-endorphin and possesses a POMC-related gene. *New Biology* **2**, 93–99.

Evans, B.D., Pohl, J., Kartsonis, N.A., and Calabrese, R.L. (1991). Identification of RFamide neuropeptides in the medicinal leech. *Peptides* **12**, 897–908.

Fricchione, G., Bilfinger, T.V., Jandorf, L., Smith, E.M., and Stefano, G.B. (1996) Surgical anticipatory stress manifests itself in immunocyte desensitization: evidence for autoimmunoregulatory involvement. *International Journal of Cardiol*ogy **53**, S65-S73.

Genedani, S., Bernardi, M., Ottaviani, E., Franceschi, C., Leung, M.K., and Stefano, G.B. (1994) Differential modulation of invertebrate hemocytes motility to CRF, ACTH and its fragments. *Peptides* **15**, 203–206.

Hernandez-Presa, M.A., Bustos, C., Ortego, M., Tunon, J., Renedo, G., Ruiz-Ortega, M., and Egido, J. (1997) Angiotensin-converting enzyme inhibition prevents arterial nuclear factor kappa B activation, monocyte chemoattractant protein-I expression, and macrophage infiltration in a rabbit model of early accelerated atherosclerosis. *Circulation* **95**, 1532–1541.

Iwai, N., Inagami, T., Ohmichi, N., and Kinoshita, M. (1996) Renin is expressed in rat macrophage/monocyte cells. *Hypertension* **27**, 399–403.

Kastin, A.J., Hahn, K., Zadina, J.E., Banks, W.A., and Hackler, L. (1995) Melanocyte sstimulating hormone release-inhibiting factor-1 (MIF-1) can be formed from Tyr-MIF-1 in brain mitochondria but not in brain homogenate. *Journal of Neurochemistry* **64**, 1855–1859.

Kastin, A.J., Hahn, K., Banks, W.A., and Zadina, J.E. (1998) Regional differences in the metabolism of Tyr-MIF-1 and Tyr-W-MIF-1 by rat brain mitochondria. *Biochemical Pharmacol*ogy **55**, 33–36.

Kim, J.A., Berliner, J.A., and Nadler, J.L. (1996) Angiotensin II increases monocyte binding to endothelial cells. *Biochemical and Biophysical Research Communications* **226**, 862–868.

Kitazono, T., Padgett, RC, Armstrong, M.L., Tomkins, P.K., and Heistad, D.D. (1995) Evidence that angiotensin II is present in human monocytes. *Circulation* **91**, 1129–1134.

Kuhlman, J.R., Li, C., and Calabrese, R.L. (1985a) FMRFamide-like substances in the leech. Immunocytochemical localization. *Journal of Neuroscience* **5**, 2301–2309.

Kuhlman, J.R., Li, C., and Calabrese, R.L. (1985b) FMRFamide-like substances in the leech. II. Bioactivity on the heartbeat system. *Journal of Neuroscience* **5**, 2310–2317.

Laurent, V., and Salzet, M. (1995a) Isolation of a neuropeptide-degrading endopeptidase from the leech *Theromyzon tessulatum*. *European Journal of Biochemistry* **233**, 186–194.
Laurent, V., and Salzet, M. (1995b) Isolation of a renin-like enzyme from the leech *Theromyzon tessulatum*. *Peptides* **8**, 1351–1358.
Laurent, V., and Salzet, M. (1996a) Metabolism of angiotensins by head membranes of the leech *Theromyzon tessulatum*. *FEBS Letters* **384**, 123–127.
Laurent, V., and Salzet, M. (1996b) Metabolism of enkephalins in head membranes of the leech *Theromyzon tessulatum* by peptidases. *Regulatory Peptides* **65**, 123–131.
Laurent, V., Stefano, G.B., and Salzet, M. (1998) Leech angiotensin converting enzyme. *Annals of the New York Academy of Sciences* **839**, 500–502.
Lubics, A., Reglodi, D., Slezak, S., Szelier, M., and Lengvari, I. (1997) Co-localization of serotonin and FMRFamide-like immunoreactivities in the central nervous system of the earthworm, *Eisenia foetida*. *Acta Histochemica* **99**, 459–467.
Minakata, H., Fujita, T., Kawano, T., Nagahama, T., Oumi, T., Ukena, K., Matsushima, O., Muneoka, Y., and Nomoto, K. (1997) The leech excitatory peptide, a member of the GGNG peptide family: isolation and comparison with the earthworm GGNG peptides. *FEBS Letters* **410**, 437–442.
Oumi, T., Ukena, K., Matsushima, O., Ikeda, T., Fujita, T., Minakata, H., and Nomoto, K. (1994) Annetocin: an oxytocin-related peptide isolated from the earthworm, *Eisenia foetida*. *Biochemical and Biophysical Research Communications* **198**, 393–396.
Salzet, M., and Stefano, G.B. (1998) Evidence for an invertebrate neuroendocrine system: neuropeptide proccessing in leech-host communication. *Trends in Comparative Biochemistry and Physiology* **5**, 85–98.
Salzet, M., and Deloffre, L. (2000) PLGamide characterization and role in osmoregulation in leech brain. *Molecular Brain Research* **76**, 161–169.
Salzet, M., Vandenbulcke, F., and Verger-Bocquet, M. (1997) Structural characterization of osmoregulator peptides from the brain of leeches: IPEPYVWD and IPEPYVWD-amide. *Molecular Brain Research* **43**, 301–310.
Salzet, M., Vieau, D., and Stefano, G.B. (1999) Serpins: an evolutionarily conserved survival strategy. *Immunology Today* **20**, 541–544.
Salzet, M., Wattez, C., Baert, J.L., and Malecha, J. (1993a) Biochemical evidence of angiotensin II-like peptides and proteins in the brain of the Rhynchobdellid leech *Theromyzon tessulatum*. *Brain Research* **631**, 247–255.
Salzet, M., Bulet, P., Van Dosselaer, A., and Malecha, J. (1993b) Isolation, structural of characterization and biological function of a lysine-conopressin in the central nervous system the Pharyngobdellid leech *Erpobdella octoculata*. *European Journal of Biochemistry* **217**, 897–903.
Salzet, M., Bulet, P., Wattez, C., and Malecha, J. (1994) FMRFamide-related peptides in the sex segmental ganglia of the Pharyngobdellid leech *Erpobdella octoculata*. *European Journal of Biochemistry* **221**, 269–275.
Salzet, M., Bulet, P., Wattez, C., Verger-Bocquet, M., and Malecha, J. (1995a) Structural charcterization of a diuretic peptide from the central nervous system of the leech *Erpobdella octoculata*. Angiotensin II amide. *Journal of Biological Chemistry* **270**, 1575–1582.
Salzet, M., Bulet, P., Verger-Bocquet, M., and Malecha, J. (1995b). Isolation and structural characterization of enkephalin-related peptides in the brain of the Rhynchobdellid leech *Theromyzon tessulatum*. *FEBS Letters* **357**, 187–191.
Salzet, M., Bulet, P., Weber, W.-M., Clauss, W., Verger-Bocquet, M., and Malecha, J. (1996a) Structural characterization of a novel neuropeptide from the central nervous system of the leech *Erpobdella octoculata*: the Leech Osmoregulator Factor. *Journal of Biological Chemistry* **271**, 7237–7243.
Salzet, M., Verger-Bocquet, M., Bulet, P., Beauvillain, J.C., and Malecha, J. (1996b) Purification, sequence analysis and cellular localization of a prodynorphin-derived peptide related to the α neo-endorphin in the rhynchobdellid leech *Theromyzon tessulatum*. *Journal of Biological Chemistry* **271**, 13191–13196.
Salzet, M., Salzet, B., Cocquerelle, C., Verger-Bocquet, M., Pryor, S., Laurent, V., and Stefano, G.B.

(1997a) Biochemical and molecular characterization of ACTH, its precursor and receptor in the leech *Theromyzon tessulatum*: morphine increases ACTH levels. *Journal of Immunology* **159**, 5400–5411.

Salzet, M., Salzet, B., Sautiere, P., Lesage, L., Beauvillain, J.C., Bilfinger, T.V., Rialas, C., Bjenning,C., and Stefano, G.B. (1998) Isolation and characterization of a leech neuropeptide in rat brain. *Molecular Brain Research*. **55**, 173–179.

Satake, H., Takuwa, K., Minakata, H., and Matsushima, O. (1999) Evidence for conservation of the vasopressin/oxytocin superfamily in Annelida. *Journal of Biological Chemistry* **274**, 5605–5611.

Seidah, N.G., Day, R., Marcinkiewicz, M., and Chretien, M. (1998) Precursor of convertases: an evolutionary ancient, cell-specific, combatorial mechanism yielding diverse bioactive peptides and proteins. *Annals of the New York Academy of Sciences* **839**, 9–24.

Shipp, M.A., Stefano, G.B., D'Adamio, L., Switzer, S.N., Howard, F.D., Sinisterra, J.I., Scharrer, B., and Reinhertz, E.L. (1990) Downregulation of enkephalin-mediated inflammatory response by CD10/neutral endopeptidase 24.11. *Nature* **347**, 394–396.

Shipp, M.A., Stefano, G.B., Switzer, S.N., Griffin, J.D. and Reinherz, E. (1991) CD10 (CALLA)/neutral endopeptidase 24.11 modulates inflammatory peptide-induced changes in neutrophil morphology, migration, and adhesion proteins and is itself regulated by neutrophil activation. *Blood* **78**, 1834–1841.

Smith, E.M. (1994) Neuropeptides and immunoregulation. In: B. Scharrer, E.M. Smith and G.B. Stefano (eds.) *Neuropeptides in Neuroimmunology*. Spinger-Verlag, Berlin, pp. 28–40.

Smith, E.M., Hughes, T.K., Hashemi, F., and Stefano, G.B. (1992) Immunosuppressive effects of corticotropin and melanotropin significance in human deficiency virus infection. *Proceedings of the National Academy of Sciences, USA* **89**, 782–86.

Soeparwata, R., Hartman, A., Frerichmann, U., Stefano, G.B., Scheld, H.H., and Bilfinger, T.V. (1996) Aprotinin diminishes inflammatory processes. *International Journal of Cardiology* **26** (Suppl.), S55-S63.

Stefano, G.B. (1999) Substance abuse and HIV-gp120: are opiates protective? *Archivum Immunologiae et Therapie Experimentalis (Warszawa)* **47**, 99–106.

Stefano, G.B., and Scharrer, B. (1991) A possible immunoregulatory function for (Met)-enkephalin-Arg 6-Phe7 involving human and invertebrate granulocytes. *Journal of Neuroimmunology* **31**, 97–103.

Stefano, G.B., and Smith, E.M. (1996) Adrenocorticotropin: a central trigger in immune responsiveness: tonal inhibition of immune activation. *Medical Hypotheses* **46**, 471–478.

Stefano, G.B., and Salzet, M. (1999) Invertebrate opioid precursors: Evolutionary conservation and the significance of enzymatic processing. *International Review of Cytology* **187**, 261–286.

Stefano, G.B., Smith, D.M., Smith, E.M., and Hughes, T.K. (1991) In: K.S. Kits, H.H. Boer and J. Joosse (eds.) *Molluscan Neurobiology*. North Holland Publishing Company, Amsterdam, pp. 206.

Stefano, G.B., Scharrer, B., Bilfinger, T.V., Salzet, M., and Fricchione, G.L. (1996) A novel view of opiate tolerance. Advances in Neuroimmunology **6**, 265–277.

Stefano, G.B., Scharrer, B., Smith, E.M., Hughes, T.K., Magazine, H.I., Bilfinger,T.V., Hartman, A.R., Fricchione, G.L., Liu,Y., and Makman, M.H. (1996) Opioid and opiate immunoregulatory processes. *Critical Reviews in Immunology* **16**, 109–143.

Takei, Y., Hasegawa, Y., Watanabe, T.X., Nakajima, K., and Hazon, N. (1993) A novel angiotensin I isolated from elasmobranch fish. *Journal of Endocrinology* **139**, 281–285.

Turner, A., Leung, M.K., and Stefano, G.B. (1994) Peptidases of significance in neuroimmunoregulation. In: B. Scharrer, E.M. Smith and G.B. Stefano (eds.) *Neuropeptides in Neuroimmunology*. Springer-Verlag, Berlin, pp. 152–169.

Vandenbulcke, F., Laurent, V., Verger-Bocquet, M., Stefano, G.B., and Salzet, M. (1997) Biochemical identification and ganglionic localization of leech angiotensin-converting enzyme. *Molecular Brain Research* **49**, 229–237.

Walker, R.J. (1992) Neuroactive peptides with an RFamide or Famide carboxyl terminal.. *Comparative Biochemistry and Physiology* **102C**, 213–222.

Wenning, A., and Calabrese, R.L. (1995) An endogenous peptide modulates the activity of a sensory neurone in the leech *Hirudo medicinalis*. *Journal of Experimental Biology* **198**, 1405–1415.
Zerbst-Boroffka, I., Bazin, B., and Wenning, A. (1997) Chloride secretion drives urine formation in leech nephridia. *Journal of Experimental Biology* **200**, 2217–2227.

CHAPTER 10

ION CHANNELS OF MICROBES

CHRISTOPHER P. PALMER, ANN BATIZA, XIN-LIANG ZHOU, STEPHEN H. LOUKIN, YOSHIRO SAIMI and CHING KUNG

Laboratory of Molecular Biology, University of Wisconsin – Madison, Winconsin 53706, U.S.A.

Summary

Electrophysiology and the study of ion channels has generally been the province of neurobiologists. However, since the application of patch clamp techniques to microbial membranes, it is evident that microbes possess a variety of ion channel activities. Microbial genome sequencing projects have revealed that, in fact, nearly all microbes that have been surveyed possess ion channel homologues. For some of these microbial ion channel genes, the corresponding ion channel activity has been identified, notably the mechanosensitive ion channel from *Escherichia coli* (MscL), the K^+ ion channel homologue from *Streptomyces lividans* (KscA) and the dual-pore K^+ channel from *Saccharomyces cerevisiae* (Tok1). This has allowed a more detailed examination of the structure/function of these ion channels. More remarkably, due to the greater ease of purification and crystallization of microbial ion channels than mammalian ion channels, the crystal structures of both MscL and KscA have been determined. For KscA, this has given great insights into the structure of the potassium ion filter in all K^+ channels. An often-neglected question concerns the function of ion channels in microbes. For MscL, this question has begun to be answered with the discovery of MscS, which functions in parallel with MscL to protect cells from osmotic stress. However, the function of KscA and Tok1 and many other ion channel homologues has as yet to be discovered. It is fascinating that the evolution of human sentience is based upon a molecule that is present in even simple forms of life.

1. Introduction

Electrophysiology, the detection of electrical currents in organisms, is largely a discipline of vertebrate physiology. Consequently, much attention has been paid to voltage-gated channels that underlie the generation and propagation of action potentials and ligand-gated channels that mediate synaptic transmission (Albright *et al.*, 2000). In contrast, microbiology's history has largely not included electrophysiology. However, vertebrate physiology and microbiology have finally met at two technical fronts. First, the advent of patch clamp (Hamill *et al.*, 1981) removed the constraint of having to work electrically with only large cells. This made possible the direct examination of ion conductances in the membranes of microbes. As reviewed in Saimi *et al.* (1999), unitary conductances have now been recorded from *Dictyostelium* (slime mould), *Chlamydomonas* (a green flagellate), *Paramecium* (a ciliate), *Neurospora* (bread mould), *Uromyes* (a parasitic bean rust fungus), budding and fission yeasts,

I. Fairweather (ed.), Cell Signalling in Prokaryotes and Lower Metazoa, 325–345.
© 2004 *Kluwer Academic Publishers. Printed in Great Britain.*

Escherichia coli and *Streptomyces* (Gram-negative bacteria), *Bacillus* (a Gram-positive bacterium) and *Haloferax* (an archaeon). Secondly, the large and rapidly increasing genome sequence information from microbes has revealed genes similar to those for ion channels in animals. Now the complete genome sequences of several bacteria, archaea, and the budding yeast (*Saccharomyces cerevisiae*) and many other partially or nearly completed microbial genomic sequences are available for scrutiny.

Both the patch-clamp survey and the examination of genomes show that ion channels and putative ion channel homologues are widespread among microbes. These findings suggest that all cellular life forms, except perhaps certain parasites, have ion channels. Cells are surrounded by a membrane that encloses the cytoplasm; however, the cell is not a closed system. Nutrients, wastes and water have to enter and/or exit the cell. The different membrane macromolecules that broker their traffic, therefore, are ancient evolutionary inventions, presumably dating back to primordial cells. Among these must have been equivalents of the modern proteinaceous pumps and/or exchangers that rely on metabolic energy to maintain gradients across the membrane. The molecules we call channels are not these pumps. Rather, they are devices that allow ion flow using the gradients created by pumps.

We would like to stress that current knowledge of microbial channels is rather limited. The two survey methods for detecting channels in microorganisms are flawed. Surveying genomes using primary sequence information is obviously limited by what can be recognized as a potential channel. At the moment, only sequence homologues to well-studied animal channels can be identified. Distant relatives of animal channels, as well as channels unique to various microbial groups, will not be picked up. On the other hand, the patch clamp can only survey surfaces that can form gigaseals with the pipette. Furthermore, only voltage and stretch force can conveniently be varied across the patches, and the chemistry of the buffers bathing the two sides of patches is usually simplified to ease analysis. Therefore, activities of channels gated by ligands or unknown gating principles frequently go undetected.

2. Channels are Gated Pores

When open, a channel produces an aqueous pathway through which different kinds of particles flow passively. For charged particles, that is, ions, the concentration gradient is compounded by the voltage gradient to form the driving force for passage through the "pore". Because flow is passive, but nonetheless driven by the electrochemical gradient, the number of particles processed through the membrane tends to be large. An ordinary channel protein can transfer some millions of particles per second, in contrast to a common enzyme's ability to turn over only thousands of substrate molecules per second. Channels are so efficient as gradient-draining, passive pores that most cells do not need and cannot afford large numbers of each channel species. Therefore, collecting

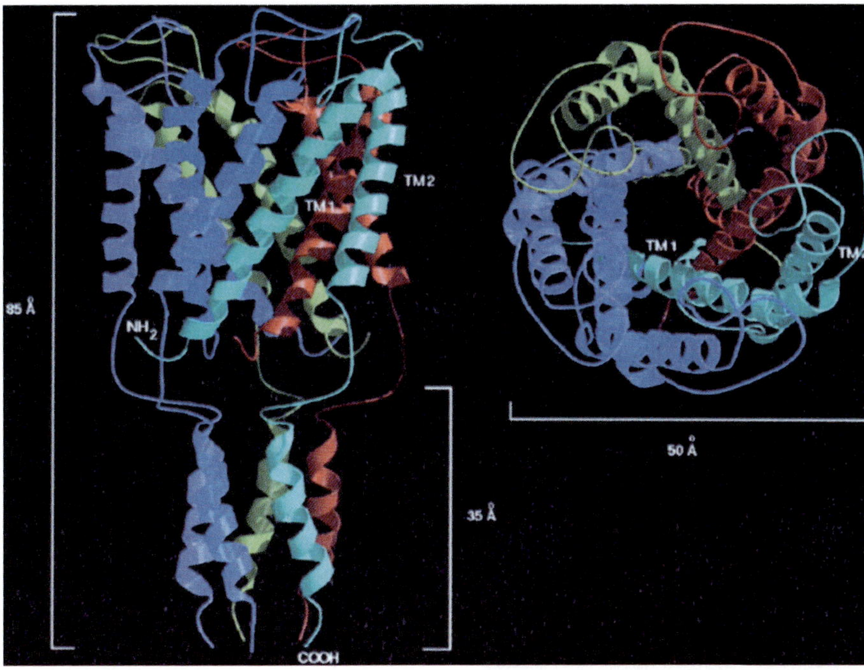

Figure 1. Structure of the homopentameric mechanosensitive channel from *Mycobacterium tuberculosis*. Side (left) and top (extracellular; right) views of Tb-MscL were rendered using BOBSCRIPT and RASTER-3D. The monomers within the channel are individually coloured. The NH_2- and COOH-terminal ends of the cyan monomer are indicated and the dimensions of the channel are shown.

Reprinted with permission from "Structure of the MscL homolog from *Mycobacterium tuberculosis*: a gated mechanosensitive ion channel." Chang, G., Spencer, R.H., Lee, A.T., Barclay, M.T., and Rees, D.C. Science **282**, 2220–2226, 1998. Copyright (2000) American Association for the Advancement of Science.

enough pure channel proteins from animal sources for biochemical or structural studies has always been a challenge. Current interest in prokaryotic ion channels derives largely from the fact that such channel proteins can be produced in milligram quantities in *E. coli*, making crystallographic studies possible (Chang *et al.*, 1998; Doyle *et al.*, 1998).

A particle-passing open pore, like a leak in a beaker or short in a battery, wastes energy. Therefore channels have "gates" and the gates are usually closed. The gate of each type of channel opens to certain specific stimulation; therefore, channels can be sorted into three general categories based on their gating principle. There are: ligand-gated channels that respond to the binding of external ligands (*e.g.* neurotransmitters), internal second messengers (Ca^{2+}, cyclic nucleotides), or neighboring proteins (*e.g.* G proteins); voltage-gated channels that respond to changes in the transmembrane electric potential (voltage); and mechanosensitive channels that respond to membrane tension

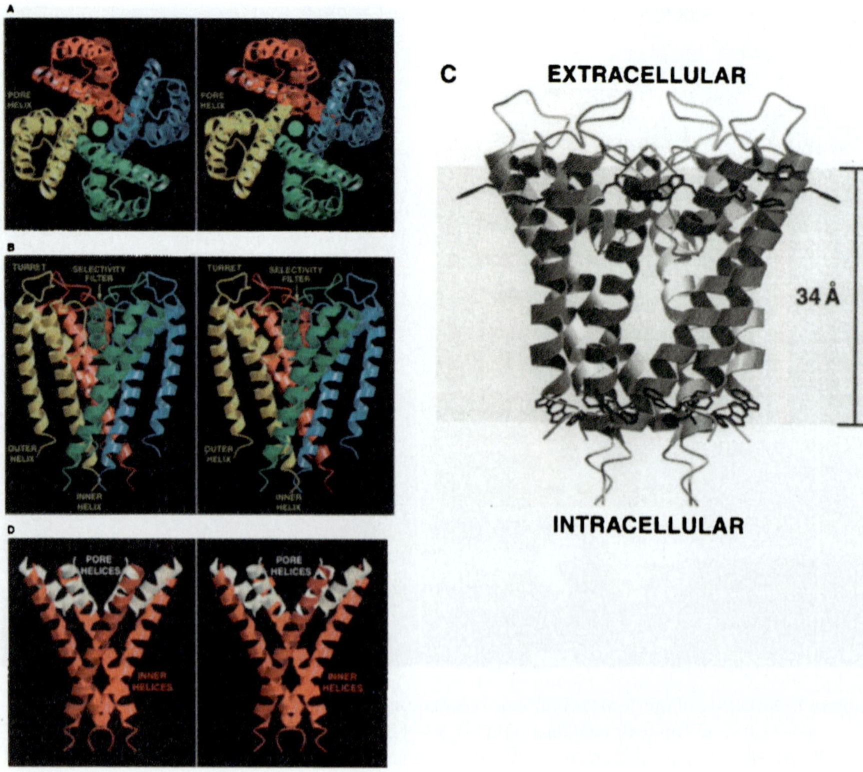

Figure 2. Structure of KcsA K^+ channel (A) Stereoview of a ribbon representation illustrating the three-dimensional fold of the KcsA tetramer viewed from the extracellular side. The four subunits are distinguished by colour. (B) Stereoview from another perspective, perpendicular to that in (A). (C) Ribbon representation of the tetramer as an integral-membrane protein. Aromatic amino acids on the membrane-facing surface are displayed in black. (D) Inverted teepee architecture of the tetramer. These diagrams were prepared with MOLSCRIPT and RASTER-3D.

Reprinted with permission from "The Structure of the Potassium Channel: Molecular Basis of K^+ Conduction and Selectivity." Doyle, D.A., Cabral, J.M., Pfuetzner, R.A., Kuo, A., Gulbis, J.M., Cohen, S.L., Chait, B.T., and MacKinnon, R. *Science* **280**, 69–77, 1998. Copyright (2000) American Association for the Advancement of Science.

(stretch). Thus, channels are "information molecules" that transduce the arrival of a ligand, a voltage change, or a stretch force, into the flow of ions across the membrane. They are the darlings of neuroscientists because they underlie neuronal activity.

All channel pores have selectivity, allowing the passage of certain particles and not others. Even the rather indiscriminate pore of MscL (below) does not permeate large proteins, nucleic acids and certainly not ribosomes. Most channels studied pass only water and certain small ions, while some have exquisitely precise "filters". Thus, there are K^+ channels, Na^+ channels and Ca^{2+} channels, for example, which favour the passage of one type of ion and virtually exclude the passage of some others.

The metaphors used by electrophysiologists to describe various parts of channels ("pores", "gates" and "filters") have recently been given flesh by the X-ray crystallographic structures of two channel molecules, MscL and KcsA, both from bacteria (Figs. 1 and 2). Both are homo-oligomeric integral membrane proteins. KcsA is a tetramer of subunits containing 160 amino acids each. Each subunit has two transmembrane α helices, an N-terminal M1 helix and a C-terminal M2 helix. MscL is a pentamer made up of 136 amino acid subunits, each also with two transmembrane domains. However, M2 is the inner helix forming the centre of the KcsA channel, while M1 is the inner helix in the MscL channel. These inner helices tilt respectively 25° for KcsA (Doyle *et al.*, 1998) and 28° for MscL (Chang *et* al., 1998). In both cases, the inner helices line the aqueous pathway and come together at the cytoplasmic face like a funnel. At the bottom of the funnel is a constriction, that is, the closed gate (see below). Gate opening refers to the transition to a new protein conformation in which the constriction is relaxed. KcsA, being a K^+-selective channel, possesses a second narrowing of the pathway on top of the gate towards the outer mouth of the channel. This is the K^+ filter, formed by four peptide loops each connecting the M1 and M2 of a subunit (see below). We will review current knowledge of microbial channels that are believed to have similar structures based upon their putative TM (transmembrane) α helical sequences. Specialized channels with β-barrel structures found in the outer membrane of Gram-negative bacteria (porins) and mitochondria (VDAC) will not be included here.

Ideally, an ion channel should be understood in terms of molecular structure and molecular activity. Activity is usually registered as the unitary conductance, the electric current driven by the electrochemical gradient through a single pore. The channel is the molecular structure that creates the pore. It is made up of subunit proteins encoded by one or more corresponding genes. At present, most of the unitary conductances seen by patch-clamping microbes have not been matched to their channel proteins or subunit genes. Conversely, most of the channels predicted from the genome sequences have no electrical description. To date, microbial channels that have both sequence information and biophysical characterization are few. They include: MscL, a mechanosensitive channel of large conductances in *E. coli* and other bacteria (Fig. 1) (Sukharev *et al.*, 1997); MscS, a mechanosensitive channel of smaller conductance in *E. coli* and other bacteria (Levina *et al.*, 1999); KcsA, a 2-TM tetrameric K^+ channel of *Streptomyces lividans* (Fig. 2) (Schrempf *et al.*, 1995; Doyle *et al.*, 1998); GluR0, a glutamate-gated K^+ channel from *Synecocystes* (Chen *et al.*, 1999); Tok1, a K^+ channel in the budding yeast (Gustin *et al.*, 1986; Ketchum *et al.*, 1995); and Mid1, a mating-related mechanosensitive cation channel of yeast (Kanzaki *et al.*, 1999).

3. K^+ Channels

K^+-selective channels are the most commonly identified channel in microbes, in part because their sequences are easily recognized. Higher eukaryotic K^+-channel homologues have now been detected in the genomes of bacteria, archaea, yeasts and protists. The first K^+-channel gene cloned and expressed was *Shaker* from *Drosophila* (Papazian *et al.*, 1987), which translates into a subunit protein with six transmembrane α helices. Each channel contains four such subunits. (Na^+- and Ca^{2+}-specific channels have a similar transmembrane motif, except that the four "subunits" are covalently linked). Analyses of the *Shaker* protein and its homologues made evident that the TXGYGD sequence between the 5th and 6th helices (M5 and M6) is the crux of K^+ filtration. Milkman (1994) recognized a likely K^+-channel gene in *E. coli*, *Kch*, in the neighbourhood of the *Trp* operon he studied. Similarly, the *KcsA* gene was first cloned and recognized on the basis of the filter sequence to be a K^+-channel homologue by Schrempf *et al.* (1995) in the bacterium, *Streptomyces lividans*. Doyle *et al.* (1998) transformed, expressed and purified the KcsA protein using *E. coli* and proceeded to solve its crystal structure at 3.2 Å resolution (Fig. 2). The most interesting feature of this structure is the 12 Å-long K^+-selective filter cradled within the pore. Here, the carbonyl oxygens of three amino acids (GYG of the canonical filter sequence) from each of the four subunits coordinate the passage of a K^+ ion. However, the smaller Na^+ ion cannot establish all of these contacts. This optimal fit, therefore, is the physical basis for ion selectivity. Using site-directed spin-labelling, Perozo, Cortes and Cuello (1999) showed that transition to the open state, that is, gating, correlates with rotations of both transmembrane helices in each subunit and the movement of the M2's to the periphery to relax the constriction at the cytoplasmic end of the channel.

In the bacterial and archaeal genomes completed to date, numerous open reading frames are known to contain the TXGYGD sequence exactly or with minor variations. In addition, more than twenty open reading frames locate this K^+ filter sequence, often called "P", between hydrophobic sequences long enough to traverse the membrane as α helices (Fig. 3) (Derst and Karschin, 1998). The simplest eukaryotic K^+ channel subunit comprises two transmembrane α helices with the arrangement of *M1-P-M2*. KcsA is an example of this arrangement, as are inward rectifier K^+ channels of animals. There is also the arrangement of *M1-M2-M3-M4-M5-P-M6*, first found in the *Drosophila* Shaker K^+ channel. In many cases, as revealed in the genome of the nematode, *Caenorhabditis elegans*, two "P" sequences are found in the same peptide in an *M1-P1-M2-M3-P2-M4* arrangement (Wang *et al.*, 1999). Finally, a K^+ channel of an *M1-M2-M3-M4-M5-P1-M6-M7-P2-M8* arrangement has been found in the yeasts *Saccharomyces cerevisiae* (Tok1, Ketchum *et al.*, 1995; Zhou *et al.*, 1995) and *Candida albicans* (Loukin, unpublished results). Extensive molecular dissection and cysteine accessibility studies have shown that M4 is an important part of the voltage sensor in voltage-gated channels (Jan and Jan, 1997). Here,

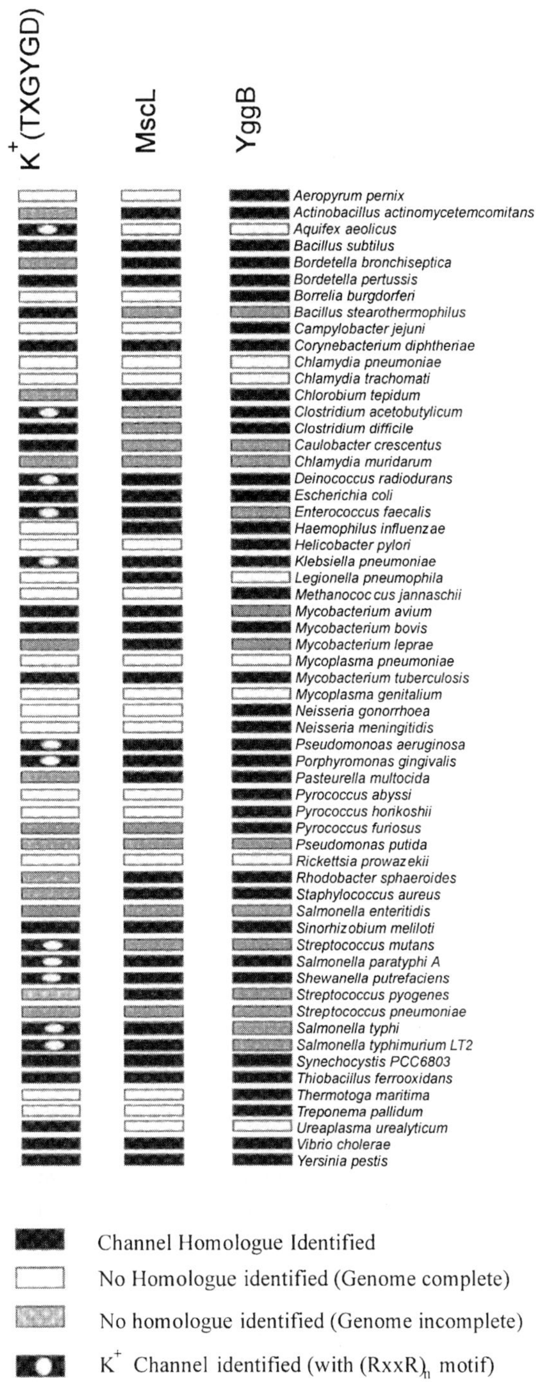

Figure 3. Ion channel homologue in prokaryotes. Homologues were identified by searching the TIGR microbiol genome database with known ion channel sequence motifs

basic amino acids are spaced three residues apart on the M4 helix, making one face of the helix positively charged. This pattern of (RXX)n, where n = 4–6, is found in the sequences of several putative prokaryotic K^+ channels. Whether or not these are voltage-gated channels, or channels at all, is yet to be shown.

As described above, several open reading frames that can be conceptually translated into proteins with K^+-channel subunit features are found in microbial genomes. However, possible activities of many of the corresponding channels have not been examined with patch clamp analysis, planar lipid bilayer single channel analysis, radiotracer fluxes through channels in liposomes, or other methods. On the other hand, K^+-specific currents through unknown channel proteins have been observed through intracellular recording from large cells of *Neurospora* (Levina *et al.*, 1995), *Dictyostelium* (Muller and Hartung, 1990) and from giant *Characious* algae (Laver and Walker, 1991). Currently, the only K^+-specific conductances for which the corresponding genes have been identified are: KcsA of the Gram-negative bacterium, *Streptomyces*; GluR0 of the photosynthetic bacterium, *Synechocystis* PCC 68803; Tok1 of the yeasts, *Saccharomyces* and *Candida*; and possibly PAK11 of the ciliate, *Paramecium tetraurelia*. All of these channels are discussed below.

KcsA heterologously expressed in *E. coli* requires low pH at the periplasmic side for its activation (Cuello *et al.*, 1998). The GluR0 sequence has regions homologous to the glutamate-binding domains of vertebrate glutamate receptor/channels (Schrempf *et al.*, 1995). Examined electrophysiologically upon expression in *Xenopus* oocytes, GluR0 indeed forms a channel gated by external glutamate. Unlike the less selective cation channel of the vertebrate glutamate receptor, the *Synechocystis* channel is K^+-specific. The sequence of GluR0 has the canonical TVGYGD, as well as other features of the K^+-specific KscA above. The Tok1 current of yeast was first reported from a patch-clamp survey of the yeast spheroplast surface (Gustin *et al.*, 1986) and later examined also by its heterologous expression in *Xenopus* oocytes (Ketchum *et al.*, 1995; Loukin *et al.*, 1997). Detailed analyses showed that this outwardly rectifying channel is, strictly speaking, not a voltage-gated channel. Its activation depends on the K^+ electromotive force across the membrane likely by means of both a traditional gating mechanism at the cytoplasmic end of M6 and M8 and subtle structural changes in the K^+ filter (Loukin *et al.*, 1997; Loukin and Saimi, 1999). *Paramecium*'s PAK11 has a membrane motif and filter sequence reminiscent of K^+ channels of the Shaker superfamily (Ling *et al.*, 1998). Though its corresponding K^+ current has not been isolated or expressed heterologously, expression of *PAK11* transgenes significantly reduces several distinct K^+ currents under a voltage clamp (Ling, Haynes and Saimi, unpublished results). While the mechanism of this phenomenon is not yet understood, it is possible that *PAK11* indeed encodes a K^+-channel subunit and that its transgenic expression somehow silences the expression of both *PAK11* and homologues in the same nucleus (Ruiz *et al.*, 1998).

4. "Chloride" Channels

Strictly speaking, Cl^- channels should be called anion channels because they allow permeation of both Cl^- and other small anions. The name Cl^- channel is used in the literature because Cl^- is the major small anion in animal cells and currents through this channel are carried mainly by Cl^- ions. However, there are very few Cl^- ions in bacteria such as *E. coli* (Kellenberger, 1996). In animal cells, Cl^- conductances stabilize the membrane potential and the resting level is near the Cl^- equilibrium potential (E_{Cl}). In addition, Cl^- channels have been implicated in other specialized functions in animals, such as volume regulation (Jentsch and Gunther, 1997). In giant fresh-water algae, the E_{Cl} is positive and Cl^--based action potentials are readily observed. But because Cl^- content in and around these algae differs greatly depending on species and niches, one needs to be cautious in extrapolating our knowledge of Cl^- channel functions in animals to those in algae.

CLC-0 from the electric organ of the *Torpedo* ray was the first voltage-dependent Cl^- channel cloned through its expression in oocytes (Jentsch, Steinmeyer and Schwarz, 1990). Many animal homologues, and especially mammalian homologues, have since been recognized. A *CLC-O*-like sequence *GEF1* is found in budding yeast (Huang, Chuat and Galibert, 1994). Its removal leads to an iron-remedial petite phenotype (Greene *et al.*, 1993). However, an anion conductance has not been reported in various electrophysiological investigations of the yeast plasma membrane and *GEF1*-knock-out mutant spheroplasts retain their constitutive K^+ and MS conductances with no other obvious deficit (Zhou and Loukin, unpublished results). Several attempts have been made to express this gene in heterologous systems also without success. Interestingly, *CLC-0*-like open-reading frames can be found in bacteria and archaea. For example, *E. coli* has a sequence that is about 20% identical to *CLC-0*. The putative *E. coli* channel has recently been expressed and purified and, upon reconstitution into liposomes, found to pass $^{36}Cl^-$ (Maduke, Pheasant and Miller, 2000). Kubalski (1995) has reported a 100-pS unitary anion conductance on the surface of giant *E. coli* syncytia. It is not known whether this conductance corresponds to the putative *CLC-0*-like sequence.

5. Mechanosensitive (MS) Channels

In the 1980's, following the detection of stretch-activated conductances in animal cells (Brehm, Kulberg and Moody-Corbett, 1984; Guharey and Sachs, 1984), researchers used the patch clamp to survey the surfaces of a few key microbes for possible stretch-activated ion-channel conductances. The MS channel activities of budding yeast (Gustin *et al.*, 1988) and *E. coli* (Martinac *et al.*, 1987) were first detected in this manner. Since then, MS channels have been found by patch clamp in a variety of microbes, including gram-positive (Zoratti and Petronilli.,1988) and photosynthetic bacteria (Moe, Blount and

Kung, 1998); the fission yeast, *Schizosaccharomyces pombe* (Zhou and Kung, 1992); the bean rust fungus, *Uromyces appendiculatus* (Zhou *et al.*, 1991); and in archea (Le Dain *et al.*, 1998; Martinac and Kloda, 2000).

A patch-clamp survey of the plasma membrane of budding yeast, *S. cerevisiae* revealed an MS unitary conductance of about 40 pS (Gustin *et al.*, 1988). This conductance is presumed to be constitutive, since it has consistently been observed in the spheroplasts of a variety of wild-type and mutant strains (Gustin *et al.*, 1988). The open probability of this channel increases dramatically with suction applied to the patch, responding to membrane stretch force and not to changes in pressure perpendicular to the membrane (Gustin *et al.*, 1988). However, the gene responsible for this activity has not been found. In order to find this gene, a patch-clamp survey of mutants having an ion- or osmosis-related phenotype and mutants missing a gene predicted to have a channel sequence has recently been conducted. Out of more than 50 mutants surveyed, none lacked this MS conductance (Zhou *et al.*, unpublished results). While the molecular identity of this conductance remains obscure, Kanzaki and colleagues recently reported that the *MID1* product accounts for a 32-pS, cation-specific MS conductance required for Ca^{2+} influx during mating (Kanzaki *et al.*, 1999). It will be interesting to see if this protein also accounts for the BAPTA-sensitive increase in cytosolic calcium upon hypotonic shock (Batiza, Schulz and Masson, 1996). However, spheroplasts from the *mid1* knockout strain nonetheless retain the constitutive 40-pS MS conductance (Zhou, Palmer and Kung, unpublished results).

While the surface of the yeast protoplast is large enough to form a seal at the mouth of the patch-clamp pipette, the surface of an individual bacterial protoplast is too small. Therefore, one must first make bacteria much larger by inhibiting septum formation during cell division (Ruthe and Adler, 1985). The *E. coli* MS channels detected in such spheroplasts (Martinac *et al.*, 1987) remain functional, even after disruption of the cells and reconstitution into artificial liposomes (Delcour *et al.*, 1989, based upon a similar protocol by Criado and Keller, 1987). More highly purified channels, derived from a detergent extract of the membranes can also be functionally reconstituted into artificial liposomes, so long as the detergent is removed. Detecting channel activity in this manner after various protein purification procedures led to identification of the protein generating the mechanosensitive channel of large conductance (MscL) and cloning of the *MscL* gene (Sukharev *et al.*, 1994). Tagged and purified recombinant MscL protein reconstituted into liposomes of defined lipid composition remains perfectly functional, clearly indicating that the stretch force is transmitted to the channel protein through the lipid bilayer (Hase, Le Dain and Martinac, 1995; Sukharev *et al.*, 1997). Genetic, biophysical and biochemical characterizations indicate that each subunit of this multimeric protein crosses the membrane twice as an alpha helix, with both ends in the cytoplasm (Sukharev *et al.*, 1994; Blount *et al.*, 1996; Arkin *et al.*, 1998). Sieving and conductance studies indicate that the opening of this channel must be huge (Cruikshank *et al.*, 1997; Ajouz *et al.*, 1998; Sukharev *et al.*, 1999).

Simultaneously imaging the patch curvature and recording the channel activities, Sukharev and colleagues (1999) derived an estimate of the energy difference between the open and the closed states and from that the in-plane expansion of 6.5 nm^2 upon opening.

In a herculean effort, Chang, Spencer and coworkers (1998) solved the crystal structure of the MscL homologue, *Mycobacterium tuberculosis* at a resolution of 3.5 Å. The structure of MscL (Fig. 1) reveals a homopentamer in a closed configuration. The pentamer protrudes above the lipid plane with the loops between the M1s and M2s capping the surface and then dipping into the external vestibule. All transmembrane helices are slanted with the five M1 helices coming together at their cytoplasmic ends. The closed channel thus approximates a funnel that narrows toward the cytoplasmic end. Polar residues near the bottom of the funnel precede a completely hydrophobic, bell-shaped cavity, closed off at both ends. Given the fact that the closed *M. tuberculosis* channel is configured so that the outer helices are largely sandwiched between the inner helices and lipid (Chang *et al.*, 1998), and given the enormous expansion which must accommodate the 3nS conductance described above for *E. coli* MscL, all ten helices must surround the lumen upon opening (Cruikshank *et al.*, 1997; Sukharev *et al.*, 1999). Relatively weak hydrophobic interactions help to stabilize the compact closed configuration, especially within the presumably closed-off hydrophobic pocket that forms the gate. However, stabilization of the closed state is not derived primarily from these interactions. Rather, this stabilization derives from the energetic barrier to moving contentedly buried hydrophobic surfaces throughout the channel into contact with the channel lumen (Batiza, Rayment and Kung, 1999; Moe, Levin and Blount, 2000). Mutations in the M1 helical face of *E. coli* MscL that by analogy forms the lining of the constricted cavity and the funnel just above, frequently result in easily gated channels and compromised cells (Ou *et al.*, 1998; Batiza, Rayment and Kung, 1999; Yoshimura *et al.*, 1999, 2001). Similarly, replacing glycine 22 with all other amino acids revealed that hydrophobic substitutions at residue 22 in the wall just below the upper constriction of this hydrophobic cavity are tolerated, while hydrophilic substitutions are not (Yoshimura *et al.*, 1999). Channels with either a hydrophilic residue 22 (Yoshimura *et al.*, 1999) or attachment of a hydrophilic methanethiosulfanate reagent at residue 22 (Yoshimura et al., 2001) prefer a substate upon opening. These results suggest that, in this substate, residue 22 is a hydrophilic environment such as the channel lumen while the environment of residue 22 in the closed state is hydrophobic (Yoshimura *et al.*, 1999, 2001).

Several bacterial genomes contain MscL sequence homologues. When several were subcloned and expressed in *E. coli* (in the *MscL* knockout background) each homologue expressed a conductance that is clearly mechanosensitive (Moe, Blount and Kung, 1998). The only residues which could substitute for wild-type Gly 22 in an osmotic shock loss-of-function assay (see section 6 below) in the *mscL*$^-$yggB$^-$ background were serine and alanine (Batiza and Kung, 2000; Batiza and Kung, unpublished data). Interestingly, alanine is the only other amino acid allowed by Mother Nature in 35 MscL sequence homologues in various prokaryotes (Moe, Blount and Kung, 1998; Spencer, Chang and Rees, 1999; Maurer *et al.*, 2000). While G22A MscL is the hardest to open of all the G22X MscL channels produced, G22S MscL has a gating threshold close to that of the smaller mechanosensitive channel MscS (Yoshimura *et al.*, 1999). Therefore, in the loss-of-function assay, G22S MscL may be complementing *YggB*, which contributes to MscS activity (Levina *et al.*, 1999). Notably, all three of these amino acids (glycine, alanine and serine) are likely occupants of helix-helix junctions (Javadpour *et al.*, 1999), which in fact is the position of the analogous residue in the Tb-MscL crystal structure (Chang *et al.*, 1998).

While the sequences of *E. coli* and *M. tuberculosis* imply structural similarity, their sequences are only 37% identical (Maurer *et al.*, 2000). In this regard, it is notable that the *M. tuberculosis* channel is much harder to open than the *E. coli* channel in both *E. coli* spheroplasts and azolectin liposomes. (Moe, Levin and Blount, 2000). In order to test the channel's structural similarity, comparable mutations were generated. Mutations in *E. coli* MscL that ease channel gating and kill *E. coli* (Ou *et al.*, 1998; Yoshimura *et al.*, 1999), presumably when the channels open inappropriately (Blount, Schroeder and Kung, 1997; Ou *et al.*, 1998; Yoshimura *et al.*, 1999), were tested by altering the analogous residues in the *M. tuberculosis* channel (Moe, Levin and Blount, 2000; Maurer *et al.*, 2000). Many of these mutant *M. tuberculosis* channels gated as expected with less stretch, but not with the ease of the *E. coli* mutant channels (Moe, Levin and Blount, 2000). Also, phenotypes were frequently milder in the *E. coli* cells carrying mutant *M. tuberculosis* MscLs (Moe, Levin and Blount, 2000). However, both results are expected, given the initially high gating threshold of the wild-type channel. Noted anomalies were the unexpected good health of cells containing charged substitutions at residues analogous to Gly 22 and Gly 30 in *E. coli*, while other analogous mutations did in fact kill cells (Maurer *et al.*, 2000). Therefore, the *E. coli* and *M. tuberculosis* MscL structures are likely to be quite similar, although they have important differences that affect the gating threshold. Quite possibly, the differences are within the constricted gate, but are definitely outside this area as well. In particular, site-directed mutagenesis highlights a hydrogen bond that provides for subunit interactions within the loop of the *M. tuberculosis* channel to stabilize the closed state (Maurer *et al.*, 2000). While loop cleavage stabilizes the open state of the *E. coli* channel (Ajouz *et al.*, 2000), the intersubunit hydrogen bond is not found in the *E. coli* channel (Maurer *et al.*, 2000).

E. coli also exhibits a mechanosensitive conductance of about 1 nS (Martinac *et al.*, 1987; Sukharev *et al.*, 1993; Berrier *et al.*, 1996). Though this is a smaller conductance than that of MscL and therefore is called MscS, this conductance is still very large compared to those of most eukaryotic channels (10–100 pS). Recently, Levina *et al.* (1999) showed that deletion of *YggB* and *KefA* removes MscS activity completely, with *YggB* playing the major role, and tag-purified YggB was recently reconstituted and shown to have MscS activity (Sukharev and Booth, 2000).

YggB and *KefA* are similar and both genes encode predicted membrane proteins with no homology to *MscL*. However, homologues can be found in a variety of bacteria and archaea. Interestingly, one homologue in the archeon, *Methanococcus jannashii* (MJ0170) is also weakly similar to *MscL* and has MS activity when expressed in *E. coli* on their reconstitution into liposomes (Martinac and Kloda, 2000).

6. Functions of Mechanosensitive Channels

Martinac *et al.* (1987) proposed an osmoregulatory role for MS channels in the very first report of their activity. The lack of an osmotic phenotype of the MscL knockout mutant was at first puzzling and a redundant function was invoked. That redundant function is apparently performed by YggB (which contributes to MscS), since genetically removing both MscL and YggB causes cells to lyse upon hypotonic shock (Levina *et al.*, 1999). Therefore, MS channels mediate the effects of a sudden osmotic downshock, such as being caught in a rainstorm. For decades we have known that *E. coli* releases its K^+, amino acids, sugars, ATP and even small proteins upon transfer into media of low osmolarity (Britten and McClure, 1962; Tsapis and Kepes, 1977; Berrier *et al.*, 1992). We now believe that MS channels are the conduits through which these osmolites are released (Blount, Schroeder and Kung, 1997; Ajouz *et al.*, 1998; Levina *et al.*, 1999; Berrier *et al.*, 2000). In fact, heterologously expressing *mscL* in a marine bacterium rescues it from osmotic lysis (Nakamura *et al.*, 1999). Note that after the near-complete release of osmolites, bacteria are presumably in a form of suspended animation and not dead, as they become metabolically active within minutes upon the return to nutrient media (Britten and McClure, 1962).

7. Possible Functions of K^+ and Anion Channels

The current focus of traditional channel biologists in working with microbial channels is not towards understanding the functions they serve in microbes. Therefore, scarcity of research in microbial channel functions, except for the MS channels above, confines the following discussion largely to the level of hypothesis. Attempts to elucidate the functions of ion channels have been disappointing, possibly because they provide for functional redundancy as demonstrated with MscS and MscL (Levina *et al.*, 1999). Alternatively, their activity may be required only under extremely stressful conditions so far not tested in the laboratory. Knocking out *MscL* or *Kch* leads to no clear phenotypes in *E. coli*; knocking out *TOK1* creates a phenotype evident only in a contrived condition in the laboratory (Saimi, unpublished results). All well-studied animal ion channels are rapid signal transducers, as most likely are the bacterial MS channels. If this rule can be extended to the rest of microbial channels, clues to their physiological functions should come from the signals they transduce and the ions they conduct.

Cation and anion channels may regulate osmolarity in microbes. All extant cells maintain a high [K^+] in their cytoplasm to neutralize the charges of organic acids and biopolymers and to maintain an outwardly directed turgor pressure. Opening a K^+ channel results in a K^+ efflux if the resting membrane potential is more positive than the equilibrium potential of K^+, E_K. One possibility is that such K^+ channels act as bleed valves to release transiently high osmotic pressure. Such a role would predict that these K^+ channels might be directly or indirectly activated by the osmotic force. MS K^+-specific channels are known in certain neurons (Kim, 1992), but not yet in microbes. However, using K^+ channels alone to lower the turgor is also unlikely. Because K^+ is charged, its efflux will polarize the membrane quickly towards the E_K. Given the small specific capacitance of the lipid bilayer, the amount of K^+ ejected would be relatively small before equilibrium was reached, even for a small cell with its large surface-to-volume ratio. For a bacterium, a polarization from 0 mV to −180 mV reduces the $[K^+]_{in}$ by less than 0.1 mM. Given that bacteria usually already rest with a high internal negativity and harbour over 100 mM of K^+, it seems unlikely that K^+ channels alone can act as osmotic adjusters. However, if a counter-ion flux is postulated, for example, a coupled K^+-Cl^- flux, so that the K^+ flux becomes non-electrogenic, the total flux can certainly be used to adjust osmolarity. Anion channels in bacteria may play such a role.

K^+ or Cl^- channels in animals are most often used to regulate the membrane potential, taking advantage of the principle set forth above, that is, a small ion flux is sufficient to produce a significant depolarization of the membrane. In prokaryotes, fungi and plants, the membrane potential is generated by pumping H^+ electrogenically. Unless there is a significant standing K^+ permeability, K^+ channels cannot be the major determinant of the resting potential. Though action potentials of these cells are not usually considered, there are many circumstances in which their membranes are discharged. They include common laboratory manoeuvres such as electroporation or the application of uncouplers or ionophores, as well as natural damages caused by agents, physical or biological (for example, colicins). In microbes, releasing most if not all of their osmolites through the MS channels does not kill them, nor does freezing or lyophilizing. Microbes surviving changing environments may have evolved mechanisms to recharge their membranes using a source of potential energy held in reserve. The K^+ gradient may well be such a reserve. Upon the loss of the inwardly directed H^+ gradient and even the loss of cytoplasmic metabolic nutrients, an efflux through K^+ channels driven by the pre-stored outwardly directed K^+ concentration gradient could reset the membrane potential. The restored cytoplasmic negativity could then be used to draw in H^+ and nutrients that symport with H^+. In this scenario, the cell would release its K^+ ions for nutrient molecules and gain energy in the exchange, after the latter were catabolized.

The above discussion suggests that K^+ channels evolved to deal with emergencies encountered in changeable environments. Another possibility is that K^+ channels have evolved as an additional route to take up K^+. Yeast made K^+-auxotrophic by the deletion of their native K^+ uptake mechanisms can be complemented by a variety of K^+ channels, including the inwardly-rectifying K^+ channels, KAT1 and AKT1 from the mustard, *Arabidopsis* (Anderson *et al.*, 1992, Sentenac *et al.*, 1992). Such a yeast K^+ auxotroph can also be complemented by the overexpression of yeast's native K^+ channel, *TOK1* (Fairman, Zhou and Kung, 1999). *Kch* did not appear in unbiased screens for the overexpression of genes that may rescue a severe *E. coli* K^+ auxotroph (Palmer and Saimi, unpublished results). Note that uptake of K^+ passively through a pore can only occur when the E_K is less negative than the membrane potential established by the H^+ pump.

8. The Use of Microbial Channels in Research

Microbial channels offer two types of experiments that cannot be carried out easily with animal channels: crystallography and forward genetics. Besides the usual difficulties in handling membrane proteins, collecting enough pure proteins of one channel species is a great challenge for crystallographers. Animal channels expressed in *E. coli* are often denatured and sequestered in inclusion bodies. The discovery of channels native to bacteria circumvented these problems, as attested by the crystal structures of MscL and KscA (Figs. 1 and 2). With regard to forward genetics, after randomly mutagenizing a channel gene, one can correlate the altered residues (molecular genotype) with altered channel biophysics (molecular phenotype). Such experiments require transformation *en masse*, replica plating, cell-phenotype selection and a large number of clonal cell lines (colonies). The numbers required are such that these experiments cannot easily be carried out in animals, even in model animals such as the fly or the roundworm. Such an experiment carried out on MscL revealed that hydrophilic substitutions of a residue within the hydrophobic wall of the closed gate promoted easy opening and conferred a severe growth phenotype (Ou *et al.*, 1998). Further analysis confirmed that all possible substitutions at this position correlated with hydrophilicity of the substitution with a lower gating threshold and cell death. In addition, substate opening was correlated with hydrophilic substitutions (Yoshimura *et al.*, 1999, 2001). A comparable experiment with Tok1 revealed that the "post-P" region (the cytoplasmic ends of the helices following each of the two "P" loops) likely forms the gate of K^+ channels (Loukin *et al.*, 1997), a conclusion in agreement with other evidence (Holmgren, Shin and Yellen, 1998).

9. Diversity, Evolution and Perspective

What seems curious is not only that ion channels, the princes of the nervous system, are present, but also that there seem to be so many different kinds of them in microbes. As reviewed above, we find among them K^+-specific, anion-specific and ion-nonspecific channels, as well as mechanically gated, ligand-gated and voltage-gated channels.

Furthermore, different microbial species seem to harbour different types of channels (Fig. 3). MscL has homologues in many, though not all, bacteria, but they are not convincingly evident in archaea and are not present in eukaryotes. The Tok1 K^+ channel with its *M1-M2-M3-M4-M5-P1-M6-M7-P2-M8* arrangement has so far been found only in *Saccharomyces* and *Candida*, but cannot be found in the largely completed genome of the fission yeast, *Schizosaccharomyces pombe.* The GluR0 of *Synechocystis* is the only known ligand-gated channel reported so far in prokayrotes. Bacteria whose completed genomes contain neither *MscL*, *YggB* nor K^+-channel-like sequences tend to be symbiotic or parasitic.

Given a cell's need to communicate with its environment, there is little doubt that channels are very ancient molecular devices. The simplest explanation for their being found in *Bacteria*, *Archaea* and *Eukarya*, is that they have appeared before the divergence of these domains. Though it is impossible to reconstruct "*the ur-channel*", conservation suggests that even early channels already had gates and filters. This scattered presence and the structural diversity among modern microbial species seem to suggest that certain features were retained in different microbes to deal with special challenges in their own special niches. Channels are presumably rapid signal transducers that monitor the environment. Presumably they are not used for the types of signal propagation, nor for the regulation of motilities based upon myosin, dynein or kinesins, that are strictly eukaryotic inventions. The widespread occurrence of K^+-channel homologues, but not animal Ca^{2+}- or Na^+-channel homologues, in bacterial and archaeal genomes further suggests that ion channels evolved long before functions such as secretion, contractile motility control and excitation, amongst others, with which we usually associate ion channels.

The alternative to early origin would be lateral transfers at more recent times in which channel genes from eukaryotes, say, became incorporated into prokaryotic genomes. This is unlikely to be a general mechanism. For example, putative channels with K^+ filters and S4-like voltage sensors are found in different branches of the *Eubacteria* and *Archaea* domains requiring multiple lateral transfers by this hypothesis. Lateral transfers likely occurred extensively among the early cells, but this is generally not considered a major mechanism in later evolution.

The greatest diversity of life forms on earth is among the microbes. Our survey here shows that most, if not all, microbes have some, but not all types of ion channels. The ability to express microbial channels heterologously in *E. coli* cells or in *Xenopus* oocytes, to patch clamp microbes directly or upon reconstitution, as well as the ease in genetic and molecular-biological manipulations guarantee progress in our understanding in the near future. The current momentum toward using microbes as models in the anthropocentric pursuit of animal-channel research will no doubt carry us further in understanding the molecular structure and activities of these "model channels" (Miller, 2000). The greatest challenge will be in understanding the physiological functions of ion channels in the microbes themselves. This will require a multidisciplinary effort making use of knowledge and techniques developed in electrophysiology and microbiology, as well as genetics and molecular biology.

Acknowledgements

Work in our laboratory has been supported by grants from NIH.

References

Ajouz, B., Berrier, C., Garrigues, A., Besnard, M., and Ghazi, E. (1998) Release of thioredoxin via the mechanosensitive channel MscL during osmotic downshock of *Escherichia coli* cells. *Journal of Biological Chemistry* **273**, 26670–26674.

Ajouz, B., Berrier, C., Besnard, M., and Ghazi, A. (2000) Contributions of the different extramembranous domains of the mechanosensitive ion channel MscL to its response to membrane tension. *Journal of Biological Chemistry* **275**, 1015–1022.

Albright, T. D., Jessell, T.M., Kandel, E.R., and Posner, M.I. (2000) Neural science: a century of progress and the mysteries that remain. *Neuron* **25**, S1-S55.

Anderson, J.A., Huprikar, S.S., Kochian, L.V., Lucas, W.J., and Gaber, R.F. (1992) Functional expression of a probable *Arabidopsis thaliana* potassium channel in *Saccharomyces cerevisiae. Proceedings of the National Academy of Sciences, USA* **89**, 3736–3740.

Arkin, I. T., Sukharev, S.I., Blount, P., Kung, C., and Brunger, A.T. (1998) Helicity, membrane incorporation, orientation and thermal stability of the large conductance mechanosensitive ion channel from *E. coli*. *Biochimica et Biophysica Acta* **1369**, 131–140.

Batiza, A.F., and Kung, C. (2000) The hydrophobicity of residue 22 is not sufficient for G22X MscL functioning upon hypotonic shock. *Biophysical* Journal **78**, 473A.

Batiza, A. F., I. Rayment, I., and Kung, C. (1999) Channel gate! Tension, leak and disclosure. *Structure with Folding and Design* **7**, R99–R103.

Batiza, A. F., Schulz, T., and Masson, P.H. (1996) Yeast respond to hypotonic shock with a calcium pulse. *Journal of Biological Chemistry* **271**, 23357–23362.

Berrier, C., M. Besnard, M., Ajouz, B., Coulombe, A., and Ghazi, A. (1996) Multiple mechanosensitive ion channels from *Escherichia coli*, activated at different thresholds of applied pressure. *Journal of Membrane Biology* **151**, 175–187.

Berrier, C., Coulombe, A., Szabo, I., Zoratti, M., and Ghazi, A. (1992) Gadolinium ion inhibits loss of metabolites induced by osmotic shock and large stretch-activated channels in bacteria. *European Journal of Biochemistry* **206**, 559–565.

Berrier, C., Garrigues, A., Richarme, G., and Ghazi, A. (2000) Elongation factor Tu and DnaK are transferred from the cytoplasm to the periplasm of *Escherichia coli* during osmotic downshock presumably via the mechanosensitive channel MscL. *Journal of Bacteriology* **182**, 248–251.

Blount, P., S. I. Sukharev, S.I., Moe, P.C., Schroeder, M.J., Guy, H.R., and Kung, C. (1996) Membrane topology and multimeric structure of a mechanosensitive channel protein of *Escherichia coli*. *EMBO Journal* **15**, 4798–4805.

Blount, P., Schroeder, M.J., and Kung, C. (1997) Mutations in a bacterial mechanosensitive channel change the cellular response to osmotic stress. *Journal of Biological Chemistry* **272**, 32150–32157.

Bonini, N. M., Gustin, M.C., and Nelson, D.L. (1986) Regulation of ciliary motility by membrane potential in *Paramecium*: a role for cyclic AMP. *Cell Motility and the Cytoskeleton* **6**, 256–272.

Brehm, P., Kullberg, K., and Moody-Corbett, F. (1984) Properties of nonjunctional acetylcholine

receptor channels on innervated muscle of *Xenopus laevis. Journal of Physiology* (Lond) **350**, 631–648.
Britten, R.J., and McClure, F.T. (1962) The amino acid pool in *Escherichia coli. Bacteriological Reviews* **26**, 292–335.
Chan, C. W.M., Saimi, Y., and Kung, C. (1999) A new multigene family encoding calcium-dependent calmodulin-binding membrane proteins of *Paramecium tetraurelia. Gene* **231**, 21–32.
Chang, G., Spencer, R.H., Lee, A.T., Barclay, M.T., and Rees, D.C. (1998) Structure of the MscL homolog from *Mycobacterium tuberculosis*: a gated mechanosensitive ion channel. *Science* **282**, 2220–2226.
Chen, G. Q., Cui, C., Mayer, M.L., and Gouaux, E. (1999) Functional characterization of a potassium-selective prokaryotic glutamate receptor. *Nature* **402**, 817–821.
Criado, M., and Keller, B.U. (1987) A membrane fusion strategy for single-channel recordings of membranes usually non-accessible to patch-clamp pipette electrodes. *FEBS Letters* **224**, 172–176.
Cruickshank, C.C., Minchin, R.F., LeDain, A.C., and Martinac, B. (1997) Estimation of the pore size of the large-conductance mechanosensitive ion channel of *Escherichia coli. Biophysical Journal* **73**, 1925–1931.
Cuello, L.G., Romero, J.G., Cortes, D.M., and Perozo, E. (1998) pH-Dependent gating in the *Streptomyces lividans* K^+ channel. *Biochemistry* **37**, 3229–3236.
Delcour, A. H., Martinac, B., Adler, J., and Kung, C. (1989) Voltage-sensitive ion channel of *Escherichia coli. Journal of Membrane Biology* **112**, 267–275.
Delcour, A. H., Martinac, B., Adler, J., and Kung, C. (1989) Modified reconstitution method used in patch-clamp studies of *Escherichia coli* ion channels. *Biophysical Journal* **56**, 631–636.
Derst, C., and Karschin, A. (1998) Evolutionary link between prokaryotic and eukaryotic K^+ channels. *Journal of Experimental Biology* **201**, 2791–2799.
Doyle, D. A., J. Morais Cabral, J.M., Pfuetzner, R.A., Kuo, A.L., Gulbis, J.M., Cohen, S.L., Chait, B.T., and MacKinnon, R. (1998) The structure of the potassium channel: molecular basis of K^+ conduction and selectivity. *Science* **280**, 69–77.
Fairman, C., Zhou, X., and Kung, C. (1999) Potassium uptake through the TOK1 K^+ channel in the budding yeast. *Journal of Membrane Biology* **168**, 149–157.
Greene, J. R., Brown, N. H., Didomenico, B.J., Kaplan, J., and Eide, D.J. (1993) The *GEF1* gene of *Saccharomyces cerevisiae* encodes an integral membrane protein; mutations in which have effects on respiration and iron-limited growth. *Molecular and General Genetics* **241**, 542–553.
Guharay, F., and Sachs, F. (1984) Stretch-activated single ion channel currents in tissue-cultured embryonic chick skeletal muscle. *Journal of Physiology* (Lond) **352**, 685–701.
Gustin, M. C., Martinac, B., Saimi, Y., Culbertson, M.R., and Kung, C. (1986) Ion channels in yeast. *Science* **233**, 1195–1197.
Gustin, M. C., Zhou, X.L., Martinac, B., and Kung, C. (1988) A mechanosensitive ion channel in the yeast plasma membrane. *Science* **242**, 762–765.
Hamill, O. P., Marty, A., Neher, E., Sakmann, B., and Sigworth, F.J. (1981) Improved patch-clamp techniques for high-resolution current recording from cells and cell-free membrane patches. *Pflugers Archiv* **391**, 85–100.
Hase, C. C., Le Dain, A.C., and Martinac, B. (1995) Purification and functional reconstitution of the recombinant large mechanosensitive ion channel (MscL) of *Escherichia coli. Journal of Biological Chemistry* **270**, 18329–18334.
Holmgren, M., Shin, K.S., and Yellen, G. (1998) The activation gate of a voltage-gated K channel can be trapped in the open state by an intersubunit metal bridge. *Neuron* **21**, 617–621.
Huang, M. E., Chuat, J.C., and Galibert, F. (1994) A voltage-gated chloride channel in the yeast *Saccharomyces cerevisiae. Journal of Molecular Biology* **242**, 595–598.
Jan, L.Y., and Jan, Y.N. (1997) Cloned potassium channels from eukaryotes and prokaryotes. *Annual Review of Neuroscience* **20**, 91–123.
Javadpour, M. M., Eilers, M., Groesbeek, M., and Smith, S.O. (1999) Helix packing in polytopic membrane proteins: role of glycine in transmembrane helix association. *Biophysical Journal* **77**, 1609–1618.

Jentsch, T. J., Steinmeyer, K., and Schwarz, G. (1990) Primary structure of *Torpedo marmorata* chloride channel isolated by expression cloning in *Xenopus oocytes*. *Nature* **348**, 510–514.
Jentsch, T.J., and Gunther, W. (1997) Chloride channels: an emerging molecular picture. *Bioessays* **19**, 117–26.
Kanzaki, M., Nagasawa, M., Kojima, I., Sato, C., Naruse, K., Sokabe, M., and Iida, H. (1999) Molecular identification of a eukaryotic, stretch-activated nonselective cation channel [published errata appear in *Science* **285**, p. 1493, 1999 and *Science* **288**, p. 1347, 2000]. *Science* **285**, 882–886.
Kellenberger, E. (1996) Structure and function at the subcellular level. In: F.C. Neidhart (ed.) *Escherichia and Salmonella, Cellular and Molecular Biology*. Vol. 1. ASM Press, Washington, D.C., pp. 17–28.
Kellenberger, S., Scheuer, T., and Catterall, W.A. (1996) Movement of the Na^+ channel inactivation gate during inactivation. *Journal of Biological Chemistry* **271**, 30971–30979.
Ketchum, K. A., Joiner, W.J., Sellers, A.J., Kaczmarek, L.K., and Goldstein, S.A.N. (1995) A new family of outwardly rectifying potassium channel proteins with two pore domains in tandem. *Nature* **376**, 690–695.
Kim, D. (1992) A mechanosensitive K^+ channel in heart cells. Activation by arachidonic acid. *Journal of General Physiology* **100**, 1021–1040.
Kubalski, A. (1995) Generation of giant protoplasts of *Escherichia coli* and an inner- membrane anion selective conductance. *Biochimica et Biophysica Acta* **1238**, 177–182.
Laver, D. R., and Walker, N.A. (1991) Activation by Ca^{2+} and block by divalent ions of the K^+ channel in the membrane of cytoplasmic drops from *Chara australis*. *Journal of Membrane Biology* **120**, 131–139.
Le Dain, A. C., Saint, N., Kloda, A., Ghazi, A., and Martinac, B. (1998) Mechanosensitive ion channels of the archaeon *Haloferax volcanii*. *Journal of Biological Chemistry* **273**, 12116–12119.
Levina, N. N., Lew, R.R., Hyde, G.J., and Heath, I.B. (1995) The roles of Ca^{2+} and plasma membrane ion channels in hyphal tip growth of *Neurospora crassa*. *Journal of Cell Science* **108**, 3405–3417.
Levina, N., Totemeyer, S., Stokes, N.R., Louis, P., Jones, M.A., and Booth, I.R. (1999) Protection of *Escherichia coli* cells against extreme turgor by activation of MscS and MscL mechanosensitive channels: identification of genes required for MscS activity. *EMBO Journal* **18**, 1730–1737.
Ling, K.-Y., Vaillant, B., Haynes, W.J., Saimi, Y., Kung, C. (1998) A comparison of internal eliminated sequences in the genes that encode two K^+ channel isoforms in *Paramecium tetraurelia*. *Journal of Eukaryotic Microbiology* **45**, 459–465.
Loukin, S. H., and Saimi, Y. (1999) $K^{(+)}$-dependent composite gating of the yeast $K^{(+)}$ channel, TOK1. *Biophysical Journal* **77**, 3060–3070.
Loukin, S. H., Vaillant, B., Zhou, X.L., Spalding, E.P., Kung, C., and Saimi, Y. (1997) Random mutagenesis reveals a region important for gating of the yeast K^+ channel YKC1. *EMBO Journal* **16**, 4817–4825.
Maduke, M., Pheasant, D.J., and Miller, C. (2000) Overexpression, functional activity, and homodimeric architecture of a prokaryotic CLC. *Biophysical Journal* **78**, 264A.
Martinac, B., and Kloda, A. (2000) Mechanosensitive (MS) ion channel of the archaeon *Methanococcus jannashii* with structural and functional homology to bacterial (MS) channels. *Biophysics Journal* **78**, 22A.
Martinac, B., Buechner, M., Delcour, A.H., Adler, J., and Kung, C. (1987) Pressure-sensitive ion channel in *Escherichia coli*. *Proceedings of the National Academy of Sciences, USA* **84**, 2297–2301.
Maurer, J.A., Elmore, D.E., Lester, D.E., and Dougherty, D.A. (2000) Comparing and contrasting *E. coli* and *M. tuberculosis* mechanosensitive channels (MscL). New gain of function mutations in the loop region. *Journal of Biological Chemistry* **29**, 22238–22244.
Milkman, R. (1994) An *Escherichia coli* homologue of eukaryotic potassium channel proteins [see comments]. *Proceedings of the National Academy of Sciences, USA* **91**, 3510–3514.
Miller, C. (2000) Ion channel surprises: prokaryotes do it again! *Neuron* **25**, 7–9.
Moe, P. C., Blount, P., and Kung, C. (1998) Functional and structural conservation in the mechanosensitive channel MscL implicates elements crucial for mechanosensation. *Molecular Microbiology* **28**, 583–592.

Moe, P. C., Levin, G., and Blount, P. (2000) Correlating a protein structure with function of a bacterial mechanosensitive channel. *Journal of Biological Chemistry* **275**, 31121–31127.

Muller, U., and Hartung, K. (1990) Properties of three different ion channels in the plasma membrane of the slime mold *Dictyostelium discoideum*. *Biochimica et Biophysica Acta* **1026**, 204–212.

Nakamaru, Y., Takahashi, Y., Unemoto, T., and Nakamura, T. (1999) Mechanosensitive channel functions to alleviate the cell lysis of marine bacterium, *Vibrio alginolyticus*, by osmotic downshock. *FEBS Letters* **444**, 170–172.

Ou, X., Blount, P., Hoffman, R.J., and Kung, C. (1998) One face of a transmembrane helix is crucial in mechanosensitive channel gating. *Proceedings of the National Academy of Sciences, USA* **95**, 11471–11475.

Papazian, D.M., Schwarz, T.L., Tempel, B.L., and Jan, L.Y. (1987) Cloning of genomic and complementary DNA from *Shaker*, a putative potassium channel gene from *Drosophila*. *Science* **237**, 749–753.

Perozo, E., Cortes, D.M., and Cuello, L.G. (1999) Structural rearrangements underlying K^+-channel activation gating. *Science* **285**, 73–78.

Ruiz, F., Vayssie, L., Klotz, C., Sperling, L., and Madeddu, L. (1998) Homology-dependent gene silencing in *Paramecium*. *Molecular Biology of the Cell* **9**, 931–943.

Ruthe, H. J., and Adler, J. (1985) Fusion of bacterial spheroplasts by electric fields. *Biochimica et Biophysica Acta* **819**, 105–113.

Saimi, Y., Loukin, S.H., Zhou, X.-L., Martinac, B., and Kung, C. (1999) Ion channels in microbes. *Methods in Enzymology* **294**, 507–524.

Schrempf, H., Schmidt, O., Kummerlen, R., Hinnah, S., Muller, D., Betzler, M., Steinkamp, T., and Wagner, R. (1995) A prokaryotic potassium ion channel with two predicted transmembrane segments from *Streptomyces lividans*. *EMBO Journal* **14**, 5170–5178.

Sentenac, H., Bonneaud, N., Minet, M., Lacroute, F., Salmon, J.M., Gaymard, F., and Grignon, C. (1992) Cloning and expression in yeast of a plant potassium ion transport system. *Science* **256**, 663–665.

Spencer, R. H., Chang, G., and Rees, D.C. (1999) 'Feeling the pressure': structural insights into a gated mechanosensitive channel [see comments]. *Current Opinion in Structural Biology* **9**, 448–454. [Published erratum appears in *Current Opinion in Structural Biology* **9**, 650–651, 1999.]

Sukharev, S., and Booth, I.R. (2000) Characterization of the small mechanosensitive channel of *E. coli* (MscS) reconstituted in liposomes. *Biophysical Journal* **78**, 22A.

Sukharev, S. I., Blount, P., Martinac, B., Blattner, F.R., and Kung, C. (1994) A large-conductance mechanosensitive channel in *E. coli* encoded by MscL alone. *Nature* **368**, 265–268.

Sukharev, S. I., Blount, P., Martinac, B., and Kung, C. (1997) Mechanosensitive channels of *Escherichia coli*: the MscL gene, protein, and activities. *Annual Review of Physiology* **59**, 633–657.

Sukharev, S. I., W. J. Sigurdson, W.J., Kung, C., and Sachs, F. (1999) Energetic and spatial parameters for gating of the bacterial large conductance mechanosensitive channel, MscL. *Journal of General Physiology* **113**, 525–539.

Tsapis, A., and Kepes, A. (1977) Transient breakdown of the permeability barrier of the membrane of *Escherichia coli* upon hypoosmotic shock. *Biochimica et Biophysica Acta* **469**, 1–12.

Wang, Z.W., Kunkel, M.T., Wei, A., Butler, A., and Salkoff, L. (1999) Genomic organization of nematode 4TM K^+ channels. *Annals of the New York Academy of Sciences* **868**, 286–303.

Yoshimura, K., Batiza, A., and Kung, C. (2001) Chemically charging the pore constriction opens the mechanosensitive channel MscL. *Biophysical Journal* **80**, 2198–2206.

Yoshimura, K., Batiza, A., Schroeder, M., Blount, P., and Kung, C. (1999) Hydrophilicity of a single residue within MscL correlates with increased channel mechanosensitivity. *Biophysical Journal* **77**, 1960–1972.

Zhou, X.-L., and Kung, C. (1992) A mechanosensitive ion channel in *Schizosaccharomyces pombe*. *EMBO Journal* **11**, 2869–2875.

Zhou, X.-L., Stumpf, M.A., Hoch, H.C., and Kung, C. (1991) A mechanosensitive channel in whole cells and in membrane patches of the fungus *Uromyces*. *Science* **253**, 1415–1417.

Zhou, X.-L., Vaillant, B., Loukin, S.L., Kung, C., and Saimi, Y. (1995) *YKC1* encodes the depolarization-activated K^+ channel in the plasma membrane of yeast. *FEBS Letters* **373**, 170–176.
Zoratti, M., and Petronilli, V. (1988) Ion-conducting channels in a gram-positive bacterium. *FEBS Letters* **240**, 105–109.

CHAPTER 11

BACTERIAL SIGNAL TRANSDUCTION: TWO-COMPONENT SIGNAL TRANSDUCTION AS A MODEL FOR THERAPEUTIC INTERVENTION

LENORE A. PELOSI, KWASI A. OHEMENG and JOHN F. BARRETT

Bristol-Myers Squibb Pharmaceutical Research Institute, Wallingford, Connecticut, USA

Summary

Bacterial signal transduction is a complex and poorly understood process that enables bacteria to grow, divide, withstand physiological insult and exert metabolic changes that allow them to adapt to and survive in their environment. These systems permit the organism to respond to biochemical or physical signals from its surroundings and within itself, in order to accomodate a dynamically changing environment. One of the major signal transduction systems in bacteria is the two-component signal transduction system, composed of hundreds of functionally diverse, but structurally-related, protein-pair motifs which facilitate the transfer of phosphate groups from ATP to key effector proteins. It is the level of phosphorylation of two-component signal-transducing proteins which ultimately controls gene expression in the bacterium. Acting upon demand, this unique histidyl-aspartyl phosphotransfer system enables the bacterium to quickly alter gene expression for processes under "stress", allowing the bacterium to survive in virtually any natural environment. As the story of two-component signal transduction unfolds, it is becoming increasingly clear that these systems play a major part in allowing organisms to survive in the host as opportunistic pathogens. Although not a single microbial signalling pathway has been successfully exploited for drug intervention to date, an awakening of the understanding of their potential will lead to opportunity and success. Although primarily a "non-essential" process in most bacteria, the two-component signal transduction system represents tremendous potential for pharmaceutical intervention, as the understanding of the *in vivo* "essentiality" of many of these two-component signal transduction systems for pathogenicity continues to develop.

1. Introduction to Signal Transduction in Bacteria

The existence of all microbial life depends on the coordinated expression of hundreds of genes simultaneously, within multiple sensor/signal transduction systems of varying receptor or catalytic motifs. The process of communication in bacteria is complex considering the size of the organism's genome; prokaryotes may have genomes as small as 0.5 Mb (*Mycoplasma*) to over 6 Mb (*Escherichia coli, Pseudomonas aeruginosa*), as compared to lower eukaryotes like *Aspergillus nidulans* with 31 Mb, *Candida elegans* with 100 Mb, *Drosophila* with 180 Mb and man, with 3,400 Mb of DNA (Database of Genome Sizes, @http://www.cbs.dtu.dk/databases/DOGS/index.html). Of the most thoroughly studied and highly visible signal transduction systems are the two-component signal transduction systems (TCSTS), which regulate multiple

I. Fairweather (ed.), Cell Signalling in Prokaryotes and Lower Metazoa, 347–402.
© 2004 *Kluwer Academic Publishers. Printed in Great Britain.*

physiological processes in prokaryotes and lower eukaryotes, including motility, osmolarity, cell cycle functions, virulence, and toxin/antibiotic resistance (see Table 1) (reviewed in Egger, Park and Inouye, 1997; Mizuno, 1998; Dutta, Quin and Inouye, 1999; Fabret, Feher and Hoch, 1999; Frosco and Barrett, 1999; Perraud, Weiss and Gross, 1999; Pirrung, 1999; Wallis, 1999a; Hoch, 2000a). The high degree of conservation of key structural and catalytic features among families of TCSTS proteins, and the lack of mammalian homologues contribute to the appeal of TCSTS as antimicrobial targets (Barrett and Hoch, 1998; Frosco and Barrett, 1999; Wallis, 1999a). Added to these advantageous features is the recent discovery of two-component regulatory systems that are "essential" *in vitro* (Fabret and Hoch, 1998; Quon *et al.*, 1998; Martin *et al.*, 1999).

In general terms, the two components of the prototypical TCSTS, typically referred to as the "sensor" histidine kinase and the "response regulator" components, may occur on a single molecule, or on two distinct molecules. Phosphorelays are created upon layering of multiple TCSTS modules (Alex and Simon, 1994; Mizuno, 1997; Wurgler-Murphy and Saito, 1997). The "sensor" histidine kinase (HK) component traverses the cell membrane, "senses" that a particular signalling event has occurred and relays that information to the cell by inducing autophosphorylation of a ubiquitously conserved histidine residue. Trans-phosphorylation of the "response regulator" (RR) component by the phosphorylated "sensor" component occurs at a conserved aspartate residue. Phosphorylation alters the secondary structure of the RR and affects the transcription-regulating capacity of the RR toward its specific target genes. Certain conserved structural motifs of the TCSTS superfamily of proteins, and their molecular organization, have been used to group TCSTS proteins into sub-families; these features characterize the molecular basis of the specificity of kinase/response regulator interaction (Fabret, Feher and Hoch, 1999).

While protein phosphorylation is a ubiquitously used mechanism of signal transduction and protein regulation in both prokaryotes and eukaryotes, the architecture of the two-component signal transduction system diverges significantly from that of other signal transduction pathways used in prokaryotes and from that comprising most eukaryotic signalling systems. It is hoped that these differences will permit selective intervention with novel, safe and effective anti-microbial agents.

1.1. AN OVERVIEW OF OPTIONS AND PROCESSES

There are numerous signal transduction systems in bacteria, among them the phosphotransferase system of sugar uptake (PTS), the ATP/GTP-dependent serine/threonine or tyrosine kinase system, and two-component signal transduction systems (Bourret, Borkovich and Simon, 1991; Brown and Firtel, 1998; Cozzone, 1998; Goudreau and Stock, 1998; Hellingwerf *et al.*, 1998; Stulke and Hillen, 1998; Bassler, 1999; Pirrung, 1999). Each system, with a commonality in catalytic process, but with varying levels of homology (Chang and

Table 1. Compendium of TCSTS in Bacteria

TCSTS	Organism where identified	Function (general)	Process
algR/agrD	*Pseudomonas aeruginosa*	capsule formation	alginate capsule production
cpxR/cpxA	*Escherichia coli*	capsule formation	multiple responses to environmental stress
hrpR/hrpS	*Pseudomonas*	capsule formation	
rcsB/rcsC	*Escherichia coli*	capsule formation	exopolysaccharide biosynthesis
rmpA/rmpB	*Escherichia coli*	capsule formation	colanic acid production
viaA/viaB	*Citrobacter freudii*	capsule formation	Vi antigen production
agrA/agrB	*Staphylococcus aureus*	exotoxin	polycistronic control of multiple exotoxins
agrA/agrC	*Staphylococcus aureus*	virulence	global virulence regulator
barA/sirA	*Escherichia coli*	virulence	pilus adherence-virulence
basS/basR	*Escherichia coli*	virulence	virulence
bvgS/bvgA	*Bordatella pertussis*	virulence	virulence
CaNikI	*Candida albicans*	virulence	phenotypic switching and virulence
CgSlnI	*Candida glabrata*	virulence	phenotypic switching and virulence
cheY/cheA-cheB	*Salmonella typhimurium*	virulence	motility-virulence
colR/colS	*Pseudomonas fluorescens*	virulence	colonization
devR/devS	*Mycobacterium tuberculosis*	virulence	heterocyst development
evgS/evgA	*Escherichia coli*	virulence	putative virulence
Exo	*Pseudomonas aeruginosa*	exotoxin	exoenzymes
expS/expA	*Erwinia carotovora*	virulence	production of extracellular, plant cell wall degrading enzymes
fimZ	*Escherichia coli*	virulence	putative fimbrial expression
finG/finN	*Salmonella typhimurium*	virulence	flagella-virulence
hrpB	*Pseudomonas solanacearum*	virulence	pathogenicity
ipaA-ipaB-ipaD	*Shigella flexneri*	virulence	mammalian entry
Mga	*Group A streptococci*	virulence	M-protein virulence
mxiD	*Shigella flexneri*	virulence	OMPs for secretion of Ipa invasion
envZ/ompR	*Escherichia coli*	virulence	osmolarity, virulence OMPs
Exo	*Pseudomonas aeruginosa*	virulence	exoenzymes
ntrA/ntrC	*Klebsiella pneumoniae*	virulence	urease production

Table 1. (Continued.)

TCSTS	Organism where identified	Function (general)	Process
o552/o226	*Escherichia coli*	virulence	putative ipa gene expression
o565/o226	*Escherichia coli*	virulence	putative fimbrial expression
phoP/phoQ,	*Salmonella typhimurium*	virulence	cationic antimicrobial peptide
pilA/pilB	*Neisseria gonorrhoeae*	virulence	pilus expression
pilG	*Pseudomonas aeruginosa*	virulence	twitching motility
pilS/pilR	*Pseudomonas aeruginosa*	virulence	fimbria and pilus formation
pmrA/pmrB	*Salmonella enterica*	virulence	polymyxin resistance
prfA	*Listeria monocytogenes*	virulence	transcription of virulence genes via RNAP
risA	*Bordetella bronchiseptica*	virulence	survival within host, resistance to oxidative stress
senX3/regX3,	*Mycobacterium bovis*	virulence	
Spv	*Salmonella typhimurium*	virulence	Salmonella plasmid virulence
ssrA-ssrB	*Salmonella typhimurium*	virulence/transport	regulates a type III secretion system needed for replication inside macrophages and systemic infection
toxS/toxR	*Vibrio cholera*	exotoxin	toxin, hemolysin production
uvrC/uvrY	*Escherichia coli*	virulence/secondary metabolism	biosynthesis of antifungal products
vacB	*Shigella flexneri*	virulence	virulence plasmid
virA	*Agrobacter tumefaciens*	virulence	host recognition, transformation
virR	*Clostridium perfringens*	virulence	regulation of collagenase, perfringolysis O, and hemagglutini
comD/come	*Streptococcus pneumoniae*	competence	competence
comP/coma	*Bacillus subtilis*	competence	competence
cheY/cheA-cheB	*Salmonella typhimurium*	environmental resistance	motility-virulence
copS/copR	*Pseudomonas sp.*	environmental resistance	Copper resistance
czcR/czcS	*Pseudomonas aeruginosa*	environmental resistance	cadmium resistance
degS/degU	*Bacillus subtilis*	environmental resistance	survival under salt stress, degradative enzyme and competence

Table 1. (Continued.)

TCSTS	Organism where identified	Function (general)	Process
f452/f239	*Escherichia coli*	environmental resistance	heavy metal homeostasis
f480/f227	*Escherichia coli*	environmental resistance	putative copper response
kdpD/kdpE	*Escherichia coli*	environmental resistance	potassium transport
pcoR/pcoS	*Pseudomonas aeruginosa*	environmental resistance	cobalt resistance
silR/silS	*Salmonella typhimurium*	environmental resistance	silver resistance
soxS/soxR	*Escherichia coli*	environmental resistance	oxidative stress
arcB/arcA	*Escherichia coli*	metabolism	respiratory control
atoS/st°C	*Escherichia coli*	metabolism	acetoacetate metabolism
citA/citB	*Klebsiella pneumoniae*	metabolism	induction of the citrate fermentation genes
creC/creB	*Escherichia coli*	metabolism	catabolite repression
divJ	*Caulobacter crescentus*	metabolism	cell division
frzE	*Myxococcus xanthus*	metabolism	fruiting body formation
hydH/hydG	*Escherichia coli*	metabolism	labile hydrogenase
kinA/ Spo0F	*Bacillus subtilis*	metabolism	sporulation
kinB/Spo0F	*Bacillus subtilis*	metabolism	sporulation
kinC	*Bacillus subtilis*	metabolism	orphan
phoR/phoB	*Salmonella typhimurium*	metabolism	phosphate regulation
phoR/phoP	*Bacillus subtilis*	metabolism	alkaline phosphatase and phosphodiesterase
resE/resD	*Bacillus subtilis*	metabolism	anaerobic respiration
rssB	*Escherichia coli*	metabolism	Sigma-S degradation
torS/torR	*Escherichia coli*	metabolism	trimethylamin metabolism
uhpB/uhpA	*Escherichia coli*	metabolism	hexose phosphate uptake
yjdH/yjdG	*Escherichia coli*	metabolism	putative citrate metabolism
fixL/fixJ	*Rhizobium leguminosarum*	nitrogen regulation	expression of nitrogenase
narX/narL	*Escherichia coli*	nitrogen regulation	nitrate regulation
narQ/narP	*Escherichia coli*	nitrogen regulation	nitrate regulation
ntrB/ntrC (NRII/NRI)	*Escherichia coli*	nitrogen regulation	nitrate regulation
absA1/absA2	*Streptomyces coelicolor*	secondary metabolism	streptomycin antibiotic synthesis
baeS/baeR	*Escherichia coli*	secondary metabolism	putative secondary metabolism

Table 1. (Continued.)

TCSTS	Organism where identified	Function (general)	Process
gacA/uvrC	*Pseudomonas fluorescens*	secondary metabolism	global regulator of 2′ metabolism
gacS/gacA	*Pseudomonas fluorescens*	secondary metabolism	hydrogen cyanide production, possibly involved in biocontrol by suppressing fungal growth
lemA/gacA	*Pseudomonas fluorescens*	secondary metabolism	pyrrolnitrin biosynthesis (anti-fungal secondary metabolite)
sapK/sapR	*Lactobacillus sake*	secondary metabolism	production and immunity to sakacin antibiotics
spaK/spar	*Bacillus subtilis*	secondary metabolism	regulates lantibiotic subtilin production
espA/espB	*Staphylococcus aureus*	unknown	regulation of timing of sporulation
f496/f444	*Escherichia coli*	unknown	
o449/o219	*Escherichia coli*	unknown	
rstB/rstA	*Escherichia coli*	unknown	
ybdK/ybdJ	*Bacillus subtilis*	unknown	non-essential
ycbA/ycbB	*Bacillus subtilis*	unknown	non-essential
ycbM/ycbL	*Bacillus subtilis*	unknown	non-essential
yccG/yccH	*Bacillus subtilis*	unknown	non-essential
yclK/yclJ	*Bacillus subtilis*	unknown	non-essential
ydfH/ydfI	*Bacillus subtilis*	unknown	non-essential
yesM/yesN	*Bacillus subtilis*	unknown	non-essential
yfiJ/yfiK	*Bacillus subtilis*	unknown	non-essential
yhcY/yhcZ	*Bacillus subtilis*	unknown	non-essential
ykoH/ykoG	*Bacillus subtilis*	unknown	non-essential
ykrQ	*Bacillus subtilis*	unknown	non-essential, orphan
ykvD	*Bacillus subtilis*	unknown	non-essential, orphan
yocF/yocG	*Bacillus subtilis*	unknown	non-essential
yojN	*Escherichia coli*	unknown	
yrkQ/yrkP	*Bacillus subtilis*	unknown	non-essential
ytsB/ytsA	*Bacillus subtilis*	unknown	non-essential
yufL/yufM	*Bacillus subtilis*	unknown	non-essential
yvcQ/yvcP	*Bacillus subtilis*	unknown	non-essential
yvfT/yvfU	*Bacillus subtilis*	unknown	non-essential
yvqB/yvqA	*Bacillus subtilis*	unknown	non-essential
yvqE/yvQC	*Bacillus subtilis*	unknown	non-essential
yvrG/yvrF	*Bacillus subtilis*	unknown	non-essential
ywpD	*Bacillus subtilis*	unknown	orphan, non-essential
yxdK/yxdJ	*Bacillus subtilis*	unknown	non-essential
yxjM/yxjL	*Bacillus subtilis*	unknown	non-essential
yycF/yycG	*Bacillus subtilis*	unknown	regulation of cell permeability

Table 1. (Continued.)

TCSTS	Organism where identified	Function (general)	Process
ciaH/ciaR	*Streptococcus pneumoniae*	resistance	cefotaxime resistance
cpxR/cpxA	*Escherichia coli*	resistance	low-level aminoglycoside resistance
phoP/phoQ	*Salmonella typhimurium*	resistance	cationic antimicrobial peptide
pmrA/pmrB	*Salmonella enterica*	resistance	polymyxin resistance
rprX/rprY	*Bacteriodes fragilis*	resistance	porin-mediated tetracycline resistance
vanS/vanR	*Enterococcus faecalis*	resistance	vancomycin resistance
vncS/vncR	*Streptococcus pneumoniae*	resistance	antibiotic tolerance via ABC transport of a death signal peptide P27
smeS/smeR	*Stenotrophomonas maltophilia*	resistance	multidrug efflux
lytS/lytT	*Bacillus subtilis*	lysis/ transport	rate of autolysis
lytS/lytR	*Staphylococcus aureus*	lysis/ transport	controls autolysis via murein hydrolase production and export
pnpR/pnpS	*Streptococcus pneumoniae*	lysis/ transport	protein export
yehU/yehT	*Escherichia coli*	lysis/ transport	putative autolysin response
ydbF/ydbG (dctB/dctD)	*Bacillus subtilis*	transport	regulates genes for C4-dicarboxylate transport
citS/citT	*Bacillus subtilis*	transport	Mg^{+2}/citrate transport
feuP/feuQ	*Rhizobium leguminosarum*	transport	iron uptake
feuP/feuQ	*Brucella suis*	transport	iron uptake

Meyerowitz, 1994), enables the bacterium to respond appropriately to a variety of favourable and unfavourable conditions, allowing quick and efficient adaptation to an ever-changing environment. These related, class-specific regulatory systems co-exist and act inter-dependently to modulate the activities of key proteins.

In *E. coli*, for example, chemotaxis is not only regulated by a specific TCSTS (CheA/CheY) (Parkinson and Kofoid, 1992), but also by a phophoenolpyruvate/carbohydrate phosphotransferase (PTS) system (Lux *et al.*, 1999). The state of phosphorylation of a PTS Regulation Domain (PRD) is responsible for its ability to interact with many anti-terminators and transcriptional activators, creating a crucial link between metabolism and regulation of catabolic genes (Stulke and Hillen, 1998). Despite similar chemical mechanisms that utilize histidine as carrier for phosphate groups, comparison of the PTS and TCSTS active sites reveals a great disparity in their tertiary structures (McEvoy and Dahlquist, 1997).

Many prokaryotes also harbour a third type of phosphorylating system, the nucleoside triphosphate-dependent serine/threonine or tyrosine kinase system, which closely resembles the classic eukaryotic phosphorylating machinery (Cozzone, 1998). Genomic analysis of the cyanobacterium, *Synechocystis* sp. strain PCC provided evidence of a potential relationship among signal transduction systems, as many of them turn out to be clustered in the genome and some appear in tandem with two-component system genes as well (Zhang, Gonzalez and Phalip, 1998).

The interplay of these three types of regulatory systems (Cozzone, 1998; Goudreau and Stock, 1998), together with the evolutionary shuffling of signalling domains which has given rise to new pairs of reaction partners (Reizer and Saier, 1997), and the development of "unorthodox" multistep His-Asp phosphorelays occurring either within or between distinct proteins (Appleby, Parkinson and Borret, 1996?), create for the organism tremendous potential for regulatory complexity and versatility. The possibility has also been debated that cross-talk may exist in some rare instances *in situ* between non-cognate proteins of TCSTS pairs, especially in those organisms in which very high numbers of signalling proteins are found, e.g. *E. coli* and *B. subtilis*, each with over 30 TCSTS (Wanner, 1992; Mizuno, 1997; Ninfa *et al.*, 1993).

Signal transduction systems exist to manage the extraordinarily complex network of signal/response scenarios that bacteria have adapted to in order to exist in a hostile environment, including a few "essential" processes. TCSTS have an unusual phosphorelay, which may signal some special significance in the bacteria, although the most reasonable explanation for their evolution is that these systems have arisen from a primordial signalling system designed to allow the microorganism to exist and adapt to a range of external conditions in dynamic flux, which constantly challenge their survival and/or propagation. The ability to quickly and specifically detect and respond to a changing environment provides obvious ecological advantages to the organism.

2. Structure/Biochemistry of TCSTS

Although the reactions that signal transduction proteins catalyze and their regulatory consequences may be similar, the structural and mechanistic differences between prokaryotic HKs and serine/threonine and tyrosine kinases are considerable, and the potential for toxicity related to non-specific inhibition of host kinases is low.

The most basic model (Fig. 1) consists of a pair of proteins possessing signalling domains that are conserved among members of their respective families, but which lack homologous mammalian counterparts (Stock *et al.*, 1995; Barrett and Hoch, 1998). Signal-induced, Mg^{+2}-dependent ATP-binding, and autophosphorylation at a conserved histidine residue occurs on the

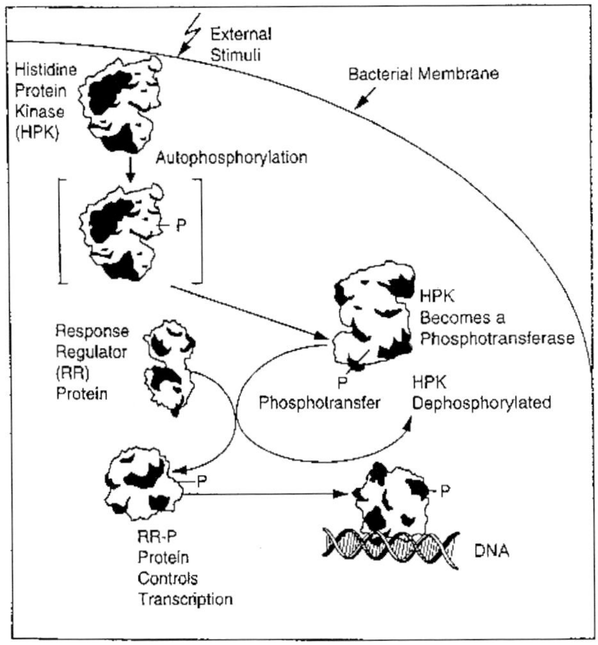

Figure 1. A typical two-component signal transduction pathway. Reprinted, with permission, from Antimicrobial Agents and Chemotherapy, Volume 42.

"sensor" histidine kinase (HK) component, which is generally an integral membrane protein (Stock *et al.*, 1995; Levit, Liu and Stock, 1998; Dutta, Quin and Inouye, 1999). Autophosphorylation elicits changes in secondary structure which render the kinase a suitable substrate for phosphotransfer to an aspartyl residue of the cognate "receiver" or response regulator (RR) component (Varughese *et al.*, 1998; Stock, 1999; Williams and Stewart, 1999). Phosphorylation activates the transcription-regulating function of the DNA-binding effector domain of the RR. In the absence of signal, a phosphatase activity residing in the HK component dephosphorylates and inactivates the response regulator, keeping target gene transcription turned off (Hoch, 2000a). The mechanism of dephosphorylation appears to be the displacement of the phosphorylated histidine side chain with water and requires close proximity of the kinase histidine and response regulator aspartate residue (Hsing and Silhavy, 1997). Response regulators also possess an autophosphatase activity which further ensures that the proteins are not permanently activated (Blat *et al.*, 1998; Zapf *et al.*, 1998). Other phosphatases may also contribute to this regulatory network (reviewed in Parkinson and Kofoid, 1992). However, rarely in nature does one find a TCSTS as simple as the prototypical two-component system model described above.

The current state of understanding of TCSTS protein structure, cellular organization, and signal transduction mechanism provides the foundation upon which strategies designed to interfere with their function may be built.

2.1. THE HISTIDINE KINASE COMPONENT

Two-component signalling follows the general model of signal transducing proteins wherein the signal "sensor" or "receptor" consists of a periplasmic sensing domain and a cytoplasmic signalling domain. Although structural information about the histidine kinase component of TCSTS is rapidly evolving, its N-terminal membrane-spanning character renders the intact sensor molecule less accessible to structural analysis than the RR component. Even within families, periplasmic signalling domains are generally unrelated in length, sequence and number of membrane-spanning domains, in contrast to the C-terminal cytoplasmic domain, which contains several conserved motifs (Fig. 2). Given the diversity in types of stimuli facing the organism, the variety in types of sensing domains is not surprising. Generally, however, the sensor

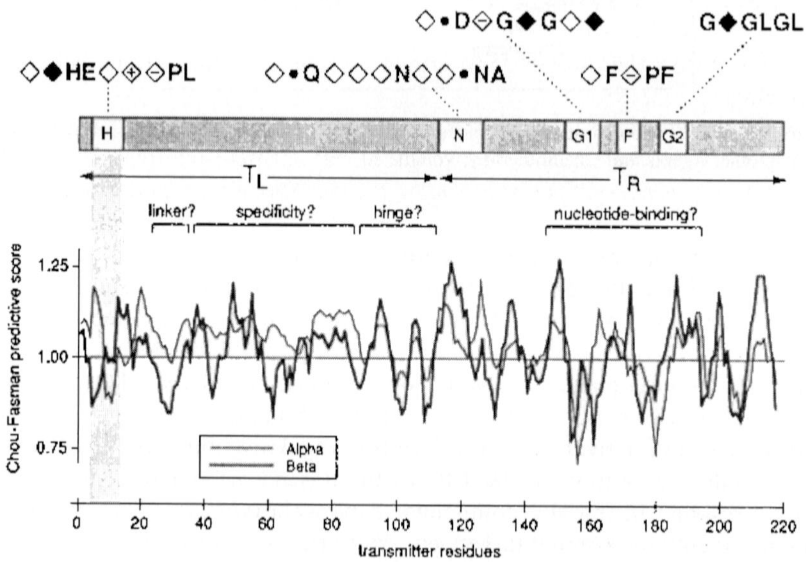

Figure 2. Structural features of orthodox histidine kinase signal transmitters. Five sequence motifs characteristic of transmitters are shown at the top. The H box comprises the Dimerization Histidine transphosphorylation (DHp) domain, or in Class II kinases, the Histidine-containing phosphotransfer (HPt) domain. The N, G1, F, and G2 boxes comprise the catalytic (CA) nucleotide-binding domain. Letters indicate amino acids conserved in >70% of aligned transmitter sequences at that position. Diamond symbols represent positions at which at least 50% of the amino acids belong to the same chemical family: white, nonpolar (I, L, K, V); black, polar (A, G, P, S, T); plus sign, basic (H, K, R); minus sign, acidic or amidic (D, E, N, Q). Black dots indicate wild card positions with less than 50% family conservation. An aggregate secondary structure prediction and possible functions of key regions are also shown. Reprinted, with permission, from the Annual Review of Medicine, Volume 26 © 1992 by Annual Reviews @ www.AnnualReviews.org.

is anchored in the membrane by at least two helical segments (Stock, Ninfa and Stock, 1989). In some cases the signal is well characterized, and direct periplasmic binding sites have been identified. For example, Ca^{++} and Mg^{+2} bind directly to the periplasmic domain of PhoQ, which regulates expression of virulence genes in *S. typhimurium* (Waldburger and Sauer, 1996; Goudreau and Stock, 1998).

Sensing may also occur indirectly, through the transmembrane domain. The Cpx system of *E. coli*, for example, is sensitive to membrane phosphatidylethanolamine content (Mileykovskay and Dowhan, 1997). The periplasmic region of the *E. coli* EnvZ kinase, which is involved in osmoregulation, can be partially deleted or replaced (Leonardo and Forst, 1996), suggesting that perhaps membrane perturbations elicited by changes in osmolarity are the signal for this system. It has been suggested that the signal for Van R/S, the vancomycin resistance TCSTS of *E. faecalis*, may also be related to membrane composition (Lai and Kirsch, 1996; Ulijasz, Grenader and Weisblum, 1996). The observation that membrane disruptive agents trigger many TCSTS (Lai and Kirsch, 1996; Ulijasz, Grenader and Weisblum, 1996; Ruzin and Novick, 2000), and the role of the essential TCSTS of *S. aureus* in cell permeability (Martin *et al.*, 1999), support a model in which the integrity of the cell membrane is of paramount importance for the proper positioning of the transmembrane helices for signalling.

TCSTS which sense intracellular signals include the redox and energy-sensitive systems, whose sensors often contain flavoprotein-binding sites or PAS motifs (Gong *et al.*, 1998; Goudreau and Stock, 1998; Taylor and Zhulin, 1999). Linking the "signalling" domain and the "transmitter" domain is a 50-amino acid alpha helical region commonly found in multi-domain signalling proteins (Aravind and Ponting, 1999). Those sensors which do not have such a HAMP (domain present in histidine kinases, adenylyl cyclases, methyl-accepting proteins and *p*hosphatases) (Aravind and Ponting, 1999), or Type P periplasmic signal transducing linker region, may possess a short, highly charged stretch linking the input and output domains (Williams and Stewart, 1999). It is thought that such sensors are capable of responding to non-periplasmic signals. It is postulated to function in transduction of signal in homodimeric receptors by regulating phosphorylation and subsequent transmembrane helix conformational changes (Williams and Stewart, 1999).

Although a novel four helix bundle motif characterizes histidine kinases and distinguishes them from serine/threonine or tyrosine kinases, differences in subdomain organization differentiate the various classes of HKs, and reflect differences in localization, function, and regulatory mechanism (Dutta, Quin and Inouye, 1999). HKs of the most abundant Class I type feature a cytoplasmic transmitter domain that consists of a catalytic ATP-binding domain (CA) and a dimerization histidine transphosphorylation domain (DHp) (Tanaka *et al.*, 1998; Dutta, Quin and Inouye, 1999). The NMR (Tanaka *et al.*, 1998) and solution (Tomamuri *et al.*, 1999) structures of the transphosphorylating catalytic domain of the *E. coli* osmosensor EnvZ demonstrate a kinase fold distinct from that of serine/threonine and tyrosine kinase, with similarities to heat shock protein 90 and DNA gyrase B. The kinase domain is an independently folding unit, which binds Mg^{+2} and ATP and catalyzes phosphorylation of a histidine residue on a second molecule of HK via formation of a homodimer. The four helix bundle motif, which is made up of two alpha helices from each of the two subunits (Varughese *et al.*, 1998; Tomomuri *et al.*, 1999), allows the ATP-binding catalytic domain to interact with the dimerization domain and phosphorylate the specific histidine residue on the opposing subunit (Yang and Inouye, 1991).

The Class II HKs lack a DHp domain, but instead contain a histidine-containing phosphotransfer (HPt) sub-domain (Dutta, Quin and Inouye, 1999). There is significant conservation of the HPt sub-domain between prokaryotic and eukaryotic kingdoms, as evidenced by the crystal structure of Ypd1p from *S. cerevesiae* (Xu, Nguyen and West, 1999). In the case of this hybrid phosphorelay protein, this four helix bundle forms within a single polypeptide and functions in the actual phosphotransfer reaction, not just dimerization (Song *et al.*, 1999). In the *E. coli* Che A kinase, the HPt feature functions in dimerization but not in phosphotransfer (Bilwes *et al.*, 1999). The sub-domains are defined by characteristic signature sequences. Within the transphosphorylating catalytic (CA) domain are several highly conserved motifs referred to as N, G1, F, and G2 boxes (Fig. 2), which together provide a nucleotide binding surface within the active site (Parkinson and Kofoid, 1992). The glycine-rich G2 box is common to many other kinases and nucleotide binding proteins (Stock, Ninfa and Stock, 1991). Likewise, the G1 box contains a conserved DXG motif also found in serine/threonine and tyrosine protein kinases, where it is involved in coordinating Mg(II), while the N and F motifs represent features which may be unique to the prokaryotic kinases.

Within the N box, the conserved asparagine has been shown to be required for Mg^{+2}-dependent ATP binding and therefore for kinase activity, whereas the first conserved glycine of the G2 box is not involved in ATP-binding, but is required for phosphoryl-transfer to the target histidine residue (Stewart *et al.*, 1998). Alanine-scanning mutagenesis of surface residues of Spo0F has identified those residues likely to be involved in recognition and binding between Spo0F and Spo0B, two TCSTS proteins involved in regulation of sporulation in *B. subtilis*. Assignment of pre-and post-phosphorylation involvement has been made for these surface residues (Tzeng and Hoch, 1997).

Between the hydrophobic, membrane receptor domain and the kinase domain lies the highly conserved H box, within which falls the phosphorylation target histidine residue. Its exact placement is variable among kinase families (Stock, Ninfa and Stock, 1991; Dutta, Quin and Inouye, 1999; Stock, 1999). Mutagenesis of the consensus residues of the H box has confirmed their importance in autophosphorylation and/or phosphatase activities (Stock, Ninfa and Stock, 1991). In the Class I HKs, the H box is directly linked to the other four boxes, which function in ATP-binding and catalysis of autophosphorylation. Phosphorylation occurs *in trans* within homodimer pairs, from the CA sub-domain of one unit to the DHp sub-domain of another (Yang and Inouye, 1991). Subsequently, the catalytic domain must vacate the histidine site in order to accommodate binding of the response regulator. Both monomers of the HK may play a role in the specific recognition and positioning of the cognate response regulator for phosphotransfer, as is suggested by work showing that contact is made between both subunits of Spo0F and the area surrounding the active site aspartyl residue of its cognate response regulator (Varughese *et al.*, 1998). Proximity between kinase and response regulator is also important for the response regulator-modifying phophatase activity of the kinase (Hsing and Silhavy, 1997).

Despite the functional division of all kinases into dimerization and transphosphorylation sub-domains, and Mg^{+2} ATP-binding and protein substrate-binding sub-domains, the organization of these domains is quite different in HK compared to serine/threonine and tyrosine protein kinases. For histidine protein kinases, the order of the conserved motifs, beginning at the amino-terminus, is H/N/G1/F/G2, while the shared motifs in non-histidine kinases appear in the opposite orientation: G2/G1/H (Stock *et al.*, 1995). Furthermore, the products of HK are phosphoesters, whereas serine/threonine and tyrosine kinases yield phosporamidates (Stock *et al.*, 1995).

2.2. THE RESPONSE REGULATOR COMPONENT

Response regulators (RRs) are defined by a sequence of approximately 120 amino acids and five dispersed but highly conserved residues form the active site; the residues generally include the phosphorylation target aspartyl residue, lysine, threonine and other residues (Volz, 1993). Like the HK component, RRs may occur either as a single domain protein, or as one unit of a multi-domain protein. A typical RR consists of a receiver domain and varying numbers of effector domains, joined by a flexible linker.

Crystal structure coordinates are available for CheY, the response regulator in the chemotaxis TCSTS in *E. coli* (Bellsolell *et al.*, 1996), as are the coordinates of its cognate histidine kinase, CheA (Bilwes *et al.*, 1999). Structures for NarL (Baikalov *et al.*, 1996) and PhoB (Sola *et al.*, 1999), two other members of the CheY family of response regulators, have also been resolved. This family is characterized structurally by 5 α-helices surrounded by a 5-stranded β-sheet and functionally by regulating transcription from promoters recognized by the major form of RNA polymerase holoenzyme (Volz, 1995).

For example, NarL, the RR for the nitrate-dependent regulation of electron transport and anaerobic fermentation genes, consists of a 131 amino acid N-terminal receiver, which forms the characteristic α_5/β_5 structure (Baikalov *et al.*, 1996). The C-terminal 62 amino acids form a compact helix-turn-helix motif characteristic of many DNA binding proteins. A flexible 13 amino acid linker joins these two domains, which are juxtaposed in the unphosphorylated form of NarL in such a way as to preclude DNA binding. Other examples in the CheY superfamily of RRs is CheY itself (which binds its protein target, the flagellar motor protein, directly and does not have a DNA-binding effector domain), OmpR (with one DNA-binding region) and NtrC, with two DNA-binding domains (Baikalov *et al.*, 1996).

Multiple alignments and comparative modelling analysis have revealed strong conservation in many regions of response regulators, including structurally important hydrophobic core residues, outer shell "contact" residues, certain hydrophobic yet solvent-accessible residues thought to be involved in Mg^{+2} binding, residues found in the loops surrounding the phosphorylated Asp site and other regions hypothesized to be functionally important (Volz, 1993). Variability in RR sequence increases with distance from the site of phosphorylation. These hypervariable regions are thought to be important for promoting the specificity that characterizes HK/RR pair interactions (Varughese *et al.*, 1998). Molecular recognition and contact itself may also contribute mechanistically in some way to response regulator activation (Welch *et al.*, 1998).

An interesting but uncommon conserved feature of the CheY family of RRs is a KPF turn featuring a cis-peptide linkage between lysine and proline (Volz, 1995). It had been hypothesized that the rigidity this linkage confers to the loop connecting β5 to α5 is important in propagating structural changes initiated upon activation when the lysine backbone configuration changes. Phosphorylation of the key aspartate residue of the RR was proposed to cause breakage of a salt bridge between aspartate and the lysine in question. This allows the adjacent proline to assume a more trans-like conformation, permitting phenylalanine to escape the RR pocket, and presumably allowing it to flip to within binding distance of its effector molecule. This trigger site model provided the basis for small molecule docking of Spo0F based on its alignment with CheY at Johnson and Johnson (Podlogar *et al.*, 1997). The KPF motif is conserved in 95% of the response regulators aligned (Volz, 1995). The following discussion will show how more recent structural studies have confirmed many of these early hypotheses.

Mutations in CheY active site residues on the β1/α1 loop and at the N-terminus of α1 were found to affect the mechanism by which movement is transmitted, without affecting the ability to bind Mg^{+2} (Bellsolell *et al.*, 1996). These mutations also implicated Lys109 in propagating conformational change from the active site to the functional surface of the molecule (Bellsolell *et al.*, 1996). More recently, structural information for the phosphorylated form of CheY (Halkides *et al.*, 2000) has shown movement of the side chain of this active site base, Lys109, as well as a predominantly buried conformation of the side chain of Tyr106. More subtle changes which are further removed from the active site involve the α4 and α5 helices (Halkides *et al.*, 2000).

Phosphorylated RRs possess a high free energy of hydrolysis which permits large and energetically favourable conformational changes (Stock *et al.*, 1995). The structural elucidation of the phosphorylated form of the *B. stearothermophilus* Spo0A regulator domain has helped to shed some light on the mechanism by which phosphorylation allows activation of the RR output domain (Lewis *et al.*, 1999). Spo0A~P, the first phosphorylated RR whose structure was determined, has shown that the phosphoryl group is coordinated with a nearby divalent cation, and is hydrogen-bonded with amino acid side chain and backbone atoms (many of which are highly conserved), as well as with water (Lewis *et al.*, 1999). The presence of the phosphoryl group causes a significant reorientation of the side chain of a highly conserved active site threonine (Thr84) and concomitant changes in residues making contact with the active site-distal face of this threonine, including conserved phenylalanine (Phe103), lysine (Lys106) and proline (Pro107) residues (Lewis *et al.*, 1999). It is hypothesized that these conformational changes destabilize interdomain interactions, allowing dis-inhibition of the output domain. NMR studies with the Spo0F response regulator of the same sporulation TCSTS in *B. subtilis* have shown that conformational changes are subsequently propagated to neighbouring backbone residues (Feher and Cavanaugh, 1999). The model generated from this work also identified conserved active site pocket threonine and lysine residues, and many buried residues of the hydrophobic core, as well as residues contributing to structural features such as the $(\alpha/\beta)_5$ fold and the β4–α4 loop, as undergoing conformational changes important for activating the RR as a transcription factor (Feher *et al.*, 1998; Feher and Cavanaugh, 1999).

Despite the degree of homology used to group the RR into a single superfamily, they vary greatly in the stability of their phosphorylated forms; stability seems to be appropriately matched to biological function (Zapf *et al.*, 1998). CheY, for example, undergoes dephosphorylation in seconds, which is roughly the amount of time the bacterium needs to change its direction of swimming, while Spo0F, which regulates the prolonged process of initiation of sporulation, autodephosphorylates over several hours (Zapf, Hoch and Whiteley, 1996). The rate of autophosphatase activity is thought to be attributable to the nature of the residues adjacent to the active aspartate residue, especially at position Asp56 (Zapf *et al.*, 1998). Interestingly, the conserved threonine and lysine residues implicated in conformational changes are also involved in phosphatase activity (Blat *et al.*, 1998; Zapf *et al.*, 1998).

In addition to the potentially critical RR mechanistic targets affecting the ability to undergo conformational change, or to dephosphorylate, one can also target the intermolecular "contact" residues involved in the transfer of phosphate from HK to RR. A class of inhibitor has been identified (which acts by interfering with the ability of a RR (VanR, of the VanR/VanS vancomycin resistance-inducing TCSTS of *E. faecalis)* to accept a phosphoryl group from its cognate HK (VanS) (Ulijasz and Weisblum, 1999). It may be possible to design very specific inhibitory agents of this kind which could target a particular TCSTS pair rather than the common catalytic features shared by the entire class of proteins.

2.3. THE CONTEXT OF TCSTS: NETWORKING AND PHOSPHORELAYS

The level of phosphorylation of two-component system proteins can be viewed as a delicate balancing of the equilibrium between active and inactive forms of the proteins, with multiple opportunities for allosteric input for both positive and negative regulation. Feeding into any individual system is information from the multi-protein, dynamically rearranging context of the membrane in which the kinases exist.

Counteracting the kinase and transferase activites of the HK and RR proteins of TCSTS are phosphatase activities which may reside either in the two-component proteins themselves (mentioned above), or on separate proteins altogether, where they create additional opportunities for expanding the spectrum of sensory cues that may influence physiological decisions. In *B. subtilis*, for example, the transition between the vegetative state and the sporulation state is under the control of the key RR and transcription factor Spo0A, whose phosphorylation state varies in response to numerous metabolic and environmental signals (Perego *et al.*, 1994). The extent of its phosphorylation is dependent upon: the activities of KinA and KinB, the two HKs responsible for feeding into the sporulation phosphorelay in response to various signals; the level of phosphorylation of the secondary messenger response regulator Spo0F resulting from KinA/KinB activity; the level of competing phosphatase activity in KinA and/or KinB; the availability of Spo0B, the phosphoprotein phosphotransferase, which transfers phosphate between Spo0F and Spo0A; the autophosphatase activities of the two RRS of this relay, Spo0F and Spo0A; and the activity of a group of protein aspartate phosphatases which complement the phosphatase activities resident in the kinases (Hoch, 1993; Tzeng and Hoch, 1997; Tzeng *et al.*, 1998).

These kinase-independent phosphatases, members of the RapA and RapB group, act specifically on Spo0F$\sim$P, the secondary messenger, in order to control its capacity to relay the phosphate signal further along this phosphorelay via Spo0B-mediated transfer to Spo0A. The RapA proteins are in turn controlled by the ComP/ComA TCSTS and the RapB proteins by the transcriptional regulator AbrB (Tzeng *et al.*, 1998). Although both of these systems play roles in competence development and stationary phase-related processes, the signals to which they respond differ. Therefore, different Rap protein aspartate phosphatases are transcriptionally controlled by different environnmental conditions, and they contribute additional layers of sensory input to the sporulation "decision"; such input may either reinforce or compete with the signals impacting the initiating kinase activity (Tzeng *et al.*, 1998).

The networking of multiple TCSTS which has been shown in *Bacillus* (Perego *et al.*, 1994; Tzeng *et al.*, 1998), is likely to occur in other organisms as well. One can envision a number of potential models of cross- and reciprocal-regulation. For example, transcription of the genes regulated by any given TCTS, for example, TCSTS *A* may be necessary to allow the sensing of the signal for TCSTS *B*. Activation of TCSTS *B* may induce transcription of a phosphatase which deactivates the kinase of TCSTS *A*, prevents *A*'s response regulator from further transcriptional activation and, in addition to the phosphatase activity of kinase A, provides an additional layer of control on that protein, as well as on itself. A third TCSTS, *C*, may induce a different phosphatase which inactivates the *B* pathway kinase and allows the *A* pathway to proceed even after the *B* pathway has been activated (Barrett and Hoch, 1998). This is one permutation of the model demonstrating temporal co-dependence and reciprocal control, and allows input from multiple input signals. Input from other global regulators of gene transcription converge into this highly sophisticated web which orchestrates cellular metabolism.

The proper cellular distribution of signalling proteins also affects their ability to elict the appropriate response. A case illustrating this point is the recently characterized cell-cycle-dependent localization of the DivJ kinase of *C. crescentus* (Wheeler and Shapiro, 1999). This bacterium undergoes asymmetric cell division, which yields a sessile stalked cell and a motile swarmer cell, requiring further differentiation before entering S phase. Multiple TCSTS in *C. crescentus* contribute to the coordination of asymmetric differentiation with cell cycle progression, including the essential global respose regulator CtrA (Quon, Marczynski and Shapiro, 1996); CckA (Jacobs *et al.*, 1999); PleC; DivJ; and DivK, another essential RR (Hecht *et al.*, 1995). It has been shown that DivK, in response to a signal at the G1-to-S phase transition, localizes to the flagellar pole in swarmer and pre-divisional cells, but is dispersed throughout stalked cells (Wheeler and Shapiro, 1999). PleC, an antagonistic HK to DivJ, may provide the signal at the G1-to-S transition.

The ability to control the local concentration of signalling proteins within the cell in response to temporal influences adds new dimensions to the complexity of the bacterial communication network. At the same time, it can account for some measure of the remarkable specificity and lack of *in sit*u "cross-talk" between non-cognate HK/RR pairs observed for these systems. Our growing understanding of the biochemical basis of the specificity of protein-protein interactions will undoubtedly have a profound impact on our ability to disrupt the balance of TCSTS for therapeutic intervention.

2.4. MULTIPLICITY/REDUNDANCY OF SYSTEMS

Sequence analysis of hundreds of histidine kinases and response regulators has revealed distinct subfamilies in both groups and a significant correlation between kinase and response regulator sub-families, suggesting that the various sub-classes have evolved independently of one another (Grebe and Stock, 1999; Stock, 1999). The evolution of multiple, seemingly redundant systems must have conferred some benefit, presumably a fine-tuning of the signal/transduced response relationship, to those organisms where TCSTS abound. Approximately thirty distinct TCSTS pathways exist in *E. coli* alone, including five composite (chimeric) kinase/response regulator proteins (Mizuno, 1997); thirty-six histidine kinases/34 response regulators have been found in *B. subtilis* (Fabret, Feher and Hoch, 1999). The *S. pneumoniae* genome has revealed thirteen distinct TCSTS protein pairs and one unpaired response regulator (Throup *et al.*, 2000).

Most, if not all, of the processes controlled by TCSTS fall into several broad and somewhat overlapping categories of survival strategies (see Table I). One such strategy enables the organism to actively seek out more favourable conditions through chemotaxis and motility (Mutoh and Simon, 1986; Stock, Ninfa and Stock, 1991; Winans, 1991). Another route allows it to compensate for a nutrient deficiency through nutrient uptake (Mueller *et al.*, 1991), sporulation (Burbulys, Trach and Hoch, 1991), control of cell division (Hecht *et al.*, 1995; Quon, Marczynski and Shapiro, 1996; Carballes *et al.*, 1999), or metabolic changes (Mueller, Bukusoglu and Sonenshein, 1992; Stretton and Goodman, 1998; Taylor and Zhulin, 1998; Msadek, 1999). Protein folding and degradation are also controlled by signals triggered by metabolic requirements (Jones *et al.*, 1997; Pogliano *et al.*, 1997). Another set of responses allows the organism to regulate its metabolism in response to the osmolarity of its surroundings (Maeda, Wurgler-Murphy and Saito, 1994), or its respiration in response to the availabilty of oxygen (Sun *et al.*, 1996; Nakano *et al.*, 1997). Another permits it to acquire the genetic elements necessary to survive in challenging environments, for example competence (Dubnau *et al.*, 1994; Guenzi *et al.*, 1994), or acquisition of antibiotic resistance (Cohen, McMurry and Levy, 1988; Joris *et al.*, 1990; Arthur, Molinas and Courvalin, 1992; Rasmussen and Kovacs, 1993;Bruns, Bruns and Pulverer, 1997; Li and Park, 1999), or tolerance determinants (Novak *et al.*, 1999a).

The expression of many virulence genes is TCSTS-modulated. Seven of the thirteen *S. pneumoniae* TCSTS, although not essential *in vitro*, are essential for growth and survival in a mouse RTI model (Throup *et al.*, 2000). The gene encoding the synthesis of the protective alginate capsule in *P. aeruginosa* (Roychoudhury, Akai and Chakrabarty, 1992), and the cell surface glycosaminoglycan hyaluronic acid in *S. pyogenes* (Bernish and Van deRijn, 1999), important in early stages of colonization of upper respiratory tract infection, are both regulated by two-component signal transduction pathways. Negative regulation of the transcription of multiple virulence operons in Group A streptococci is also a TCSTS-mediated process (Federle, McIver and Scott, 1999). In *C. jejuni*, a common agent of animal and human diarrheal disease, a TCSTS has been implicated in temperature-dependent growth and colonization in an intestinal tract infection model of that disease (Bras *et al.*, 1999). In *Salmonella*, a Ca^{+2}/Mg^{+2}-modulated TCSTS regulates expression of genes of the bacterial cell envelope which promote entrance and intracellular survival, increase resistance to anti-microbial peptides and decrease host cytokine production (Ernst, Guina and Miller, 1999). Production of the exotoxin perfringolysin O, as well as that of enzymes required for its transport, is a TCSTS-modulated process in *C. perfringens* (Shimizu *et al.*, 1994). Host recognition is mediated by VirA of *A. tumefaciens* (Winans *et al.*, 1994); adherence via the BarA/SirA system of *E. coli* (Zhang and Normark, 1996); enzymatic metabolism of host extracellular matrix components and adhesion to and penetration of host cell surfaces are also TCSTS-modulated events (Zhang and Normark, 1996), as is transcription of the ToxR regulon in *V. cholera* (Wong *et al.*, 1998).

Invasion occurs in response to the ColR/ColS TCSTS of *P. fluorescens* (Dekkers *et al.*, 1999) or the toxR/toxS system of *V. cholera* (Miller, DiRita and Mekalanos, 1989); and resistance to oxidative and chemotype-related stresses are also TCSTS-mediated (Guenzi *et al.*, 1994; Lyristis *et al.*, 1994; Haldiman *et al.*, 1997; Ernst, Guina and Miller, 1999; McFarlane *et al.*, 1999). The role that TCSTS play in regulating secondary metabolism of antimicrobial entities (Blumer *et al.*, 1992; Horinouchi and Beppu, 1992; Laville *et al.*, 1992; Stutzman-Engwall, Otten and Hutchinson, 1992; Beppu, 1995; Chang *et al.*, 1996) further extends the list of processes under the TCSTS domain and speaks to the central function of these entities in monitoring the economic expenditure of the organism's energy on evasive and protective processes.

Considering the gamut of potential stimuli to which an organism may be exposed, one can clearly see the necessity for either redundant but highly specialized signalling systems, or for a conservative but highly networked system of phosphorelay proteins capable of an expanded level of specificity and control. Layering and integration of these regulatory phosphoproteins permit an enhanced complexity in processing a broader spectrum of input signals and in determining the degree and duration of response (Soncini and Groisman, 1996; Fabret, Feher and Hoch, 1999; Hoch, 2000a).

2.5. CROSS-TALK?

Despite the conserved signalling domains shared among members of the HK and the RR families, there is apparently a remarkable lack of *in vivo* cross-talk between the HK component of one pathway and the RR component of another. The molecular basis of the specificity of signal transduction has not been elucidated, though it has been shown that structural analysis has suggested that specific molecular recognition domains may be important in the lack of cross-recognition and activation (Haldiman *et al.*, 1997; Bilwes *et al.*, 1999). TCSTS are clearly effective in accomplishing the precise shuttling of phosphoryl groups to convey physiological information to the necessary centres of response.

"Cross-regulation", which may include covalent modification or non-TCSTS-mediated effector binding, is distinguished from "cross-talk", which is defined as phosphotransfer between non-cognate sensor histidine kinase/response regulator pairs. From a genetic perspective, the genes for two-component proteins are commonly linked, and are often found as part of an operon containing the structural or target genes that the system modulates (Wanner,1992). Proximity of functionally related genes suggests a close relationship between their expression and activity. Yet there also exist "orphan" sensor or regulator proteins for which a partner has not yet been identified; for example, an unpartnered response regulator P31489 has been found in *S. pneumoniae* (Throup *et al.*, 2000) and the Aer sensor kinase found in *E. coli* is unpaired (Nichols and Harwood, 2000). It is not clear at present whether such unpaired genes are actually lacking cognate partner components, and so must integrate and converge with other signal transduction pathways, as does the Aer kinase (Taylor and Zhulin, 1998), or whether their partners are yet to be identified.

It has been proposed that, depending on whether a regulator is an activator or a repressor, sensor mutants should display an uninducible or constitutive phenotype, respectively, if *in vivo* cross-talk does not occur (Wanner, 1992). Mutants of the PhoR sensor/kinase gene do not result in a null phenotype (Wanner, Wilmes and Young, 1988). PhoB, for example, can be regulated by two kinases, both PhoR and CreC (Amemura *et al.*, 1990). Similarly, the response regulator Spo0A can be regulated by two kinases, KinA/KinB with a common secondary messenger Spo0F (Hoch, 1993). The Nar two-component regulatory system, consisting of the dual sensors NarX and NarQ and the dual response regulators NarL and NarP, controls the expression of various anaerobic respiratory and fermentation pathway genes (Stewart, 1994). Both NarX and NarQ are known to detect the two environmental signals nitrate and nitrite; either of the two sensor-transmitter proteins, NarX and NarQ, can activate the response-regulator proteins, NarL and NarP (Chiang, Cavicchioli and Gunsalus, 1997). However, differential stimulation of autokinase activity has been shown, at least for NarX, by nitrate and nitrite; the ability to respond to or discriminate between nitrate and nitrite being determined by a single residue (Williams and Stewart, 1997).

Evidence both for and against cross-talk exists for the vancomycin resistance-conferring TCSTS, VanR/VanS, which may exists in either of two forms in *E. faecium*: Type A, which is responsible for resistance to both vancomycin and teicoplanin, and Type B, which is limited to vancomycin resistance. Mutants in the sensor component of the VanR/VanS TCSTS of type B ($VanS_B$) can be compensated by over-expression of a wild type copy of the non-cognate sensor of type A ($VanS_A$) (Arthur, Depardieu and Courvalin, 1999). Yet under conditions where $VanR_A$ and $VanR_B$ activate their cognate promoters, neither activates the heterologous promoters, PvanR, PvanH, or PvanY, arguing against the interchangeability of type A and B two-component regulatory switches in vancomycin-resistant enterococci (Silva *et al.*, 1998). The ability of these highly related kinases to "cross-talk" when placed under stoichiometric pressure *in vitro* is of questionable import to the natural functioning of this TCSTS.

$VanR_A$ can also be activated by the non-partner kinase PhoR *in vitro* (Silva *et al.*, 1998). Likewise, $VanS_A$ can phosphorylate the non-partner response regulator PhoB, but the HK exhibits a 10,000-fold preference for transfer to its cognate RR, $VanR_A$, over PhoB (Fisher *et al.*, 1996).

In these examples where cross-talk can be demonstrated *in vitro*, the physiological relevance of the findings is uncertain. Spatial and temporal relationships within the living cell must certainly impact on *in situ* interactions and outcomes, and should be considered when interpreting the experimental results. Differential binding affinities and substrate turnover rates are also very likely to complement the specific molecular recognition mechanisms intrinsic to the proteins in determining specificity.

3. TCSTS in Other Organisms

Previously thought to be strictly prokaryotic vehicles of phosphotransfer, TCSTS have also been found in lower eukaryotes, specifically some fungi, including *N. crassa* (Oda and Hasunuma, 1994; Alex, Borkovich and Simon, 1996), *C. albicans* (Nagahashi *et al.*, 1998; Srikantha and Soll, 1998a; Calera and Calderone, 1999), *C. glabrata* (Srikantha and Soll, 1998b) and *S. cerevesiae* (Maeda *et al.*, 1994); in the slime mold, *D. discoideum* (Brown and Firtel, 1998); and in plants, where *A. thaliana* represents the most extensively characterized species (Susuki *et al.*, 1998; Imamura *et al.*, 1999). Eukaryotes appear to use the multi-step phosphorelay model more so than do prokaryotes, possibly suggesting a greater suitability for the signalling needs in these more complex organisms. Further conclusions about the differences in the choice of primary mode of signal transduction between prokaryotes and eukaryotes requires a better understanding of their specific roles in the organism and how differences in cellular organization may affect function (Perraud, Weiss and Gross, 1999). In all organisms in which they have been studied, the carefully paired "sensor/receiver" and "response regulator/effector" proteins of TCSTS comprise the foundation of a relatively sophisticated mechanism for regulating gene expression.

3.1. TCSTS IN LOWER EUKARYOTES

The Sln1 protein (Sln1p) of *S. cerevesiae*, for which a homologue exists in *C. albicans*, is a hybrid kinase/ response regulator which serves an essential function by modulating key events in response to changes in osmolarity. In *S. cerevesiae*, Sln1p appears to modulate a bifurcated pathway of transcription controls (Fassler *et al.*, 1997). One branch of this pathway regulates a downstream mitogen-activated protein kinase cascade (Map kinases) and may represent an evolutionary link between prokaryotic and eukaryotic models of phosphate-mediated regulatory mechanisms. Under conditions of low osmolarity, the N-terminal domain of Sln1 undergoes autophosphorylation at a conserved histidine residue, and subsequent phospho-transfer to an aspartate in the C-terminus. In one branch of the Sln1 pathway, phosphorylated Sln1p appears to increase gene expression mediated by the transcription factor Mcm1p (Yu, Deschenes and Fassler, 1995), which regulates such vital processes as cell wall synthesis, membrane integrity and cell-cycle regulation. The response regulator for this Sln1p signalling pathway is Skn7p (Li *et al.*, 1998). Alternatively, Sln1p may transfer the phosphate group to a second intermediate histidine kinase, Ypd1p, which in turn phosphorylates the first in a series of three Map protein kinases, Ssk1p, Ssk2p, and Ssk22p (Posas *et al.*, 1996). These Map protein kinases comprise the Hog1p Map kinase cascade. The phosphorylated form of Ssk1p is incapable of further activating the other two Map kinases and therefore exerts a negative regulation of the Hog1 Map kinase pathway in response to low osmolarity (Posas *et al.*, 1996). The same osmotic signal efficiently utilizes a common chimeric protein (Sln1p) to sense the change in environment and initiate the necessary chain of events in each.

A second chimeric osmosensing signal transducing protein, Nik1p, has been found in *C. albicans* (Nagahashi *et al.*, 1998); this protein has homologues in *N. crassa*, where it is involved in regulation of hyphal development (Alex, Borkovich and Simon, 1996). Disruption of either *Nik1*, *Sln1* or a third histidine kinase gene, *CaHK1*, impaired hyphal formation and attenuated the virulence of *C. albicans* in a mouse systemic candidiasis model (Yamada-Okabe *et al.*, 1999). Since invasion of the host is accompanied by the conversion from unicellular morphology to a filamentous (hyphae, pseudohyphae) growth form, all three of the characterized TCSTS from *C. albicans* and other fungi appear to be related to the pathogenic process (Calera and Calderone, 1999).

There are at least three other distinct MAPK cascades in budding yeast which transduce a variety of intracellular signals, including the mating-pheromone response, pseudohyphal/invasive growth, cell wall integrity and high osmolarity adaptation (Posas, Takekawa and Saito, 1998). Although there is extensive sharing of signalling elements among the MAPK pathways, little undesirable cross-talk occurs between various cascades, probably due to the formation of multi-protein signalling complexes.

The slime mold, *D. discoideum* utilizes the two-component mechanism of signal transduction to regulate the central processes of development and fruiting body formation via its effect on the degradation of cAMP (Aubry and Firtel, 1999; Thomason, Traynor and Kay, 1999). A hybrid kinase/response regulator, DokA, has also been implicated in the regulation of cellular responses to changing osmolarity (Schuster *et al.*, 1996).

3.2. TCSTS IN HIGHER EUKARYOTES

Despite their abundance in bacteria (Table 1), relatively few TCSTS have been found in eukaryotes. TCSTS with the characteristic His-Asp phosphorelay mechanism have emerged as very important signalling systems in plants, where they are responsible for responding to a wide variety of stimuli, including ethylene, cytokinin and osmolarity (Sakakibara, Taniguchi and Sugiyama, 2000; Urao, Yamaguchi-Shinozaki and Shinozaki, 2000). At least 10 histidine kinases and 14 response regulators have been identified in *A. thaliana* (Susuki *et al.*, 1998; Imamura *et al.*, 1999), along with a number of phosphorelay proteins which have been implicated in regulating aging and stress responses (Chang and Shockey, 1999; D'Agostino and Kieber, 1999).

Plant phytochromes, photoreceptors which are able to effect chromophore changes in their ligand, had been previously proposed to be histidine kinase paralogs based on comparative protein sequence alignments and biochemical cross-talk experiments. These photo-sensory kinases possess a serine/threonine specificity, possibly representing divergence from a prokaryotic ancestor after duplication of the transmitter module (Quail, 1997). These kinases show an ability to cross-talk with the response regulator substrate of the cyanobacterial phytochrome Cph1 (Yeh and Lagarias, 1998), reflecting the redundancy of these systems and possible common ancestral origin.

A family of mitochondrial protein kinases, previously suspected as being relatives of the prokaryotic TCSTS histidine kinase family, have been recently characterized as serine kinases (Harris *et al.*, 1997). These kinases play pivotal regulatory roles, functioning either at the interface between glycolysis and the citric acid cycle, or in the pathway for the disposal of branched chain fatty acids.

Parallels have also been drawn between prokaryotic response regulators and the G-protein family of signal-transducing proteins in humans (Artymiuk *et al.*, 1990; Chen *et al.*, 1990). A search for effectors of RhoA, a Ras-related G protein, has revealed an interaction domain in heterotrimeric G-protein beta subunits which is conserved in the yeast response regulator protein Skn7. (Alberts *et al.*, 1998). Mechanistic similarities have also been demonstrated between RapB phosphatase-driven hydrolysis of Spo0F ~ P and the GTPase-activating, protein-assisted GTP hydrolysis of Ras; both involve conformational changes in the substrate which result in increased autodephosphorylation (Tzeng *et al.*, 1998).

The discovery of eukaryotic systems resembling the prokaryotic His-Asp two-component model of signal transduction, and the ubiquitous distribution of the serine/threonine and tyrosine kinases and G-protein-like phosphorelay systems have erased the distinction between the two kingdoms that had previously been drawn along signal transduction kinase lines. While phosphotransfer is a common element of all known mechanisms of signal transduction, there seem to exist sufficient opportunities for exploitation of mechanistic differences between two-component His-Asp phosphotransfer and the eukaryotic models.

4. TCSTS as Drug Targets

Signal transduction inhibitors are expected to interrupt the mechanism(s) by which bacteria "sense" the stress signals which mobilize genes in response to their environment and to affect the bacterium's ability not only to survive in the host, but also to respond to the normal signals that would induce the events necessary for infection. "Essential" targets, which are defined by researchers as those gene products that are necessary for survival of a bacterium in the laboratory, are believed to make terrific potential drug targets, as an inhibitor against an "essential" gene would be expected to cripple, if not kill, the growing bacterium.

The anti-TCSTS therapeutic approach, however, did not start as a single drug target approach, but rather as an anti-virulence effort, as the initial reports from Eli Lilly (Roychoudhury, Sakai and Chakrabarty, 1992; Roychoudhury *et al.*, 1993) and The R.W. Johnson Pharmaceutical Research Institute (Barrett and Hoch, 1998; Barrett *et al.*, 1998) did not appear to focus on "essential" gene targets. And the leading laboratories in the world (Ninfa and Bennett,

1991; Stock, Ninfa and Stock, 1991; Hoch, 1993) did not emphasize the "essential" nature of the TCSTS. It was only later that the potential for targeting "essential" genes arose. So where lies the interest in targeting these "non-essential" TCSTS?

Until recently, the many processes associated with two-component modulation were closely linked to, yet not essential for, survival *in vitro*. The conditions under which essentiality is determined in the laboratory conditions are very different from those that the organism will encounter in nature. A defect in an organism's ability to express collagenase or exotoxins, for example, or to adapt to dramatically different nutritional conditions, may have no effect on its ability to grow under optimized laboratory culture conditions. Yet such organisms may be unable to establish infection in their hosts and clearance by the host's immune system would render them harmless.

4.1. LINKS TO SURVIVAL YET NOT "ESSENTIAL" GENES: ROLE OF TCSTS IN RESISTANCE

Despite all our efforts to eliminate the threat bacteria pose to our health, these organisms have, over time, adapted to the pharmacological challenges we have placed in their paths. Four basic mechanisms of resistance are: drug inactivation; target modification; decreased uptake; and efflux. By way of example, inactivation of β-lactams by β-lactamases is the primary mechanism of resistance to this type of antibiotic, especially in *E. coli*. Other examples of inactivation include glycoside inactivating enzyme, chloramphenicol acetyl transferases, macrolide esterases or phosphotransferases, and host renal dipeptidase (which is counteracted by peptidase inhibitors) (Russell and Day, 1996). To date, no β-lactamases have been discovered in *S. pneumoniae* or *H. pylori*, nor has any other antibiotic inactivating mechanism. Strategies for combating resistance by inactivation include finding inhibitors of the inactivation enzymes, developing more stable antibiotics, or preventing TCSTS-mediated induction of the modifying enzymes. Target modification as a resistance mechanism also occurs in resistance to β-lactams and vancomycin, for example, where the normal cellular binding sites for the drugs are altered to reduce their affinity for the drug ligand, thus lowering the achievable dose of inhibitor per cell.

Of the many mechanisms of resistance, perhaps the most troublesome are efflux and uptake, since they are broad-stroke evasion strategies which often complement the more chemotype-specific resistance routes of target modification or drug inactivation. For example, β-lactam resistance can also be accomplished through altered uptake/efflux mechanisms in *P. aeruginosa*. Other antibiotics eliciting this response are the tetracyclines, chloramphenicol, quinolones, aminoglycosides (which decrease permeability indirectly through enzyme modification), macrolides and trimethoprim (Williams, 1996). The repertoire of mechanisms that bacteria have developed to circumvent or compensate the effects of antibiotics in order to survive speaks to the tenacity and resilience of these organisms. This flexibility will almost certainly continue to challenge any novel antimicrobial therapy and therefore is an attractive target itself. The following brief review will show that TCSTS are the agents of this flexibility in a growing number of cases.

3.1.1. *Vancomycin Resistance*

Since the initial emergence of glycopeptide-resistant *E. faecium* in 1986, the incidence of its occurrence has risen at an alarming rate: currently, 52% of all enterococcal isolates show some level of vancomycin resistance (Huycke, Sahm and Gilmore, 1998). Since these vancomycin-resistant enterococcal strains (VRE) are often multiply resistant to other classes of antibiotics, the mortality rate is ~60% for these patients. Roughly 90% of resistance of this type is mediated by the *vanA* determinant, which comprises five genes on two coordinately expressed operons. Transposable element TN1546, for example, confers inducible, high-level resistance, often a precedent of constitutive resistance. Experimentally, resistance is transferable to *S. aureus*, but the progression of Van_A-type resistance has not yet occurred in nature. Although intermediate vancomycin resistance has been reported in *S. aureus* and as a sub-population of methicillin-resistant *S. aureus* (MIC to vancomycin 8–16 μg/ml), none of these strains harboured the *vanA*, *vanB* or *vanC* genes from VRE (Kinoshita and Nakamachi, 2000). The need to address vancomycin resistance is clear and urgent and this is but one of many surging antibiotic resistance problems worldwide.

The VanR/S two-component system mediates type A, high-level, inducible resistance to the glycopeptide antibiotics vancomycin and teicoplanin and is one of the most extensively characterized of the TCSTS-mediated resistance models. In this system, the sensor histidine kinase, VanS, phosphorylates the response regulator, VanR, in the presence of an as yet unidentified signalling event (Bugg, 1991; Wright, Holman and Walsh, 1993; Haldiman *et al.*, 1997). VanR ~ P modulates the transcription of three genes whose protein products are involved in the synthesis of an altered peptidoglycan target with a lower affinity for vancomycin. Of the three types of vancomycin resistance (Dougherty and Barrett, 1999), VanA-mediated resistance is by far the most clinically important in resistant enterococcal isolates (Huycke, Sahm and Gilmore, 1998). The VanB resistance phenotype differs from VanA in conferring resistance to vancomycin, but not to teicoplanin, due to a different ordering of the structural genes in the operon regulated by the $VanR_B/VanS_B$ two-component system (Arthur *et al.*, 1995). The VanS sensor protein, in fact, was shown to be non-essential in transcriptional activation of *van*H/A/X, indicating that in the absence of signal, VanS negatively controls VanR (Arthur *et al.*, 1996). In the absence of VanS, activation may occur through acetyl phosphate, if levels are high enough in the cell, or through a heterologous HK (Arthur *et al.*, 1996). The nature of the VanR/S inducing signal is not completely understood, but induction is not related to any structural feature of the glycopeptide itself, since antibiotics affecting the biosynthesis of peptidoglycan are also able to induce the system (Lai and Kirsch, 1996). Hydrolytic enzymes which damage peptidoglycan cross-links have similar inducing effects (Ulijasz, Grenader and Weisblum, 1996). It is possible that the build-up of peptidoglycan precursors, or the ratio of cross-linked to non-cross-linked peptidoglycan molecules may trigger the system.

4.1.2. *β-Lactam Resistance*

Methicillin-resistant *S. aureus* (MRSA) comprises one of the largest and clinically most important groups of drug-resistant nosocomial infections affecting U.S. patients. Resistant isolates harbour the complementary resistance mechanisms of both target modification, through low affinity penicillin-binding proteins (PBPs), and drug inactivation through β-lactamases. The BlaR sensory-transducer, shown in *B. licheniformis* to be required for β-lactamase inducibility, bears topographic resemblance to the signal transducing agents of chemotaxis (Joris *et al.*, 1990). This plasma membrane-embedded penicillin-binding protein is thought to transduce signal, elicited by penicillin binding, via a four-alpha-helix bundle in a manner analogous to that of the aspartate receptor Tar. (Hardt *et al.*, 1997). Induction of β-lactamase is under the control of a similar regulatory system in MRSA (Bruns, Bruns and Pelverer, 1997), where it is an attractive target for suppressing methicillin resistance.

Competence and penicillin-resistance in *S. pneumoniae* are also subject to TCSTS modulation (Guenzi *et al.*, 1994). In addition to the production of modified PBPs with a decreased affinity for penicillin, resistance to β-lactams also occurs through a PBP-independent mechanism which is mediated by the CiaH/CiaR two-component system (Guenzi *et al.*, 1994). A single point mutation in the kinase *cia*H conferred both cefotaxime resistance and transformation deficiency, which could not be complemented by the addition of exogenous competence factor (Giammariaro, Sicard and Gasc, 1999). Mutations in the *S. pneumoniae ciaH* kinase have other effects, including the ability to form protoplasts and susceptibilty to lysis with deoxycholate (Giammariaro, Sicard and Gasc, 1999). It has been suggested that the Cia TCSTS mediates a PBP-independent resistance mechanism which affects the biosynthesis of cell wall components at a step prior to the biosynthetic functions of PBPs (Hakenbeck *et al.*, 1999).

4.1.3. *Tetracycline Resistance*

A possible mechanism of tetracycline resistance was suggested by experiments demonstrating low level tetracycline resistance in *E. coli* when an operon containing the RprX/RprY TCSTS from *B. fragilis* was over-expressed (Rasmussen and Kovacs, 1993; Rasmussen, 1994). The over-expression of these TCSTS proteins resulted in an increase in the level of outer membrane porin protein, OmpC, and a decrease in the level of OmpF. The decreased expression of OmpF correlated with, and may be responsible for, the observed tetracycline resistance in these cells. It has been suggested that the RprX/RprY TCSTS interferes at the transcriptional level with the normal regulation of OmpF, which is normally controlled by the EnvZ/OmpR TCSTS (Rasmussen and Kovacs, 1993).

4.1.4. *Polymyxin B, Aminoglycoside and Multiple Drug Resistance*

Part of the adaptive response of *P. aeruginosa* to Mg^{+2} starvation is overexpression of the outer membrane protein, OprH and increased resistance to the polycationic antibiotic polymyxin B. Genetic analysis revealed a linkage of *opr*H, *pho*P and *pho*Q into an operon that was demonstrated to be under the joint control of PhoP-PhoQ TCSTS proteins and Mg^{+2} ion concentration (McFarlane *et al.*, 1999).

Low-level aminoglycoside resistance in *E. coli* also appears to be related to the integrity of the Cpx two-component system (Mileykovskaya and Dowhan, 1997), whose primary function in this organism is believed to be regulation of conjugative plasmid expression.

Recent data suggests that a signal transduction pathway may also be involved in the induction of multiple antibiotic resistance in *E. coli* (Li and Park, 1999). MppA is a periplasmic binding protein whose mutation leads to increased resistance to a wide spectrum of antibiotics, overproduction of at least nine inner membrane proteins, as well as the transcriptional activator MarA, and reduced production of OmpF (Li and Park, 1999). Since all of these phenotypic changes require the presence of the *mar* operon (Cohen, McMurry and Levy, 1988), it is proposed that MppA is linked to *mar*A expression, either directly or indirectly, through a sensor kinase or phosphorelay mechanism. The existence in MarA of a putative response regulator domain containing the signature aspartate and lysine residues supports this hypothesis. If MppA does, in fact, exert its control on multiple drug resistance *via* signal transduction, it would define another subset of attractive targets for therapeutic intervention.

Recently, a multidrug efflux system in *S. maltophilia* has also been reported to be regulated by a two-component switch, SmeS/SmeR (Li, Zhang and Poole, 2000). Having plagued the current therapies by creating pathways whereby toxic agents are pumped out of the cell, efflux represents another potential target falling under the TCSTS umbrella of regulated processes. An inhibitor in this area would represent a major breakthrough for antimicrobial therapy.

4.1.5. *Antibiotic Tolerance*

A new VncS/VncR TCSTS-mediated antibiotic tolerance phenomenon is responsible for the recent failure of vancomycin therapy against clinical isolates of *S. pneumoniae* (Gilmore and Hoch, 1999). Vancomycin tolerance has recently been reported in *S. pneumoniae*. Clinical isolates of this strain, which is responsible for meningitis, pneumonia and otitis media, and which causes over one million deaths per year, have turned up with a mutation in the histidine kinase of the newly identified VncS/VncR two-component system controlling multiple antibiotic tolerance (Novak *et al.*, 1999a). A single defect in the "sensor" component of this TCSTS protects the organism from this treatment of last resort, and further underscores the pivotal role such signal transducers play in ensuring the organism's survival under adverse conditions.

Antibiotic tolerance is a phenomenon separable from resistance, but which is believed to be its precursor (Novak *et al.*, 1999a, 2000). The VncS/VncR TCSTS is believed to act normally *via* a quorum-sensing mechanism to trigger a number of cell death pathways, including autolysis, but is also apparently required for the bactericidal activity of antibiotics. Antibiotic treatment apparently stimulates the same lytic response, the common element being a small "death signal peptide", P27, which is coordinately expressed along with the three subunits of an ABC-transporter gene also required for antibiotic-induced killing (Novak *et al.*, 2000). Interestingly, five of the thirteen TCSTS gene pairs found in *S. pneumoniae* are also found adjacent to genes encoding ABC transporters (Throup *et al.*, 2000), which may suggest a potential involvement in either the import or export of their related signalling entities.

4.2. LINKS TO SURVIVAL YET NOT "ESSENTIAL" GENES: ROLE OF TCSTS IN PATHOGENICITY

Several TCSTS have been shown to be essential for growth of the bacterium in the host, though not necessarily for growth *in vitro* (i.e., on laboratory media). These systems offer opportunities for bacteriostatic agents which might render the pathogen harmless, especially those that are involved in virulence and pathogenicity, which require the precise temporal and spatial control of new gene expression for successful colonization in the host.

The stimuli inducing these virulence factors must signal the transition from environmental reservoir to the host and may include changes in pH, temperature, osmolarity, calcium, magnesiium, iron, oxygen and carbon dioxide (Dziejman and Mekalonos, 1995). Organisms have developed the adaptive processes to recognize specific host cell types and even sub-cellular location (e.g. endosome, lysosome, cytoplasm) before initiating their own growth.

Table 1 provides a sampling of some of the virulence determinants controlled by TCSTS. The diversity of systems reflects the range of anatomical sites, signals and responses used to accomplish each infective process. While *V. cholera*, *B. pertussis* and *P. aeruginosa* all produce toxins in the course of pathogenesis, the TCSTS proteins which regulate their production are distinct and are triggered by distinct signals. Even within a species, multiple TCSTS may modulate diverse functions. For example, in *Pseudomonas* spp., some of the TCSTS-mediated virulence factors utilized include PfeR/PfeS (ferric enterobactin receptor), AlgRI/AlgRII (alginate capsule), PilS/PilR (fimbriae expression), CopS/CopR (copper resistance) and HrpR/HrpS (capsule synthesis) (reviewed in Dziejman and Mekalonos, 1995).

S. aureus produces a wide variety of exotoxins, hemolysins and enterotoxins; a secreted nuclease, protease and lipase; a variety of cell-wall-associated proteins including Protein A, collagen-binding protein and fibronectin-binding protein;

as well as mammalian matrix protein receptors. The coordinated production of these proteins is accomplished in this organism by the global accessory gene regulator (*agr*) locus, which embodies a unique and complex TCSTS.

The *agr* locus comprises two divergent operons, one of which forms a cell density- or quorum-sensing TCSTS unit (*agr*A and *agr*C) (Lina *et al.*, 1998). Environmental pH is thought to modulate Agr regulation (Regassa and Betly, 1992). Expression of *agr*ABCD results in autocrine-like transcription from its own promoter, P2, and from the divergent promoter, P3 (Novick *et al.*, 1993). An auto-inducing octapeptide ligand (AgrD) binds the HK AgrC, which is believed to transmit information through the RR AgrA, resulting in production of an RNAIII transcript from the divergent promoter P3. This transcript is the effector of the agr system, causing down-regulation of the expression of exponential-phase genes (Protein A, coagulase) and up-regulation of post-exponential proteins (enterotoxins, haemolysins, proteases) (Janzon and Arvidson, 1990; Novick *et al.*, 1993; Patel *et al.*, 1992). AgrA may also be involved in regulating genes whose transcription is not mediated by RNAIII (Abdelnour, 1993). Agr-deleted mutants were unable to migrate to joint cavities through the blood in a bacterial arthritis model and were instead recovered in the spleen (Abdelnour *et al.*, 1993). Apparently, the Agr proteins have an important role in allowing *S. aureus* to penetrate blood vessels into joints. An interesting observation has been reported that may explain the ability of one strain to exclude others from subsequent infection of the same tissue. It has been reported that the agr-encoded, auto-inducing ligand produced in some strains inhibits the expression of agr in other strains where the amino acid sequence of the peptide differs markedly (Ji, Beavis and Novick, 1997). Thus, the *agr* regulatory system has a built-in hypervariability mechanism for cross-inhibition of competing pathogens.

In *S. typhimurium*, the PhoP/PhoR TCSTS regulates up to 40 loci in response to divalent cations (Groisman, Heffron and Solomon, 1992). Distinct binding sites exist for Ca^{+2} and Mg^{+2} on the PhoQ sensor protein (Vescovi *et al.*, 1997). Though Mg^{+2} was more efficient than Ca^{+2} in repressing transcription from this TCSTS, an avirulent mutant resulting from a point mutation in PhoQ possessed an altered affinity for Ca^{+2}, but not for Mg^{+2}, suggesting that both cations are required for regulation (Vescovi *et al.*, 1997).

The role of PhoP/PhoQ in virulence in this organism was established when a transposon-generated, virulence-attenuated mutant was mapped to the PhoP locus (Fields, Groisman and Heffron, 1989). This system is unique in that both null and constitutive mutations attenuate virulence, although different virulence phenotypes are affected in each case (Garcia *et al.*, 1994). Approximately 75% of the loci whose transcription is regulated by PhoP are not involved in virulence (Garcia *et al.*, 1994). One of the targets of the PhoP/PhoQ system is pmrCAB, which encodes a TCSTS involved in resistance to polymyxin B (Soncini and Groisman, 1996). Of the 75 loci whose transcription was regulated by PhoP/PhoQ, 7 were found to be mediated through *pmr*A/*pmr*B. While transcription from pmr-dependent genes was modulated by both Mg^{+2}

(PhoP/PhoQ) and pH (PmrA/PmrB), transcription from *pmr*-independent genes could not be modulated by Mg^{+2} (Soncini and Groisman, 1996). This cascade, allowing reaction to a broader range of cues, may play an indirect role in pathogenicity, by conferring resistance to the cationic peptides of phagocytic cells (Groisman *et al.*, 1989). PhoP/PhoQ-dependent loci include *pag*G, an outer membrane protein required for macrophage survival, and *pgr*H, which plays a role in invasion beyond entry into intestinal epithelial cells, among many others, some of which have yet to be characterized (reviewed in Groisman and Heffron, 1995). The PhoP/PhoQ TCSTS is capable of auto-regulation, providing yet another level of control of this massive regulon (Soncini, Vescovi and Groisman, 1995).

4.3. LINKS TO SURVIVAL: "ESSENTIAL" GENES

An essential TCSTS has been reported in *B. subtilis (yycF/G)* (Fabret and Hoch, 1998), for which strong homologues exist in other gram-positive organisms, including clinically important staphylococcal strains. In particular, a strong *S. aureus* homologue has been identified, with evidence that it may play a role in regulating the permeability of the cell membrane (Martin *et al.*, 1999). Both the HK and RR components of this TCSTS were shown to be essential (Fabret and Hoch, 1998; Martin *et al.*, 1999). Temperature-sensitive mutants in the *yyc*F, RR component in *B. subtilis* resulted in the cells appearing to have lost their cytoplasmic content at the non-permissive temperature (Fabret and Hoch, 1998). A temperature-sensitive regulatory mutant in *yyc*F in *S. aureus* gave results which suggest a defect in branched-chain fatty acid biosynthesis, as well as an increased susceptibility to macrolides (Martin *et al.*, 1999). Studies of the homologous system in *S. pneumoniae* (Wallis, 1999b) showed that only the RR is essential, suggesting that an alternative kinase may be involved in this pathway.

A TCSTS involved in cell cycle control in *C. crescentus* was also shown to be essential for the organism's viability (Quon, Marczynski and Shapiro, 1996). The CtrA RR of this system binds and controls transcription from a number of cell cycle-regulated promoters, all of which share a nine bp recognition motif. Some of these promoters are responsible fpr driving transcription of genes involved in DNA replication and methylation and in flagellar biogenesis. *Ctr*A itself is regulated both transcriptionally and via phosphorylation and is proposed to be part of a phosphorelay coordinating the timing of cell cycle events. It appears to be pivotal in the direct coupling of chromosome replication with the cell cycle (Quon *et al.*, 1998).

Systematic knockout experiments have demonstrated the presence of just one *in vitro* "essential" gene in *S. pneumoniae*, but multiple *in vivo* "essential" TCSTS genes (Throup *et al.*, 2000). Three essential response regulators have been identified in *Heliocobacter pylori* (Beier and Frank, 2000) and one essential TCSTS, *mtrA*/*mtrB*, has been found in *Mycobacterium tuberculosis* (Zahrt and Deretic, 2000).

As antibacterial targets, TCSTS not only meet the traditional criteria of essentiality but, in serving as a crude nervous system for the microorganism, they regulate a collection of individually non-essential functions whose simultaneous inhibition may cripple the cell by disarming its ability to respond appropriately to its natural host environment

4.4. THEORETICAL CONSIDERATION OF TCSTS INTERVENTION

Two-component systems offer several potential sites for interference (see Fig. 3, Table 2), including the "sensing" of the signalling condition or ligand, Mg^{+2}-dependent ATP-binding, autophosphorylation of the conserved histidine residue of the HK, phosphotransfer from HK → RR (involving intermolecular contact formation and phosphate-releasing and -accepting activities), the phosphatase activity of the HK, without which the system would be in a perpetual "on" state, and the intermolecular response regulator/DNA interaction. The predominant eukaryotic signal transduction systems employ ATP-dependent serine/threonine and tyrosine kinases rather than histidine kinases. With the exception of certain mitochondrial enzymes, the prokaryotic ATP-binding site of HKs differs from that of most eukaryotic proteins (Harris *et al.*, 1995). One might anticipate limited risk of human toxicity considering the multiple novel opportunities for drug intervention, including both highly conserved catalytic activities and specific recognition events.

What impact could design and use of a TCSTS inhibitor have on the growing or non-growing bacterium? The potential for therapeutic intervention of a TCSTS may be rich in opportunity, albeit with great initial risk. Due to the high degree of homology in both the carboxyl terminus and amino terminus of the histidine kinase and response regular, respectively (Volz, 1995; Podlogar

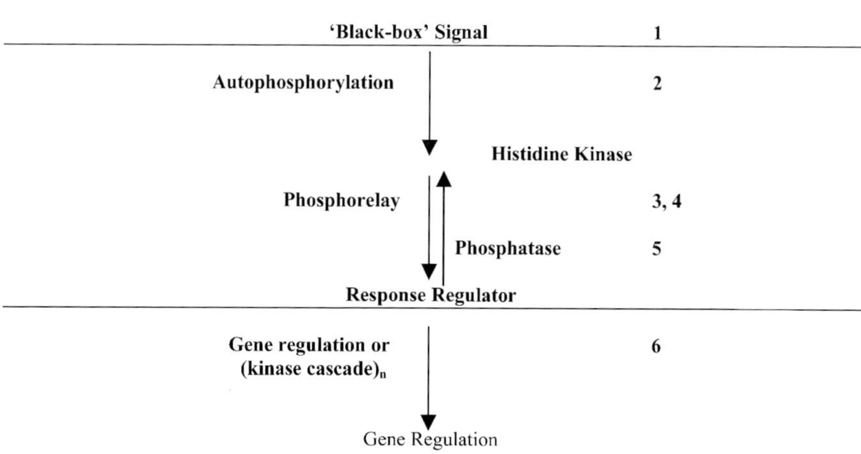

Figure 3. Potential sites for therapeutic intervention

Table 2. Potential therapeutic intervention sites and expected outcomes

Site	Target	Molecular mechanism	Spectrum potential	Therapeutic outcome
1	Signal initiation	Unknown	Broad	Bactericidal/ Bacteristatic
2	Autophosphorylation	Auto-catalytic Phosphorylate for ATP donor	Broad	Bactericidal
3	HK :: RR interaction	Non-covalent association	Broad	Bactericidal
4	Phosphorylation	Transfer of H-poly to Asp	Broad	Bactericidal
5	Dephosphorylation of RR	Reversal of "D" by phosphorylation activity	Broad	Bactericidal
6	2nd round phosphorylation or direct gene regulation	Transfer of H-PO_4 to Asp	Selective	Bactericidal/ Bacteristatic

HK – histidine kinase
RR – response regulator

et al., 1997; Lewis *et al.*, 1999), it is widely speculated that in targeting a particular TCSTS, one may find inhibitors that cross-react with other two-component systems (Volz, 1993; Domagala *et al.*, 1998; Macielag *et al.*, 1998).

The potential broad range of utilities of TCSTS inhibitors includes potential use as: independent antimicrobial agents for monotherapy; potentiators for treating resistance; synergistic bactericidal combination or anti-virulence therapy; or sub-MIC use in prophylaxis (Barrett and Hoch, 1998; Frosco and Barrett, 1999; Wallis, 1999a). Each usage of TCSTS inhibitors may lead to the inhibition of growth, inhibition of pathogenicity, bacteriostatic inhibition of bacteria and/or bactericidal activity against the bacterium.

If the primary mechanism of resistance to TCSTS inhibitors is not a generalized strategy to prevent accumulation of the offending agent, then multiple genetic events may be required to compensate for the pathways affected. Though it is difficult to predict the nature of resistance to agents which do not yet exist, loss of sensitivity to the signal is one potential mechanism that may be selected. Yet the transient nature of TCSTS-mediated processes suggests that another system for braking or activating the system would have to be engaged, in order to minimize the metabolic expense and other consequences of a loss of temporal control of gene expression. TCSTS inhibitors may also select an enhanced promiscuity between components of non-cognate systems, yet the cell must somehow accomplish this without losing the signal specificty of the affected TCSTSs. While discussion of possible resistance routes may be premature, these scenarios illustrate the potential results of selective pressure that might be elicited to counter TCSTS inhibition.

Whether inhibition is bactericidal or through combination therapy, reduced resistance emergence of monotherapy may be possible by inhibiting multiple targets with a single drug, where the chance of resistance emergence becomes exponentially additive. That is, whereas the rate of resistance emergence may be 1 in 10^5 to 1 in 10^9 in bacteria with monotherapy, targeting two or more targets should make the likelihood of a double mutation event in the range of 1 in 10^{10} to 1 in 10^{18}, assuming completely independent events are required for simultaneous single-step mutations. Minimally, resistance in a multi-step process, sequentially rather than simultaneously, would have a reduced rate of resistance, somewhere between the two aforementioned rates.

One of the other approaches to inhibition of these TCSTS may be to eliminate expression of the TCSTS at the gene level. Down-regulation of TCSTS expression would permit intervention at an earlier point than inhibition at the protein (HK/RR) level, by small molecule pharmacophores. This type of therapy, i.e. the inhibition of gene regulation of TCSTS, rather than the production of any single TCSTS, may provide a "broad-spectrum" inhibition of multiple TCSTS.

Unknown resistance mechanisms, toxicities and possible synergies and antagonisms that TCSTS inhibitors may possess should not deter efforts in this or in any novel target drug discovery effort.

4.4.1. *Monotherapy: Antibacterial Agent*

Since many of the key features of His → Asp signalling pathways are conserved across systems (and across species), an inhibitor has the potential to hit many TCSTS targets (and organisms). The end result of inhibiting several non-essential TCSTS may be the same as inhibiting a single essential TCSTS. The outcome of inhibition of any single TCSTS essential gene would be expected to cause a bactericidal event in the bacterium.

Likewise, agents active against one or more "non-essential" TCSTS may also be useful monotherapies. Structural data suggests that one may be able to design both narrow spectrum (narrow target) agents or broad spectrum agents, depending on what part of the conserved binding site in the HK/RR one targets with an inhibitor. Thus, it may be possible to design a single agent that targets multiple TCSTS, which would provide just the fourth example of an antimicrobial targeting more than one target (the others being aminoglycosides, β-lactams and quinolones).

A subset of the "non-essential" TCSTS are the ones regulating virulence factor expression. As listed in Table 1, there are numerous virulence targets in bacteria that could be the next vaccine candidate or non-antibiotic target in bacteria. Although not fully appreciated in the clinical community, it is theoretically possible to inhibit bacterial pathogenesis *in vivo* by inhibiting a "non-essential" target or process that is "essential" to the microbe in the host itself.

4.4.2. *Potentiation: Adjunct Therapy*

The involvement of TCSTS in antibiotic resistance is an emerging and expanding scenario. Strategies to address VanA-mediated vancomycin resistance in *E. faecium*, for example, involve interference with its induction by disabling the two-component system controlling the genetic machinery that permits adaptation to the drug. This TCSTS-vancomycin resistance operon is also on a promiscuous transposon, Tn*1545* (Arthur *et al.*, 1995), which has the potential for movement throughout the Gram-positive world. Down-regulation of *vanS*/*vanR*, suppressing the production of the downstream cell wall modifying enzymes VanA and VanH, should restore vancomycin susceptibility as implied from the work of several laboratories (Bugg *et al.*, 1991; Arthur *et al.*, 1995; Harris *et al.*, 1995; Lai and Kirsch, 1996; Haldiman *et al.*, 1997). Another TCSTS-controlled antibiotic resistance factor system involves a complex regulatory network in *S. pneumoniae*, in which competence and penicillin-susceptibility are tied to the *cia*H/*cia*R TCSTS system (Guenzi *et al.*, 1994; Novak *et al.*, 1999a). Similar rationale applies when considering other TCSTS-mediated resistance mechanisms as they continue to be elucidated.

While the ultimate goal may be to identify a broad spectrum antibiotic that recognizes one of the common elements of the TCSTS model, a VanR/VanS-specific inhibitor may also be of interest and considered for development as an adjunct therapy, or potentiator for vancomycin, rendering a resistant pathogen susceptible to co-administered glycopeptide (Barrett and Hoch, 1998; Frosco and Barrett, 1999).

4.4.3. *Other Adjunct Therapies: Anti-Adaptive Response Intervention*

The inhibition of multiple TCSTS, even though "non-essential" in themselves, may cripple the bacterium sufficiently to cause a synergistic bactericidal event in the bacterium, in much the same way that proven drug combinations, such as the mixture of bacteriostatic streptogramins that make up Synercid®, become synergistically bactericidal. Broad-target TCSTS inhibitors may expand the spectrum of activity of a given antibiotic by inhibiting a virulence or resistance factor that may reduce the single agent antibacterial *in vitro* activity, or breadth of coverage.

Anti-virulence, non-antibiotic TCSTS agents may provide exceptionally valuable adjunct therapy in the clinic, without neccesarily adding selective pressure to the bacterium. For example, the use of an anti-virulence agent in cystic fibrosis patients, along with a potent antibacterial (i.e. ciprofloxacin), may allow for greater ciprofloxacin penetration in the infection site by reducing the production of the virulence factor, alginate capsule, by *P. aeruginosa.*

TCSTS as drug targets not only provide a handle on essential and non-essential signal-inducible metabolic processes and the acquisition and expression of resistance genes, but also on the expression of virulence factors. Interference with TCSTS-mediated attachment of the bacteria to the host mucosa provides a unique set of opportunities for prophylactic TCSTS inhibition. Similarly, inhibitors of TCSTS-mediated invasion of the host tissue, which determine bacterial survival and propagation, would provide therapeutic benefit since they would open an alternative "retreat corridor" through which the bacterium would be forced to find another environment in which to settle.

Because only the ability to establish infection and not survival *per se* would be threatened by such agents, anti-virulence TCSTS inhibitors may perhaps mitigate the genetic back-lash of resistance mutations to lethal agents.

4.4.4. *Sub-MIC Prophylaxis: Single Agent Versus the Lethal Target*

Key to design of TCSTS agents is an understanding of what is the "lethal target" in a given bacterium. With this understanding in hand, the sensitivity of different TCSTS in a single bacterium could determine how one can position the TCSTS inhibitor. For example, designing an anti-VanRS adjunct therapy, anti-vancomycin-resistance agent to restore vancomycin susceptibility in VRE may have its non-antibiotic affect at one drug concentration and its "lethal target" inhibition in the same bacterium at a higher level. This is similar to sub-MIC-type effects of antibiotics on bacteria, well documented in the clinic and laboratory, but without appreciation of the non-killing action of such agents.

4.5. ANTI-TCSTS THERAPY AS MEETING THE UNMET MEDICAL NEED

The World Health Organization has recently warned that the increasingly drug-resistant infections in both wealthy and developing nations threaten to make once-treatable diseases incurable (Neergaard, 2000). The U.S. Department of Health and Human Services (DHHS) announced in June of 2000 a "Draft Public Health Action Plan to Combat Antimicrobial Resistance". The Action Plan represents an unprecedented blueprint for the comprehensive and coordinated efforts of federal agencies, including the Centers for Disease Control (CDC), The Food and Drug Administration (FDA), The National Institutes of Health (NIH), the Agency for Healthcare Research and Quality (AHRQ), the Department of Agriculture (USDA), the Department of Defense (DoD), the Department of Veterans Affairs (DVA), the Environmental Protection Agency (EPA) and other agencies, to address the emergence of antimicrobial resistance, focussing on the issues of surveillance, prevention and control, research and product development.

The problem of antibiotic resistance has created a need for agents with novel mechanisms of action. Ironically, TCSTS not only provide a novel group of targets, but the recent literature points to an increasing awareness of TCSTS in mediating antibacterial resistance. TCSTS inhibitors as pharmaceuticals may not only fulfill the need for antibacterials with novel mechanisms of action, but also have the potential to cripple the organism's ability to evade the antibacterial effects of other agents. As the details unfold regarding TCSTS participation in resistance and virulence, we expand our potential to combat these adaptive responses, either prophylactically with agents useful as adjunct treatment with known antibiotics, or with anti-TCSTS agents that are antimicrobial in their own right.

5. Known Inhibitors of TCSTS

Inhibitors of TCSTS from both natural and synthetic sources have been recently identified in the literature (both in Journals and as Patents). Some of these include natural product inhibitors, including unsaturated fatty acids such as oleic acid (Strauch, de Mendoza and Hoch, 1992). The fatty acids were reported as being non-competitive inhibitors of ATP-dependent autophosphorylation of the histidine protein kinase, KinA, which is involved in the regulation of sporulation in *B. subtilis* (Strauch, de Mendoza and Hoch, 1992). Strauch and co-workers reported that oleic acid inhibits the formation of KinA-phosphate (inhibition of autophosphorylation of KinA in the presence of ATP) in a concentration-dependent manner; this inhibition was found to occur regardless of whether the cognate response regulator is present (Spo0F), but it did not inhibit bacterial growth.

Genistein, an isoflavone originally isolated from a *Pseudomonas* sp., inhibits tyrosine and histidine kinases, but not serine/threonine kinases (Huang *et al*, 1992). Inhibition of a *S. cerevesiae* HK by genistein is non-competitive with respect to both ATP and protein substrate (Huang *et al.*, 1992). The complete resistance of rat microsomal histidine kinase to genistein points to the non-cognate nature of the proteins from these two sources, and the possibility for successful differential inhibition of the microbial protein (Motojima and Goto, 1994).

One of the earliest reports of TCSTS inhibition surfaced from a relatively obscure source in plant microbiology. Two natural products had been reported to inhibit the activity of a simple two-component system which allows *A. tumefaciens* to infect plants through injury or wounds and induce crown gall tumors in dicotyledonous plants (Lee *et al.*, 1995). Acetosyringone and/or other phenolic compounds produced by wounded tobacco cells have been reported to be the environmental stimuli that activate this two-component system (Winans, 1991; Lee *et al.*, 1995). Bromoacetosyringone thoroughly inhibits expression of genes regulated by the two-component system at concentrations that do not affect *A. tumefaciens* growth (McLean, 1994). Bromoacetosyringone binds irreversibly to the TCSTS and does not affect other inducible genes.

Another example of a natural product inhibitor of TCSTS is glycerol monolaurate, and/or its hydrolytic product lauric acid, which is reported to inhibit both the activation of virulence factor production (i.e. toxic shock toxin) and induction of β-lactamase in *S. aureus*, and to prevent and reverse vancomycin resistance in *E. faecalis* (Ruzin and Novick, 2000). Glycerol monolaurate had no effect, however, on erythromycin-inducible macrolide resistance in *S. aureus*, which does not involve a signal transduction pathway. The implication is that glycerol monolaurate and/or lauric acid effects these results through a non-specific mechanism of signal transduction interference, versus a target-based HK or RR inhibition.

The biological surfactant polidocanol (PDO), an amphiphilic dodecyl polyethyleneoxide ether, non-competitively "re-sensitizes" methicillin-resistant

S. aureus (MRSA) to methicillin, presumably at the level of the membrane-embedded β-lactam sensory transduction protein, BlaR1, which possesses a membrane topography similar to that of hybrid two-component proteins (Bruns, Bruns and Pelverer, 1997).

The Agr quorum-sensing TCSTS of staphylococci, which controls expression of several surface proteins and exoproteins, including many virulence factors, is stimulated by a thiolactone-containing octapeptide pheromone, the exact sequence of which varies among strains (Ji, Beavis and Novick, 1995; Novick *et al.*, 1995). As cells enter post-exponential phase, surface protein gene expression is repressed and secretory protein expression is activated by this global regulator, whose intracellular effector is a regulatory RNA, RNAIII (Novick *et al.*, 1995). Synthetic analogues of the *Staphylococcus epidermidis* pheromone have been shown to be effective inhibitors of *agr*-activated virulence factor production in *S. aureus* (Otto *et al.*, 1999).

Several synthetic compounds have been identified and published over the past decade as inhibitors of TCSTS. The initial pharmaceutical company-associated inhibitors that regulate the expression of virulence factors in bacteria pathogenic for humans had been reported for the *P. aeruginosa* two-component system that controls the alginate biosynthetic pathway (Roychoudhury *et al.*, 1993). Alginate is a virulence factor/product synthesized by *P. aeruginosa* and provides a protective exopolysaccharide layer outside the cell wall of the bacterium that plays an important role in the pathogenesis of this microorganism in cystic fibrosis patients (Roychoudhury, Sakai and Chakrabarty, 1992). The synthesis of alginate in patients' lungs is suppressed by a two-component system, AlgR2/AlgR1 (Roychoudhury *et al.*, 1993). Several compounds were identified as potent inhibitors of *alg*D promoter activity (Fig. 4).

Four compounds were initially reported by Roychoudhury *et al.* (1993). Compounds **1** and **3** inhibited the autophosphorylation of AlgR2, the autophosphatase activity of AlgR2, and the phosphotransfer of the phosphate to AlgR1, but they did not inhibit the binding of AlgR1 to the *alg*D promoter (Roychoudhury *et al.*, 1993). In contrast, compounds **2** and **4** did not inhibit AlgR2 kinase activity or the phosphorylation of AlgR1, but compounds **2** and **4** did inhibit the binding of AlgR1 to the *alg*D promoter. Noteworthy is the observation that inhibitors such as these are not antibacterial agents, but have the potential for use as adjunct therapy in combination with an antibacterial agent (capable of bacteristatic or bactericidal activity). From a therapeutic standpoint, theoretically, an inhibitor of the alginate two-component system could render the bacterium susceptible to a separate bactericidal antibacterial agent and to phagocytosis (normally inhibited by the *P. aeruginosa* capsule) by weakening the protective alginate barrier.

The halophenyl isothiazolone **1** also attenuates phosphotransfer from VanS$\sim$P to VanR by affecting the ability of VanR ability to accept the phosphate group (Ulijasz and Weisblum, 1999). Phosphorylated VanS accumulated in the presence of this inhibitor, which had no effect (in this system) on the kinase activity of the HK component. The finding that a compound can have

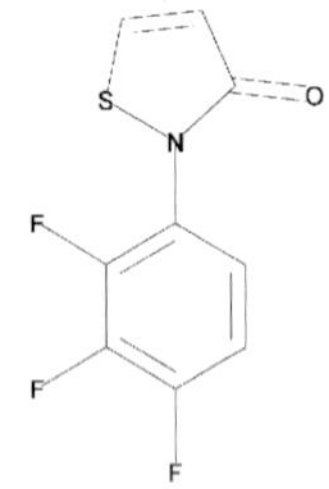

1

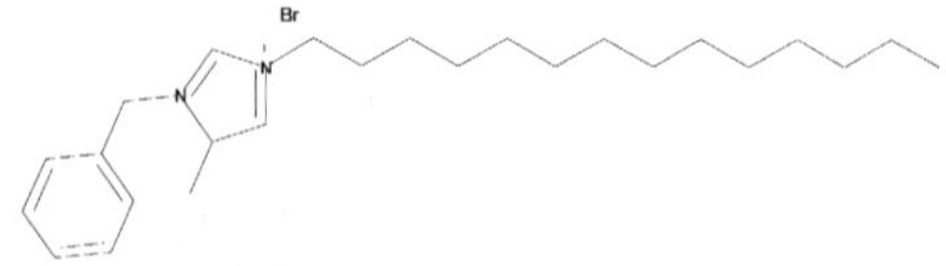

2

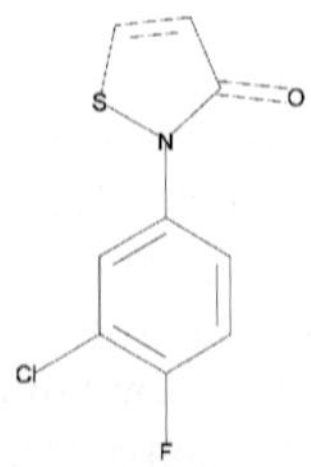

3

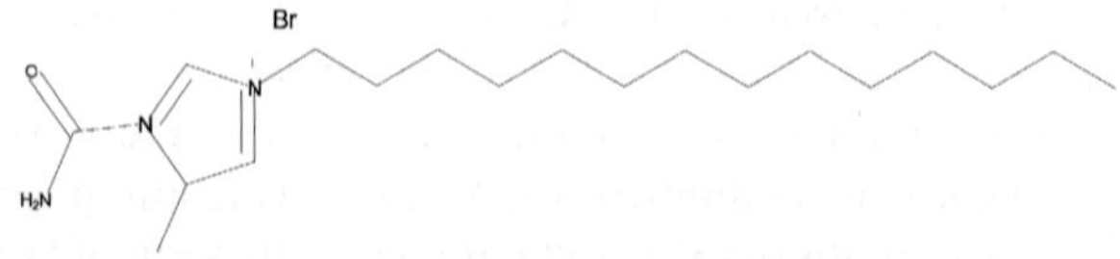

4

Figure 4. Chemical structures of the anti-alginate two-component kinase inhibitors.The autophosphorylation and autophosphatase activities of AlgR2, and phosphotransfer to AlgR1 were all inhibited by compounds 1 and 3, while inhibitors 2 and 4 did not affect these activities. Inhibitors 2 and 4 instead have their effect on the DNA-biding properties of AlgR1, inhibiting AlgD promoter activation at the level of the protein:DNA interaction.

different mechanisms of action against different TCSTS is significant because it implies a greater pharmacologic challenge to the organism in terms of resistance development. It also supports the hypothesis that it may be possible to identify inhibitors that are selective for a given TCSTS without affecting others.

Domagala *et al.* (1998) have reported the structure-activity relationship of two-component system inhibitors of gram-positive organisms for several series of diphenol-methane compounds that possess antibacterial activity (compound **5** series [e.g., **5a** and **5b**]: see Fig. 5).

These compounds were identified and initially characterized as inhibitors against the NR_{II} kinase (Domagala *et al.*, 1998). The MICs of these compounds for gram-positive bacteria (*B. subtilis, S. aureus* and *S. pyogenes*) were found to

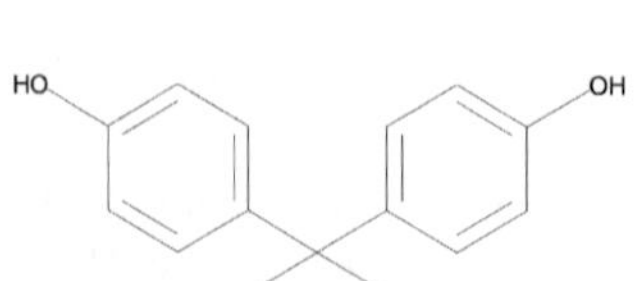

5

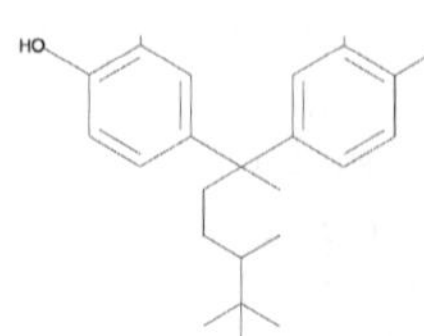

5a

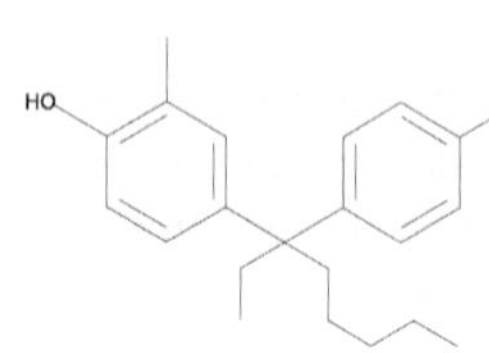

5b

Figure 5. The bis-phenolic class of TCSTS inhibitors.

range from 1 to 4 μg/ml. Likewise, the MICs for methicillin-resistant *S. aureus* were as low as 4 μg/ml; these compounds were also active at 2 μg/ml against enterococci. It was shown that the phenolic nucleus is essential for inhibitory activities (Domagala *et al.*, 1998) in those compounds. Overall, there was approximately an 85 to 90% correlation between the 50% inhibitory concentration for the inhibition of NR_{II} kinase activity and the MICs for gram-positive bacteria (Domagala *et al.*, 1998).

Researchers at the R.W. Johnson Pharmaceutical Research Institute have reported at least six chemotypes (Fig. 6) that possess two-component system inhibitory activity and antibacterial activities (Frechette *et al.*, 1997; Licata *et al.*, 1997; Macielag *et al.*, 1997; Melton *et al.*, 1997; Podlogar *et al.*, 1997; Urbanski *et al.*, 1997; Ohemeng and Nguyen, 1998; Weidner-Wells *et al.*, 1997; and Xiang, Urbanski and Chen, 1997). These compounds were mostly active against gram-positive bacteria and demonstrated a mechanism-based inhibition of two-component systems based on *in vitro* reporter gene systems. Among those highlighted, the triphenylalkyl derivatives (compound 6; Weidner-Wells *et al.*, 1997), cyclohexenes (compound 7; Xiang, Urbanski and Chen, 1997), salicylanilides (compound 8; Macielag *et al.*, 1997), benzoxazines (compound 9; Frechette *et al.*, 1997) and the benzimidazoles (compound 10; Ohemeng and Nguyen, 1998) were reported as inhibitors, with supportive data including

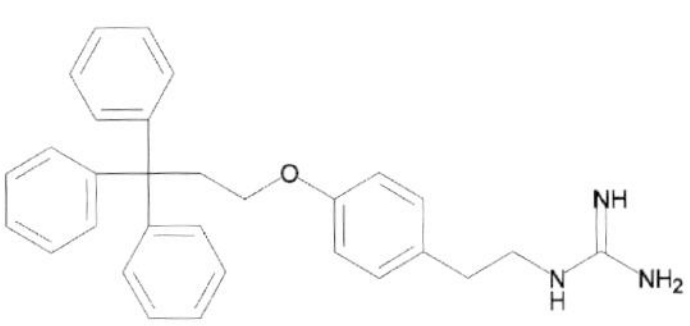

6

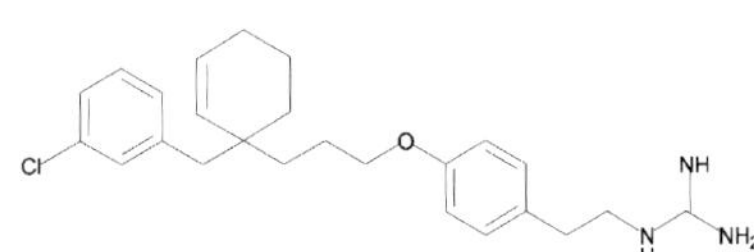

7

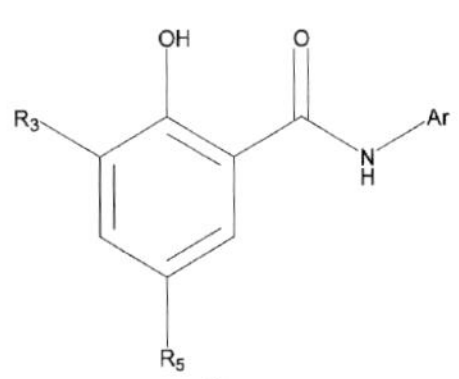

8

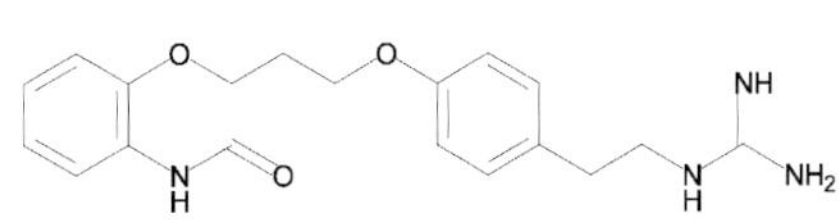

9

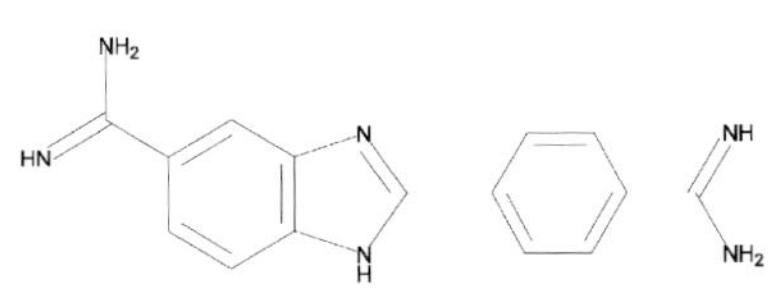

10

Figure 6. Structures of various chemical classes of TCSTS inhibitors. Inhibitor 6 represents the triphenylalkyl or trityl derivatives of hydrophobic tyramines; inhibitor 7 represents the cyclohexene class; inhibitor 8 represents the salicylanilides; inhibitor 9 represents the benzoxazines; inhibitor 10 represents the benzimidazoles.

MICs, target-based inhibition data demonstrating *in vitro* activity against KinA/Spo0F and/or NR_{II}/NR_{I} (as well as other two-component systems), kill-curves, compound ligand- binding data, *in silico* structural data (Podlogar *et al.*, 1997) and *in vivo* data (Melton *et al.*, 1997).

Compound 7, a general representative of the cyclohexene series (Urbanski *et al.*, 1997), has MICs in the range of 1 to 4 μg/ml for gram-positive bacteria and is rapidly bactericidal. The salicylanilides, based on the antihelminthic drug closantel, inhibited both KinA/Spo0F and NR_{II}/NR_{I} and possessed potent antibacterial activity *in vitro*. MICs were reported in the range of 0.5 to 4 μg/ml, with rapid bacterial killing (Macielag *et al.*, 1997). Interpretation of the susceptibility data for the salicylanilides is complicated by the recognized activity of these compounds in uncoupling oxidative phosphorylation (Hlasta *et al.*, 1998), as well as the potential for other non-specific inhibition. The benzoxazines have been shown to be potent antibacterial agents (Frechette *et al.*, 1997), but were limited in their *in vivo* activity due to serum-binding effects (Licata *et al.*, 1997; Melton *et al.*, 1997). Recently, a series of natural streptopyrrole compounds from *Streptomyces rimosus* were found to inhibit the *E. coli* nitrogen regulator HK, NR_{II}, with 9–80 μM potency and to exhibit antimicrobial activity against a range of bacteria and fungi (Trew *et al.*, 2000). The triphenylalkyl derivative series (Weidner-Wells *et al.*, 1997), closely related to the cyclohexenes, were also potent agents against gram-positive bacteria. Similarly, the benzimidazoles were reported as potent enzyme inhibitors (Ohemeng and Nguyen, 1998).

Several series were reported as having, in molecularly engineered reporter gene systems, *in vitro* "proof-of-principle" activity, indicating that they have activity against a specific two-component system in the whole bacterium (Macielag *et al.*, 1997). For example, the salicylamide series possessed activity in an *in situ* VanH::luciferase reporter assay system, a modification of the *van*R/*van*S TCSTS in *E. faecium*, which is an indication of specific two-component system inhibition at sub-MIC levels in the whole bacterium (Macielag *et al.*, 1998). Similarly, inhibitors of the type exemplified by compound **6**, triphenylalkyl derivatives of hydrophobic tyramines, show a concentration-dependent inhibition of a Taz-1 reporter, even at concentrations of drug having no effect on growth (Barrett and Hoch, 1998). Taz-1 is a modification of the EnvZ/OmpR TCSTS of *E. coli*, which functions in modifying porin expression in response to changes in osmolarity. It is important to remember that, while all of these compounds inhibit both bacterial growth and TCSTS, the results to date prevent assignment of a causal relationship between inhibition of two-component systems and growth inhibition, or the distinction between TCSTS-related inhibition and secondary mechanism(s) which may come into play in the whole cell bacterium.

Novel sources of inhibitors of TCSTS are actively being explored using combinatorial chemistry technology to quickly and easily prepare large numbers of related variants of a desired core structure (Roychoudhury *et al.*, 1998). Antibacterial inhibitors of CheA, the chemotactic HK sensor of the CheA/CheY TCSTS, were identified in libraries of N-acetylated, C-amidated D-amino acid hexapeptides, but the mechanism of killing has not been determined (Roychoudhury *et al.*, 1998).

While some degree of membrane perturbation is characteristic of most inhibitors identified to date, it is still unclear whether membrane disruption is a prerequisite to acquiring access to membrane-associated HKs, or whether loss of membrane integrity is the combined consequence of successfully jamming multiple signalling pathways. A closer examination of mode of action was undertaken for a group of HK inhibitors belonging to classes depicted in compounds **5**–**10**. Twenty-three out of the 24 compounds evaluated either significantly affected *S. aureus* membrane integrity, or caused haemolysis of equine erythrocytes. Eleven out of twelve compounds evaluated inhibited DNA, RNA and protein synthesis >80% after a short exposure to the inhibitors (Hilliard *et al.*, 1999). It is thought that this is due to a general effect on either uptake or retention of small molecule precursors, rather than specific effects on each of the biosynthetic pathways. With the exception of the salicylanilides, correlation was seen between membrane effects and bactericidal activity; the lack of such correlation for compounds related to closantel may be accounted for by multiple mechanisms of action contributing simultaneously to antibacterial activity (Hilliard *et al.*, 1999).

A simplified explanation of these results is that the aromatic and planar nature of many of the inhibitors found to date may interfere with the 4-helix bundle that forms the active unit of the HK dimer, thus interrupting phosphotransfer by physical and not chemical means (Hoch, 2000a,b). The inability to generate resistant mutants to some of these inhibitors lends further credence to this hypothesis (Domagala, 2000). But membrane disruption alone does not define a TCSTS inhibitor; among a group of 45 commercially available detergents, only 12 inhibited the KinA/Spo0F enzyme pair, with an SAR pointing to aliphatic chain length and the nature of the charged group as the specific detergent-like features which appear to be required for inhibition (Hilliard *et al.*, 1999).

The lack of success to date in identifying specific, selective antimicrobial TCSTS inhibitors that are not also general membrane perturbation agents does not invalidate these systems as targets. It is hoped that inhibitors with mechanisms of action not based on hydrophobic interactions can be identified by adjusting screening methods to accommodate our expanding grasp of the two-component signal transduction mechanism and how these systems impact the bacterial cell.

Clearly, any discovery effort must include an early evaluation of specificity. Some experts in the field feel that perhaps the best approach to finding more specific inhibitors of TCSTS might be *via* a whole cell approach, targeting either essential systems or those that fill a key biological niche, such as virulence or resistance. While *in vitro* biochemical assays permit detailed characterization of inhibition at a molecular level, they ignore the interplay of parallel cellular processes which may either also be critically affected by the inhibitors or influence the inhibitor's ability to interact with the TCSTS target, i.e. through efflux, compound inactivation or cellular compartmentalization, for example. Although mechanim of action may be difficult to assess in the whole cell, the living organism may be the most appropriate context for finding effective inhibitors of these complex communication modules, since the integration of their physiological functions with cellular processes may still be incompletely defined. "Proof of principle" for TCSTS will no doubt continue to be a challenge, as it requires a demonstration of "sufficient" evidence for TCSTS mechanism-based growth inhibition in the context of the whole cell network of TCSTS, where the potential to simultaneously affect numerous key physiological processes exists.

The on-going discovery and characterization of essential TCSTS (Fabret and Hoch, 1998; Quon *et al.*, 1998; Martin *et al.*, 1999) and of TCSTS involved in the regulation of pathogenesis and inducible or constitutive resistance, reflect the pre-eminence of TCSTS in key cellular decision processes and provide support for targeting these important physiological switches for intervention. TCSTS inhibitors have the potential to affect either multiple or select targets and species and may enjoy a prolonged effective life span as a result of their novelty and mode of action. Such qualities contribute to their appeal for targeted rational drug design.

6. Patent Position on Intellectual Property of TCSTS

An indication of the level of interest and faith that the scientific community, including both academia and industry, has in the potential value of TCSTS, lies in the volume of patent activity around these biological entities. Recent advances in sequencing technologies have uncovered a wealth of information, particularly in bacterial genomes, where one can search the entire genetic composition of several organisms to identify related proteins and utilize the homologies among them to speculate on evolutionary and physiological relationships among signal transduction pathways. The interest in protecting the intellectual property value in sequences resembling the two-component protein paradigm has evolved naturally out of the availability of new sequence data, particularly in pathogenic organisms such as *S. pneumoniae* (Wallis, 1998; Black, 1999; Wallis *et al.*, 1999; Zalacain and Brown, 1999), *S. aureus* (Wallis and Hodgson, 1997; Wallis, Shilling and Warren, 1998; Hodgson and Wallis, 1999) and *C. albicans* (Alex *et al.*, 1999; Srikantha and Soll, 1998a). Several of these patents have issued based on homologies with TCSTS proteins of known function.

As the physiological roles of the proteins encoded by these new sequences become clear, patent protection is also being sought for the related proteins involved in a particular pathway, and for methods of screening for inhibitors (Rasmussen, 1994; Selitrenikoff, 1998; Srikantha and Soll, 1998a,b; Choi, Calera and Calderone, 1999). The recent elucidation of the mechanism of antibiotic tolerance, for example, and the role of the signal transduction pathway

(VncS/VncR) that results in the synthesis and transport of the operative anti-bacterial peptide illustrates this point (Novak and Toumanen, 1999). The patent describes the utility of mutants, instrumental in the discovery of the mechanism of antibiotic-induced killing, for discovery of novel antibiotics effective against penicillin and/or vancomycin-tolerant strains. In addition, patents for genes showing a high degree of homology to the recently discovered "essential" TCSTS of *B. subtilis* (*yycF/G*), i.e. the *S. pneumoniae* homologue (Wallis *et al.*, 1999) and the *S. aureus* EspA histidine kinase of the EspA/EspB TCSTS (Benton *et al.*, 1999), describe specific screening methods for antibacterial inhibitors. Indeed, patents have even been issued describing generic high throughput *in vitro* and whole cell assays for histidine kinase activity with and without reference to a particular enzyme (Selitrennikoff, 1998; Hoch and Huang, 1997).

Although no anti-TCSTS agents have yet been brought to clinic, a number of potent kinase inhibitors have been identified and protected, some of which have shown anti-bacterial activity. They include triphenylalkyl-agents (Demers *et al.*, 1999), diaryl-agents (Kanojia *et al.*, 1997), cyclohexenyl and cyclohexyl derivatives (Barrett *et al.*, 1998b) and benzoxazines (Beach and Frechette, 1997). It is believed that non-specific hydrophobic interaction with the target, possibly within the membrane, is responsible for the inhibition detected to date (Hoch 2000a,b). A better understanding of how signals are transduced might afford a clearer approach to the problem of finding mechanism-based inhibitors that do not interfere with membrane integrity.

Healthy efforts to protect biological and chemical discoveries in the field of bacterial signal transduction provide evidence that, despite the many challenges intrinsic to these targets, efforts persist to penetrate the surface of the TCSTS barrier. As legal vehicles for the disclosure and protection of inventions intended to promote and encourage further invention, patents reflect not only the continuous and accelerating expansion of our knowledge base in the field of prokaryotic signal transduction, but also the growing appreciation for their potential exploitability as therapeutic targets. The rapid pace of discovery challenges the legal system to make appropriate and scientifically sound judgements which serve this goal.

7. Summary and Conclusion

Two-component signal transduction in prokaryotes is a biological phenomenon about which we know surprisingly little considering the number and scope of central cellular functions these systems participate in. Recent discoveries of a role in quorum sensing and in an expanding repertoire of virulence systems seem only to add fuel to the fire that has inspired the already impressive body of work developed in the relatively short time since the discovery of this mode of inter- and intra-cellular communication. Characterization of the common structural motifs, such as the ATP-binding and phosphotransferase domains of histidine kinases and the regulator and effector domains of response regulators, has allowed identification and classification of an expanding number of newly identified sequences into the "superfamily" of two-component signal transduction proteins. In many cases it has also permitted speculation about biological pathways and relationships without specific evidence of a biochemical mechanism. We can make very few (if any) general (or specific) statements about the mechanism of two-component signal transduction, including the nature of the signal in most cases, how signalling occurs, the subsequent phosphotransfer between kinase and response regulator and activation of the output domain. The information we do have, however, suggests that at least some of these events must be conserved among systems and that targeted disruption of such events would severely disable an organism's adaptive and infective capability. Alternately, identification of those unique features which render a particular system or group of systems selectively susceptible, e.g. essentiality, antibiotic resistance or virulence, could facilitate the pursuit of agents having a more limited target niche, but equal value in controlling microbial infections.

TCSTS may provide a genuine intervention opportunity like no other target class to date, able to be used an single agent antibacterial or adjunct therapy. Moreover, this class of agent may provide only the third example of antibacterial therapeutic agents which target more than one target in a given bacterium (the other examples being quinolones and certain β-lactams), thus potentially providing an advantage in suppression of resistance emergence. Clearly there is much work to be done to understand (and therefore outmanoeuvre) the microbial two-component signal transduction process, but the impressive progress made to date in this relatively young field is a credit to researchers who recognize the potential impact such an understanding would have, not only on anti-microbial drug discovery, but also on our understanding of the even larger puzzle of eukaryotic signal transduction.

References

Abdelnour, A., Arvidson, S., Bremell, T., Ryden, C., and Tarkowski, A. (1993) The accessory gene regulator (agr) controls *Staphylococcus aureus* virulence in a murine arthritis model. *Infection and Immunity* **61**, 3879–3885.

Alberts, A.S., Bouquin, N., Johnston, L.H., and Treisman, R. (1998) Analysis of RhoA-binding proteins reveals an interaction domain conserved in heterotrimeric G protein beta subunits and the yeast response regulator protein Skn7. *Journal of Biological Chemistry* **273**, 8616–8622.

Alex, L.A., and Simon, M.I. (1994) Protein histidine kinases and signal transduction in prokaryotes and eukaryotes. *Trends in Genetics* **10**, 133–138.

Alex, L.A., Borkovich, K.A., and Simon, M.I. (1996) Hyphal development in *Neurospora crassa*: involvement of a two-component histidine kinase. *Proceedings of the National Academy of Sciences, USA* **93**, 3416–3421.

Alex, L.A., Simon, B., Melvin, I., Selitrennikoff, C., and Agnan, J. (1999) Fungal histidine kinases involved in osmoregulation, the genes encoding them and their use as targets for antifungals. U.S. Patent No. 5939306.

Amemura, M., Makino, K., Shinagawa, H., and Nakata, A. (1990) Cross-talk to the phosphate regulon of *Escherichia coli* by PhoM protein: PhoM is a histidine kinase and catalyzes phosphorylation of PhoB and PhoM open reading frame 2. *Journal of Bacteriology* **172**, 6300–6307.

Appleby, J.L., Parkinson, J.S., and Bourret, R.B. (1996) Signal transduction via the multi-step phosphorelay: not necessarily a road less traveled. *Cell* **86**, 845–848.

Aravind, L., and Ponting, C.P. (1999) The cytoplasmic helical linker domain of receptor histidine kinase and methyl-accepting proteins is common to many prokaryotic signaling proteins. *FEMS Microbiology Let*ters **17776**, 111–116.

Arthur, M., Molinas, C., and Courvalin, P. (1992) The VanS-VanR two-component regulatory system controls synthesis of depsipeptide peptidoglycan precursors in *Enterococcus faecium* BM4147. *Journal of Bacteriologuy* **174**, 2582–2591.

Arthur, M., Depardieu, F., and Courvalin, P. (1999) Regulated interactions between partner and non-partner sensors and response regulators that control glycopeptide resistance gene expression in enterococci. *Microbiology* **145**, 1849–1858.

Arthur, M., Depardieu, F., Holman, T., Wu, Z., Wright, Walsh, C.T., and Courvalin, P. (1995) Regulation of glycopeptide resistance genes of enterococcal transposon Tn1546 by the VanR-VanS two-component regulatory system. In: J.A. Hoch and T.J. Silhavy (eds.) *Two-Component Signal Transduction*. ASM Press, Washington, D.C., pp. 387–400.

Arthur, M., Depardieu, F., Reynolds, P., and Courvalin, P. (1996) Quantitative analysis of the metabolism of soluble cytoplasmic peptidoglycan precursors of glyopeptide-resistant enterococci. *Molecular Microbiology* **21**, 33–44.

Artymiuk, P.J., Rice, D.W., Mitchell, E.M., and Willet, P. (1990) Structural resemblance between the families of bacterial signal-transduction proteins and of G-proteins revealed by graph theoretical techniques. *Protein Engineering* **4**, 39–43.

Aubry, L., and Firtel, R. (1999) Integration of signaling networks that regulate *Dictyostelium* differentiation. *Annual Review of Cell and Developmental Biology* **15**, 469–517.

Baikalov, I., Schroder, I., Kaczor-Grzeskowiak, M., Grzeskowiak, K., Gunsalas, R.P., and Dickerson, R.E. (1996) Structure of the *E. coli* response regulator NarL. *Biochemistry* **35**, 11053–11061.

Barrett, J.F., and Hoch, J.A. (1998) Two-component signal transduction as a target for microbial anti-infective therapy. *Antimicrobial Agents and Chemotherapy* **42**, 1529–1536.

Barrett, J.F., Goldschmidt, R.M., Lawrence, L.E., Foleno, B., Chen, R., Demers, J.P., Johnson, S., Konojia, R., Fernandez, J., Bernstein, J., Licata, L., Donetz, A., Huang, S., Hlasta, D.J., Macielag, M.J., Ohemeng, K., Freschette, R., Frosco, M.B., Klaubert, D.H., Whitekey, J.M., Wang, L., and Hoch, J.A. (1998a) Antibacterial agents that inhibit two-component signal transduction systems. *Proceedings of the National Academy of Sciences, USA* **95**, 5317–5322.

Barrett, J.F., Chen, R.H., Urbanski, M., and Xiang, M. (1998b) New cyclohexene compounds are histidine kinase inhibitors-used to treat bacterial infections by inhibiting autophosphorylation of kinase A and transphosphorylation of Spo0F. U.S. Patent No. 5,854,242.

Bassler, B.L. (1999) How bacteria talk to one another: regulation of gene expression by quorum sensing. *Current Opinion in Microbiology* **2**, 582–587.

Beach, M., and Frechette, R. (1997) New antibacterial benzoxazine and pyrido-oxazine derivatives-used to inhibit bacterial histidine kinase and are active against bacteria resistant to some antibiotics. U.S. Patent No. 5707990.

Beier, D., and Frank, R. (2000) Molecular characterization of two-component systems of *Helicobacter pylori*. *Journal of Bacteriology* **182**, 2068–2076.

Bellsolell, L., Cronet, P., Macielag, M., Serrano, L., and Coll, M. (1996) The three dimensional structure of two mutants of the
signal transduction protein CheY suggests its molecular activation mechanism. *Journal of Molecular Biology* **257**, 116–128.

Benton, B., Malouin, F., Martin, P.K., Schmid, M.B., and Sun, D. (1999) *Staphylococcus aureus* protein kinase gene *espB* and response regulator gene *espA* and methods for screening for antibacterial agents and for treating bacterial infections. PCT International Application WO 9932657.

Beppu, T. (1995) Signal transduction and secondary metabolism: prospects for controlling productivity. *Trends in Biotechnology* **13**, 264–269.

Bernish, B., and Van de Rijn, I. (1999) Characterization of a two-component system in *Streptococcus pyogenes* which is involved in regulation of hyaluronic acid production. *Journal of Biological Chemistry* **274**, 4786–4793.

Bilwes, A.M., Alex, L.A., Crane, B.R., and Simon, M.I. (1999) Structure of CheA, a signal transducing His kinase. *Cell* **96**, 131–141.

Black, M.T. (1999) A *Streptococcus pneumoniae* homolog of the lytS histidine kinase gene of *Bacillus subtilis* and the development of novel antibiotics. European Patent Application EP 900845.

Blat, Y., Gillespie, B., Bren, A., Dahlquist, F.W., and Eisenbach, M. (1998) Regulation of phosphatase activity in bacterial chemotaxis. *Journal of Molecular Biology* **284**, 1191–1199.

Blumer, C., Heeb, S., Pessi, G., and Haas, D. (1999) Global GacA-steered control of cyanide and exoprotease production in *Pseudomonas fluorescens* involves specific ribosome binding sites. *Proceedings of the National Academy of Sciences, USA* **96**, 14073–14078.

Bourret, R.B., Borkovich, K.A., and Simon, M.I. (1991) Signal transduction pathways involving protein phosphorylation in prokaryotes. *Annual Review of Biochemistry* **60**, 401–441.

Bras, A.M., Chaterjee, S., Wren, B.W., Newell, B.G., and Ketley, J.M. (1999) A novel *Campylobacter jejuni* two-component regulatory system important for temperature-dependent growth and colonization. *Journal of Bacteriology* **181**, 3298–3302.

Brown, J.M., and Firtel, R.A. (1998) Phosphorelay signalling: new tricks for an ancient pathway. *Current Biology* **8**, R662–R665.

Bruns, O., Bruns, W., and Pulverer, G. (1997) Regulation of beta-lactamase synthesis as a novel site of action for suppression of methicillin resistance in *Staphylococcus aureus*. *Zentralblatt fuer Bakteriologie* **285**, 413–430.

Bugg, T.D.H., Wright, G.D., Dutka-Malen, S., Arthur, M., Courvalin, P., and Walsh, C.T. (1991) Molecular basis for vancomycin resistance in *Enterococcus faecium* BM4147:biosynthesis of a depsipeptide peptidoglycan precursor by vancomycin resistance proteins VanH and VanA. *Biochemistry* **30**, 10408–10415.

Burbulys, D., Trach, K.A., and Hoch, J.A. (1991) Initiation of sporulation in *B. subtilis* is controlled by a multicomponent phosphorelay. *Cell* **64**, 545–552.

Calera, J.A., and Calderone, R. (1999) Histidine kinase, two-component signal transduction proteins of *Candida albicans* and the pathogenesis of candidosis. *Mycoses* **43** (Suppl. 2), 49–53.

Carballes, F., Bertrand, C., Bouche, J.P., and Cam, K. (1999) Regulation of *Escherichia coli* cell division genes ftsA and ftsZ by the two-component system rcsC-rcsB. *Molecular Microbiology* **34**, 442–450.

Chang, C., and Meyerowitz, E.M. (1994) Eukaryotes have "two-component" signal transducers. *Research in Microbiology* **145**, 481–486.

Chang, C., and Shockey, J.A. (1999) The ethylene-response pathway: signal perception to gene regulation. *Current Opinion in Plant Biology* **2**, 352–358.

Chang, H.M., Chen, M.Y., Shieh, Y.T., Bibb, M.J., and Chen, C.W. (1996) The cutRS signal transduction system of *Streptomyces lividans* represses the biosynthesis of the polyketide antibiotic actinorhodin. *Molecular Microbiology* **21**, 1075–1085.

Chen, J.M., Lee, G., Murphy, R.B., Brandt, R.P., and Pincus, M.R. (1990) Comparisons between the three-dimensional structures of the chemotaxis protein CheY and the normal Gly 12-p21 protein. *International Journal of Peptide and Protein Research* **36**, 1–6.

Chiang, R.C., Cavicchioli, R., and Gunsalus, R.P. (1997) 'Locked-on' and 'locked-off' signal transduction mutations in the periplasmic domain of the *Escherichia coli* NarQ and NarX sensors affect nitrate- and nitrite-dependent regulation by NarL and NarP. *Molecular Microbiology* **24**, 1049–1060.

Choi, G.H., Calera, J.A., and Calderone, R.A. (1999) *Candida albicans* histidine kinase and CaHK-1 gene and detection and inhibition of the enzyme. PCT International Application WO 9902700.

Cohen, S.P., McMurry, L.M., and Levy, S.B. (1988) marA locus causes decreased expression of OmpF porin in multiple-antibiotic-resistant (Mar) mutants of *Escherichia coli*. *Journal of Bacteriology* **170**, 5416–5422.

Cozzone, A.J. (1998) Post-translational modification of proteins by reversible phosphorylation in prokaryotes. *Biochimie* **80**, 43–48.

D'Agostino, I.B., and Kieber, J.J. (1999) Phosphorelay signal transduction: the emerging family of plant response regulators. *Trends in Biochemical Sciences* **24**, 452–456.

Dekkers, L.C., Bloemendaal, C.J.P., de Weger, L.A., Wijffelman, C.A., Spaink, H.P., and Lugtenberg, B.J.J. (1998) A two-component system plays an important role in the root-colonizing ability of *Pseudomonas fluorescens* strain WCS365. *Molecular Plant-Microbe Interactions* **11**, 45–56.

Demers, J.P., Johnson, S., Weidner-Wells, M.A., Kanojia, R.M., Fraga, S.A., and Klaubert, D. (1999) Preparation of triphenylalkyl antimicrobial agents. U.S. Patent No. 5874436.

Domagala, J.M., Alessi, D., Cummings, M., Gracheck, S., Huang, L., Huband, M., Johnson, G., Olson, E., Shapiro, M., Singh, R., Song, Y., Van Bogelen, R., Vo, D., and Wold, S. (1998) Bacterial two-component signalling as a therapeutic target in drug design. Inhibition of NRII by the diphenolic methanes (bisphenols). *Advances in Experimental Medicine and Biology* **456**, 269–286.

Domagala, J.M. (2000) Signal transduction as a therapeutic target. Symposium 172/A at the 100th American Society for Microbiology Annual Meeting.

Dougherty, T., and Barrett, J.F. (1999) Glycopeptide antibiotics:A historical overview and current perspectives. *Current Opinion in Anti-Infective and Investigational Drugs* **1**, 18–25.

Dubnau, D,. Hahn, J., Roggiani, M., Piazza, F., and Weinrauch, Y. (1994) Two-component regulators and genetic competence in *Bacillus subtilis. Research in Microbiology* **145**, 403–411.

Dutta, R., Quin, L., and Inouye, M. (1999) Histidine kinases: diversity of domain organization. *Molecular Microbio*logy **34**, 633–640.

Dziejman, M., and Mekalanos, J.J. (1995) Two component signal transduction and its role in the expression of virulence factors. In: J.A. Hoch and T.J. Silhavy (eds.) *Two-Component Signal Transduction*. ASM Press, Washington, D.C., pp. 305–317.

Egger, L.A., Park, H., and Inouye, M. (1997) Signal transduction via the histidyl-aspartyl phosphorelay. *Genes to Cells* **2**, 167–184.

Ernst, R.K., Guina, T., and Miller, S.I. (1999) How intracellular bacteria survive: surface modifications that promote resistance to host innate immune responses. *Journal of Infectious Diseases* **179** (Supplt. 2), S326–S330.

Fabret, C., and Hoch, J.A. (1998) A two-component signal transduction system essential for growth of *Bacillus subtilis*: Implications for anti-infective therapy. *Journal of Bacteriology* **180**, 6375–6383.

Fabret, C., Feher, V.A., and Hoch, J.A. (1999) Two-component signal transduction *in Bacillus subtilis*: how one organism sees its world. *Journal of Bacteriology* **181**, 1975–1983.

Fassler, J.S., Gray, W.M.,.Malone, C.L., Tao, W., Lin, H., and Deschenes, R.J. (1997) Activated alleles of yeast SlnI increase MCM-1-dependent reporter gene expression and diminish signalling through the Hogl osmosensing pathway. *Journal of Biochemistry* **271**, 13365–13371.

Federle, M.J., McIver, K.S., and Scott, J.R. (1999) A response regulator that represses transcription of several virulence operons in the group A streptococcus. *Journal of Bacteriology* **181**, 3649–3657.

Feher, V.A., Tzeng, Y.L., Hoch, J.A., and Cavanagh, J. (1998) Identification of communication networks in Spo0F: a model for phosphorylation-induced conformational change and implications for activation of multiple domain bacterial response regulators. *FEBS Letters* **425**, 1–6.

Feher, V.A., Cavanagh, J. (1999) Millisecond-timescale motions contribute to the function of the bacterial response regulator protein Spo0F. *Nature* **400**, 289–293.

Fields, P.I., Groisman, E.A., and Heffron, F. (1989) A Salmonella locus that controls resistance to microbicidal proteins from phagocytic cells. *Science* **243**, 1059–1062.

Fisher, S.L., Kim, S.K., Wanner, B.L., and Walsh, C.T. (1996) Kinetic comparison of the specificity of the vancomycin resistance VanS for two response regulators, VanR and PhoB. *Biochemistry* **35**, 4732–4740.

Frechette, R.F., Beach, M., Bernstein, J., Dow, T., Foleno, B., Johnson, C.E., Loeloff, M., Weidner-Wells, M.A., Werblood, H., and Barrett, J.F. (1997) Novel benzoxazine derivatives with inhibitory activity against bacterial two-component signal transduction systems. Abstr. 152, p. 44. In: Proceedings of the 214th ACS National Meeting. Abstract 152, p. 44.

Frosco, M.B., and Barrett, J.F. (1999) Bacterial two-component systems as antimicrobial drug discovery targets. *Drug News and Perspectives* **12**, 293–299.

Garcia Vescovi, E., Soncini, F.C., Groisman, E.A. (1994) The role of the PhoP/PhoQ regulon in *Salmonella* virulence. *Research in Microbiology* **45**, 473–480.

Giammarinaro, P., Sicard, M., and Gasc, A.M. (1999) Genetic and physiological studies of the CiaH-CiaR two-component signal-transducing system involved in cefotaxime resistance and competence of *Streptococcus pneumoniae. Microbiology* **145**, 1859–1869.

Gilmore, M.S., and Hoch, J.A. (1999) Antibiotic resistance. A vancomycin surprise. *Nature* **399**, 524–525 and 527.

Gong, W., Hao, B., Mansy, S.S., Gonzales, G., Gilles-Gonzales, M.A., and Chan, M.K. (1998) Structure of a biological oxygen sensor: a new mechanism for heme-driven signal transduction. *Proceedings of the National Academy of Sciences, USA* **95**, 15177–15182.

Goudreau, P.N., and Stock, A.M. (1998) Signal transduction in bacteria: molecular mechanisms of stimulus-response coupling. *Current Opinion in Microbiology* **1**, 160–169.

Grebe, T.W., and Stock, J.B. (1999) The histidine kinase superfamily. *Advances in Microbial Physiology* **41**, 139–127.

Groisman, E.A., and Heffron, F. (1995) Regulation of Salmonella virulence by two-component regulatory systems. In: J.A. Hoch and T.J. Silhavy (eds.) *Two-Component Signal Transduction.* ASM Press, Washington, D.C., pp. 319–349.

Groisman, E.A., Heffron, F., and Solomon, F. (1992) Molecular genetic analysis of the *Escherichia coli* phoP locus. *Journal of Bacteriology* **174**, 486–491.

Groisman, E.A., Chiao, E., Lipps, C.J., and Heffron, F. (1989) *Salmonella typhimurium phoP* virulence gene is a transcriptional regulator. *Proceedings of the National Academy of Sciences, USA* **86**,7077–7081.

Guenzi, E., Gasc, A.M., Sicard, M.A., and Hakenbeck, R. (1994) A two-component signal-transducing system is involved in competence and penicillin susceptibility in laboratory mutants of *Streptococcuss pneu*moniae. *Molecular Microbiology* **12**, 505–515.

Hakenbeck, R., Grebe, T., Zahner, D., and Stock, J.B. (1999) beta-lactam resistance in *Streptococcus pneumoniae*: penicillin-binding proteins and non-penicillin-binding proteins. *Molecular Microbiology* **33**, 673–678.

Haldiman, A., Fisher, S.L., Daniels, L.L., Walsh, C.T., and Wanner, B.L. (1997) Transcriptional regulation of the *Enterococcus faecium* BM4147 vancomycin resistance gene cluster by he VanR-VanS two-component regulatory system in *Escherichia coli* K-12. *Journal of Bacteriology* **179**, 5903–5913.

Halkides, C.J., McEvoy, M.M., Casper, E., Matsumura, P., Volz, K., and Dahlquist, F.W. (2000) The 1.9 A resolution crystal structure of phosphono-CheY, an analogue of the active form of the response regulator, CheY. *Biochemistry* **39**, 5280–5286.

Hardt, K., Joris, B., Lepage, S., Brasseur, R., Lampen, J.O., Frere, JM., Fink, A.L., and Ghuysen, J.M. (1997) The penicillin sensory transducer, BlaR, involved in the inducibility of beta-lactamase synthesis in *Bacillus licheniformis* is embedded in the plasma membrane via a four-alpha-helix bundle. *Molecular Microbiology* **23**, 935–944.

Harris, R.A., Popov, K.M., Zhao, Y., Kedishvili, N.Y., Shimomura, Y., and Crabb, D.W. (1995) A new family of protein kinases – the mitochondrial protein kinases. *Advances in Enzyme Regulation* **35**, 147–162.

Harris, R.A., Hawes, J.W., Popov, K.M., Zhao, Y., Shimomura, Y., Sato, J., Jaskiewicz, J., and Hurley, T.D. (1997) Studies on the regulation of the mitochondrial alpha-ketoacid dehydrogenase complexes and their kinases. *Advances in Enzyme Regulation* **37**, 271–293.

Hecht, G.B., Lane, T., Ohta, N., Sommer, J., and Newton, A. (1995) An essential single domain response regulator required for normal cell division and differentiation *in Caulobacter crescentus. EMBO Journal* **14**, 3915–3924.

Hellingwerf, K.J., Crieland, W.C., Teixeira de Mattos, M.J., Hoff, W.D., Kort, R., Verhamme, D.T., Avignone-Rossa, C. (1998) Current topics in signal transduction in bacteria. *Antonie van Leeuwenhoek* **74**, 211–227.

Hilliard, J.J., Goldschmidt, R.M., Licata, L., Baum, E.Z., and Bush, K. (1999) Multiple mechanisms of action for inhibitors of histidine kinases from bacterial two-component systems. *Antimicrobial Agents and Chemotherapy* **43**, 1693–1699.

Hlasta, D.J., Demers, J.P., Foleno, B.D., Fraga-Spano, S.A., Guan, J., Hillard, J.J., Macielag, M.J., Ohemeng, K.A., Shepard, C.N.,

Sui, Z., Webb, G.C., Weidner-Wells, M.A., Werblood, H., and Barrett, J.F. (1998.) Novel inhibitors of bacterial two-component systems with gram positive activity: pharmacophore identification based on the screening hit closantel. *Bioorganic and Medicinal Chemistry Letters* **8**, 1923–1928.

Hoch, J.A. (1993) The phosphorelay signal transduction pathway in the initiation of *Bacillus subtilis* sporulation. *Journal of Cellular Biochemistry* **51**, 55–61.

Hoch, J.A. (2000a) Two-component and phosphorelay signal transduction. *Current Opinion in Microbiology* **3**, 165–70.

Hoch, J.A. (2000b) Signal transductiona as a therapeutic target. Symposium 172/A at 100th American Society for Microbiology Annual Meeting.

Hoch, J.A., and Huang, S. (1997) Screening methods for the identification of novel antibiotics. U.S. Patent No. 5747276.

Hodgson, J.E., and Wallis, N.G. (1999) Two-component signal transduction response regulator from *Staphylococcus aureus*. U.S. Patent No. 6001600.

Horinouchi, S., and Beppu, T. (1992) Regulation of secondary metabolism and cell differentiation in Streptomyces: A-factor as a microbial hormone and the AfsR protein as a component of a two-component regulatory system. *Gene* **115**, 167–172.

Hsing, W., and Silhavy, T.J. (1997) Function of conserved histidine-243 in phosphatase activity of EnvZ, the sensor for porin osmoregulation in *Escherichia coli*. *Journal of Bacteriology* **179**, 3729–3735.

Huang, J., Nasr, M., Kim, Y., and Matthews, H.R. (1992) Genistein inhibits protein histidine kinase. *Journal of Biological Chemistry* **267**, 15511–15515.

Huycke, M.M., Sahm, D.F., and Gilmore, M.S. (1998) Multiple-drug resistant enterococci: the nature of the problem and an agenda for the future. *Emerging Infectious Diseases* **4**, 239–249.

Imamura, A., Hanaki, N., Nakamura, A., Suzuki, T., Taniguchi, M., Kiba, T., Ueguchi, C., Sugiyama, T., and Mizuno, T. (1999) Compilation and characterization of *Arabidopsis thaliana* response regulators implicated in His-Asp phosphorelay signal transduction. *Plant and Cell Physiology* **40**, 733–742.

Jacobs, C., Domian, I.J., Maddock, J.R., and Shapiro, L. (1999) Cell cycle-dependent polar localization of an essential bacterial histidine kinase that controls DNA replication and cell division. *Cell* **97**, 111–120.

Janzon L, and Arvidson S. (1990) The role of the delta-lysin gene (hld) in the regulation of virulence genes by the accessory gene regulator (agr) in *Staphylococcus aureus*. *EMBO Journal* **9**, 1391–1399.

Ji, G., Beavis, R.C., and Novick, R.P. (1995) Cell density control of staphylococcal virulence mediated by an octapeptide pheromone. *Proceedings of the national Academy of Sciences, USA* **92**, 12055–12059.

Ji, G., Beavis, R., and Novick, R.P. (1997) Bacterial interference caused by autoinducing peptide variants. *Science* **276**, 2027–2030.

Jones, C.H., Danese, P.N., Pinket, J.S., Silhavy, Y.J., and Hultgren, S.J. (1997) The chaperone assisted membrane release and folding pathway is sensed by two signal transduction systems. *EMBO Journal* **16**, 6394–6406.

Joris, B., Ledent, P., Kobayashi, T., Lampen, J.O., and Ghuysen, J.M. (1990) Expression in *Escherichia coli* of the carboxy terminal domain of the BLAR sensory-transducer protein of *Bacillus licheniformis* as a water-soluble Mr 26,000 penicillin-binding protein. *FEMS Microbiology Letters* **58**, 107–113.

Kanojia, M., Demers, J.P., Hlasta, D.J., Johnson, S.G., and Klaubert, D.H. (1997) Preparation of diaryl antimicrobial agents. U.S. Patent No. 5773469.

Kinoshita, S., and Nakamachi, Y. (2000) Detection methods for drug-resistant bacteria in routine examination-VISA. *Rinsho Byori* Supplement **111**, 69–74.

Lai, M.H., and Kirsch, D.R. (1996) Induction signals for vancomycin resistance encoded by the vanA gene cluster *Enterococcus faecium*. *Antimicrobial Agents and Chemotherapy* **40**, 1645–1648.

Laville, J., Voisard, C., Keel, C., Maurhofer, M., Defago, G., and Haas, D. (1992) Global control in *Pseudomonas fluorescens* mediating antibiotic synthesis and suppression of black root rot of tobacco. *Proceedings of the National Academy of Sciences, USA* **89**, 1562–1566.

Lee, Y.W., Jin, S., Sim, W.S., and Nester, E.W. (1995) Genetic evidence for direct sensing of phenolic compounds by the VirA protein of *Agrobacterium tumefaciens. Proceedings of the National Academy of Sciences, USA* **92**, 12245–12249.

Leonardo, M.R., and Forst, S. (1996) Re-examination of the role of the periplasmic domain of EnvZ in sensing of osmolarity signals in *Escherichia coli. Molecular Microbiology* **22**, 405–413.

Levit, M.N., Liu, Y., and Stock, J.B. (1998) Stimulus response coupling in bacterial chemotaxis: receptor dimers in signalling arrays. *Molecular Microbiology* **30**, 459–466.

Lewis, R.J., Brannigan, J.A., Muchova, K., Barak, I., and Wilkinson, A.J. (1999) Phosphorylated aspartate in the structure of a response regulator protein. *Journal of Molecular Biology* **294**, 9–15.

Li, H., and Park, J.T. (1999) The periplasmic murein peptide-binding protein MppA is a negative regulator of multiple antibiotic resistance in *Escherichia coli. Journal of Bacteriology* **181**, 4842–4847.

Li, S., Ault, A., Malone, C.L., Raitt, D., Dean, S., Johnston, L.H., Deschenes, R.J., and Fassler, J.S. (1998) The yeast histidine protein kinase, Sln1p, mediates phosphotransfer to two response regulators, Ssk1p and Skn7p. *EMBO Journal* **7**, 6952–6962.

Li, X.Z., Zhang, L., and Poole, K. (2000) Identification and characterization of a multidrug efflux system Sme ABC, regulated by a two-component regulatory system, SmeSR in *Stenotrophomonas maltiphilia.* The 100th American Society for Microbiology Meeting, Absract A33.

Licata, L., Melton, J.L., Fernandez, J.A., Frechette, R.F., Beach, M., Webb, G., Lawrence, L.E., Barrett, J.F., and Frosco, M. (1997) *In vitro* characterization of a novel class of antibacterial agents that inhibit bacterial two-component systems. In: Proceedings of the 37th Interscience Conference on Antimicrobial Agents and Chemotherapy. American Society for Microbiology, Washington, D.C. Abstract F-226, p. 87.

Lina, G., Jarraud, S., Ji, G., Greenland, T., Pedraza, A., Etienne, J., Novick, R.P., and Vandenesch, F. (1998) Transmembrane topology and histidine protein kinase activity of AgrC, the *agr* signal receptor in *Staphylococcus aureus. Molecular Microbiology* **30**, 655–662.

Lux, R., Munasinghe, V.R., Castellano, F., Lengeler, J.W., Corrie, J.E., and Khan, S. (1999) Elucidation of a PTS-carbohydrate chemotactic signal pathway in *Escherichia coli* using a time-resolved behavioral assay. *Molecular Biology of the Cell* **4**, 1133–1146.

Lyristis, M., Bryant, A.E., Sloan, J., Awad, M.M., Nisbet, I.T., Stevens, D.L., and Rood, J.I. (1994) Identification and molecular analysis of a locus that regulates extracellular toxin production in *Clostridium perfringens. Molecular Microbiology* **12**, 761–777.

Macielag, M.J., Bernstein, J., Demers, J.P., Dow, T., Foleno, B., Goldschmidt, R., Guan, J. Hilliard, J.J., Hlasta, D.J., Johnson, C.E., Johnson, S.G., Kanojia, R.M., Loeloff, M., Ohemeng, K.A., Russell, R.K., Sheppard, C.M., Fraga-Spano, S.A., Sui, Z., Weidner-Wells, M.A., Werblood, H., and Barrett, J.F. (1997) In: Proceedings of the 37th Interscience Conference on Antimicrobial Agents and Chemotherapy. American Society for Microbiology, Washington, D.C. Abstract F-228.

Macielag, M.J., Demers, J.P., Fraga-Spano, S.A., Hlasta, D.J., Johnson, S.G., Kanojia, R.M., Russel, R.K.l., Sui, Z., Weidner-Wells, M.A., Werblood, H., Foleno, B.D., Goldschmidt, R.M., Loeloff, M.J., Webb, G.C., and Barrett, J.F. (1998) Substituted salicylanilides as inhibitors of two-component regulatory systems in bacteria. *Journal of Medicinal Chemistry* **41**, 2939–2945.

Maeda, T., Wurgler-Murphy, S.M., and Saito, H. (1994) A two-component system that regulates an osmosensing MAP kinase cascade in yeast. *Nature* **369**, 242–245.

Martin, P.K., Li, T., Sun, D., Biek, D.P., and M Schmid, M. (1999) Role in cell permeability of an essential two-component system in *Staphylococcus aureus. Journal of Bacteriology* **181**, 3666–3673.

McEvoy, M.M., and Dahlquist, F.W. (1997) Phosphohistidines in bacterial signaling. *Current Opinion in Structural Biology* **7**, 793–797.

McFarlane, E.L., Kwasnicka, A., Ochs, M.M., and Hancock, R.E. (1999) PhoP-PhoQ homologues in *Pseudomonas aeruginosa* regulate expression of the outer-membrane protein OprH and polymyxin B resistance. *Molecular Microbiology* **34**, 305–316.

McLean, B.G., Greene, E.A., and Zambryski, P.C. (1994) Mutants of *Agrobacterium* VirA that activate *vir* gene express in the absence of the inducer actosyringone. *Journal of Biological Chemistry* **269**, 2645–2651.

Melton, J.L., Fernandez, J.A., Frechette, R.F., Beach, M., Licata, L., Mehta, D.B., Barrett, J.F., and Frosco, M. (1997) *In vivo* activity of a novel series of antibacterial agents that inhibit the bacterial two-component signal transduction system. In: Proceedings of the 37th Interscience Conference on Antimicrobial Agents and Chemotherapy. American Society for Microbiology, Washington, D.C. Abstract F-229, p. 87.

Mileykovskaya, E., and Dowhan, W. (1997) The Cpx two-component signal transduction pathway is activated in *Escherichia coli* mutant strains lacking phosphatidylethanolamine. *Journal of Bacteriology* **179**, 1029–1034.

Miller, V.L., DiRita, V.J., and Mekalanos, J.J. (1989) Identification of toxS, a regulatory gene whose product enhances toxR-mediated activation of the cholera toxin promoter. *Journal of Bacteriology* **171**, 1288–1293.

Mizuno, T. (1997) Compilation of all genes encoding two-compound phosphotransfer signal transducers in the genome of *Escherichia coli. DNA Research* **4**, 161–168.

Mizuno T. (1998) His-Asp phosphotransfer signal transduction. *Journal of Biochemistry***123**, 5555–5563.

Motojima, K., and Goto, S. (1994) Histidyl phosphorylation and dephosphorylation of P36 in rat liver extract. *Journal of Biological Chemistry* **269**, 9030–9037.

Msadek, T. (1999) When the going gets tough: survival strategies and environmental signaling networks in *Bacillus subtilis. Trends in Microbiology* **7**, 201–207.

Mueller, J.P., Bukusoglu, G., and Sonenshein, A.L. (1992) Transcriptional regulation of *Bacillus subtilis* glucose starvation-inducible genes: control of gsiA by the ComP-ComA signal transduction system. *Journal of Bacteriology* **174**, 4361–4373.

Mueller, J.P., Mathiopoulos, C., Slack, F.J., and Sonenshein, A.L. (1991) Identification of *Bacillus subtilis* adaptive response genes by subtractive differential hybridization. *Research in Microbiology* **142**, 805–813.

Mutoh, N., and Simon, M.I. (1986) Nucleotide sequence corresponding to five chemotaxis genes in *Escherichia coli. Journal of Bacteriology* **165**, 161–166.

Nagahashi, S., Mio, T., Ono, N., Yamada-Okabe, T., Arisawa, M., Bussey, H., and Yamada-Okabe, H. (1998) Isolation of *CaSLN1* and *CaNIK1*, the genes for osmosensing histidine kinase homologues, from the pathogenic fungus *Candida albicans. Microbiology* **144**, 425–432.

Nakano, M.M., Dailly, Y.P., Zuber, P., and Clark, D.P. (1997) Characterization of anaerobic fermentative growth of *Bacillus subtilis*: identification of fermentation end products and genes required for growth. *Journal of Bacteriology* **179**, 6749–6755.

Neergaard, L.(2000) WHO issues antibiotic alert. *AP Health Headlines*, June 12, 2000. The U.S. Department of Health and Human Services (DHHS).

Nichols, N.N., and Harwood, C.S. (2000) An aerotaxis transducer gene from *Pseudomonas putida. FEMS Microbiology Letters* **182**, 177–183.

Ninfa, A.J., and Bennett, R.L. (1991) Identification of the site of autophosphorylation of the bacterial protein kinase/phosphatase NRII. *Journal of Biological Chemistry* **266**, 6888–6893.

Ninfa, E.G., Atkinson, M.R., Kamberov, E.S., and Ninfa, A.J. (1993) Mechanism of autophosphorylation of *Escherichia coli* nitrogen regulator II (NRII or NtrB): trans-phosphorylation between subunits. *Journal of Bacteriology* **175**, 7024–7032.

Novak, R., and Toumanen, E.I. (1999) *Pneumococcus*-inhibiting peptide antibiotics, ABC transporter and two-component signal transduction system of genes of *Streptococcus* and uses thereof. PCT International Application WO 9957281.

Novak, R., Henriques, B., Charpentier, E., Normark, S., and Tuomanen, E. (1999a) Emergence of vancomycin tolerance in *Streptococcuss pneumoniae. Nature* **399**, 590–593.

Novak, R., Cauwels, A,, Charpentier, E., Tuomanen, E. (1999b) Identification of a *Streptococcus pneumoniae* gene locus encoding proteins of an ABC phosphate transporter and a two-component regulatory system. *Journal of Bacteriology* **181**, 1126–1133.

Novak, R., Charpentier, E., Braun, J.S., and Tuomanen, E. (2000) Signal transduction by a death signal peptide: uncovering the mechanism of bacterial killing by penicillin. *Molecules and Cells* **5**, 49–57.

Novick, R.P., Ross, H.F., Projan, S.J., Kornblum, J., Kreiswirth, B., and Moghazeh, S. (1993)

Synthesis of staphylococcal virulence factors is controlled by a regulatory RNA molecule. *EMBO Journal* **12**, 3967–3975.

Novick, R.P., Projan, S.J., Kornblum, J., Ross, H.F., Ji, G., Kreiswirth, B., Vandenesch, F., and Moghazeh, S. (1995) The *agr* P2 operon: an autocatalytic sensory transduction system in *Staphylococcus aureus*. *Molecular and General Genetics* **248**, 446–458.

Oda, K., and Hasunuma, K. (1994) Light signals are transduced to the phosphorylation of 15 kDa proteins in *Neurospora crassa*. *FEBS Letters* **345**, 162–166.

Ohemeng, K.A., and Nguyen, V.N. (1998) 2-substituted phenyl-benzimidazole antibacterial agents. PCT International Application WO 98-US18586 980904.

Otto, M., Sussmuth, R., Vuong, C., Jung, G., and Gotz, F. (1999) Inhibition of virulence factor expression in *Staphylococcus aureus* by the *Staphylococcus epidermidis* agr pheromone and derivatives. *FEBS Letters* **450**, 257–262.

Parkinson, J.S., and Kofoid, E.C. (1992) Communication modules in bacterial signaling proteins. *Annual Review of Genetics* **26**, 71–112.

Patel, A.H., Kornblum, J., Kreiswirth, B., Novick, R., and Foster, T.J. (1992) Regulation of the protein A-encoding gene in *Staphylococcus aureus*. *Gene* **114**, 25–34.

Perego, M., Hanstein, C., Welsh, K.M., Djavakhishvili, T., Glaser, P., and Hoch, J.A. (1994) Multiple protein-aspartate phosphatases provide a mechanism for the integration of diverse signals in the control of development in *B. subtilis*. *Cell* **79**, 1047–1055.

Perraud AL, Weiss V, and Gross R. (1999) Signalling pathways in two-component phosphorelay systems. *Trends in Microbiology* **7**, 115–120.

Pirrung, M. (1999) Histidine kinases and two-component signal transduction systems. *Chemistry and Biology* **6**, R167–R175.

Pogliano, J., Lynch, A.S., Belia, D., Liu, E.C., and Beckwith, J. (1997) Regulation of *Escherichia coli* cell envelope proteins involved in protein folding and degradation by the Cpx two-component system. *Genes and Development* **11**, 1169–1182.

Podlogar, B.L., Leo, G.C., McDonnell, P.A., Loughney, D.A., Caldwell, G.W., and Barrett, J.F. (1997) Protein structure determination using a combination of comparative modeling and NMR spectroscopy. Application to the response regulator protein Spo0F. *Journal of Medicinal Chemistry* **40**, 3453–3455.

Posas, F., Takekawa, M., and Saito, H. (1998) Signal transduction by MAP kinase cascades in budding yeast. *Current Opinion in Microbiology* **1**, 175–182.

Posas, F., Wurgler-Murphey, S.M., Maeda, T., Witten, E.A., Thai, T.C., and Saito H. (1996) Yeast HOG1 MAP kinase cascade is regulated by a multistep phosphorelay mechanism in the SLN1-YPD1-SSK1 "two-component" osmosensor. *Cell* **86**, 865–875.

Quail, P.H. (1997) The phytochromes: a biochemical mechanism of signaling in sight? *Bioessays* **19**, 571–579.

Quon, K.C., Marczynski, G.T., and Shapiro, L. (1996) Cell cycle control by an essential bacterial two-component signal transducing protein. *Cell* **84**, 83–93.

Quon, K.C., Yang, B., Domian, I.J., Shapiro, L., and Marczynski, G.T. (1998) Negative control of bacterial DNA replication by a cell cycle regulatory protein that binds at the chromosome origin. *Proceedings of the National Academy of Sciences, USA* **95**, 120–125.

Rasmussen, B.A. (1994) Cloning and identification of a two-component signal transducing regulatory system from *Bacterioides fragilis*. European Patent Application EP 612849.

Rasmussen, B.A., and Kovacs, E. (1993) Cloning and identification of a two-component signal-transducing regulatory system from *Bacteroides fragilis*. *Molecular Microbiology* **7**, 765–776.

Regassa LB, and Betley, MJ. (1992) Alkaline pH decreases expression of the accessory gene regulator (agr) in *Staphylococcus aureus*. *Journal of Bacteriology* **174**, 5095–5100.

Reizer, J., and Saier, M.H. Jr. (1997) Modular multidomain phosphoryl transfer proteins of bacteria. *Current Opinion in Structural Biology* **7**, 407–415.

Roychoudhury, S., Sakai, K., and Chakrabarty, A.M. (1992) AlgR2 is an ATP/GTP-dependent protein kinase involved in alginate synthesis by *Pseudomonas aeruginosa*. *Proceedings of the National Academy of Sciences, USA* **89**, 2695–2663.

Roychoudhury, S., Zielinski, N.A., Ninfa, A.J., Allen, N.E., Jungheim, L.N., Nicas, T.I., and

Chakrabarty, A.M. (1993) Inhibitors of two-component signal transduction systems:inhibition of alginate gene activation in *Pseudomonas aeruginosa. Proceedings of the National Academy of Sciences, USA* **90**, 965–969.

Roychoudhury, S., Blondelle, S.E., Collins, S.M., Davis, M.C., McKeever, H.D., Houghten, R.A., and Parker, C.N. (1998) Use of combinatorial library screening to identify inhibitors of a bacterial two-component signal transduction kinase. *Molecular Diversity* **4**, 173–182.

Russell, A.D., and Day, M.J. (1996) Antibiotic and biocide resistance in bacteria. *Microbios* **85**, 45–65.

Ruzin, A., and Novick, R.P. (2000) Equivalence of lauric acid and glycerol monolaurate as inhibitors of signal transduction in *Staphylococcus aureus. Journal of Bacteriologyl* **182**, 2668–2671.

Sakakibara, H., Taniguchi, M., and Sugiyama, T. (2000) His-Asp phosphorelay signaling: a communication avenue between plants and their environment. *Plant Molecular Biology* **42**, 273–278.

Schuster, S.S., Noegel, A.A., Oehme, F., Gerisch, G., and Simon, M.I. (1996) The hybrid histidine kinase DokA is part of the osmotic response system of *Dictyostelium. EMBO Journal* **15**, 3880–3889.

Selitrennikoff, C.P. (1998) Histidine kinase assay and its use for screening putative inhibitors. PCT International Application WO 9844148.

Shimizu, T., Ba-Thein, W., Tamaki, M., and Hayashi, H. (1994) The virR gene, a member of a class of two-component response regulators, regulates the production of perfringolysin O, collagenase, and hemagglutinin in *Clostridium perfringens. Journal of Bacteriology* **176**, 1616–1623.

Silva, J.C., Haldimann, A., Prahalad, M.K., Walsh, C.T., and Wanner, B.L. (1998) *In vivo* characterization of the type A and B vancomycin-resistant enterococci (VRE) VanRS two-component systems in *Escherichia coli*: a nonpathogenic model for studying the VRE signal transduction pathways. *Proceedings of the National Academy of Sciences, USA* **95**, 11951–11956.

Sola, M., Gomis-Ruth, F.X., Serrano, L., Gonzales, A., and Coll, M. (1999) Three-dimensional structure of the transcription factor PhoB receiver domain. *Journal of Molecular Biology* **285**, 675–687.

Soncini, F.C., and Groisman, E.A. (1996) Two-component regulatory systems can interact to process multiple environmental signals. *Journal of Bacteriology* **178**, 6796–6801.

Soncini, F.C., Vescovi, E.G., and Groisman, E.A. (1995) Transcriptional autoregulation of the *Salmonella typhimurium* phoPQ operon. *Journal of Bacteriology* **177**, 4364–4371.

Song, H.K., Lee, J.Y., Lee, M.G., Moon, J., Min, K., Yang, J.K., and Suh, S.W. (1999) Insights into eukaryotic multistep phosphorelay signal transduction revealed by the crystal structure of Ypd1p from *Saccharomyces cerevisiae. Journal of Molecular Biology* **293**, 753–761.

Srikantha, T., and Soll, D.R. (1998a) *Candida albicans* two-component hybrid kinase gene, *CaNik1*, and use for screening antifungal candidate compounds. PCT International Application WO 9855654.

Srikantha, T., and Soll, D.R (1998b) *Candida glabrata* two-component hybrid kinase gene, *CgSln1*, and use for screening antifungal candidate compounds. PCT International Application WO 9855655.

Stewart, R.C., VanBruggen, R., Ellefson, D.D., and Wolfe, A.J. (1998) TNP-ATP and TNP-ADP as probes of the nucleotide binding site of CheA, the protein histidine kinase in the chemotaxis signal transductipn pathway of *Escherichia coli. Biochemistry* **37**, 12269–12279.

Stewart, V. (1994) Regulation of nitrate and nitrite reductase synthesis in enterobacteria. *Antonie Van Leeuwenhoek* **66**, 37–45.

Stock, J.B. (1999) Signal transduction: Gyrating protein kinases. *Current Biology* **9**, R364–R367.

Stock, J.B., Ninfa, A.J., and Stock, A.M. (1989) Protein phosphorylation and regulation of adaptive responses in bacteria. *Microbiological Reviews* **53**, 450–490.

Stock, J.B., Surette, M.G., Levit, M., and Park, P. (1995) Two-component signal transduction systems: structure-function relationships and mechanisms of catalysis. In: J.A. Hoch and T.J. Silhavy (eds.) *Two-Component Signal Transduction.* ASM Press, Washington, D.C., pp. 25–51.

Strauch, M.A., de Mendoza, D., and Hoch, J.A. (1992) Cis-unsaturated fatty acids specifically inhibit a signal-transducing protein kinase required for initiation of sporulation in *Bacillus subtilis. Molecular Microbiology* **6**, 2909–2917.

Stretton, S., and Goodman, A.E. (1998) Carbon dioxide as a regulator of gene expression in microorganisms. *Antonie Van Leeuwenhoek* **73**, 79–85.

Stulke, J., and Hillen, W. (1998) Coupling physiology and gene regulation in bacteria: the phosphotransferase sugar uptake system delivers the signals. *Naturwissenschaften* **85**, 583–592.

Stutzman-Engwall, K.J., Otten, S.L., and Hutchinson, C.R. (1992) Regulation of secondary metabolism in *Streptomyces* spp. and overproduction of daunorubicin in *Streptomyces peucetius. Journal of Bacteriology* **174**, 144–154.

Sun, G., Sharkova, E., Chesnut, R., Birkey, S., Duggan, M.F., Sorokin, A., Pujic, P., Ehrlich, S.D., and Hulett, F.M. (1996) Regulators of aerobic and anaerobic respiration in *Bacillus subtilis. Journal of Bacteriology* **178**, 1374–1385.

Suzuki, T., Imamura, A., Ueguchi, C., and Mizuno, T. (1998) Histidine-containing phosphotransfer (HPt) signal transducers implicated in His-to-Asp phosphorelay in *Arabidopsis. Plant and Cell Physiology* **39**, 1258–1268.

Tanaka, T., Saha, S.K., Tomomuri, C., Ishima, R., Liu, D., Tong, K.I., Park, H., Dutta, R., Qin, L., Swindells, M.B., Yamazaki, T., Ono, A.M., Kainosho, M., Inouye, M., and Ikura, M. (1998) NMR structure of the histidine kinase domain of the *E. coli* osmosensor EnvZ. *Nature* **396**, 88–92.

Taylor, B.L., and Zhulin, I.B. (1998) In search of higher energy: metabolism-dependent behaviour in bacteria. *Molecular Microbiology* **28**, 683–690.

Taylor, B.L., and Zhulin, I.B. (1999) PAS domains: internal sensors of oxygen, redox potential, and light. *Microbiology and Molecular Biology Reviews* **63**, 479–506.

Thomason, P., Traynor, D., and Kay, R. (1999) Taking the plunge. Terminal differentiation in *Dictyostelium. Trends in Genetics* **15**, 15–19.

Throup, J.P., Koretke, K.K., Bryant, A.P., Ingraham, K.A., Chalker, A.F., Ge, Y., Marra, A., Wallis, N.G., Brown, J.R., Holmes, D.J., Rosenberg, M., and Burnham, M.K. (2000) A genomic analysis of two-component signal transduction in *Streptococcus pneumoniae. Molecular Microbiology* **35**, 566–576.

Tomomuri, C., Tanaka, T., Dutta, R., Park, H., Saha, S.K., Zhu, Y., Ishima, R., Liu, D., Tong, K.I., Kurokawa, H., Qian, H., Inouye, M., and Ikura, M. (1999) Solution structure of the homodimeric core domain of *Escherichia coli* histidine kinase EnvZ. *Nature Structural Biology* **6**, 729–734.

Trew, S.J., Wrigley, S.K., Pairet, L., Sohal, J., Shanu-Wilson, P., Hayes, M.A., Martin, S.M., Manohar, R.N., Chicarelli-Robinson, M.I., Kau, D.A., Byrne, C.V., Wellington, E.M., Moloney, J.M., Howard, J., Hupe, D., and Olson, E.R. (2000) Novel streptopyrroles from *Streptomyces rimosus* with bacterial protein histidine kinase inhibitory and antimicrobial activities. *Journal of Antibiotics (Tokyo)* **53**, 1–11.

Tzeng, Y.L., and Hoch, J.A. (1997) Molecular recognition in signal transduction:the interaction surfaces of the Spo0F response regulator with its cognate phosphorelay proteins revealed by alanine scanning mutagenesis. *Journal of Molecular Biology* **272**, 200–212.

Tzeng, Y.L., Feher, V.A., Cavanagh, J., Perego, M., Hoch, J.A. (1998) Characterization of interactions between a two-component response regulator, Spo0F, and its phosphatase, RapB. *Biochemistry* **37**, 16538–16545.

Ulijasz, A.T., and Weisblum, B. (1999) Dissecting VanRS signal transduction pathway with specific inhibitors. *Journal of Bacteriology* **181**, 627–631.

Ulijasz, A.T., Grenader, A., and Weisblum, B. (1996) A vancomycin-inducible LacZ reporter system in *Bacillus subtilis*: induction by antibiotics that inhibit cell wall synthesis and by lysozyme. *Journal of Bacteriology* **178**, 6305–6309.

Urao, I., Yamaguchi-Shinozaki, I., and Shinozaki, I. (2000) Two-component systems in plant signal transduction. *Trends in Plant Sciences* **5**, 67–74.

Urbanski, M.J., Xiang, M.A., Foleno, B.D., Bernstein, J.I., Lawrence, L., Goldschmidt, R.M., Barrett, J.F., Frechette, R., and Chen, R. (1997) Novel cyclohexene derivatives with histidine protein kinase inhibitory activity-pontential new antibacterial agents. In: Proceedings of the 214th ACS National Meeting. Abstract 270, p. 68.

Varughese, K.I., Madhusudan, C.E.G., Zhou, X.Z., Whiteley, J.M., and Hoch, J.A. (1998) Formation of a novel four helix bundle and molecular recognition sites by dimerization of a response regulator phosphotransferase. *Molecules and Cells* **2**, 48–493.

Vescovi, E.G., Ayala, Y.M., Di Cera, E., and Groisman, E.A. (1997) Characterization of the bacterial sensor protein PhoQ. Evidence for distinct binding sites for Mg^{2+} and Ca^{2+}. *Journal of Biological Chemistry* **272**, 1440–1443.

Volz, K. (1993) Structural conservation in the CheY superfamily. *Biochemistry* **32**, 11741–11753.

Volz, K. (1995) Structural and functional conservation in response regulators. In: J.A. Hoch and T.J. Silhavy (eds.) *Two-Component Signal Transduction*. ASM Press, Washington, D.C., pp. 53–64.

Waldburger, C.D., and Sauer, R.T. (1996) Signal detection by the PhoQ sensor-transmitter. Characterization of the sensor domain and a response-impaired mutant that identifiesligand-binding determinants. *Journal of Biological Chemistry* **271**, 26630–26636.

Wallis, N. (1998) *Streptococcus pneumoniae* homologs of the SapR gene and protein of *Streptococcus mutans*. European Patent Application EP 885963.

Wallis, N.G. (1999a) Bacterial two-component signal transduction systems as drug targets. *Current Opinion in Anti-Infective Investigational Drugs* **1**, 428–434.

Wallis, N.G. (1999b) U.S. Patent #05910472.

Wallis, N.G., and Hodgson, J.E. (1997) Novel two-component signal transduction system (TCTS) response regulator of *Staphylococcus aureus*. PCT International Application WO 9723620.

Wallis, N.G., Shilling, L.K., and Warren, R.L. (1998) Cloning, sequence, and expression of histidine kinase gene from *Staphylococcus aureus*. European Patent Application.EP 857787.

Wallis, N., Ingraham, K., Holmes, D., Zalacain, M., Throup, J., Biswas, S., and Ge, Y. (1999) Nucleic acid encoding *Streptococcus pneumoniae* response regulator. European Patent Application EP 900846.

Wanner, B.L. (1992) Is cross-regulation by phosphorylation of two-component response regulator proteins important in bacteria? *Journal of Bacteriology* **174**, 2053–2058.

Wanner, B.L., Wilmes, M.R., and Young, D.C. (1988) Control of bacterial alkaline phosphatase synthesis and variation in an *Escherichia coli* K-12 phoR mutant by adenyl cyclase, the cyclic AMP receptor protein, and the phoM operon. *Journal of Bacteriology* **170**, 1092–1102.

Weidner-Wells, M.A., Bernstein, J., Chen, R.H.K., Demers, J.P., Foleno, B., Fraga-Spano, S.A., Hlasta, D.J., Johnson, S.G., Kanojia, R.M., Klaubert, D.H., Sheppard, C.M., and Barrett, J.F. (1997) Triphenylalkyl derivatives as inhibitors of bacterial two-component regulatory systems. In: Proceedings of the 214th ACS National Meeting. Abstract 147, p. 43.

Welch, M., Chinardet, N., Mourey, L., Birck, C., and Samama, J.P. (1998) Structure of the CheY-binding domain of histidine kinase CheA in complex with CheY. *Nature Structural Biology* **5**, 25–29.

Wheeler, R.T., and Shapiro, L. (1999) Differential localization of two histidine kinases controlling bacterial cell differentiation. *Molecules and Cells* **4**, 683–694.

Williams, J.B. (1996) Drug efflux as a mechanism of resistance. *British Journal of Biomedical Sciences* **53**, 290–293.

Williams, S.B., and Stewart, V. (1997) Discrimination between structurally related ligands nitrate and nitrite controls autokinase activity of the NarX transmembrane signal transducer of *Escherichia coli* K-12. *Molecular Microbiology* **26**, 911–925.

Williams, S.B., and Stewart, V. (1999) Functional similarities among two-component sensors and methyl-accepting chemotaxis proteins suggest a role for the linker region amphipathic helices in transmembrane signal transduction. *Molecular Microbiology* **33**, 1093–1102.

Winans, S.C. (1991) An *Agrobacterium* two-component regulatory system for the detection of chemicals released from plant wounds. *Molecular Microbiology* **5**, 2345–2350.

Winans, S.C., Mantis, N.J., Chen, C.Y., Chang, C.H., and Han, D.C. (1994) Host recognition by the VirA, VirG two-component regulatory proteins of *agrobacterium tumefaciens*. *Research in Microbiology* **45**, 461–473.Wong, S.M., Carroll, P.A., Rahme, L.G., Ausubel, F.M., and Calderwood, S.B. (1998) Modulation of expression of the ToxR regulon in *Vibrio cholerae* by a member of the two-component family of response regulators. *Infection and Immunity* **66**, 5854–5861.

Wright, G.R., Holman, T.R., and Walsh, C.T. (1993) Purification and characterization of VanR and the cytosolic domain of VanS: a two-component regulatory system required for vancomycin resistance in *Enterococcus faecium* BM4147. *Biochemistry* **32**, 5057–5063.

Wurgler-Murphy, S.M., and Saito, H. (1997) Two-component signal transducers and MAPK cascades. *Trends in Biochemical Sciences* **22**, 172–176.

Xiang, M.A., Urbanski, M.J., and Chen, R.H.K. (1997) Synthesis of novel cyclohexene derivatives and potential antibacterial Agents. In: Proceedings of the 214th ACS National Meeting. Abstract 153, p. 45.

Xu, Q., Nguyen, V., and West, A.H. (1999) Purification, crystallization and preliminary X-ray diffraction analysis of the yeast phosphorelay protein YPD1. *Acta Crystallographica. Section D. Biological Crystallography* **55**, 291–293.

Yamada-Okabe, T., Mio, T., Ono, N., Kashima, Y., Matsui, M., Arisawa, M., and Yamada-Okabe, H. (1999) Roles of three histidine kinase genes in hyphal development and virulence of the pathogenic fungus *Candida albicans*. *Journal of Bacteriology* **181**, 7243–7247.

Yang, Y., and Inouye, M. (1991) Intermolecular complementation between two defective mutant signal-transducing receptors of *Escherichia coli*. *Proceedings of the National Academy of Sciences, USA* **88**, 11057–11061.

Yeh, K.C., and Lagarias, J.C. (1998) Eukaryotic phytochromes: light-regulated serine/threonine protein kinases with histidine kinase ancestry. *Proceedings of the National Academy of Sciences, USA* **95**, 13976–13981.

Yu, G., Deschenes, R.J., and Fassler, J.S. (1995) The essential transcription factor, Mcm1, is a downstream target of Sln1, a yeast "two-component" regulator. *Journal of Biological Chemistry* **270**, 8739–8743.

Zahrt, T.C., and Deretic, V. (2000) An essential two-component signal transduction system im *Mycobacterium tuberculosis*. The 100th American Society for Microbiology Annual Meeting. Abstract U-16.

Zalacain, M., and Brown, J.R. (1999) A *Streptococcus pneumoniae* homologue of the phoR histidine kinase gene of *Bacillus subtilis* and development of novel antibiotics. European Patent Application EP 913478.

Zapf, J.W., Hoch, J.A., and Whiteley, J.M. (1996 A phosphotransferase activity of the *Bacillus subtilis* sporulation protein Spo0F that employs phosphoramidate substrates. *Biochemistry* **35**, 2926–2933.

Zapf, J., Madhusudan, C.E.G., Hoch, J.A., Varughese, K.I., and Whiteley, J.M. (1998) A source of response regulator autophophatase activity: the critical role of a reidue adjacent to the Spo0F autophosphoylation active site. *Biochemistry* **37**, 7725–7732.

Zhang, C.C., Gonzalez, L., and Phalip, V. (1998) Survey, analysis and genetic organization of genes encoding eukaryotic-like signaling proteins on a cyanobacterial genome. *Nucleic Acids Research* **26**, 3619–3625.

Zhang, J.P., and Normark, S. (1996) Induction of gene expression in *Escherichia coli* after pilus-mediated adherence. *Science* **273**, 1234–1236.

INDEX

a cells 4
α cells 4
a loci 14, 15
a-factor GPCR Ste2p 4
A-factor signalling 44–5, 53–4
ACE enzymes 317
acetate 80–1
acetosyringone 382
N-acetylation 263–4
acetylcholine 145, 216, 252–7
ACTH *see* adrenocorticotrophic hormone
Activating Transcription Factor *see* Atf1
acyl HSLs 32, 40, 51–2
acylated homoserine lactones 45–52
adenylyl/ate cyclase 2, 7, 14, 81, 84, 154–5, 170
adjunct therapy 379–81
adrenocorticotrophic hormone (ACTH) 317, 318
AF *see Ascaris suum*, neuropeptides
agr locus 375
Agr quorum-sensing TCSTS 383
Agrobacter tumefaciens 382
AI-2-type factors 52–3
aldicarb resistance 214, 216, 217
alginate 383, 384
allogeneic competitors 104–8
alloinduction 30–1
allorecognition 105–8
allorejection 105–8
alpha subunits 196–8, 203–7, 209–13
amidation of neuropeptides 126–9
amino acid sequences
  annelid excitatory peptide precursor 316
  Anthopleura elegantissima neuropeptide preprohormone 123, 124
  ciliate pheromones 78–9
  leech proenkephalins 313–14
  nematode neuropeptides 277–9
  neuropeptides 128
aminoglycoside resistance 373
ammonium ions 80
amoscanate 180
anatomy of nematodes 244–7
angiotensin II-immunoreactive cells 306
angiotensin-like peptides 305–11
anion channels 333, 337–9
annelids 303–20
anthelmintic drugs 179–80, 268
Antho-RFamide 125, 126
Anthopleura elegantissima 107, 115–35
anthozoans 115–35
anti-adaptive response intervention 380–1
antibiotics 370–4, 381
anticoagulants 320
APEASPFIRFamide 290
aprotinin 319
arachidonic acid 95–6, 159–60
arylaminopyridazines 271
Ascaris
  A. galli, neuropeptides 284
  A. lumbricoides 244, 247, 249, 251, 263, 275
  A. suum
    APEASPFIRFamide 290
    benzoquinonium 253, 255
    forskolin 285
    GABA 269, 270
    5-hydroxytryptamine 258–61
    KPNFIRFamide 288–9
    KPSFVRFamide 289–90
    KSAYMFRamide 286–8
    KYSALMFamide 273
    muscle cell, pilocarpine 254
    nervous system 244–6
    neuromuscular innervation 248, 249, 250
    neuropeptides 277, 279–86
    PF-1 effects 287
    philanthotoxin 256
    potassium channel blockers 288
  dopamine/noradrenaline levels 263

asexual sporulation 10
Atf1 (Activating Transcription
  Factor 1) 8
autoinduction 30–1, 39
avermectins 265–7
AVPGVLRFamide 282–4
avr-15 gene 266–7
axis specification 94–5
axon guidance 223
azole derivatives 270–1

b loci 14, 15
Bacillus subtilis 35, 36–9, 362, 376–7
bacteria *see* prokaryotes
bacteriocin 40–1
behaviour control 195–231
benzimidazoles 385, 386
benzoquinonium (BNZQ) 253, 255
benzoxamines 385, 386
beta lactams 370
beta subunits 196–7
biogenic amines 263
bioluminescence 108
Biomphalaria 148
biosynthesis 120–6
biotin 80–1
bis-phenols 384
Blepharisma 77
BNZQ *see* benzoquinonium
bromoacetosyringone 382
Brugia pahangi 273
gamma-butyrolactones 53–5

C-signalling 60–1
Caenorhabditis elegans 195–231
  dopamine 263
  electropharyngeogram 251
  glutamic acid 265–6
  heterotrimeric G proteins 195–222
  5-hydroxytryptamine 260–2
  neuropeptides 277, 278–9, 282
  non-FaRPs 290–1
  small GTP-binding proteins 222–9
calcitonin-gene-related peptide
  (CGRP) 273
calcium-ATPase activity 80–1
Calliactis parasitica 126–9, 130
calmodulin 151, 162–3
cAMP 80–1, 101, 150, 155–6
Candida albicans 368
carbohydrate metabolism 172–3
β-catenin 101–2
Caulobacter crescentus 351, 363, 376
Cdc24p membrane localized protein 5, 7
Cdc42 5, 6, 222–3
cDNA sequences 125
cell division 198
cell membrane 79–80, 338–9, 387
cell surface receptors 149–50
cell-to-cell signalling 27–63
cellular distribution 363
CePAK 227–8
cercaria larvae 149
CGRP *see* calcitonin-gene-related peptide
chelerythrine 286
chemotaxis 80–2, 198, 210–13, 221, 353
CheY family of response regulators 360
Chlamydomonas 325
chloride ions 269–70, 333
ciliates 73–86
citrulline 274, 275
classical transmitters 252–72
cma genes 58
cnidarians 91–109, 115–35
coexpression 129–34
competence-inducing peptide
  pheromones 36–9
ComX 36, 37
concentration gradients 326
conformational changes 360, 361
contact-requiring prokaryotic cell-to-cell
  signalling 59–62
corals 115–35
CP2 gene codes 129
cross-talk in TCSTS 365–6
Cryptococcus neoformans 10–12
CTX 83–4
cyanobacteria 41
cyclohexenes 385, 386
cysticercosis 142
cytokines 78, 165
cytoplasmic pheromone receptors 33

dauer larvae 199, 211–12, 221
dbcAMP 286
DE-2 motoneurone 268
dec-4 gene 205, 218–19
defaecation cycles 219

dephosphorylation 361
dgk-1/sag-1 genes 213–20
DHp *see* dimerization histidine transphosphorylation domain
diacylglycerol (DAG) 100, 159–60, 216, 217
diacylglycerol kinase 214
diBrcGMP 286
Dictyostelium 83, 325, 332
differentiation 93
diffusible extracellular factors 56–7
see also pheromones
Dig1p/Dig2p proteins 7
dihydroinvermectin 271, 272
Dileptus anser 77
dimerization histidine transphosphorylation (DHp) domain 356, 358, 359
dimethylphenylpiperazinium (DMPP) 253
Diphyllobothrium dendriticum 142
Dirofilaria immitis 273
diversity 339–41
DivJ kinase 351, 363
DMPP *see* dimethylphenylpiperazinium
DNA 125, 149, 362–3
dopamine 174, 262–5
dorsal nerve cord 244, 245
downstream events 104
Drosophila melanogaster 126, 128, 134, 223, 330
drug resistance 370–4
drug targets 369–88
DSF diffusible extracellular factor 56–7
Dugesia tigrina 143, 144

eat-16 gene 213–20
Echinococcus granulosus 142
ecological role 107
effector molecules 82–5
EGF *see* epidermal growth factor
egg-laying
C. elegans 217–18
egl-8 gene 205, 214, 215, 217
egl-10 gene 198, 199, 201–3
egl-19 gene 218
egl-30 gene 198, 203–5, 213–20
dopamine in nematodes 264
eicosanoids 95–6, 180
Eirene viridula 97, 99
Eisenia foetida 310, 311, 312
electropharyngeograms (EPGs) 251
electrophysiological studies 247–52, 271, 312–13, 325–41
endocrine systems 142
see also neuroendocrine systems
β-endorphin-like immunoreactivity 75–7
endosporulation 36–9
Enterococcus faecalis 41–2, 353, 366
enzyme-linked receptors 149, 150, 151–2, 163–6
enzymes 121, 316–19
EPGs *see* electropharyngeograms
epidermal growth factor (EGF) 144
Erpobdella octoculata 307–8, 309, 310
Escherichia coli 351, 352, 353, 359
MscL 328, 329, 331, 335–7
'essential' TCSTS genes 369–70, 374–7, 379
Eudendrium racemosum 97
eukaryotes 51, 367–9
Euplotes raikovi 77–80
evolution 78–9, 106–7, 115, 303, 339–41, 363
excitable tissues 243–92
excitatory peptide precursor sequences 316
exocytosis 228–9
exopolysaccharide synthesis regulation 55–6

Far1p 7
FaRPs *see* FMRFamide-related peptides
Fasciola hepatica 142, 153–63
fibrils 62
fibroblast growth factor (FGF) 144
filamentation response elements (FREs) 6
'filters', ion channels 328
fissioning *see* scissiparity
flatworms *see* platyhelminths
flp genes 219–20, 276, 278–9, 284, 289
fluorescence studies 262
FMRFamide-related peptides (FaRPs) 144, 178, 220, 276–91
folate 80–1
foot formation 92, 93, 96–7
forskolin 285, 286

free-living flatworms 143–4
FREs *see* filamentation response elements
fruiting body formation 44–5, 59–60
fungi 1–19
furanones 51
FUS3 6, 7
Fuz7 MAPKK 16

G proteins
  α subunits 1–2, 7, 13, 15
  C. elegans 195–231
  family 369
  fungal pheromone response 1–19
  heterotrimeric 1–2, 80–5, 195–222
  linked receptors 149, 150–1
  mediated pathways 153–63
  protozoa 83
  signalling pathways 213–20
GABA *see* gamma-aminobutyric acid
gamma subunits 196–7
gamma-aminobutyric acid (GABA) 216, 267, 269–72, 289, 291
gamma-butyrolactones 53–5
'gamones' *see* pheromones
ganglia 244–5
GAP 164
gated pores 326–9
GbG moiety 4, 5, 6
GDVPGVLRFamide 282–4
GEF *see* Guanine nucleotide Exchange Factor
gene expression 48–9, 195–231, 347–90
genistein 382
genome sequencing 326
genome size 347
GFP *see* green fluorescent protein
GFP reporter gene 209, 210
Gierer-Meinhardt model 93
gliding motility 61–2
glucose 7, 8
glutamate 80, 145, 177–8, 265–9
glutamine signalling 57
GLW-amides 103
glycerol monolaureate 382
glycogen-synthase kinase (GSK3) system 101
glycoprotein hormones 134
gna-1/2/3 12–13
goa-1 gene 199–201, 213–20
gpa genes 4, 7, 210–12, 215
gpb genes 11, 208–9
Gram-negative bacteria 29, 31–2, 46–7
Gram-positive bacteria 31, 39
granule-containing sensory-motor-interneurons 133
green fluorescent protein (GFP) 108
growth factors 144
gsa-1 gene 198, 205–9
GSK3 *see* glycogen-synthase kinase
GTP 82, 83–5
guanine nucleotide exchange factors 5, 223–5
guanylyl cyclase 84
guinea pig proencephalins 313
GYRIFamide 161, 178

HA *see* head activator peptide
Haemadipsa sylvestris 320
Haemonchus contortus 267, 272, 277
haploid fruiting 10
head activator (HA) peptide 92, 93–4, 95
head ganglia 244–5
head inhibitors 96
Heady gene 94–5
heart surgery 319
helix-turn-helix motif (HTH) 48
12-R-HETE 95–6
heterocyst formation 41
heterotrimeric G proteins 1–2, 83–5, 195–222
higher eukaryotes 368–9
Hirudinaria manillensis 320
Hirudo 309, 311, 313, 315, 320
histidine kinase (HK) 348, 355–9, 362, 368, 377–9
homologies 197, 198, 203–9
hormone-like molecules 145–6
hormones 166–9, 317, 318
  see also endocrine systems
HPLC purification 120
HPPT *see* m-hydroxyphenylpropyl-trimethylammonium
HSN ablation 201
5-HT *see* 5-hydroxytryptamine
HTH *see* helix-turn-helix motif
Hydra
  evolutionary aspects 115–16
  life cycle control 102–4

neuropeptides structures 116–20
periodic patterns 99
preprohormones 131, 132
signalling systems 91–109
'hydra peptide project' 96
Hydra-KVamides 118, 120
Hydra-LWamides 118, 120
Hydra-RFamides 129, 131
Hydractinia 105–8
H. echinata 93, 97, 98, 99, 120
metamorphosis 102, 103, 104
hydric balance 312–13
3-hydroxypalmitic acid methyl ester (PAME) 55
m-hydroxyphenylpropyltrimethyl-ammonium (HPPT) 253
5-hydroxytryptamine (5-HT, serotonin)
C. elegans 200–1, 203, 215, 216, 217–18, 260–2
nematodes 258–62
platyhelminths 143, 145, 154–5, 170–3, 176
hydrozoans 115–20, 129, 131, 132
Hym-330 97
Hymenolepis diminuta 145, 158, 162, 173

immune response 51, 104–8, 304, 317, 318, 319
IMP chemorepellent 81
in situ hybridization 132, 133
inhibitors 96, 382–8
inositol phospholipid pathway 150–1
inositol trisphosphate (IP_3) 158, 204, 205, 218–19
intellectual property aspects 388–9
interactions analysis 213–20
intercellular signalling 27–63
intracellular targets 34–5
ion channels 149, 151, 169–70, 325–41
itr-1 gene 205, 218–19
ivermectin 265–6, 267, 271–2

Janus kinase (JaK) cytoplasmic tyrosine

kinases 152
KcsA channel molecule 328, 329, 330
KHEYLRFamide 279–82
KinA/KinB histidine protein kinases 362, 382
KNEFIRFamide 279–82
KPNFIRFamide 288–9
Kpp2 MAPK 16–17
KPSFVRFamide 289–90
KSAYMFRamide 286–8
KSS1 6
KYSALMFamide 273

lactate 80–1
lauric acid 382
Leech OsmoRegulator Factor (LORF) 315–16
leeches 305–12, 313–14, 317, 318, 320
let genes 219, 222, 225–7
levamisole 179
lfe-1 gene 205, 218–19
life cycle 102–4
ligand-gated ion channels 327
light-emitting substances 108
lin-3 gene 219
lipid modification 4
LORF *see* Leech OsmoRegulator Factor
lower eukaryotes 367–8
Lumbricus terrestris 312
luminescence regulation 45–6
LuxR family of proteins 47, 48–9, 51–2
Lymnaea stagnalis 126
lysozyme 82

M-factor 9
magA/B/C Ga subunits 13
Magnaporthe grisea 13–14
mammals 134, 197, 198, 203–9
MAP kinases 164
MAPK cascade 1–19
MAPK kinase kinase (MAPKKK) 2–3, 5, 6
marine cnidarians 108
marine vibrios 45–6
mat locus 4, 9, 13
mating 1–19, 79–80, 119
'mating type substances' *see* pheromones
mechanosensitive channel of large conductance *see* MscL
mechanosensitive (MS) channels 327, 333–7
mel-11 gene 225–7
membrane (cell) 79–80, 338–9, 387
metamorphosin (MMA) 120, 123–4

metamorphosis 102–4
metazoa 115–34, 141–81
methicillin-resistant *Staphylococcus aureus* (MRSA) 372, 385
methyl donors 99
N-methyl-scopolamine 257
microbes *see* prokaryotes
microbial biofilms 49–50
mig-2 gene 199
mitochondrial protein kinases 369
mitogen-activated kinases (MAPK) 85
Mitrocomella polydiademata 104
MMA *see* metamorphosin
molecular structure 271, 328, 329, 335
monoamines 143, 145, 173–7
monotherapy 379
morphine 76
morphogens 91–102
motility 171–2, 177–8
motoneurones 244, 245, 246, 251–2, 268, 270, 271
mRNA 149
MRSA *see* methicillin-resistant *Staphylococcus aureus*
MS *see* mechanosensitive channels
MscL channel molecule 328, 329, 331, 334, 335–7
MscS channel molecule 329, 336
MT A-1/2/3 12
multiple drug resistance 373
multiplicity/redundancy 363–5
muscimol 270
muscle system 244–52
Mycobacterium tuberculosis 328, 335, 336
myosin phosphatase regulatory subunits 225–7
myristoylation 4
myxobacteria 42–5, 59–61
Myxococcus xanthus 43–5, 59–61

N-acetylation 263–4
NADPH diaphorase 272–3, 274
naloxone-reversible effects 76
nematocytes 106
nematodes 195–231, 243–92
NEP *see* neutral endopeptidase
Nereis 310, 311, 312
nervous systems 115, 116, 142, 143, 146, 244–7
neuroendocrine systems 303–20
neurohormones 93, 103, 134
neuromuscular systems 178–80, 198, 244–6, 248–50, 268
neuropeptide F (NPF) 144
neuropeptides 115–35, 146–7, 220, 275–91, 303–20
Neurospora crassa 12–13, 325, 332
neurotransmission 143–7, 214–20, 243–92
neutral endopeptidase 24.11 (NEP) 317
nicotinic receptors 253–7
Nik1 protein 368
nitric oxide 145, 272–5
nitrogen 7, 8, 10
NLP-1/2 neuropeptides 291
non-classical opioid peptides 315
'non-essential' TCSTS 379, 380
non-FaRPs 290–1
noradrenaline 262–5
Notoplana acticola 143
NPY receptor 221–2

octopamine 262–5
ODR proteins 212–13
olfaction 199
oligopeptide signal molecules 30–42
opioids 74–7, 313–15
osmolarity changes 367–8
oxytocin/vasopressin peptides 311–12

P-factor 9
p21 Activated Kinase (PAK) family 5, 227–8
PAF-E.rac (proportion-altering factor) 97–8
PAK *see* p21-activated kinase
PAL *see* peptidyl-α-hydroxyglycine α-amidating lyase
palmitoylation 4
PAME *see* 3-hydroxypalmitic acid methyl ester
Panagrellus redivivus 277
Paramecium spp. 77, 80–1, 84, 325, 332
parasitic flatworms 145–7
patch clamps 325, 326, 333–4
patents, TCSTS 388–9
pathogenicity 374–6

Pathogenicity MAP Kinase 1 *see PMK1*
pattern formation 91–102
PBPs *see* penicillin-binding proteins
PCs *see* proprotein convertases
PDO *see* polidocanol
pedin 97
penicillin-binding proteins (PBPs) 372
peptide pheromones 33–42
peptidergic nervous system 146
peptides 96–7, 178
peptidyl-α-hydroxyglycine α-amidating lyase (PAL) 121–2, 124, 126, 129, 130
peptidyl-glycine α-hydroxylating monooxygenase (PHM) 121–2, 124, 126, 127–30
Perinereis vancaurica 310
periodic patterns 99
periplasmic binding sites 356–9
PF neuropeptides 284–9
PGF *see* plerocercoid factor
pharmacological studies 252, 253–7, 261, 264–75, 280, 285–6
pharyngeal pumping 198, 200, 203, 264, 267, 287
pheromone response elements (PREs) 6
pheromones
  amino acid sequences 78–9
  Cryptococcus neoformans 10–12
  definition 28–9
  Euplotes 77–80
  fungal 1–19
  Hydra 102
  Magnaporthe grisea 13–14
  Neurospora crassa 12–13
  prokaryotic 27–9, 33–42
  receptors 79, 80–1
  Saccharomyces cerevisiae 3–7
  Schizosaccharomyces pombe 3, 7–10
  signalling mechanism summary 32
  Ustilago maydis 14–17
philanthotoxin 256
PHM *see* peptidyl-glycine α-hydroxylating monooxygenase
PhoP/PhoR TCSTS 375–6
phorbol esters 161
phosphatidylinositol-PKC system 100–1
phosphodiesterase 157
phosphokinase A (PKA) 8, 16
phospholipase C (PLC) 157–8
phospholipase Cβ (PLCβ) 203–5
phosphorelays 361–3
phosphorylation 348
phosphotransferase system (PTS) 348, 353–4
photo-sensory kinases 368
pilocarpine 254–5
PKA *see* phosphokinase A
plant pathogens 55–7
plant phytochromes 368
platyhelminths
  adenylate cyclase 154–5, 170
  arachidonic acid 159–60
  calmodulin 151, 162–3
  cAMP 155–6
  carbohydrate metabolism 172–3
  diacylglycerol 159–60
  enzyme-linked receptor pathways 149, 150, 151–2, 163–6
  evidence for signalling systems 153–70
  G protein-mediated pathways 153–63
  ion channels 149, 151, 169–70
  monoamines and regeneration 143, 145, 173–7
  motility 171–2, 177–8
  nervous systems 142, 143, 146
  neuromuscular system 178–80
  phosphodiesterase 157
  phospholipase C 157–8
  protein kinase C 151, 160–1, 177
  protein kinase A 150, 156–7, 176
  receptor serine/threonine kinases 152, 165–6
  receptor tyrosine kinase 151–2, 163–4
  retinoids 166–9
  rhodopsin 163
  serotonin receptors 170–1
  signalling mechanisms 141–81
  steroid hormones 166–9
  thyroid hormones 166–9
  transmitter substances 143–7
  tyrosine kinase-associated receptors 164–5
PLC *see* phospholipase C
plerocercoid factor (PGF) 146
PMK1 14
polarized cell growth 5, 6, 7
polidocanol (PDO) 382–3

polymyxin B resistance 373
POMC *see* pro-opiomelanocortin
population density 30–1
potassium channels 288, 329, 330–2, 335, 336, 337–9, 340
preprohormones 118, 121–3, 129–34
PREs *see* pheromone response elements
prf1 pheromone response factor 15
primary structure 79
pro-Dyn *see* prodynorphin
pro-Enk *see* proenkephalin
pro-opiomelanocortin (POMC) 314, 318
prodynorphin (pro-Dyn) 314
proenkephalin (pro-Enk) 313–14
prohormone convertases 121
prokaryotes
  acylated homoserine lactones 45–52
  AI-2-type factors 52–3
  autoinduction 30–1
  bacteriocin 40–1
  contact-requiring 59–62
  drug resistance 370–4
  Enterococcus faecalis 41–2
  genome size 347
  glutamine signalling 57
  intercellular signalling 27–63
  intracellular targets 34–5
  ion channels 325–41
  myxobacteria 42–5
  pheromones 27–9, 33–42
  phosphotransferase system 348, 353–4
  quorum sensing 30–1
  serine/threonine and tyrosine kinase phosphorylating system 354
  signal molecules diversity 31–2
  signal transduction 347–90
  survival strategies 364
  TCSTS-controlled virulence determinants 349–53, 374
  two-component phosphorelay systems 34, 35–42
proportion-altering factor PAF-E.rac 97–8
proprotein convertases (PCs) 318–19
protein kinase C (PKC) 151, 160–1, 177, 217–18
protein kinase A (PKA) 85, 150, 156–7, 176
protein phosphorylation 348
Proteus 57
protozoa 73–86
PTS *see* phosphotransferase system
PTX 83–4

quorum sensing 30–1, 40

RAB-3 function 228–9
Rac 222–8
Ralstonia solanacearum 55–6
RapA/RapB proteins 362
RAS gene 11, 317
Ras pathway 198–9, 219, 222
reaction-diffusion systems 91–2
receptor serine/threonine kinases 152, 165–6
receptor tyrosine kinases 151–2, 163–4
receptors
  5-HT in nematodes 260–2
  annelid neuroendocrine 304
  platyhelminths 170–1
  protozoa 74–7, 83–5
  signal molecules 149
reef-building hydrocorals 107
regeneration 142, 173–7
Rennilla köllikeri 124, 125
RER *see* rough endoplasmic reticulum
research 339
residues conservation 360
resistance to antibiotics 370, 372–3, 381
'response regulator' (RR) component 348, 355, 359–61, 377–9
retinoic acid 99
retinoids 166–9
RFamides 312–13
RGS proteins 203, 204
Rho 222–3, 225–7
rhodopsin 163
ribbon structures 78
RNAIII transcript 375
rough endoplasmic reticulum (RER) 120
RR *see* 'response regulator' component

Sabellastarte magnifica 312
Saccharomyces 3–7, 83, 367
SADPNFLRFamide 284–6
SAEPFGTMRFamide 290
salicylanilides 385, 386, 387
Salmonella typhimurium 364, 375–6

SAM *see* sterile alpha motif
scaffolding protein 5, 6
Schistosoma mansoni 146, 147, 148, 149, 153–63
Schizosaccharomyces pombe 3, 7–10, 334, 340
scissiparity cycle 143–4
scyphozoans 118
SDPNFLRFamide 284–6
sea anemone *see Anthopleura elegantissima*; *Calliactis parasitica*
sea pansy *see Renilla köllikeri*
second messenger molecules 150
secretion 228–9
selectivity 327
self-organization 91–2
'sensor' histidine kinase component 348, 355–9, 362, 368, 377–9
sequence motifs 356–9
serine/threonine and tyrosine kinase phosphorylating system 354
serotonin *see* 5-hydroxytryptamine
seven-pass transmembrane receptors (7-TMRs) 196, 197, 220–1
sexual development 1–19
Shaker protein 330
shmoo formation 5, 7
SIF *see* stolon-inducing factor
signalling molecules 31–2, 50–1, 74–82, 304–15
signalling switches 82–5
Sln1 protein 367
Smad pathway 152
smrab gene 153
spawning 102
specificity 6
Spirometra mansonoides 146
SpoOA regulator domain 360–1, 362
SpoOF regulator domain 382
sporulation pathway 37–9
SST2 protein 202
Staphylococcus aureus 39, 374–5
stationary phase 58–9
STATS 152
Ste proteins 4, 5, 6, 8, 11
Stentor coeruleus 74, 75–7
stereoisomers 270
sterile alpha motif (SAM) 9
steroid hormones 166–9
stigmalone 44–5
Stigmatella aurantiaca 44–5
stolon-inducing factor (SIF) 98
Streptococcus pneumoniae 28, 39, 353, 372, 373, 376, 380
Streptomyces 53–5
structural studies 116–20, 354–66
sub-MIC prophylaxis 381
supra-oesophageal ganglia 305–6
survival strategies 364
swarming motility 57

Taenia 142
Taz-1 reporter 386
TCSTS *see* two-component signal transduction systems
teicoplanin resistance 371
Telodorsagia circumcincta 267
tetracycline resistance 372–3
Tetrahymena 74–7, 81–2
theophylline 286
therapeutic intervention 377–81
Theromyzon tessulatum 305–6, 308, 309, 310–14, 320
thyroid hormones 166–9
7-TMRs *see* seven-pass transmembrane receptors
transcription regulation 362–3
transmembrane domain 357
transmitters *see* neurotransmitters
transplantation studies 92
trifluoperazine (TFP) 162–3
tubocurarine 257
turbellarians 144
Turing, Alan 91–2
two-component phosphorelay systems 34, 35–42
two-component signal transduction systems (TCSTS) 347–90
TXGYGD sequence 330, 331
tyramines 385
tyrosine kinase-associated receptors 152, 164–5

uka1/2 PKA catalytic subunits 16
unc genes 216, 217, 223–5, 271
unicellular eukaryotes 73–86
Uromyces appendiculatus 325, 334
Ustilago maydis 14–17